Nuclear Hormone Receptors

Nuclear Hormone Receptors
Molecular Mechanisms, Cellular Functions, Clinical Abnormalities

Edited by

MALCOLM G. PARKER

*Molecular Endocrinology Laboratory,
Imperial Cancer Research Fund,
London, UK*

ACADEMIC PRESS
Harcourt Brace Jovanovich, Publishers

London   San Diego   New York
Boston   Sydney   Tokyo   Toronto

This book is printed on acid-free paper

ACADEMIC PRESS LIMITED
24/28 Oval Road
LONDON NW1 7DX

United States Edition published by
ACADEMIC PRESS INC.
San Diego, CA 92101

Copyright © 1991 by
ACADEMIC PRESS LIMITED

All Rights Reserved
No part of this book may be reproduced in any form by photostat, microfilm, or by any other means, without written permission from the publishers

British Library Cataloguing in Publication Data

Is available

ISBN 0-12-545072-9

Filmset by Bath Typesetting Ltd, Bath, Avon
Printed in Great Britain by St Edmundsbury Press Ltd, Bury St Edmunds, Suffolk

Contents

List of Contributors   xiii
Preface   xvii

1. Overview of the Nuclear Receptor Family   1
E. V. Jensen

  1.1. Intracellular receptors   1
      1.1.1. The nuclear receptor family   1
      1.1.2. Hormone–receptor complexes   3
  1.2. Hormone–receptor interaction in target cells   4
      1.2.1. Receptor transformation   4
      1.2.2. Receptor localization   6
  1.3. Status of the two-step mechanism   8
  1.4. Summary   10
      Acknowledgement   10
      References   10

2. The Oestrogen Receptor: From Perception to Mechanism   15
S. Green and P. Chambon

  2.1. Historical perspective   15
  2.2. The primary amino acid sequence   16
  2.3. The hormone-binding domain   17
  2.4. The DNA-binding domain   18
  2.5. Interaction between the oestrogen receptor and the 90-kD heat-shock protein   21
  2.6. The oestrogen receptor contains a nuclear localization signal   22
  2.7. Oestrogen response elements are not always perfectly palindromic   23
  2.8. The oestrogen receptor binds to a palindromic oestrogen response element as a dimer   24
  2.9. Synergy between multiple oestrogen response elements and/or the binding sites of other transcription factors   26
  2.10. The oestrogen receptor contains multiple distinct regions important for transcriptional activation   27
  2.11. Mechanisms of oestrogen antagonism   29
  2.12. How does the oestrogen receptor activate gene transcription?   32
      References   33

3. Structure and Function of the Glucocorticoid Receptor   39
M. Danielsen

  3.1. Introduction   39

3.2.		The N-terminal domain	39
3.3.		The DNA-binding domain	41
	3.3.1.	DNA-binding specificity	45
	3.3.2.	Role of the basic region in nuclear localization	51
3.4.		The hormone-binding domain	52
	3.4.1.	Limits of the hormone-binding site	52
	3.4.2.	Affinity labelling studies to characterize the hormone-binding site	57
	3.4.3.	Characterization of hormone-binding deficient receptors	58
	3.4.4.	Role of the hormone-binding region in transcriptional activation	59
3.5.		Phosphorylation	60
	3.5.1.	Sites of phosphorylation	60
	3.5.2.	Phosphorylation and activation	61
	3.5.3.	Hormone-induced phosphorylation *in vivo*	62
3.6.		Role of heat-shock proteins in glucocorticoid receptor function	63
	3.6.1.	Hsp90	63
	3.6.2.	Hsp90 and receptor activation	65
	3.6.3.	Hsp90 and hormone binding	66
	3.6.4.	p56/p59	67
		Acknowledgements	67
		References	68

4. Nuclear Thyroid Hormone Receptors
W. W. Chin

4.1.	Introduction	79
4.2.	Biochemistry of thyroid hormone receptors	80
4.3.	Thyroid hormone receptors are encoded by the proto-oncogene, c-*erb*A	81
4.4.	A bevy of thyroid hormone receptors	82
4.5.	Thyroid hormone response elements	85
4.6.	Functional domains of thyroid hormone receptors	86
4.7.	Function of thyroid hormone receptors	88
4.8.	Negative regulation by thyroid hormone receptors and related forms	89
4.9.	Heterodimer formation involving thyroid hormone receptors	90
4.10.	Other aspects of thyroid hormone receptor biology	90
4.11.	Clinical syndromes	94
4.12.	Summary	95
	References	96

5. The Steroid Receptor Superfamily: Transactivators of Gene Expression
S. Y. Tsai, M.-J. Tsai and B. W. O'Malley

103

5.1.	Introduction	103
5.2.	Identification and characterization of chicken ovalbumin upstream promoter transcription factor	104

5.3.	Chicken ovalbumin upstream promoter transcription factor binds specifically to apparently dissimilar promoter sequences	105
5.4.	The chicken ovalbumin upstream promoter transcription factor belongs to the steroid hormone receptor superfamily	106
5.5.	Chicken ovalbumin upstream promoter transcription factor requires S300-II for its function	108
5.6.	Chicken ovalbumin upstream promoter transcription factor and orphan receptors	108
5.7.	Progesterone receptor: its role in transactivation of target gene expression	110
5.8.	Multiple hormone response elements mediate synergistic transcriptional activation of target genes	113
5.9.	The role of ligand in steroid receptor function	115
5.10.	Conclusion and perspectives	117
	References	118

6. Characterization of Hormone Response Elements 125
E. Martinez and W. Wahli

6.1.	Introduction	125
6.2.	Hormone response elements are usually but not always palindromic DNA sequences	125
6.3.	Hormone response elements as hormone-inducible enhancers	133
6.4.	Synergistic action of hormone response elements	134
6.5.	Hormone response elements constitute a family of related DNA sequences: the problem of specificity	136
6.6.	Conclusions	145
	Acknowledgements	146
	References	146

7. Co-operative Transactivation of Steroid Receptors 155
M. Muller, C. Baniahmad, C. Kaltschmidt, R. Schüle and R. Renkawitz

7.1.	Introduction	155
7.2.	Co-operation between hormone response elements	156
7.3.	Functional co-operativity between hormone response elements and other transcription factor binding sites	157
	7.3.1. Natural genes	157
	7.3.2. Synthetic genes	160
7.4.	Functional details of co-operativity	161
	7.4.1. Arrangement of co-operative binding sites	161
	7.4.2. Stereo-specific alignment	162
7.5.	Functional synergism and DNA-binding affinity	163
	7.5.1. Two receptor binding sites	163
	7.5.2. Combination of a glucocorticoid response element with another transcription factor	164
7.6.	Synergistic domains of the receptor	164
7.7.	Conclusion	168
	7.7.1. Mechanisms of functional synergism	168

| | Acknowledgements | 170 |
| | References | 170 |

8. Repression of Gene Expression by Steroid and Thyroid Hormones — 175
I. E. Akerblom and P. L. Mellon

- 8.1. Introduction — 175
- 8.2. Molecular mechanisms involved in negative regulation of transcription by steroid hormones — 176
- 8.3. Competition between receptors and transcription factors at the level of DNA binding — 178
  - 8.3.1. Pro-opiomelanocortin gene — 179
  - 8.3.2. α-Subunit gene of the glycoprotein hormones — 179
  - 8.3.3. α-Fetoprotein gene — 184
  - 8.3.4. Prolactin gene — 184
  - 8.3.5. Negative glucocorticoid response elements — 185
- 8.4. Interactions of receptors with other transcription factors in the absence of DNA binding — 186
  - 8.4.1. Oestrogen regulation of the prolactin gene — 187
  - 8.4.2. Glucocorticoid regulation of the proliferin gene — 187
- 8.5. Competition between hormone receptors — 188
  - 8.5.1. Retinoic acid and thyroid hormone receptors — 188
  - 8.5.2. Thyroid hormone and oestrogen receptors — 189
  - 8.5.3. Competition between multiple forms of the thyroid hormone receptor — 189
- 8.6. Summary — 190
- References — 191

9. Characterization of DNA Receptor Interactions — 197
M. Beato, D. Barettino, U. Brüggemeier, G. Chalepakis, R. J. G. Haché, M. Kalff, B. Piña, M. Schauer, E. P. Slater and M. Truss

- 9.1. Introduction — 197
- 9.2. Interaction of hormone receptors with hormone responsive elements on linear DNA — 198
- 9.3. Role of the hormone ligand — 201
- 9.4. Similarities and differences between the response elements for various steroid hormones — 203
- 9.5. Interaction of steroid hormone receptors with circular DNA molecules of different topologies — 205
- 9.6. Interaction of receptors with nucleosomally organized DNA — 207
- 9.7. Mechanism of transcriptional activation — 209
- 9.8. Cell-free transcription assay — 210
- Acknowledgements — 211
- References — 211

10. The Interaction of Steroid Receptors with Chromatin — 217
G. L. Hager and T. K. Archer

- 10.1. Introduction — 217

10.2.	Nucleoprotein organization of eukaryotic DNA		218
10.3.	Receptor interactions with characterized chromatin structures		219
	10.3.1.	The MMTV paradigm	219
	10.3.2.	Formation of initiation complex	221
	10.3.3.	Transactivation by factor interactions	222
	10.3.4.	Factor exclusion by chromatin	223
	10.3.5.	Mechanism of factor exclusion	224
	10.3.6.	Mechanism of nucleosome positioning	226
	10.3.7.	Modulation of factor access by alteration of nucleosomes	228
	10.3.8.	Selective transactivating factor access	230
10.4.	Summary		231
	References		232

11. Ontogeny of Sex Steroid Receptors in Mammals 235
G. R. Cunha, P. S. Cooke, R. Bigsby and J. R. Brody

11.1	Introduction	235
11.2.	Methodological considerations	238
11.3.	Ontogeny of androgen receptors	240
11.4.	Ontogeny of oestrogen receptors in the female genital tract	243
11.5.	Ontogeny of oestrogen receptors in male mouse reproductive organs	247
11.6.	Steroid receptors in developing brain and pituitary	249
11.7.	Progesterone receptors in fetal tissues	250
11.8.	Oestrogen induction of progesterone receptors	251
11.9.	Role of cell–cell interactions in the expression of steroid hormone receptors	252
11.10.	Role of mesenchymal–epithelial interactions in hormonal response	254
11.11.	Conclusions	259
	Acknowledgements	259
	References	259

12. Retinoic Acid Receptors and Vertebrate Limb Morphogenesis 269
C. W. Ragsdale Jr. and J. P. Brockes

12.1.	Introduction		269
12.2.	Biology of retinoid action		269
12.3.	Effects on pattern formation in limb morphogenesis		271
12.4.	Retinoic acid receptors		275
	12.4.1.	Identification	275
	12.4.2.	Region A	277
	12.4.3.	Region C and retinoic acid response elements	279
	12.4.4.	Region D	280
	12.4.5.	Region E	280
	12.4.6.	Region F	281
12.5.	Functions of the retinoic acid receptors		281
12.6.	Ligand affinities of the retinoic acid receptors		282

12.7.	Localization of the retinoic acid receptors	283
12.8.	Regulation of gene expression by retinoic acid	285
12.9.	Prospects for functional studies in cells and limbs	287
12.10.	Conclusions	290
	Acknowledgements	290
	References	291

13. The Role of Steroid Hormones and Growth Factors in the Control of Normal and Malignant Breast — 297
R. Clarke, R. B. Dickson and M. E. Lippmann

13.1.	Introduction	297
13.1.1.	Steroid hormones in normal and malignant breast tissue	297
13.1.2.	Steroid hormone receptors in breast cancer	299
13.2.	The autocrine hypothesis of steroid hormone function	300
13.3.	The role of growth factors in normal and malignant breast	302
13.3.1.	Epidermal growth factor/transforming growth factor-α	302
13.3.2.	Effects of constitutive transforming growth factor-α expression in MCF-7 cells	303
13.3.3.	Insulin-like growth factor-I and insulin-like growth factor-II	304
13.3.4.	Fibroblast growth factors	306
13.3.5.	Other mitogenic growth factors and hormones	307
13.4.	Inhibitory growth factors and hormones	307
13.5.	Growth factors as mediators of steroid hormone function	308
	References	310

14. Genetic Defects of Receptors Involved in Disease — 321
M. R. Hughes and B. W. O'Malley

14.1.	Introduction	321
14.2.	The vitamin D receptor	322
14.2.1.	Background	322
14.2.2.	Hereditary generalized resistance to 1,25-dihydroxyvitamin D_3	322
14.2.3.	Families with defective receptor–DNA interaction	324
14.3.	Syndrome of androgen insensitivity	342
14.4.	Glucocorticoid receptor resistance	345
14.5.	General comments	347
	References	348

15. Avian Erythroleukaemia: Possible Mechanisms Involved in v-erbA Oncogene Function — 355
H. Beug and B. Vennström

15.1.	Introduction	355
15.2.	Avian thyroid hormone receptors	357

15.3.	v-*erb*A, an oncogene acting by aberrant transcriptional regulation?		359
15.4.	Transcriptional regulation by v-*erb*A and TRα		359
15.5.	Regulation of erythrocyte differentiation and erythrocyte gene expression by v-*erb*A, TRα and chimaeras thereof		361
	15.5.1.	v-*erb*A	361
	15.5.2.	TRα	364
	15.5.3.	Mutations that activate TRα as an oncogene	364
	15.5.4.	Repression of erythrocyte genes by v-*erb*A: direct or indirect mechanism?	366
15.6.	Significance of erythrocyte gene repression for the leukaemic erythroblast phenotype		367
15.7.	Modulation of v-*erb*A activity by phosphorylation		368
15.8.	Conclusions and open questions		370
	References		372

16. Nuclear Hormone Receptors: Concluding Remarks   377
M. G. Parker and O. Bakker

16.1.	Introduction	377
16.2.	Receptor gene structure and function	377
16.3.	Hormone binding and receptor dimerization	380
16.4.	DNA binding of nuclear hormone receptors	384
16.5.	Transcriptional activation	385
16.6.	Hormonal effects on mRNA stability	386
16.7.	Mechanism of action of hormone antagonists	387
16.8.	Final remarks	388
	References	389

Subject index   397

Contributors

Ingrid E. Akerblom Department of Biology, University of Virginia, Charlottesville, VA 22901, USA

Trevor K. Archer Hormone Action and Oncogenesis Section, Laboratory of Experimental Carcinogenesis, National Cancer Institute, N.I.H., Bethesda, MD 20892, USA

Domingo Barettino Institut für Molekularbiologie und Tumorforschung, Emil-Mannkopff Strasse 2, D-3550 Marburg, FRG

Claudia Baniahmad Max Plank Institut für Biochemie, Genzentrum, D-8033 Martinsried, FRG

Onno Bakker Molecular Endocrinology Laboratory, Imperial Cancer Research Fund, Lincoln's Inn Fields, London WC2A 3PX, UK

Hartmut Beug Institute of Molecular Pathology, Dr Bohrgasse 7, A 1030 Wein, Austria

Miguel Beato Institut für Molekularbiologie und Tumorforschung, Emil-Mannkopff Strasse 2, D-3550 Marburg, FRG

Robert Bigsby Department of Obstetrics and Gynecology, Indiana University Medical School, Indianapolis, IN 46202-5196, USA

Jeremy P. Brockes The Ludwig Institute for Cancer Research, 91 Riding House Street, London W1P 8BT, UK

Joel R. Brody Department of Anatomy, University of California, San Francisco, CA 94143, USA

Ulf Brüggemeier Institut für Molekularbiologie und Tumorforschung, Emil-Mannkopff Strasse 2, D-3550 Marburg, FRG

Georges Chalepakis Institut für Molekularbiologie und Tumorforschung, Emil-Mannkopff Strasse 2, D-3550 Marburg, FRG

CONTRIBUTORS

Pierre Chambon Laboratoire de Génétique Moléculaire des Eucaryotes du CNRS, Unité 184 de Biologie Moléculaire et de Génie Génétique de l'INSERM, Institut de Chimie Biologique, Faculté de Médecine, 11 rue Humann, 67085 Strasbourg Cedex, France

William W. Chin Division of Genetics, Department of Medicine, Brigham and Women's Hospital, Howard Hughes Medical Institute, and Harvard Medical School, Boston, MA 02115, USA

Robert Clarke Room S128, Lombardi Cancer Research Center, Georgetown University Medical Center, 3800 Reservoir Road NW, Washington, DC 20007, USA

Paul S. Cooke Department of Veterinary Biosciences, University of Illinois, 2001 South Lincoln Avenue, Urbana, IL 61801, USA

Gerald R. Cunha Department of Anatomy, University of California, San Francisco, CA 94143, USA

Mark Danielsen Department of Biochemistry and Molecular Biology, Georgetown University Medical School, 3900 Reservoir Road NW, Washington, DC 20007, USA

Robert B. Dickson Room S128, Lombardi Cancer Research Center, Georgetown University Medical Center, 3800 Reservoir Road NW, Washington, DC 20007, USA

Stephen Green Cell and Molecular Biology Department, ICI Central Toxicology Laboratory, Alderley Park, Macclesfield, Cheshire SK10 4TJ, UK

Robert J. G. Haché Institut für Molekularbiologie und Tumorforschung, Emil-Mannkopff Strasse 2, D-3550 Marburg, FRG

Gordon L. Hager Hormone Action and Oncogenesis Section, Laboratory of Experimental Carcinogenesis, National Cancer Institute, N.I.H., Bethesda, MD 20892, USA

Mark R. Hughes Institute for Molecular Genetics and the Department of Cell Biology, Baylor College of Medicine, One Baylor Plaza, Houston, TX 77030, USA

CONTRIBUTORS

Elwood V. Jensen Scholar-in-Residence, New York Hospital, Cornell Medical Center, 525 East 68th Street, New York, NY 10021, USA

Martha Kalff Institut für Molekularbiologie und Tumorforschung, Emil-Mannkopff Strasse 2, D-3550 Marburg, FRG

Christian Kaltschmidt Max Plank Institut für Biochemie, Genzentrum, D-8033 Martinsried, FRG

Marc E. Lippman Room S128, Lombardi Cancer Research Center, Georgetown University Medical Center, 3800 Reservoir Road NW, Washington, DC 20007, USA

Ernest Martinez Institut de Biologie Animale, Université de Lausanne, Bâtiment de Biologie, CH-1015 Lausanne, Switzerland

Pamela L. Mellon Regulatory Biology Laboratory, The Salk Institute for Biological Studies, 10010 N. Torrey Pines Road, La Jolla, CA 92037, USA

Marc Muller Max Plank Institut für Biochemie, Genzentrum, D-8033 Martinsried, FRG

Bert W. O'Malley Department of Cell Biology, Baylor College of Medicine, One Baylor Plaza, Houston, TX 77030, USA

Malcolm G. Parker Molecular Endocrinology Laboratory, Imperial Cancer Research Fund, Lincoln's Inn Fields, London WC2A 3PX, UK

Benjamin Piña Institut für Molekularbiologie und Tumorforschung, Emil-Mannkopff Strasse 2, D-3550 Marburg, FRG

Clifton W. Ragsdale, Jr The Ludwig Institute for Cancer Research, 91 Riding House Street, London W1P 8BT, UK

Rainer Renkawitz Max Plank Institut für Biochemie, Genzentrum, D-8033 Martinsried, FRG

Michael Schauer Institut für Molekularbiologie und Tumorforschung, Emil-Mannkopff Strasse 2, D-3550 Marburg, FRG

Roland Schüle Salk Institute for Biological Studies, 10010 N. Torrey Pines Road, La Jolla, CA 92037, USA

Emily P. Slater Institut für Molekularbiologie und Tumorforschung, Emil-Mannkopff Strasse 2, D-3550 Marburg, FRG

Mathias Truss Institut für Molekularbiologie und Tumorforschung, Emil-Mannkopff Strasse 2, D-3550 Marburg, FRG

Ming-Jer Tsai Department of Cell Biology, Baylor College of Medicine, One Baylor Plaza, Houston, TX 77030, USA

Sophia Y. Tsai Department of Cell Biology, Baylor College of Medicine, One Baylor Plaza, Houston, TX 77030, USA

Björn Vennström Karolinska Institute, Department of Molecular Biology CMB, Box 60400, S 10401 Stockholm, Sweden

Walter Wahli Institut de Biologie Animale, Université de Lausanne, Bâtiment de Biologie, CH-1015 Lausanne, Switzerland

Preface

In the past decade there has been remarkable progress in the field of endocrinology due in large part to advances in molecular and cell biology. *Nuclear Hormone Receptors* describes the family of proteins responsible for mediating the action not only of steroid and thyroid hormones but also different types of ligand such as retinoic acid and others predicted to exist but not yet discovered. In view of the similarities in receptor structure it seems likely that they function by similar mechanisms to stimulate or repress the expression of specific genes and thereby play a crucial role in cell growth and morphogenesis, as well as differentiation. It is now also clear that disruption of hormone receptor function is implicated in oncogenesis and a number of endocrine disorders and further understanding will lead to new rational therapeutic and preventive approaches.

MALCOLM PARKER

1
Overview of the Nuclear Receptor Family

E. V. JENSEN

Ben May Institute, University of Chicago, Chicago, IL 60637, USA

Current address: New York Hospital Cornell Medical Center, New York, NY 10021, USA

1.1 Intracellular receptors

1.1.1 The Nuclear Receptor Family

The nuclear receptors consist of proteins that mediate the actions of many important cell regulators. These include the steroid hormones (oestrogens, progestins, androgens, glucocorticoids, mineralocorticoids, ecdysteroids, vitamin D), thyroid hormones, and retinoids. In contrast to receptors for peptide hormones, which are located in the cell membrane and evoke a second messenger to deliver the regulatory signal, the nuclear receptors are present within the cell. After associating with their respective ligands, they eventually act as transcription factors in the cell nucleus to enhance the expression of specific genes.

Agents that act by way of nuclear receptors influence the behaviour of many tissues in vertebrate species. Often certain cells are especially responsive; these generally contain an elevated level of receptor and are called target cells. For example, the gonadal steroid hormones control growth, differentiation, and function of reproductive and accessory sex tissues: oestrogens and progestins in the female (uterus, vagina, oviduct, mammary gland) and androgens in the male (prostate, seminal vesicles, penis, chick comb). They also influence many other tissues, including skin, hair, bone, pituitary, hypothalamus, and behavioural centres in the brain. In birds, reptiles and amphibians, oestrogens induce synthesis of egg yolk proteins in the liver. Glucocorticoids promote gluconeogenesis and regulate carbohydrate metabolism in many mammalian cells, cause involution of lymphoid tissues, such as spleen, thymus and other components of the immune system, and have profound effects on brain and nerve tissues. Mineralocorticoids promote sodium transport across membranes in organs such as kidney and toad bladder. Vitamin D acts in intestinal mucosa, bone and kidney to

NUCLEAR HORMONE RECEPTORS
ISBN 0-12-545072-9

enhance calcium transport and utilization. Ecdysteroids modulate gene expression in essentially all insect cells and, in the absence of juvenile hormone, induce metamorphosis. Thyroid hormones regulate development and metabolic behaviour of a great variety of mammalian cells and are required for metamorphosis in amphibians. Retinoids are morphogenic agents, essential for the differentiation and growth of epithelial cells and for the development of embryonic structures such as limb buds.

Intracellular receptors originally were discovered for the oestrogens by the ability of target tissues to bind labelled hormone, both *in vivo* and *in vitro*. That these binding substances are true receptors was established by the correlation between the inhibition of oestradiol uptake and the reduction in uterotrophic response when diffferent amounts of oestrogen antagonist are given along with the hormone. It has since been demonstrated that the binding components of target cells participate directly in the biological actions of all members of the steroid and thyroid hormone family. Steroid hormone receptors are phosphorylated proteins containing sulphydryl groups necessary for ligand interaction; sulphydryl dependence has also been shown for thyroid receptor. In their native state, steroid receptors are complex entities, consisting of a hormone-binding protein associated with other macromolecules.

The primary structure for most of the intracellular receptors has been deduced by sequencing their corresponding cDNAs (Fig. 1). Though they vary in length from 427 amino acids for vitamin D receptor to 984 for mineralocorticoid receptor, they are composed of comparable units. Each of these intracellular receptors contains a DNA-binding domain of 66–68 amino acids (C), which shows a high degree of homology with the other members of the family, a ligand-binding region (E) showing some homology, and variable regions with little homology that contribute somehow to optimal function of the receptor. The DNA-binding domain contains two 'zinc fingers', looped structures involving chelated metal ions, that are responsible for reaction with cognate DNA in the hormone response elements of target genes. Chicken and human progestin receptors come in two sizes, A and B, the B form consisting of the A protein plus an additional unit at the N-terminus (165 amino acids in human receptor, 128 in chicken). Two forms (α, β), arising from different genes and each with submodifications, have been identified for thyroid hormone receptor, and three for retinoic acid (α, β, γ). It is interesting that the oncogene v-*erb*A is related to the thyroid hormone receptor. References to studies of receptor cloning and structure, as well as to interactions with the genome, are given in the following reviews: Ringold (1985); Yamamoto (1985); Green and Chambon (1986, 1988); Samuels *et al.* (1988); Minghetti and Norman (1988); Evans (1988, 1989); Savouret *et al.* (1989); Berg (1989); Beato (1989); Liao *et al.*

1. OVERVIEW OF THE NUCLEAR RECEPTOR FAMILY

(1989); Godowski and Picard (1989); Ham and Parker (1989); Picard *et al.* (1990); O'Malley (1990).

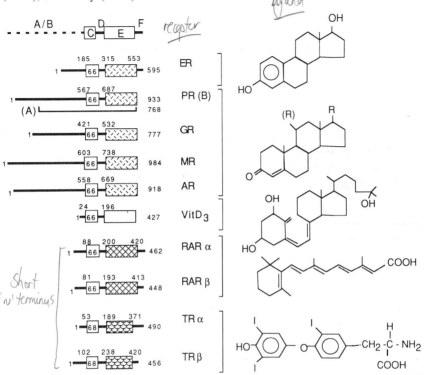

Fig. 1 The human nuclear receptor family and their ligands. The top diagram shows the six functional domains: A/B, F, modulating regions; C, DNA-binding region; D, 'hinge' region; E, ligand-binding region. Boxes indicate highly conserved domains; thin black lines are regions of low homology. The position of each domain boundary is given as the number of amino acids from the amino terminus (1). Figures in the C boxes indicate the number of amino acids in the DNA-binding domain. Receptors are: ER, oestrogen; PR, progestin [(R) = H, R = CH_3CO]; GR, glucocorticoid [(R) = OH, R = CH_2OHCO, OH]; MR, mineralocorticoid [(R) = OH, R = CH_2OHCO, plus =O at C-18]; AR, androgen [(R) = H, R = OH]; Vit D_3, vitamin D; RAR, retinoic acid; TR, thyroid hormone. It should be noted that androgen receptors in male reproductive tissues react preferentially with dihydrotestosterone, the saturated analogue of the AR ligand shown. Figure courtesy of Professor Pierre Chambon.

1.1.2 Hormone–Receptor Complexes

When target tissues are exposed to steroid hormones, *in vivo* or under physiological conditions *in vitro*, two different steroid–receptor complexes can be identified. A small portion of the labelled steroid is readily extractable and appears as a hormone–receptor complex in the supernatant or cytosol fraction of tissue homogenates. Most of the incorporated steroid is associated with a protein that is more tightly bound in the cell nucleus and requires salt concentrations of 300–400 mM for its solubilization. Before exposure to hormone, essentially all of the receptor is in the more extractable form. For gonadal and adrenal hormones, the unoccupied receptor is usually solubilized on cell disruption in hypotonic media, whereas unoccupied receptor for vitamin D requires physiological salt concentrations for its extraction into the cytosol. Addition of steroid to the supernatant fraction in the cold gives rise to the same hormone–receptor complex as that present in cytosol from previously treated animals or tissues.

Thyroid hormones and retinoic acid differ from steroid hormones in that only a nuclear form of their receptors has been identified. This phenomenon is discussed further in Section 1.2.1.

For steroid hormones, the receptor in the cytosol is a heterogeneous entity, consisting of a steroid-binding unit associated with macromolecular as well as micromolecular factors. In physiological or hypotonic salt concentrations, the cytosol steroid–receptor complex sediments at 8–10 S, but in the presence of 300 mM salt it is dissociated into a 3.5–4 S steroid-binding subunit. The complex extracted from the nuclei of hormone-treated tissues also sediments at about 4 S in salt-containing sucrose gradients, except for oestrogen receptor which sediments at about 5 S. In all cases, the nuclear complex can be distinguished from the cytosol complex by its ability to bind to nuclei, chromatin, DNA, and polyanions such as phosphocellulose.

References to the identification of intracellular receptors and their complexes are given in the following reviews: Raspé (1971); Jensen and DeSombre (1972); O'Malley and Means (1974); Liao (1975); Edelman (1975); Gorski and Gannon (1976); Yamamoto and Alberts (1976); Jost (1978); Norman et al. (1982); Pike (1985); Oppenheimer et al. (1987).

1.2 Hormone–receptor interaction in target cells

1.2.1 Receptor Transformation*

The relationship between the nuclear and cytosol forms of steroid hormone receptors has been the subject of much investigation, for it has proved to be of importance in identifying a biochemical role for the steroid. It was shown very early that the oestradiol–receptor complex bound in target cell nuclei is not produced directly but arises in a two-step process via the initial formation of the cytosol complex. That the cytosol protein is not just a carrier but actually becomes the nuclear receptor was demonstrated by its conversion to the nuclear protein on exposure of uterine cytosol to oestradiol at physiological temperatures. Only the 5 S nuclear form of the oestradiol–receptor complex binds to target cell nuclei (or to DNA) and stimulates the rate of transcription of specific genes. Thus, an important function of the steroid is to convert the native receptor to a biochemically functional state that can bind in the nucleus and enhance gene transcription. Early experiments leading to the concept of receptor transformation and the two-step interaction mechanism are described in the following articles: Gorski et al. (1968); Jensen et al. (1968); Jensen and DeSombre (1973).

Subsequent studies in many laboratories demonstrated that hormone-induced transformation of the native receptor to a form that can interact with target genes is a general phenomenon for all classes of steroid hormones, but not for thyroid hormones. Oestrogen receptor is unique in that this conversion is accompanied by a change in its sedimentation properties because the transformed receptor forms a dimer. Although other transformed receptors do not dimerize spontaneously**, progestin, glucocorticoid and thyroid receptors appear to react in a dimeric form with the palindromic response elements in their target genes. Hormone-induced transformation of the vitamin D receptor takes place at a lower temperature than that of other steroid receptors, which require temperatures near physiological for this change to proceed readily under conditions of cellular pH and ionic strength.

Several early findings suggested that the transformation process might involve the removal of some component of the native complex. Transformation is effected in the absence of hormone by ammonium sulphate precipitation, gel filtration, or dialysis, whereas conversion of the native 4 S oestrogen receptor to the 5 S form proceeds more rapidly at lower concentrations, contrary to what would be expected for a simple dimerization reaction. Better understanding of the phenomenon came with the demonstration that native receptors are associated with heat shock proteins that obscure the molecular domain responsible for DNA binding. These are lost on transformation, enabling the receptor to react with target genes.

The association with heat-shock protein requires the presence of micromolecular factors, identified in the case of glucocorticoid receptor as an endogenous metal anion and an aminophosphoglyceride. Removal of these

---

* In the first reports of hormone-induced conversion of native receptor protein to its nuclear-binding form, this phenomenon was known as receptor transformation. With the appreciation of its biological significance, many investigators began to call it activation. More recently there has been a trend back to the use of transformation, reserving the term activation for the acquisition of steroid-binding properties as by the rephosphorylation of processed receptor.

** Added in proof: recent studies using gel filtration have shown that transformed glucocorticoid receptor also exists as a dimer (Wrange et al., 1989).

by dialysis, gel filtration, or chelation eliminates the requirement for the hormone in receptor transformation. How interaction with the steroid effects unmasking of the DNA-binding site in the native receptor is not entirely clear. It has been proposed that the steroid may act directly on the metal anion centre or that it may stimulate a self-proteolytic activity inherent in the receptor protein. Detailed references to studies of receptor transformation and of the involvement of heat-shock proteins are given in the following articles: Milgrom (1981); Grody et al. (1982); Schmidt and Litwack (1982); Dahmer et al. (1984); Pratt (1987); Gustafsson et al. (1987); Bodine and Litwack (1988); Baulieu et al. (1989); Denis and Gustafsson (1989); Pratt et al. (1989); Jensen (1990).

Recent experiments (Dalman et al., 1990) have shown that receptor for thyroid hormone is synthesized directly in the DNA-binding state, not associated with heat-shock protein, and thus it does not require transformation of the type seen with steroid receptors. This finding is consistent with the failure to detect a cytosol form of the thyroid receptor. No cytosol receptor has been identified for retinoic acid, but its target tissues contain a cellular binding protein (CRABP) that may serve as a carrier to bring it to the receptor in the nucleus.

1.2.2 Receptor Localization

Because untransformed receptors for most steroid hormones usually appear in the cytosol fraction of tissue homogenates, it was originally assumed that the unoccupied receptor is a cytoplasmic protein. Thus, hormone-induced conversion of the native receptor to the transformed state was thought to be accompanied by its migration to the nucleus (receptor translocation). Although this scheme was consistent with most earlier experiments, observations were made that suggested that much of the native receptor may already be in the nucleus before exposure to hormone.

Unfilled and presumably untransformed receptors for oestrogens, progestins, mineralocorticoids and ecdysteroids are sometimes found in the nuclear fraction of tissue homogenates, whereas unoccupied vitamin D receptor remains in the nuclear pellet in media containing less than physiological salt concentration. Autoradiographic studies of target tissues, exposed to tritiated oestrogens or progestins in the cold to minimize transformation, show most but not all tissue radioactivity to be associated with cell nuclei. Immunocytochemical staining with monoclonal antibodies usually, but not always, is seen only in the nuclei of target cells for oestrogens, progestins and androgens. After enucleation with or without cytochalasin B, nearly all of the steroid-binding activity of target cells for oestrogens, progestins and glucocorticoids, but not for vitamin D, is present in the nucleoplasts rather than the cytoplasts.

The foregoing findings, more fully described and referenced in recent reviews (Walters, 1985; Jensen, 1990), have led some investigators to conclude that untransformed receptors for gonadal hormones are entirely confined to the nucleus, even though this is not true for other steroid hormones. Immunocytochemical studies with glucocorticoids, and recently mineralocorticoids (Rundle et al., 1989; Lombès et al., 1990), clearly show both nuclear and cytoplasmic staining, with the latter often predominating. Although conventional immunocytochemical methods indicate native vitamin D receptors to be largely in the nucleus (Clemens et al., 1988), improved techniques suggest that they are actually cytoplasmic (Barsony et al., 1990). Even with gonadal hormones, none of the procedures used is capable of proving that native receptors are exclusively nuclear, and there are experimental observations that are difficult to reconcile with this conclusion.

Immunocytochemistry and autoradiography have the limitation that they often underestimate cytoplasmic components. Both techniques measure concentration rather than amount, and in most target cells the cytoplasmic volume greatly exceeds the nuclear volume. For example, in a roughly spherical cell with a diameter twice that of its nucleus, if half the receptor were in the cytoplasm the staining would appear to be 88% nuclear, whereas a 20% content of cytoplasmic receptor would appear as less than 4% extranuclear and thus be undetectable. Several studies have found that prior exposure to hormone, either *in vivo* or *in vitro*, markedly increases the intensity of nuclear staining. This suggests the presence of receptor that is not seen immunochemically until it is converted to the more tightly bound form that either resists loss during experimental manipulation or is concentrated to detectable levels in the nucleus from a dilute state in the cytoplasm. It is also possible that, if they are in close proximity to other macromolecules in a 'cytoplasmic anchor' (Picard and Yamamoto, 1987; Hunt, 1989), native receptors for some steroids may not be immunoreactive until exposure to hormone disrupts this association.

The results of enucleation experiments are far from compelling, especially since their indication of a 90% nuclear localization for native glucocorticoid receptor is clearly incompatible with other experimental findings. In contrast to oestrogens, progestins and glucocorticoids, only 30% of the vitamin D receptors are found in the nucleoplasts of target cells unless the cells are first exposed to hormone, whereupon the nucleoplasts contain 90% of the receptors. Whether these conflicting results reflect the fact that the cells used in the enucleation experiments were grown in culture with fetal calf serum, known to contain oestrogens, progestins and glucocorticoids but not vitamin D, is not certain. It is known that enucleation techniques can lead to an erroneous assumption of nuclear localization, as in the case of DNA polymerase α in fetal bovine fibroblasts (Brown *et al.*, 1981).

With the foregoing uncertainties in the quantitative assignment of untransformed receptors to the nuclear compartment, and the fact that native glucocorticoid, mineralocorticoid and probably vitamin D receptors are present appreciably or even predominantly in the cytoplasm of their target cells, an exclusive nuclear localization for oestrogen, androgen and progestin receptors is open to question. Based on present knowledge, one cannot exclude the possibility that, like adrenal steroid receptors, those for gonadal hormones are also present in both cytoplasm and nucleus, but with a nuclear–cytoplasmic equilibrium more in the direction of nuclear residency. Recent studies of localization signals (Guiochon-Mantel *et al.*, 1989; Picard *et al.*, 1990) suggest a greater tendency toward nuclear binding in receptors for oestrogens and progestins than for glucocorticoids, but whether these signals are strong enough to effect exclusive nuclear localization is not established. What is evident is that, contrary to original assumptions, much of the untransformed receptor, especially for gonadal hormones, resides within the nuclear compartment, loosely bound and readily extractable on cell disruption.

1.3 Status of the two-step mechanism

With the recognition that much of the native receptor for steroid hormones is in the nucleus, it became obvious that some revision was needed of the original view that the cytosol receptor is an extranuclear entity until interaction with hormone converts it to the nuclear bound form. However, the rush to proclaim that the two-step mechanism is no longer valid and that one must search for a new model for steroid hormone action reflects misunderstanding of what is meant by 'two-step'. This term was originally proposed to indicate that the oestradiol–receptor complex, bound in target cell nuclei after hormone administration *in vivo*, is not produced directly but is derived from an initially formed complex of the steroid with the receptor protein identified in the cytosol. When it was found that the 4 S binding unit of the cytosol receptor actually becomes the 5 S nuclear receptor under the influence of hormone, and that only this transformed oestrogen–receptor complex can bind to chromatin and enhance transcriptional activity in target cell nuclei, the two-step concept took on additional meaning. In this refinement, it came to denote the hormone-induced conversion of a native receptor protein to its biochemically functional form that then reacts with target genes.

1. OVERVIEW OF THE NUCLEAR RECEPTOR FAMILY

Despite repeated emphasis on receptor transformation followed by genomic interaction as the key steps in steroid hormone action, some investigators have regarded translocation from cytoplasm to nucleus as the essence of the two-step mechanism, which it never was. Translocation is a secondary feature in that the transformed steroid–receptor complex must assume its genomic binding site from wherever the native receptor protein is originally located. If it is entirely within the nucleus, translocation will be only intranuclear, but if it is partly in the cytoplasm, intracellular translocation will also take place. These are important aspects, but they are details rather than differences in basic mechanism.

Thus, the question of where the untransformed receptor is actually located has no direct bearing on the validity of the two-step mechanism. If there is a cytoplasmic–nuclear equilibrium of receptor distribution, as there appears to be with glucocorticoid and mineralocorticoid receptors, the only refinement needed in the original concept is that the level of cytoplasmic receptor is less than previously considered, but that it is continually replenished from a nuclear pool (Fig. 2A). If it can be established that native receptors for gonadal hormones are entirely confined to the nucleus, there will be need for elucidation of how the steroid makes its way so rapidly *in vivo* from the blood transport proteins to the cell nucleus. But once the hormone is in the nucleus, the two-step mechanism, involving receptor transformation and genomic binding, would operate as previously conceived (Fig. 2B).

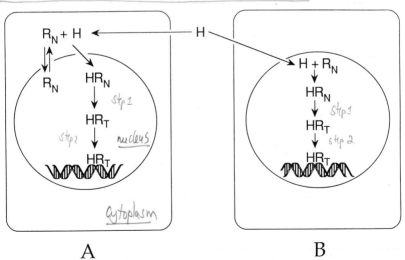

Fig. 2 Schematic representation of hormone–receptor interaction in target cell. H, hormone; R_N, native (untransformed) receptor; R_T, transformed receptor. (A) Extranuclear receptor in equilibrium with nuclear pool of loosely bound native receptor. (B) Native receptor confined to the nucleus.

On the basis of considerations here and elsewhere (Clark, 1984; Walters, 1985; Gravanis and Gurpide, 1986; Gustafsson et al., 1987; Parikh et al., 1987; Jensen, 1990), it appears that the iconoclastic fervor concerning the two-step mechanism has been a bit excessive and tends to present a distorted picture, especially to investigators new to the field. Neither of the pathways shown in Fig. 2 represents more than a modification of the original concept. As has been aptly stated, 'if steroid receptors are indeed weakly held non-histone nuclear proteins ... the change [for the "two-step" hypothesis] may be more cosmetic than substantive' (Schrader, 1984). One may reasonably conclude that, with certain refinements based on two decades of research since its original proposal, the two-step mechanism is alive and well and operating in target cells for steroid hormones.

1.4 Summary

Receptors for steroid and thyroid hormones and for retinoic acid comprise a family of proteins with many similarities in their structures and modes of action. These receptors are present within target cells where, in association with their respective ligands, they act in the nucleus to regulate transcription in responsive genes. Receptors for thyroid hormones, and probably for retinoic acid, appear to exist only in a DNA-binding state in the nucleus, but those for steroid hormones are present in an inactive modification, in some and possibly all cases partly in the cytoplasm. Interaction with the hormone transforms the native receptor to its DNA-binding form. Thus, the biological action of steroid hormones is said to involve a two-step mechanism: conversion of the native receptor to a functional transcription factor and reaction of this transformed receptor with response elements of target genes.

Acknowledgement

Preparation and publication of this overview was supported by a grant (RDP-53A) from the American Cancer Society.

References

Barsony, J., Pike, J. W., DeLuca, H. F. and Marx, S. J. (1990). Immunocytology with microwave fixed fibroblasts shows $1\alpha,25$-dihydroxyvitamin D_3 dependent rapid and estrogen dependent slow reorganization of vitamin D receptors. *J. Cell Biol.* in press.

1. OVERVIEW OF THE NUCLEAR RECEPTOR FAMILY

Baulieu, E.-E., Binart, N., Cadepond, F., Catelli, M. G., Chambraud, B., Garnier, J., Gasc, J. M., Groyer-Schweizer, G., Oblin, M. E., Radanyi, C., Redeuilh, G., Renoir, J. M. and Sabbah, M. (1989). Do receptor-associated nuclear proteins explain earliest steps of steroid hormone function? In "The Steroid/Thyroid Hormone Receptor Family and Gene Regulation" (eds J. Carlstedt-Duke, H. Eriksson and J.-Å. Gustafsson), pp. 301–318. Birkhaüser-Verlag, Basel.

Beato, M. (1989). Gene regulation by steroid hormones. *Cell* **56**, 335–344.

Berg, J. M. (1989). DNA binding specificity of steroid receptors. *Cell* **57**, 1065–1068.

Bodine, P. V. and Litwack, G. (1988). Purification and structural analysis of the modulator of the glucocorticoid–receptor complex. *J. Biol. Chem.* **263**, 3501–3512.

Brown, M., Bollum, F. J. and Chang, L. M. S. (1981). Intracellular localization of DNA polymerase α. *Proc. Natl. Acad. Sci. USA* **78**, 3049–3052.

Clark, C. R. (1984). The cellular distribution of steroid hormone receptors: have we got it right? *Trends Biochem. Sci.* **9**, 207–208.

Clemens, T. L., Garrett, K. P., Zhou, X.-Y., Pike, J. W., Haussler, M. R. and Dempster, D. W. (1988). Immunocytochemical localization of the 1,25-dihydroxyvitamin D_3 receptor in target cells. *Endocrinology* **122**, 1224–1230.

Dahmer, M. K., Housley, P. R. and Pratt, W. B. (1984). Effects of molybdate and endogenous inhibitors on steroid receptor inactivation, transformation and translocation. *Annu. Rev. Physiol.* **46**, 67–81.

Dalman, F. C., Koenig, R. J., Perdew, G. H., Massa, E. and Pratt, W. B. (1990). In contrast to the glucocorticoid receptor, the thyroid hormone receptor is translated in the DNA binding state and is not associated with hsp90. *J. Biol. Chem.* **265**, 3615–3618.

X Denis, M. and Gustafsson, J.-Å. (1989). The $M_r \approx 90,000$ heat shock protein: an important modulator of ligand and DNA-binding properties of the glucocorticoid receptor. *Cancer Res.* **49**, 2275s–2281s.

Edelman, I. S. (1975). Mechanism of action of steroid hormones. *J. Steroid Biochem.* **6**, 147–159.

Evans, R. M. (1988). The steroid and thyroid hormone receptor superfamily. *Science* **240**, 889–895.

Evans, R. M. (1989). Molecular characterization of the glucocorticoid receptor. *Recent Prog. Horm. Res.* **45**, 1–27.

Godowski, P. J. and Picard, D. (1989). Steroid receptors—how to be both a receptor and a transcription factor. *Biochem. Pharmacol.* **38**, 3135–3143.

Gorski, J. and Gannon, F. (1976). Current models of steroid hormone action: a critique. *Annu. Rev. Physiol.* **38**, 425–450.

Gorski, J., Toft, D., Shyamala, G., Smith, D. and Notides, A. (1968). Hormone receptors: studies on the interaction of estrogen with the uterus. *Recent Prog. Horm. Res.* **24**, 45–80.

Gravanis, A. and Gurpide, E. (1986). Enucleation of human endometrial cells: nucleo-cytoplasmic distribution of DNA polymerase α and estrogen receptor. *J. Steroid Biochem.* **24**, 469–474.

Green, S. and Chambon, P. (1986). A superfamily of potentially oncogenic hormone receptors. *Nature* **324**, 615–617.

Green, S. and Chambon, P. (1988). Nuclear receptors enhance our understanding of transcription regulation. *Trends Genet.* **4**, 309–314.

Grody, W. W., Schrader, W. T. and O'Malley, B. W. (1982). Activation, transformation, and subunit structure of steroid hormone receptors. *Endocr. Rev.* **3**, 141–163.

Guiochon-Mantel, A., Loosfelt, H., Lescop, P., Sar, S., Atger, M., Perrot-Applanat, M. and Milgrom, E. (1989). Mechanisms of nuclear localization of the progesterone receptor: evidence for interaction between monomers. *Cell* **57**, 1147–1154.

Gustafsson, J.-Å., Carlstedt-Duke, J., Poellinger, L., Okret, S., Wikström, A.-C., Brönnegård, M., Gillner, M., Dong, Y., Fuxe, K., Cintra, A., Härfstrand, A. and Agnati, L. (1987). Biochemistry, molecular biology, and physiology of the glucocorticoid receptor. *Endocr. Rev.* **8**, 185–234.

Ham, J. and Parker, M. G. (1989). Regulation of gene expression by nuclear hormone receptors. *Curr. Opinion Cell Biol.* **1**, 503–511.

Hunt, T. (1989). Cytoplasmic anchoring proteins and the control of nuclear localization. *Cell* **59**, 949–951.

Jensen, E. V. (1990). Molecular mechanisms of steroid hormone action in the uterus. *In* "Uterine Function: Molecular and Cellular Aspects" (eds M. E. Carsten and J. D. Miller), pp. 315–359. Plenum Publishing Corp., New York.

Jensen, E. V. and DeSombre, E. R. (1972). Mechanism of action of the female sex hormones. *Annu. Rev. Biochem.* **41**, 203–230.

Jensen, E. V. and DeSombre, E. R. (1973). Estrogen–receptor interaction. Estrogenic hormones effect transformation of specific receptor proteins to a biochemically functional form. *Science* **182**, 126–134.

Jensen, E. V., Suzuki, T., Kawashima, T., Stumpf, W. E., Jungblut, P. W. and DeSombre, E. R. (1968). A two-step mechanism for the interaction of estradiol with rat uterus. *Proc. Natl. Acad. Sci. USA* **59**, 632–638.

Jost, J.-P. (1978). Facts and speculation on the mode of action of the steroid-binding protein complex. *Int. Rev. Biochem.* **20**, 109–132.

Liao, S. (1975). Cellular receptors and mechanisms of action of steroid hormones. *Int. Rev. Cytol.* **41**, 87–172.

Liao, S., Kokontis, J., Sai, T. and Hiipakka, R. A. (1989). Androgen receptors: structures, mutations, antibodies and cellular dynamics. *J. Steroid Biochem.* **34**, 41–51.

Lombès, M., Farman, N., Oblin, M. E., Baulieu, E.-E., Bonvalet, J. P., Erlanger, B. F. and Gasc, J. M. (1990). Immunohistochemical localization of renal mineralocorticoid receptor by using an anti-idiotypic antibody that is an internal image of aldosterone. *Proc. Natl. Acad. Sci. USA* **87**, 1086–1088.

Milgrom, E. (1981). Activation of steroid–receptor complexes. *In* "Biochemical Actions of Hormones", Vol. 8 (ed. G. Litwack), pp. 465–492. Academic Press, New York.

Minghetti, P. P. and Norman, A. W. (1988). 1,25(OH)$_2$–Vitamin D$_3$ receptors: gene regulation and genetic circuitry. *FASEB J.* **2**, 3043–3053.

Norman, A. W., Roth, J. and Orci, L. (1982). The vitamin D endocrine system: steroid metabolism, hormone receptors, and biological response (calcium binding proteins). *Endocr. Rev.* **3**, 331–366.

O'Malley, B. (1990). The steroid receptor superfamily: more excitement predicted for the future. *Mol. Endocrinol.* **4**, 363–369.

O'Malley, B. W. and Means, A. R. (1974). Female steroid hormones and target cell nuclei. *Science* **183**, 610–620.

Oppenheimer, J. H., Schwartz, H. L., Mariash, C. N., Kinlaw, W. B., Wong, N. C. W. and Freake, H. C. (1987). Advances in our understanding of thyroid hormone action at the cellular level. *Endocr. Rev.* **8**, 288–308.

Parikh, I., Rajendran, K. G., Su, J.-L., Lopez, T. and Sar, M. (1987). Are estrogen receptors cytoplasmic or nuclear? Some immunocytochemical and biochemical studies. *J. Steroid Biochem.* **27**, 185-192.

Picard, D. and Yamamoto, K. R. (1987). Two signals mediate hormone-dependent nuclear localization of the glucocorticoid receptor. *EMBO J.* **11**, 3333-3340.

Picard, D., Kumar, V., Chambon, P. and Yamamoto, K. R. (1990). Signal transduction by steroid hormones: nuclear localization is differentially regulated in estrogen and glucocorticoid receptors. *Cell Regulation* **1**, 291-299.

Pike, J. W. (1985). Intracellular receptors mediate the biologic action of 1,25-dihydroxyvitamin D_3. *Nutr. Rev.* **43**, 161-168.

Pratt, W. B. (1987). Transformation of glucocorticoid and progesterone receptors to the DNA-binding state. *J. Cell. Biochem.* **35**, 51-68.

Pratt, W. B., Sanchez, E. R., Bresnick, E. H., Meshinchi, S., Scherrer, L. C., Dalman, F. C. and Welsh, M. J. (1989). Interaction of the glucocorticoid receptor with the M_r 90,000 heat shock protein: An evolving model of ligand-mediated receptor transformation and translocation. *Cancer Res.* **49**, 2222s-2229s.

Raspé, G. (ed.) (1971). "Schering Workshop on Steroid Hormone Receptors". Advances in the Biosciences, Vol. 7. Pergamon-Vieweg, Braunschweig.

Ringold, G. M. (1985). Steroid hormone regulation of gene expression. *Annu. Rev. Pharmacol. Toxicol.* **25**, 529-566.

Rundle, S. E., Smith, A. I., Stockman, D. and Funder, J. W. (1989). Immunocytochemical demonstration of mineralocorticoid receptors in rat and human kidney. *J. Steroid Biochem.* **33**, 1235-1242.

Samuels, H. H., Forman, B. M., Horowitz, Z. D. and Ye, Z.-S. (1988). Regulation of gene expression by thyroid hormone. *J. Clin. Invest.* **81**, 957-967.

Savouret, J. F., Misrahi, M., Loosfelt, H., Atger, M., Bailly, A., Perrot-Applanat, M., Vu Hai, M. T., Guiochon-Mantel, A., Jolivet, A., Lorenzo, F., Logeat, F., Pichon, M. F., Bouchard, P. and Milgrom, E. (1989). Molecular and cellular biology of mammalian progesterone receptors. *Recent Prog. Horm. Res.* **45**, 65-120.

Schmidt, T. J. and Litwack, G. (1982). Activation of the glucocorticoid-receptor complex. *Physiol. Rev.* **62**, 1131-1192.

Schrader, W. T. (1984). New model for steroid hormone receptors? *Nature* **308**, 17-18.

Walters, M. R. (1985). Steroid hormone receptors and the nucleus. *Endocr. Rev.* **6**, 512-543.

Wrange, Ö., Eriksson, P. and Perlmann, T. (1989). The purified activated glucocorticoid receptor is a homodimer. *J. Biol. Chem.* **264**, 5253-5259.

Yamamoto, K. R. (1985). Steroid receptor regulated transcription of specific genes and gene networks. *Annu. Rev. Genet.* **19**, 209-252.

Yamamoto, K. R. and Alberts, B. M. (1976). Steroid receptors: elements for modulation of eukaryotic transcription. *Annu. Rev. Biochem.* **45**, 721-746.

2
The Oestrogen Receptor: From Perception to Mechanism

S. GREEN[1] and P. CHAMBON[2]

[1] *Cell and Molecular Biology Department, ICI Central Toxicology Laboratory, Alderley Park, Macclesfield, Cheshire SK10 4TJ, UK*

[2] *Laboratoire de Génétique Moléculaire des Eucaryotes du CNRS, Unité 184 de Biologie Moléculaire et de Génie Génétique de l'INSERM, Institut de Chimie Biologique, Faculté de Médecine, 11 rue Humann, 67085 Strasbourg Cedex, France*

2.1 Historical perspective

The observation made almost a century ago by Beatson (1896) that female sex steroid hormones play a pivotal role in promoting mammary tumour growth generated considerable interest in the role and mechanism of oestrogen action. However, it was not until the 1960s and the pioneering work of the groups of Gorski and Jensen that the key mediator of oestrogen action, the oestrogen receptor (ER), was identified and a model for its mechanism of action proposed (Toft and Gorski, 1966; Jensen *et al.*, 1968; see Chapter 1.3). Since then, many groups have characterized and purified the ER, proposed and refined models to explain its mechanism of action, and gradually improved our understanding of how oestrogens modulate gene transcription. A further advance in our knowledge of the receptor's structure and molecular mechanism of action came with the cloning of the ER gene (Green *et al.*, 1986). This, together with the advent of recombinant DNA technology, has allowed the structure and function of the receptor to be dissected so that we now have a refined, yet incomplete, picture of how small ligands such as oestrogens are able to orchestrate the intricate pathways active during differentiation and development.

The ER is a member of a supergene family, known as the nuclear hormone receptor family, that includes receptors for steroid and thyroid hormones, vitamin D_3, and retinoic acid (Evans, 1988; Green and Chambon, 1988; see Chapter 1, Fig. 1). A number of additional members of the family have been cloned where the putative ligand remains to be identified. The ER is a ligand-activated transcription factor that modulates specific gene expression by binding to short DNA sequences, termed oestrogen response elements (ERE), located in the vicinity of oestrogen-regulated genes. The ERE is an enhancer since it exerts its action irrespective of its orientation and when positioned at a variable distance upstream or downstream from a variety of promotors (Klein-Hitpass et al., 1986; Seiler-Tuyns et al., 1986). Therefore the ER represents an inducible enhancer factor that contains regions important for hormone binding, ERE recognition and activation of transcription.

The ER is detectable in more than half of human breast tumours. Approximately two-thirds of these ER-positive tumours respond favourably to either the withdrawal of oestrogens or to anti-oestrogen therapy compared to only 5% of ER-negative tumours (Edwards et al., 1979). This correlation suggests that the ER plays an important role in breast cancer by mediating the mitogenic action of oestrogens. Similarly, oestrogens are mitogenic in the breast cancer cell line, MCF-7, that contains functional ER (Lippman et al., 1976; see Chapter 13.1). Although we have yet to understand how oestrogens act as mitogens in breast cancer, we are now beginning to uncover the complexity of oestrogen-regulated gene transcription and to identify those features of the regulated genes that confer responsiveness.

In this chapter we aim to outline our current understanding of the molecular mechanism of the ER and explore the implications of this model to our knowledge of how gene networks are regulated and how anti-oestrogens may act to antagonize oestrogen action.

2.2 The primary amino acid sequence

The human ER (hER) was cloned from cDNA libraries prepared from the MCF-7 breast cancer cell line (Walter et al., 1985; Green et al., 1986). The MCF-7 ER mRNA is 6322 nucleotides in length and encodes a 595 amino acid protein of 66 kD (Green et al., 1986). ER cDNA clones have been sequenced from a number of other species including chicken (Krust et al., 1986), mouse (White et al., 1987), rat (Koike et al., 1987), Xenopus (Weiler et al., 1987) and rainbow trout (Pakdel et al., 1989). Comparison of the primary amino acid sequence of the human and chicken ERs indicates six regions (A–F) of differing homology (Fig. 1; Krust et al., 1986). There are three highly conserved regions, A (87%), C (100%) and E (94%), and three that have been less well conserved, B (56%), D (38%) and F (41%). The DNA-binding domain (region C) and the hormone-binding domain (region E) are the most highly conserved regions. A transcriptional activation function has been detected in regions A/B and E but no function has been ascribed to regions D or F. A comparison of the primary amino acid sequence of the human, mouse, rat, chicken, Xenopus and trout ER demonstrates that the basic structure and presumably also the function, of the ER has been very highly conserved throughout evolution.

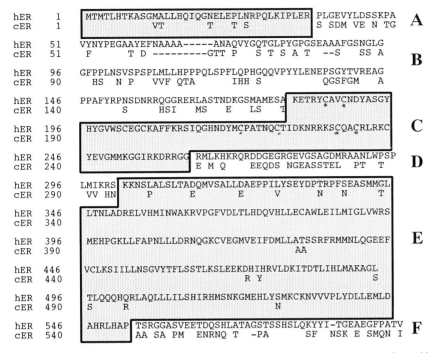

Fig. 1 Comparison of the human (hER) and chicken (cER) oestrogen receptor amino acid sequences. Only non-identical amino acids are shown in the cER sequence. The sequence is divided into six regions (A–F) based upon sequence homology between the hER and cER. The most highly conserved regions (A, B and C) are boxed and shaded.

2.3 The hormone-binding domain

Region E is a large region of approximately 250 amino acids that is hydrophobic in character (Green *et al.*, 1986). A number of ER deletion mutants generated using site-directed mutagenesis have been transiently expressed in mammalian cells and tested for their ability to bind to oestradiol (Kumar *et al.*, 1986). The integrity of a large portion of region E between amino acids 302 and 552 in the hER is required for efficient hormone binding. These results are supported by direct binding studies using the oestrogen, ketononestrol aziridine, and the anti-oestrogen, tamoxifen aziridine, which covalently bind to cysteine 530 (Harlow *et al.*, 1989). Point mutations close to the corresponding cysteine in the mouseER, namely isoleucine-518 and glycine-525 (hER amino acids 514 and 521), greatly reduced oestrogen binding although mutation of the cysteine had no effect (Fawell *et al.*, 1990). These results suggest that the C-terminal of region E is important for high-affinity oestrogen binding and that the cysteine is merely close enough to the aziridine group to react. A point mutation within region E of the MCF-7 ER cDNA that results in the replacement of the natural glycine at position 400 with a valine (Ponglikitmongkol et al., 1988) produces an unstable ER protein with an affinity for oestradiol of 3 nM at 37°C but with the wild-type affinity of 0.4 nM at 4°C (Tora et al., 1989a). A truncated receptor lacking regions A/B and C binds oestradiol with an affinity similar to that of the complete receptor (Kumar et al., 1986), suggesting that region E is an independently-folded structural domain. This is supported by the production of a 28-kD trypsin-resistant ER fragment that contains the hormone-binding domain (Katzenellenbogen et al., 1987). The hormone-binding domain may therefore represent a hydrophobic pocket. Amino acids within region E that are conserved amongst other members of the nuclear hormone receptor family may be responsible for the conformation of the pocket and some of the non-conserved amino acids may be important for ligand-specific binding. It is therefore of interest that region E is composed of five exons and that comparison of the hER with the chicken progesterone receptor (cPR) indicates that both the location and number of exons have been conserved during evolution (Huckaby et al., 1987; Ponglikitmongkol et al., 1988; see Chapter 16.2). The contribution that each of these exons makes to the structure and function of the hormone binding domain as well as a number of other functions located within region E such as the dimerization domain, interaction with heat-shock protein and activation of transcription (see below) is unknown.

2.4 The DNA-binding domain

Region C contains a core sequence of 66 amino acids (cysteine-185 to methionine-250) that is the most highly conserved region of the nuclear hormone receptor family (Fig. 2A; Green and Chambon, 1988; Evans, 1988; Beato, 1989). The corresponding region of the glucocorticoid receptor (GR) binds two atoms of zinc (Freedman et al., 1988) and is thought to fold into two DNA-binding structures (Evans and Hollenberg, 1988) that are similar to, yet clearly distinguishable from, the proposed 'zinc fingers' of the 5 S ribosomal RNA transcription factor TFIIIA (Klug and Rhodes, 1987). The putative ER zinc fingers (CI and CII; see Fig. 2A) appear to bind zinc with two pairs of cysteines rather than the pair of cysteines and histidines present in TFIIIA (Lee et al., 1989). Interestingly, like most of the TFIIIA zinc fingers, each ER zinc finger is encoded by a separate exon (Fig. 2A; Ponglikitmongkol et al., 1988). In TFIIIA, and other putative transcription factors that contain TFIIIA-like zinc fingers, the finger sequences can be

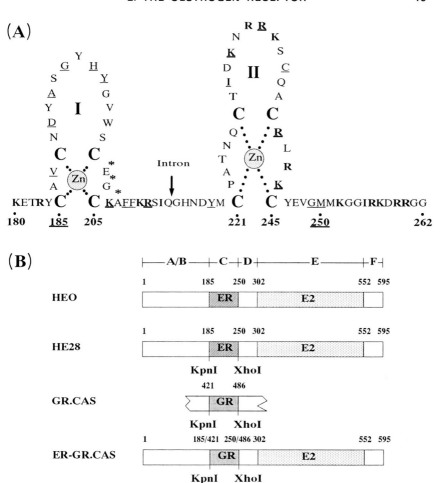

Fig. 2 The hER DNA binding domain. (A) Hypothetical structure of the hER region C (amino acids 180–262). The highly conserved core of region C (amino acids 185–250) is shown as two subregions CI (amino acids 185–215) and CII (amino acids 216–250). CI and CII each contain four highly conserved cysteine residues which may tetrahedrally coordinate zinc to form a 'zinc finger'. CI and CII are separated by an intron located within the codon of glycine-215. Amino acids that are highly conserved amongst the nuclear receptor family are underlined. Basic amino acids are shown in bold. The three amino acids within region CI that are important for the receptor to distinguish between an ERE and a GRE are denoted by asterisks. (B) Schematic structure of the chimaeric receptor ER-GR.CAS. The structures of HE0, HE28, GR.CAS and ER-GR.CAS are shown with the numbers referring to amino acid positions. The division of hER into regions A–F is shown at the top of the figure. The highly conserved core region of the DNA-binding domain of the hER (ER) and hGR (GR), together with the oestrogen-binding domain (E2) are shaded. The position of the KpnI and XhoI sites, created within region C of the hER and hGR to allow exchange of this region as a cassette (ER.CAS and GR.CAS) are indicated.

aligned to give a consensus sequence. However, the two ER zinc fingers are structurally quite distinct from one another; CI contains several hydrophobic amino acids and four invariant cysteines, whilst CII contains five invariant cysteines and is richer in basic amino acids (Fig. 2A). Together, this suggests that they could represent independent structural and functional subdomains within the DNA binding domain.

Chimaeric receptors with altered DNA binding specificity can be created by exchanging the 66 amino acid core of region C between receptors. An hER chimaera (ER-GR.CAS, Fig. 2B) that contains the two zinc fingers of the hGR, is able to activate glucocorticoid-responsive genes in the presence of oestradiol (Green and Chambon, 1987). This demonstrates that it is this core region that is responsible for specific targeting of the receptor to the hormone responsive element. Although both zinc fingers are required to bind DNA with high affinity it is the N-terminal finger (CI) that is of greater importance in determining binding specificity (Green *et al.*, 1988). Additional ER-GR chimaeras in which one or a few amino acids in CI are exchanged indicates that three amino acids at the base of this finger are important in distinguishing between the ERE and the glucocorticoid response element (GRE; Fig. 2B; Mader *et al.*, 1989; Umesono and Evans, 1989; Danielsen *et al.*, 1989; see Chapter 3.3).

Since the ER recognizes the ERE as a dimer (see below) it could be speculated that the postulated helical region of each CI finger (see Evans and Hollenberg, 1988) may contact specific bases in the major groove of each half site of the ERE palindrome. The basic amino acid-rich CII region could stablize DNA binding by contacting phosphates in the DNA backbone. An interesting possibility is that interaction between two receptor monomers could be stabilized by the coordination of zinc between some of the five invariant cysteines present in CII in a manner similar to that suggested for the HIV *tat* protein (Frankel *et al.*, 1988). It has also been suggested that a region within CII, termed the distal (D) box, may be important for protein–protein contacts in the receptor dimer that allow the correct spacing of the ERE-bound monomers (Umesono and Evans, 1989).

Although essential, the putative zinc fingers are by themselves insufficient for DNA binding (Kumar and Chambon, 1988). The sequence upstream of amino acid 179 can be deleted without loss of DNA binding activity, however, removal of sequence downstream of amino acid 261 destoys DNA binding activity and can only be restored by extending the sequence to amino acid 282. A fragment of the ER consisting of amino acids 176–282 expressed in bacteria is sufficient for specific DNA binding *in vitro* in the absence of other proteins (S. Mader, L. Tora, and P. Chambon, unpubl. res.). It is interesting to note that the region between positions 262 and 271 has been well conserved during evolution (Fig. 1) and contains a number of basic amino acids, perhaps indicating that this region also stabilizes DNA binding by contacting the DNA phosphate backbone. Interestingly, deletion of amino acids 260–270 from the full-length ER has no effect either on DNA binding or on the ability of the ER to activate transcription, whereas deletion of amino acids 250–264 prevents both (Chambraud et al., 1990). Therefore the basic region between 260 and 270 appears to be essential only in the absence of the hormone-binding domain, perhaps indicating that the presence of a strong dimerization domain located in region E (Fawell et al., 1990) compensates for this deletion.

2.5 An interaction between the oestrogen receptor and the 90-kD heat-shock protein

The unoccupied ER migrates as an 8 S form on low-salt sucrose gradients but can be dissociated with high salt to give a 4 S form. Hormone binding 'transforms' or 'activates' the receptor so that its affinity for DNA is increased and the receptor–hormone complex becomes tightly associated with the nuclear compartment. Receptor activation appears to involve several concomitant events including transformation from the 8 S form to a 4 S form; dimerization, which may represent the 5 S nuclear form of the ER (Miller et al., 1985); and increased affinity for the nuclear compartment and specific EREs.

Hormone-induced transformation of the 8 S receptor to the 4 S form appears to reflect loss of an associated 90-kD heat-shock protein, Hsp90 (Catelli et al., 1985; Chambraud et al., 1990). Although it is not clear whether the ER is associated with Hsp90 in vivo, formation of an ER–Hsp90 complex in the absence of hormone could explain why the unoccupied ER fails to bind to the ERE. Since the ER binds to the palindromic ERE more stably as a dimer (see below), then DNA binding could be reduced if interaction with Hsp90 prevents dimerization. Alternatively, or in addition, Hsp90 may mask the DNA binding domain, thereby preventing recognition of the ERE. A role for Hsp90 in preventing the unoccupied receptor from binding to its target genes is favoured by experiments demonstrating that partially purified steroid receptors that have lost Hsp90 bind to their cognate hormone response element in vitro in the absence of hormone (Bailly et al., 1986; Denis et al., 1988). In addition, deletion of the hormone-binding domain produces a truncated receptor (amino acids 1–281) that fails to interact with Hsp90 (Chambraud et al., 1990) but binds weakly to the ERE and constitutively activates transcription at a level that is 5–10% of that observed with the full-length ER (Kumar et al., 1987).

It is noteworthy that the hER functions identically when expressed in yeast or HeLa cells and that DNA binding and transactivation are hormone-dependent (Metzger et al., 1988). This suggests that if Hsp90 is responsible for the inactivity of the unoccupied receptor, then the human receptor should be able to associate with the yeast Hsp90 homologue, Hsp85 (Farrelly and Finkelstein, 1984).

Functional analyses of ER deletion mutants indicate that more than one region is involved in interaction with Hsp90 (Chambraud et al., 1990). Deletion of amino acids 250–274 produces an ER that fails to interact with Hsp90, whilst an ER with deletion of amino acids 250–264 or 260–270 interacts weakly with Hsp90 since both 4–5 S and 8–9 S forms are observed on low-salt sucrose gradients. However, this region alone is insufficient for interaction with Hsp90 since a receptor truncated from amino acid 282 migrates at 4–5 S. No specific part of region E that is important for interaction with Hsp90 has been identified. This may indicate that the primary region of interaction of the ER with Hsp90 is with amino acids 250–274 and that this interaction is stabilized at a number of sites within region E. Clearly, these results are consistent with the possibility that hormone binding alters the conformation of the hormone-binding domain sufficiently to prevent interaction of region E with Hsp90. This in turn would destabilize interaction of Hsp90 with amino acids 250–274 and, with the loss of Hsp90, the ER may then be free to dimerize and/or bind to the ERE.

2.6 The oestrogen receptor contains a nuclear localization signal

In the absence of hormone the ER is weakly associated with the nuclear compartment of the cell (King and Greene, 1984; Welshons et al., 1984). The cellular distribution of ER deletion mutants has been analysed using ER antibodies (P. Chambon group, unpubl. res.). A truncated receptor consisting of regions A/B, C and part of D (amino acids 1–281) is localized in the nucleus, whereas an ER fragment (amino acids 282–595) that contains part of region D and all of regions E and F is cytoplasmic. Additional experiments suggest that one of the regions important for interaction with Hsp90 (amino acids 250–270) is also important for nuclear localization. For example, deletion of amino acids 250–260 or 260–270 produces a receptor that is distributed between the nuclear and cytoplasmic compartments, whereas a receptor with deletion of amino acids 250–270 is exclusively cytoplasmic (P. Chambon group, unpubl. res.). Moreover, a 48 amino acid region (residues 256–303) is able to confer nuclear localization activity on β-galactosidase although amino acids 263–271 (RMLKHKRQR) were inactive (Picard *et al.*, 1990). It should be noted that this basic region is similar in sequence to the nuclear localization domain (NL1) of the glucocorticoid receptor (Picard and Yamamoto, 1987) and the progesterone receptor (Guiochon-Mantel *et al.*, 1989). These results are intriguing since they may suggest that interaction of Hsp90 with the receptor could prevent nuclear localization. Alternatively, they may indicate that interaction of the receptor with Hsp90 only occurs in the nucleus.

2.7 Oestrogen response elements are not always perfectly palindromic

The hormone-activated receptor binds to specific EREs located in the vicinity of oestrogen-regulated genes. EREs (Fig. 3; see Chapter 6) have been identified in the upstream flanking regions of several genes including the *Xenopus* (Seiler-Tuyns *et al.*, 1986; Klein-Hitpass *et al.*, 1988) and chicken (Jost *et al.*, 1984) vitellogenin genes, the chicken ovalbumin gene (Tora *et al.*, 1988), the rat prolactin gene (Waterman *et al.*, 1988), and the pS2 gene (Berry *et al.*, 1989) associated with ER-positive breast cancer (Rio *et al.*, 1987). The vitellogenin gene A2 ERE is a perfect 13 bp palindromic element consisting of two 5 bp arms separated by a 3 bp spacer (Fig. 3). The pS2 ERE differs from the vitellogenin A2 ERE by a base change in one arm and in the spacer (Fig. 3). The pS2 ERE is able to confer oestrogen-inducibility to a heterologous promoter sequence but is approximately three–four times less efficient than the perfect palindromic sequence (Berry *et al.*, 1989). This is consistent with the affinity of the ER being approximately five times lower for the pS2 ERE compared with the perfect palindrome (Kumar and Chambon, 1988). The *Xenopus* vitellogenin gene B1 contains two closely spaced 13 bp sequences (ERE-1 and ERE-2) that differ from the perfect ERE palindrome by two and one base substitutions, respectively (Fig. 3; Martinez and Wahli, 1989). Individually these imperfect palindromes confer little or no oestrogen responsiveness, but together they act as a strong ERE (see below and Martinez and Wahli, 1989; Ponglikitmongkol *et al.*, 1990).

An ERE present in the chicken ovalbumin gene is unusual in that it contains only half of the palindromic sequence and is present just upstream of the TATA box (Fig. 3; Tora *et al.*, 1988). Point mutations in this half palindrome ablate the oestrogen response, demonstrating that this half site is functional. In the presence of ER the ovalbumin gene is responsive to oestrogens in chicken embryo fibroblast (CEF) cells but not in HeLa cells.

Furthermore, *in vitro* binding studies indicate that the ER is unable to bind to a half palindrome. Together, these results suggest that interaction with a cell-specific factor(s) stabilizes ER binding to the ovalbumin gene ERE.

Xenopus vit A2	(−325)	5'	GGTCAcagTGACC	3'
Chicken vit A2	(−619)	5'	GGTCAgcgTGACC	3'
Xenopus vit B1 ERE1	(−308)	5'	AGTTAtcaTGACC	3'
Xenopus vit B1 ERE2	(−325)	5'	AGTCActgTGACC	3'
Human pS2	(−399)	5'	GGTCAcggTGGCC	3'
Rat prolactin	(−1575)	5'	TGTCActaTGTCC	3'
Chicken ovalbumin	(−41)	3'	CTGAAttgTGACC	5'

Fig. 3 Sequence of oestrogen response elements. The *Xenopus* vitellogenin A2 gene ERE is considered to represent the perfect ERE sequence. Nucleotides within other EREs that deviate from the A2 ERE are underlined. The 'arms' of the palindrome are shown in capitals. The position of the centre of the ERE relative to the transcription start site is shown to the left of the sequence. Note that all the sequences apart from the chicken ovalbumin sequence are written 5' to 3'.

2.8 The oestrogen receptor binds to a palindromic oestrogen response element as a dimer

DNA binding experiments, using the gel retardation assay, indicate that ER-containing extracts prepared from either transfected cell lines (Kumar and Chambon, 1988), yeast extracts that express the hER (Metzger *et al.*, 1988) or an *in vitro* translation system (Lees *et al.*, 1989) bind to the palindromic ERE in the presence of oestradiol. Unfortunately, the story in the absence of oestradiol is confused. With the hER expressed in yeast (*Saccharomyces cerevisiae*) DNA binding is strictly dependent upon the presence of oestradiol (Metzger *et al.*, 1988). Some DNA binding is occasionally seen in the absence of oestradiol with the hER expressed in HeLa cells (Kumar and Chambon, 1988), although this may be due to trace amounts of endogenous oestrogen in the culture medium. However the *in vitro* synthesized mouse ER binds DNA equally well in the absence or presence of hormone (Lees *et al.*, 1989). Whether this is because the conformation of the *in vitro* synthesized ER differs from that found *in vivo* (for example, the *in vitro* synthesized ER may not be associated with the heat-shock protein Hsp90) remains to be determined. It should be noted, however, that purification or physical treatment of steroid receptors (e.g. warming or high salt) allows them to bind constitutively to DNA (Willman and Beato, 1986; Bailly *et al.*, 1986; Denis *et al.*, 1988). In this respect it is of interest that the thyroid hormone receptor binds to its response element and may repress transcription in the absence of hormone (Damm et al., 1989; Sap et al., 1989) and does not appear to be associated with Hsp90 (Dalman et al., 1990). The in vivo evidence for the ER favours a model in which the receptor is not bound to the ERE in the absence of hormone and suggests that the hormone plays a major role in activating the ER to allow it to bind to DNA (see Chapter 9.3).

DNA binding experiments, using mixtures of wild-type and truncated ERs, have indicated that in the presence of oestradiol two ER monomers can dimerize in solution and bind to a palindromic ERE (Fig. 4; Kumar and Chambon, 1988). This is supported by the observation in vitro where the ER only binds weakly to an ERE that contains mutations in one half of the palindrome (Kumar and Chambon, 1988). ER dimers are relatively stable in vitro for 30 min at 25°C since mixing extracts of wild-type and truncated ER that have been pre-incubated in vivo with oestradiol produces very little heterodimer complex (Fig. 4). A greater proportion of the heterodimer complex is observed if the extracts are incubated together overnight at 4°C, suggesting that some dissociation and re-association can occur (Kumar and Chambon, 1988).

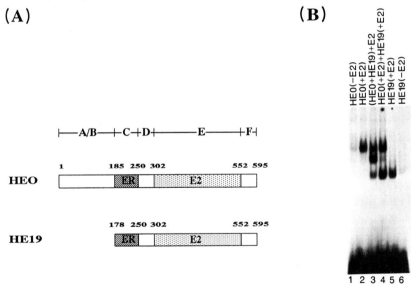

Fig. 4 The hER binds to the palindromic ERE as a dimer. (A) Schematic structure of the wild-type (HE0) and truncated (HE19) receptor with the numbers referring to the amino acid sequence. The division of hER into regions A–F is shown at the top of the figure. The highly conserved core region of the DNA-binding domain (ER) together with the oestrogen-binding domain (E2) are shaded. (B) Gel retardation assay performed using an oligonucleotide containing the vitellogenin A2 gene ERE and HeLa cell extracts containing either HE0, HE19 or both in the presence (+E2) or absence (−E2) of oestradiol (see Kumar and Chambon, 1988 for details).

Removal of the A/B region has no effect on DNA binding. However, removal of the hormone-binding domain greatly reduces the ability of the ER to bind to the palindromic ERE, suggesting the presence of a dimerization domain in this region. Similarly, this truncated receptor fails to bind to a half site ERE, suggesting that a weaker dimerization domain exists in or near to the DNA-binding domain (Kumar and Chambon, 1988). Such a dimerization domain has recently been identified in the C-terminal of region E (Fawell *et al.*, 1990). The region contains a heptad repeat that is functionally distinct from the leucine zipper motif (Landschulz *et al.*, 1988) and is conserved amongst the members of the nuclear receptor family. This may indicate that this region has a similar function in other receptors and that heterodimers may occur in some cases (see Chapter 16.3). Interestingly, the hormone-binding and dimerization activities overlap in this region and can be functionally separated by point mutations. These results indicate that *in vitro* steroid binding is not necessary to promote dimerization and support a model in which one role of the hormone is to allow dissociation of the heat-shock protein so as to unmask the dimerization domain.

2.9 Synergy between multiple oestrogen response elements and/or the binding sites of other transcription factors

The strength of a promoter is dependent upon the number and position of binding sites for *trans*-acting factors. Using minimal promoters that contain just the TATA region it is possible to examine the synergistic effect of single or tandem perfect or imperfect EREs. In this situation perfect EREs act additively and independently of their relative spacing when close to the TATA box, but synergistically and dependent upon their stereo-alignment when positioned 175 bp further upstream (Ponglikitmongkol *et al.*, 1990). The synergy of imperfect EREs, however, is dependent upon their stereo-alignment even when positioned close to the TATA box (Ponglikitmongkol *et al.*, 1990). These results differ from those observed by others (Klein-Hitpass *et al.*, 1988; Martinez and Wahli, 1989; see Chapter 6.4) where separation of the tandem imperfect EREs (ERE-1 and ERE-2) of the vitellogenin B1 gene by 2, 2.5 or 3 turns of DNA had essentially no effect when located just upstream of the herpes simplex virus thymidine kinase (tk) promoter. It is possible that the strength of the tk promoter masks or compensates for any effect of changing the stereo-alignment of the EREs.

As noted above, the presence of ERE-1 or ERE-2 alone confers little or no oestrogen-responsiveness although together they act as a strong ERE. It has been suggested that the transcriptional synergy of ERE-1 and ERE-2 may be due to co-operative binding of the ER to DNA (Martinez and Wahli, 1989). Gel retardation assays performed using a DNA fragment containing both ERE-1 and ERE-2 indicate that at higher ER concentrations occupancy of both sites is favoured over occupancy of a single site. Moreover, a DNA fragment that contains both ERE-1 and ERE-2 can prevent the formation of a complex between the ER and a single perfect ERE twenty to five times better, respectively, than competition with fragments containing either ERE-1 or ERE-2 alone (Martinez and Wahli, 1989; see Chapter 6.4). However, similar experiments performed by others (Ponglikitmongkol et al., 1990) provide no evidence for co-operative binding of the ER to tandem perfect or imperfect EREs.

The presence of binding sites for other transcription factors close to the hormone response element can increase transcription synergistically (Schule et al., 1988; Strahle et al., 1988; see Chapters 7.3 and 7.4). An additional mechanism to explain the synergistic effects of multiple *trans*-acting factor binding sites is for there to be protein–protein interactions between different *trans*-acting factors or for them to interact with common 'bridging' transcription factors that do not themselves bind DNA.

Thus although the strongest oestrogen response is seen when the promoter contains a perfect ERE, deviations from the palindrome sequence can be tolerated and are offset by multiple binding sites and the position of the ERE within the promoter.

2.10 The oestrogen receptor contains multiple distinct regions important for transcriptional activation

Once bound to the ERE the hormone–receptor complex activates transcription. Functional dissection of the hER using mutants obtained by site-directed mutagenesis indicates there to be a transcriptional activation function (TAF) within both the A/B region (TAF-1) and hormone-binding domain (TAF-2) of the receptor (Kumar et al., 1987; Metzger et al., 1988; Webster et al., 1988; White et al., 1988; Tora et al., 1989b; Lees et al., 1989) but unlike others (Waterman et al., 1988) we find no evidence for a transcriptional activation function located in the ER DNA-binding domain.

A chimaeric *trans*-activator protein (GAL-ER) constructed from the DNA-binding domain of the yeast GAL4 transcription factor and the hormone-binding domain of the hER is able to activate transcription from reporter genes that contain GAL4 binding sites (Webster et al., 1988). Importantly, transactivation is strictly hormone-dependent, suggesting that the mechanism by which the GAL-ER chimaera activates transcription may parallel that of the wild-type ER. The mechanism by which hormone binding actuates TAF-2 is poorly understood but could result from a number of mechanisms including unmasking of a pre-formed region, a conformational change in the hormone-binding domain, dimerization or phosphorylation. Interestingly, none of the five exons of the hormone-binding domain are capable of activating transcription when individually tethered to the GAL4 DNA-binding domain (Webster *et al.*, 1989). This may indicate that TAF-2 is constructed from several exons to form a structure that is recognized by the transcriptional machinery of the cell. Using mouse ER mutants that contain progressive C-terminal deletions, Lees *et al.* (1989) have identified a region between amino acid 538 and 552 (hER amino acids 534–548) that when deleted produces a receptor that binds oestradiol (K_d 0.3 nM compared to 0.1 nM for the wild-type) but only weakly activates transcription (20-fold lower than wild-type). Whether this region forms part of TAF-2 is unknown.

Truncation of the hER by removal of the hormone-binding domain produces a constitutively-active transcription factor with approximately 5% wild-type activity in HeLa or COS cells (Kumar *et al.*, 1987; Bocquel *et al.*, 1989) but with 60–70% activity in chicken embryo fibroblasts (CEF) and 100% activity in yeast (White *et al.*, 1988; Tora *et al.*, 1989b). A similarly truncated mouse ER has approximately 3% activity in 3T3 cells (Lees *et al.*, 1989). These results indicate several points: (1) TAF-2 represents the major transactivation activity of the receptor in HeLa and COS cells; (2) there is an additional transactivation function, TAF-1, contained within the truncated receptor; and (3) the activity of TAF-1 and TAF-2 is dependent upon additional cell-specific factors.

The effect of removing the A/B region (TAF-1) is dependent upon the recipient cell line and the nature of the reporter gene used to measure the activity of the receptor. In HeLa cells, removal of TAF-1 produces a receptor with wild-type activity when measured using vit-tk-CAT but only 20% activity with pS2-CAT (Kumar *et al.*, 1987). Similarly, removal of parts of the mouse ER A/B region reduces transactivation of a vitellogenin promoter 10-fold (Lees *et al.*, 1989). These differences are thought to reflect the nature of additional *trans*-acting factors that bind to the promoter regions of the reporter genes rather than to differences in the affinity of the receptor for the ERE.

Interestingly, TAF-1 and TAF-2 are structurally dissimilar from the acidic-, glutamine- or proline-rich activating regions that have been described for other transcription factors. Importantly, TAF-1 and TAF-2 are different from the acidic activating regions that have been identified in both the A/B and E regions of the glucocorticoid receptor (Godowski *et al.*, 1988;

2. THE OESTROGEN RECEPTOR

Hollenberg and Evans, 1988; see Chapters 3.2 and 3.4). TAF-1, TAF-2 or the acidic activating domain (AAD) of the herpes simplex viral gene, VP16, were functionally compared after fusion with the DNA-binding domain of either GAL4 or the ER (Tora et al., 1989b). A reporter gene containing one or two GAL4 binding sites was used to analyse the ability of these different activating regions to homosynergize. Similarly, a reporter gene containing an ERE and a GAL4 binding site was used to analyse their ability to heterosynergize. TAF-1 was unable to homosynergize so that, for example, the activity of GAL-ER(A/B) observed with the reporter gene containing two GAL4 binding sites was approximately double that seen with the reporter gene with a single site. However, TAF-1 was able to heterosynergize with either TAF-2 or AAD to give, depending upon the combination between five- and 60-fold greater activity than expected from additive effects. TAF-2 could homosynergize (2.5-fold in HeLa cells and 20-fold in CEF cells) but could not synergize with AAD, whilst AAD was able to homosynergize. These results (summarized in Fig. 5) demonstrate that the properties of these different activating regions are functionally distinct and moreover suggest that TAF-1, TAF-2 and AAD interact with different components of the transcription apparatus (see below).

	ER(A/B) (TAF-1)	ER(EF) (TAF-2)	VP16 (AAD)
ER(A/B) (TAF-1)	-	+	+
ER(EF) (TAF-2)		+	-
VP16 (AAD)			+

Fig. 5 Summary of homosynergy and heterosynergy between the two transcription activating regions (TAF-1, TAF-2) of the hER and the acidic activating domain (AAD) of the herpes simplex virus protein VP16 (see Tora et al., 1989b for details).

2.11 Mechanisms of oestrogen antagonism

Anti-oestrogens such as tamoxifen (Nolvadex), 4-hydroxytamoxifen, and the pure anti-oestrogen, ICI 164 384, antagonize oestrogen action by competing with oestrogens for receptor binding (see Wakeling and Bowler, 1988). Since these compounds bind to the receptor it is pertinent to ask whether anti-oestrogens invoke or inhibit conformational changes in the receptor in order to suppress full oestrogen activity. Using gel retardation and methylation interference assays it has been demonstrated that the ER recognizes an ERE specifically when either oestradiol or hydroxytamoxifen is bound (Kumar and Chambon, 1988; Metzger *et al.*, 1988; Lees *et al.*, 1989). However, the receptor fails to bind to DNA when bound to ICI 164 384 (V. Kumar, M. Berry and P. Chambon, unpubl. res.; S. Fawell and M. Parker, unpubl. res.; see Chapter 16.7). This suggests that *in vivo* the ER recognizes the same target genes irrespective of whether hydroxytamoxifen or oestradiol is bound but does not bind DNA when bound to ICI 164 384. It is interesting to note that in the gel retardation assay the receptor–ligand–DNA complex migrates slightly slower in the presence of hydroxytamoxifen than it does in the presence of oestradiol (Metzger *et al.*, 1988; Kumar and Chambon, 1988; Lees *et al.*, 1989), suggesting that the structure of the ER is different when bound to tamoxifen. Differences in ER structure when bound to either oestradiol or tamoxifen have also been noted when using monoclonal antibodies against the ER (Fauque *et al.*, 1985).

Although the receptor binds to an ERE equally well in the presence of either oestradiol or hydroxytamoxifen, it activates transcription less well when bound to the anti-oestrogen. For example, the ER or the GAL-ER chimaera (see above) stimulate transcription in HeLa cells in the presence of oestradiol but not in the presence of hydroxytamoxifen (Webster *et al.*, 1988). These results imply that oestrogens, but not anti-oestrogens, promote a structural change in the ER hormone-binding domain that is necessary for complete activation of a functional receptor.

Similar experiments compared the ability of the GR and GAL-GR (a chimaera consisting of the GAL4 DNA-binding domain and the GR hormone-binding domain) to activate transcription in the presence of either the synthetic glucocorticoid, dexamethasone, or the anti-glucocorticoid, RU486 (Webster *et al.*, 1988). Interestingly, RU486 behaved as a complete antagonist with GAL-GR but as a partial agonist with the intact GR, demonstrating 20% of the activity seen with dexamethasone. RU486 may act as a GR antagonist by promoting DNA binding in the absence of activation of the hormone-binding domain transactivation function (Webster *et al.*, 1988). A transactivation function, representing approximately 20% of the activity seen with the intact receptor, has been identified in the GR A/B region (Hollenberg *et al.*, 1987). Because the A/B region is absent in GAL-GR but present in the GR it was speculated that anti-hormones may function as partial agonists by promoting DNA binding, thereby allowing the A/B region of the receptor to activate transcription (Fig. 6; Green and Chambon, 1988).

The potency of tamoxifen as an oestrogen antagonist depends upon the

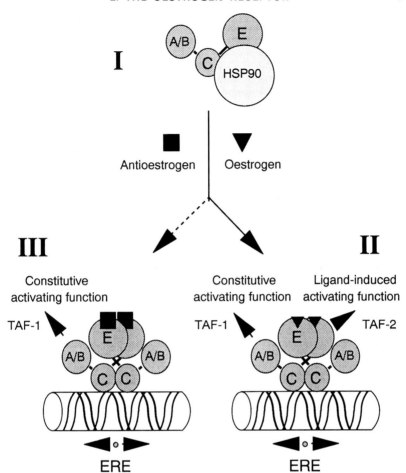

Fig. 6 Model for the mechanism of oestrogen and anti-oestrogen action. The receptor is schematically shown as three domains representing regions A/B, the DNA binding domain (C) and the hormone binding domain (E). The unoccupied receptor (I) is unable to bind strongly to the ERE possibly because of the presence of the heat-shock protein Hsp90 which may prevent dimerization and/or recognition of the ERE. Addition of oestrogen (II) promotes dimerization involving strong interaction between the hormone-binding domains (E) and weak interactions between the DNA-binding domains (C). The receptor dimer binds tightly to the ERE and activates transcription via a ligand-inducible transcription activating function (TAF-2) present in the hormone-binding domain and a constitutive transcription activating function (TAF-1) in region A/B. Anti-oestrogens, such as tamoxifen, promote dimerization and DNA binding (III) whereas others, such as the pure anti-oestrogen ICI 164 384, do not. Tamoxifen appears not to activate the oestrogen-inducible transcription activating function in region E (TAF-2) whereas the constitutive TAF-1 region remains active. The agonistic potency of tamoxifen depends upon the ability of TAF-1 to synergize with other transcription factors that bind to the promoter and therefore it is both gene and tissue specific.

tissue and response examined (Jordan, 1984; May and Westley, 1987). The ability of hydroxytamoxifen to function as a partial oestrogen agonist appears to result from its ability to promote DNA binding of an ER in which TAF-1 is active but TAF-2 is not (Fig. 6). A variable degree of agonist activity expressed by tamoxifen may be explained by a variation in the activity of TAF-1 in different species and tissues and also by the nature of the oestrogen-responsive target gene. For example, we speculate that specific transactivating factors are able to co-operate functionally with TAF-1 and that the presence of these factors will be species and tissue specific (see above). Thus it can be demonstrated that hydroxytamoxifen is a potent ER agonist and strongly activates transcription of the pS2 promoter in either CEF cells or yeast where TAF-1 accounts for a major proportion of the receptor activity (M. Berry, D. Metzger and P. Chambon, unpubl. res.). Similarly, the contribution made by TAF-1 to target gene transcription is dependent upon other transcription factors that can bind to the promoter (Nunez et al., 1989). For example, although TAF-1 is unable to homosynergize, it can synergize with transcription factors that possess acidic activation domains (see above).

The above results have important implications for the role of tamoxifen in breast cancer therapy. Circulating levels of oestrogens are believed to play an important role in promoting the growth of ER-positive breast tumours (Jordan et al., 1986). The ability of tamoxifen treatment to achieve remission in patients with oestrogen-responsive tumours is thought to be due to the role of tamoxifen as an oestrogen antagonist that competitively binds to the ER. Because tamoxifen may also act as a partial agonist it is possible that tamoxifen may partially activate transcription of some of the oestrogen-responsive genes involved in the regulation of tumour growth. Pure anti-oestrogens, such as ICI 164 384, may therefore be expected to suppress completely the expression of these genes and therefore be of greater efficacy in controlling hormone-dependent breast cancer.

2.12 How does the oestrogen receptor activate gene transcription?

The basic mechanism by which the ER stimulates an increase in transcription from target genes is unknown. Currently, a popular theory to explain enhancer function is that the binding of transcription factors to cognate DNA elements located in the vicinity of target genes promotes protein–protein interactions that favour formation of a stable initiation complex (Ptashne, 1986, 1988).

A number of yeast and mammalian transcription factors function in heterologous systems and complement each other *in vitro* (Buratowski *et al.*, 1988; Cavallini *et al.*, 1988; Metzger *et al.*, 1988; Webster *et al.*, 1988). This, together with conservation of the large subunit of RNA polymerase B from yeast to man (see Bartholomei *et al.*, 1988), suggests that the mechanism of transcription of polymerase B transcribed genes has been highly conserved during evolution. The ER is able to stimulate transcription from minimal promoters that contain a simple TATA box element (see Tora *et al.*, 1989b). Enhancer factors may therefore function by recruiting or promoting interaction of basic transcription factors such as the TATA box factor TFIID (Horikoshi *et al.*, 1989), allowing initiation of transcription. In addition to interacting with other DNA-binding proteins, the ER may interact with a number of 'bridging' transcription factors that do not themselves bind DNA. Competition amongst enhancer factors for a limiting number of such 'bridging' transcription factors could explain the inhibition of transcription (squelching) of genes by the over-expression of enhancer factors that are incapable of binding to those genes (Gill and Ptashne, 1988; Meyer *et al.*, 1989). The challenge is to understand the nature and function of these 'bridging' transcription factors by the isolation and characterization of proteins that associate with the hormone-activated receptor.

Although the molecular mechanism of oestrogen action may have been highly conserved during evolution, it is evident that physiological mechanisms have not. Thus, although the basic function of oestrogen as a female sex steroid is to regulate reproductive function, the gene network required to achieve this differs among species. This is dependent upon the presence of ERE in the promoters of oestrogen-responsive genes rather than to changes in the receptor itself. Clearly, there is greater flexibility during evolution to regulate gene networks by altering promoters that contain response elements for the ER or other transcription factors than by altering the sequences of the factors themselves. The fine tuning of the response may be achieved by altering the number, position or sequence of the response elements. In addition, the ability of different combinations of enhancer factors with some, like the ER, containing multiple transcriptional activating functions, to recruit a limiting number of tissue-specific or developmentally-regulated transcription factors in order to allow initiation of transcription, allows for a complex and intricate number of combinations to modulate spatial and temporal gene expression during development.

References

Bailly, A., Le Page, C., Rauch, M. and Milgrom, E. (1986). Sequence-specific DNA binding of the progesterone receptor to the uteroglobin gene: effects of hormone, anti-hormone and receptor phosphorylation. *EMBO J.* **5**, 3235–3241.

Beato, M. (1989). Gene regulation by steroid hormones. *Cell* **56**, 335–344.
Beatson, G. T. (1896). *Lancet* **ii**, 104–107; 162–165.
Bartholomei, M., Halden, N., Cullen, C. and Corden, J. (1988). Genetic analysis of the repetitive carboxy-terminal domain of the largest subunit of mouse RNA polymerase II. *Mol. Cell. Biol.* **8**, 330–339.
Berry, M., Nunez, A. M. and Chambon, P. (1989). Estrogen-responsive element of the human pS2 gene is an imperfectly palindromic sequence. *Proc. Natl. Acad. Sci. USA* **86**, 1218–1222.
Bocquel, M. T., Kumar, V., Stricker, C., Chambon, P. and Gronemeyer, H. (1989). The contribution of the N- and C-terminal regions of steroid receptors to activation of transcription is both receptor and cell-specific. *Nucl. Acids Res.* **17**, 2581–2595.
Buratowski, S., Hahn, S., Sharp, P. A. and Guarente, L. (1988). Function of a yeast TATA element-binding protein in a mammalian transcription system. *Nature* **334**, 37–42.
Catelli, M. G., Binart, N., Jung-Testas, I., Renoir, J.-M., Baulieu, E. E., Feramisco, J. R. and Welch, W. J. (1985). The common 90-kD protein component of nontransformed '8S' steroid receptors is a heat-shock protein. *EMBO J.* **4**, 3131–3135.
Cavallini, B., Huet, J., Plassat, J.-L., Sentenac, A., Egly, J. M., Chambon, P. (1988). A yeast activity can substitute for the HeLa cell TATA box factor. *Nature* **334**, 77–80.
Chaumbraud, B., Berry, M., Redeuilh, G., Chambon, P. and Baulieu, E. E. (1990). The formation of estrogen receptor–Hsp90 complexes involves several sites of interaction. *J. Biol. Chem.* in press.
Dalman, F. C., Koenig, R. J., Perdew, G. H., Massa, E. and Pratt, W. B. (1990). In contrast to the glucocorticoid receptor, the thyroid hormone receptor is translated in the DNA binding state and is not associated with Hsp 90. *J. Biol. Chem.* **265**, 3615–3618.
Damm, K., Thompson, C. C. and Evans, R. M. (1989). Protein encoded by v-erb A functions as a thyroid hormone receptor antagonist. *Nature* **339**, 593–597.
Danielsen, M., Hinck, L. and Ringold, G. M. (1989). Two amino acids within the knuckle of the first zinc finger specify DNA response element activation by the glucocorticoid receptor. *Cell* **57**, 1131–1138.
Denis, M., Poellinger, L., Wikström, A. and Gustafsson, J. (1988). Requirement of hormone for thermal conversion of the glucocorticoid receptor to a DNA-binding state. *Nature* **333**, 686–688.
Edwards, D. P., Chamness, G. C. and McGuire, W. L. (1979). Estrogen and progesterone receptor proteins in breast cancer. *Biochim. Biophys. Acta.* **560**, 457–486.
Evans, R. M. (1988). The steroid and thyroid hormone receptor family. *Science* **240**, 889–895.
Evans, R. M. and Hollenberg, S. (1988). Zinc fingers: gilt by association. *Cell* **52**, 1–3.
Farrelly, F. W. and Finkelstein, D. B. (1984). Complete sequence of the heat shock-inducible Hsp90 gene of *Saccharomyces cerevisiae*. *J. Biol. Chem.* **259**, 5745–5751.
Fauque, J., Borgna, J. L. and Rochefort, H. (1985). A monoclonal antibody to the estrogen receptor inhibits *in vitro* criteria of receptor activation by an estrogen and an anti-estrogen. *J. Biol. Chem.* **260**, 15 547–15 553.
Fawell, S. E., Lees, J. A., White, R. and Parker, M. G. (1990). Characterization and co-localization of steroid binding and dimerization activities in the mouse estrogen receptor. *Cell* **60**, 953–962.

Frankel, A. D., Bredt, D. S. and Pabo, C. O. (1988). Tat protein from human immunodeficiency virus forms a metal-linked dimer. *Science* **240**, 70–73.
Freedman, L. P., Luisi, B. F., Korszum, Z. R., Basavappa, R., Sigler, P. B. and Yamamoto, K. R. (1988). The function and structure of the metal coordination sites within the glucocorticoid receptor DNA binding domain. *Nature* **334**, 543–546.
Gill, G. and Ptashne, M. (1988). Negative effect of the transcriptional activator GAL4. *Nature* **334**, 721–724.
Godowski, P. J., Picard, D. and Yamamoto, K. R. (1988). Signal transduction and transcriptional regulation by glucocorticoid receptor–LexA fusion proteins. *Science* **241**, 812–816.
Green, S., Walter, P., Kumar, V., Krust, A., Bornert, J.-M., Argos, P. and Chambon, P. (1986). Human oestrogen receptor cDNA: sequence, expression and homology to v-erb-A. *Nature* **320**, 134–139.
Green, S. and Chambon, P. (1987). Oestradiol induction of a glucocorticoid-responsive gene by a chimaeric receptor. *Nature* **325**, 75–78.
Green, S. and Chambon, P. (1988). Nuclear receptors enhance our understanding of transcription regulation. *Trends Genet.* **4**, 309–314.
Green, S., Kumar, V., Theulaz, I., Wahli, W. and Chambon, P. (1988). The N-terminal DNA-binding 'zinc finger' of the oestrogen and glucocorticoid receptors determines target gene specificity. *EMBO J.* **7**, 3037–3044.
Guiochon-Mantel, A., Loosfelt, H., Lescop, P., Sar, S., Atger, M., Perrot-Applanat, M. and Milgrom, E. (1989). Mechanisms of nuclear localization of the progesterone receptor: evidence for interaction between monomers. *Cell* **57**, 1147–1154.
Harlow, K. W., Smith, D. N., Katzenellenbogen, J. A., Greene, G. L. and Katzenellenbogen, B. S. (1989). Identification of cysteine 530 as the covalent attachment site of an affinity-labeling estrogen (ketononestrol aziridine) and antiestrogen (tamoxifen aziridine) in the human estrogen receptor. *J. Biol. Chem.* **264**, 17 476–17 485.
Hollenberg, S. M. and Evans, R. M. (1988). Multiple and cooperative trans-activation domains of the human glucocorticoid receptor. *Cell* **55**, 899–906.
Hollenberg, S. M., Giguere, V., Segui, P. and Evans, R. M. (1987). Colocalization of DNA-binding and transcriptional activation functions in the human glucocorticoid receptor. *Cell* **49**, 39–46.
Horikoshi, M., Wang, C. K., Fujii, H., Cromlish, J. A., Weil, P. A. and Roeder, R. G. (1989). Cloning and structure of a yeast gene encoding a general transcription initiation factor TFIID that binds to the TATA box. *Nature* **341**, 299–303.
Huckaby, C. S., Conneely, O. M., Beattie, W. G., Dobson, A. D. W., Tsai, M. J. and O'Malley, B. W. (1987). Structure of the chromosomal chicken progesterone receptor gene. *Proc. Natl. Acad. Sci. USA* **84**, 8380–8384.
Jensen, E. V., Sujuki, T., Kawashima, T., Stumpf, W. E., Jungblut, P. W. and DeSombre, E. R. (1968). A two-step mechanism for the interaction of estradiol with rat uterus. *Proc. Natl. Acad. Sci. USA* **59**, 632–638.
Jordan, V. C. (1984). Biochemical pharmacology of antiestrogen action. *Pharmacol. Rev.* **36**, 245–276.
Jordan, V. C. (ed.) (1986). "Estrogen/Anti-estrogen Action and Breast Cancer Therapy". University of Wisconsin Press.
Jost, J.-P., Seldran, M. and Geiser, M. (1984). Preferential binding of estrogen–receptor complex to a region containing the estrogen-dependent hypomethylation site preceding the chicken vitellogenin II gene. *Proc. Natl. Acad. Sci. USA* **81**, 429–433.

Katzenellenbogen, B. S., Elliston, J. F., Monsma, F. J., Springer, P. A., Ziegler, Y. S. and Greene, G. L. (1987). Structural analysis of covalently labeled estrogen receptors by limited proteolysis and monoclonal antibody reactivity. *Biochemistry* **26**, 2364–2373.

King, W. J. and Greene, G. L. (1984). Monoclonal antibodies localize oestrogen receptor in the nuclei of target cells. *Nature* **307**, 745–747.

Klein-Hitpass, L., Schorpp, M., Wagner, U. and Ryffel, G. U. (1986). An estrogen-responsive element derived from the 5' flanking region of the *Xenopus* vitellogenin A2 gene functions in transfected human cells. *Cell* **46**, 1053–1061.

Klein-Hitpass, L., Kaling, M. and Ryffel, G. U. (1988). Synergism of closely adjacent estrogen-responsive elements increases their regulatory potential. *J. Mol. Biol.* **201**, 537–544.

Klug, A. and Rhodes, D. (1987). 'Zinc finger': a novel protein motif for nucleic acid recognition. *Trends Biochem. Sci.* **12**, 464–469.

Koike, S., Sakai, M. and Maramatsu, M. (1987). Molecular cloning and characterization of rat estrogen receptor cDNA. *Nucl. Acids Res.* **15**, 2499–2513.

Krust, A., Green, S., Argos, P., Kumar, V., Walter, P., Bornert, J.-M. and Chambon, P. (1986). The chicken oestrogen receptor sequence: homology with v-erb-A and the human oestrogen and glucocorticoid receptors. *EMBO J.* **5**, 891–897.

Kumar, V. and Chambon, P. (1988). The estrogen receptor binds tightly to its reponsive element as a ligand-induced homodimer. *Cell* **55**, 145–156.

Kumar, V., Green, S., Staub, A. and Chambon, P. (1986). Localisation of the oestradiol-binding and putative DNA-binding domains of the human oestrogen receptor. *EMBO J.* **5**, 2231–2236.

Kumar, V., Green, S., Stack, G., Berry, M., Jin, J. R. and Chambon, P. (1987). Functional domains of the human estrogen receptor. *Cell* **51**, 941–951.

Landschulz, W. H., Johnson, P. F. and McKnight, S. L. (1988). The leucine zipper: a hypothetical structure common to a new class of DNA binding proteins. *Science* **240**, 1759–1764.

Lee, M. S., Gippert, G. P., Soman, K. V., Case, D. A. and Wright, P. E. (1989). Three-dimensional solution structure of a single zinc finger DNA-binding domain. *Science* **245**, 635–637.

Lees, J. A., Fawell, S. E. and Parker, M. G. (1989). Identification of two transactivation domains in mouse oestrogen receptor. *Nucl. Acids Res.* **17**, 5477–5488.

Lippman, M. E., Bolan, G. and Huff, K. K. (1976). The effects of androgens and antiandrogens on hormone-responsive human breast cancer in long-term tissue culture. *Cancer Res.* **36**, 4595–4601.

Mader, S., Kumar, V., de Verneuil, H. and Chambon, P. (1989). Three amino acids of the oestrogen receptor are essential to its ability to distinguish an oestrogen from a glucocorticoid-responsive element. *Nature* **338**, 271–274.

Martinez, E. and Wahli, W. (1989). Cooperative binding of estrogen receptor to imperfect estrogen-responsive DNA elements correlates with their synergistic hormone-dependent enhancer activity. *EMBO J.* **8**, 3781–3791.

May, F. E. and Westley, B. R. (1987). Effects of tamoxifen and 4-hydroxytamoxifen on the pNR-1 and pNR-2 estrogen-regulated RNAs in human breast cancer cells. *J. Biol. Chem.* **262**, 15 894–15 899.

Miller, M. A., Mullick, A., Greene, G. L. and Katzenellenbogen, B. S. (1985). Characterization of the subunit nature of nuclear estrogen receptors by chemical cross-linking and dense amino acid labeling. *Endocrinology* **117**, 515–522.

Metzger, D., White, J. H. and Chambon, P. (1988). The human oestrogen receptor functions in yeast. *Nature* **334**, 31–36.
Meyer, M.-E., Gronemeyer, H., Turcotte, B., Bocquel, M.-T., Tasset, D. and Chambon, P. (1989). Steroid hormone receptors compete for factors that mediate their enhancer function. *Cell* **57**, 433–442.
Nunez, A.-M., Berry, M., Imler, J.-L. and Chambon, P. (1989). The 5′ flanking region of the pS2 gene contains a complex enhancer region responsive to oestrogens, epidermal growth factor, a tumour promoter (TPA), the c-Ha-ras oncoprotein and the c-jun protein. *EMBO J.* **8**, 823–829.
Pakdel, F., Le Guellec, C., Vaillant, C., Le Roux, M. G. and Valtaire, Y. (1989). Identification and estrogen induction of two estrogen receptors (ER) messenger ribonucleic acids in the rainbow trout liver: sequence homology with other ERs. *Mol. Endocrinol.* **3**, 44–51.
Picard, D. and Yamamoto, K. R. (1987). Two signals mediate hormone-dependent nuclear localization of the glucocorticoid receptor. *EMBO J.* **6**, 3333–3340.
Picard, D., Kumar, V., Chambon, P. and Yamamoto, K. R. (1990). *Cell. Reg.* (in press).
Ponglikitmongkol, M., Green, S. and Chambon, P. (1988). Genomic organization of the human oestrogen receptor gene. *EMBO J.* **7**, 3385–3388.
Ponglikitmongkol, M., White, J. H. and Chambon, P. (1990). Synergistic activation of transcription by the human oestrogen receptor bound to tandem responsive elements. *EMBO J.* **9**, 2221–2232.
Ptashne, M. (1986). Gene regulation by proteins acting nearby and at a distance. *Nature* **322**, 697–701.
Ptashne, M. (1988). How eukaryotic transcriptional activators work. *Nature* **335**, 683–689.
Rio, M. C., Bellocq, J. P., Gairard, B., Rasmussen, U. B., Krust, A., Koehl, C., Calderoli, H., Schiff, V., Renaud, R. and Chambon, P. (1987). Specific expression of the pS2 gene in subclasses of breast cancers in comparison with expression of estrogen and progesterone receptors and the oncogene ERBB2. *Proc. Natl. Acad. Sci. USA* **84**, 9243–9247.
Sap, J., Munoz, A., Schmitt, J., Stunnenberg, H. and Vennström, B. (1989). Repression of transcription mediated at a thyroid hormone response element by the v-erb-A oncogene product. *Nature* **340**, 242–244.
Schule, R., Muller, M., Kaltschmidt, C. and Renkawitz, R. (1988). Many transcription factors interact synergistically with steroid receptors. *Science* **242**, 1418–1420.
Seiler-Tuyns, A., Walker, P., Martinez, E., Merillat, A. M., Givel, F. and Wahli, W. (1986). Identification of estrogen-responsive DNA sequences by transient expression experiments in a human breast cancer cell line. *Nucl. Acids Res.* **14**, 8755–8770.
Strahle, U., Schmid, W. and Schutz, G. (1988). Synergistic action of the glucocorticoid receptor with transcription factors. *EMBO J.* **7**, 3389–3395.
Toft, D. O. and Gorski, J. (1966). A receptor molecule for estrogens: isolation from the rat uterus and preliminary characterization. *Proc. Natl. Acad. Sci. USA* **5**, 1574–1581.
Tora, L., Gaub, M. P., Mader, S., Dierich, A., Bellard, M. and Chambon, P. (1988). Cell-specific activity of a GGTCA half-palindromic oestrogen-responsive element in the chicken ovalbumin gene promoter. *EMBO J.* **7**, 3771–3778.
Tora, L., Mullick, A., Metzger, D., Ponglikitmongkol, M., Park, I. and Chambon, P. (1989a). The cloned human oestrogen receptor contains a mutation which alters its hormone binding properties. *EMBO J.* **8**, 1891-1896.

Tora, L., White, J., Brou, C., Tasset, D., Webster, N., Scheer, E. and Chambon, P. (1989b). The human estrogen receptor has two independent nonacidic transcriptional activation functions. *Cell* **59**, 477–487.

Umesono, K. and Evans, R. M. (1989). Determinants of target gene specificity for steroid/thyroid hormone receptors. *Cell* **57**, 1139–1146.

Wakeling, A. E. and Bowler, J. (1988). Novel antioestrogens without partial agonist activity. *J. Steroid Biochem.* **31**, 645–653.

Walter, P., Green, S., Greene, G., Krust, A., Bornert, J.-M., Jeltsch, J. M., Staub, A., Jensen, E., Scrace, G., Waterfield, M. and Chambon, P. (1985). Cloning of the human estrogen receptor cDNA. *Proc. Natl. Acad. Sci. USA* **82**, 7889–7893.

Waterman, M. L., Adler, S., Nelson, C., Greene, G. L., Evans, R. M. and Rosenfeld, M. G. (1988). A single domain of the estrogen receptor confers deoxyribonucleic acid binding and transcriptional activation of the rat prolactin gene. *Mol. Endocrinol.* **2**, 14–21.

Webster, N. J., Green, S., Jin, J. R. and Chambon, P. (1988). The hormone-binding domains of the estrogen and glucocorticoid receptors contain an inducible transcription activation function. *Cell* **54**, 199–207.

Webster, N., Green, S., Tasset, D., Ponglikitmongkol, M. and Chambon, P. (1989). The transcriptional activation function located in the hormone-binding domain of the human oestrogen receptor is not encoded in a single exon. *EMBO J.* **8**, 1441–1446.

Weiler, I. J., Lew, D. and Shapiro, D. J. (1987). The *Xenopus laevis* estrogen receptor: sequence homology with human and avian receptors and identification of multiple estrogen receptor messenger ribonucleic acids. *Mol. Endocrinol.* **1**, 355–362.

Welshons, W. V., Lieberman, M. E. and Gorski, J. (1984). Nuclear localization of unoccupied oestrogen receptors. *Nature* **307**, 747–749.

White, J. H., Metzger, D. and Chambon, P. (1988). Expression and function of the human estrogen receptor in yeast. *Cold Spring Harbor Symp. Quant. Biol.* **53**, 819–828.

White, R., Lees, J. A., Needham, M., Ham, J. and Parker, M. (1987). Structural organization and expression of the mouse estrogen receptor. *Mol. Endocrinol.* **1**, 735–744.

Willmann, T. and Beato, M. (1986). Steroid-free glucocorticoid receptors binds specifically to mouse mammary tumour virus DNA. *Nature* **324**, 688–691.

3
Structure and Function of the Glucocorticoid Receptor

M. DANIELSEN

*Department of Biochemistry and Molecular Biology,
Georgetown University Medical School, 3900 Reservoir Road, N.W.,
Washington, DC 20007, USA*

3.1 Introduction

Scarcely any tissue in the mammalian organism fails to respond in one way or another to glucocorticoids. The responses range from small changes in the activity of enzymes often involved in homeostasis to growth inhibition and cell death. The glucocorticoid receptor (GR), like other members of the steroid receptor family (Evans, 1988; Ham and Parker, 1989), consists of three structural domains, the N-terminal domain, the DNA-binding domain and the hormone-binding domain (Fig.1). The N-terminal 50 kDa of the receptor is essential for full transcriptional activity. The DNA-binding domain, comprising approximately 75 amino acids, contains two zinc fingers that confer DNA binding specificity on the receptor. It also contains a basic region involved in nuclear localization and possibly in interactions with DNA. The hormone-binding domain of approximately 30 kDa, upon binding of hormone, promotes DNA binding and transcriptional activation. In the absence of hormone, this domain is involved in suppressing receptor activity since mutants lacking this domain are constitutively active. This suppression may be due to interaction with masking proteins such as Hsp90, which appears to form a complex with the receptor only in the absence of hormone. This review will summarize our current knowledge of the structure and function of the glucocorticoid receptor and the potential role played by heat-shock proteins.

3.2 The N-terminal domain

The N-terminal domain of the GR is required for full transcriptional activity (Danielsen *et al.*, 1987; Hollenberg *et al.*, 1987; Miesfeld *et al.*, 1987) and, since it is not required for hormone binding (Rusconi and Yamamoto,

1987), receptor activation (Gehring and Arndt, 1985), dimerization (Eriksson and Wrange, 1990) or DNA recognition (Freedman et al., 1988), it has been termed the modulatory domain. There is strong evidence that the N-terminal domain of the GR contains at least one sequence that can act as a transcriptional activator. A hybrid protein in which the GR N-terminal domain is attached to the LexA DNA-binding domain is capable of activating transcription from a promoter regulated by the LexA operator (Godowski et al., 1988). Most of the transcription activating activity was located between amino acids 237 and 318 of the rat receptor (mGR 225–306), a region these authors call enh2. Similarly Hollenberg and Evans (1988) spliced the N-terminal domain of the human GR to the GAL4 DNA-binding domain, yielding a hybrid receptor that could activate transcription from a GAL4-responsive promoter. These authors mapped this activation function to between amino acids 77 and 262 of the human GR (mGR 84–270) and have called this sequence τ1. Duplication of τ1 in the human GR results in a modest (two- to four-fold) increase in receptor activity, indicating that the region can act as a transcription activator in the context of a modified GR, and that its position in the recombinant protein is not critical. The structural integrity of the N-terminal domain of the wild-type GR is important, however, since insertions of three or four amino acids, both within [at residues 120, 204, 214 of the human GR (mGR 129, 213, 223)] and outside [at residue hGR 346 (mGR 354)] enh2/τ1 inhibit transcriptional activity.

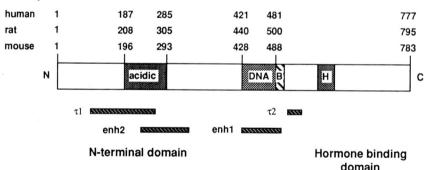

Fig. 1 Structure of the glucocorticoid receptor. The amino acid co-ordinates of the domains are shown for the mouse (Danielsen et al., 1986), rat (Miesfeld et al., 1986), and human GR (Hollenberg et al., 1985). τ1, τ2, and enh1, enh2 are transcription activation domains as defined by Hollenberg and Evans (1988) and Godowski et al. (1988). B is a basic region adjacent to the zinc fingers. The region of greatest homology (H) in the hormone-binding domain of members of the gene family is shown in more detail in Fig. 5.

Both τ1 and enh2 overlap a highly acidic region of the receptor between amino acids 196 and 293 of the mGR that has been shown to be required for near normal receptor activity (Danielsen et al., 1987). Since acidic regions of proteins can act as transcriptional activators (Sigler, 1988), it is tempting to suggest that this acidic region in the GR has the role of a transcriptional activator. However, the removal of the acidic region increases the non-specific DNA-binding properties of the receptor with the result that the difference between specific and non-specific DNA binding is reduced *in vitro* (Payvar and Wrange, 1984). It is possible that this reduced ability to distinguish between specific and non-specific sites is at least a contributory lesion in these N-terminally truncated receptors. It has been suggested that the increased non-specific DNA-binding properties of N-terminally truncated receptors is due to their increased positive charge. This is supported by an inverse relationship between positive charge and transcriptional activity in a series of N-terminal deletions (Danielsen et al., 1987). An alternative explanation has been put forward by Eriksson and Wrange (1990), who have presented evidence that loss of the N-terminal domain of the GR results in altered protein–protein contacts in the GR dimer, resulting in altered protein–DNA interactions.

Among steroid receptors the N-terminal domain is the most variable in size and amino acid composition (Evans, 1988; see Chapters 1 and 5). For instance, the rat T3 α receptor has only 52 amino acids in this domain compared to 427 in the mGR. In some steroid receptors the N-terminal domain can have dramatic differential effects on the ability of the receptor to activate transcription from specific promoters. An N-terminally truncated oestrogen receptor, for instance, can regulate the vitellogenin promoter normally, but has little activity on the pS2 promoter (Kumar et al., 1987; see Chapter 2). Similarly, the A and B forms of the progesterone receptor, which differ by 127 amino acids at the N-terminus, interact differentially with the ovalbumin and MMTV promoters (Tora et al., 1988; see Chapter 5). Perhaps the variability in the structure of the N-terminal domain gives to the receptor the ability to interact with a specific subset of transcription factors. This might help explain why the GR and progesterone, androgen and mineralocorticoid receptors (PR, AR and MR) all recognize the same response element yet bring about such diverse effects *in vivo*.

3.3 The DNA-binding domain

The three-dimensional structure of the DNA-binding domain of the GR has recently been solved using two-dimensional nuclear magnetic resonance (2D NMR) of a DNA-binding domain fragment produced in *Escherichia coli* (Härd et al., 1990) (see cover illustration). The domain contains two fingers formed by the tetrahedral co-ordination of two zinc atoms by four pairs of cysteines (Freedman et al., 1988, Fig. 2). α-Helices are located immediately adjacent to the distal side of each finger (Ser_{447}–Glu_{457}, and Pro_{481}–Gly_{492}), and are orientated perpendicular to each other. There is a type 1 and a type 2 turn near the beginning of the second finger (Arg_{467}–Cys_{470}, and Leu_{463}–Gly_{466}, respectively). In addition there is a small region of antiparallel β-sheet involving amino acids Cys_{428} and Leu_{429} and Leu_{463}–Gly_{466}, in the first finger. The DNA-binding domains of all members of the steroid and thyroid hormone receptor family are likely to have similar structures since their primary amino acid sequences are highly conserved. The cysteine and histidine co-ordinated zinc fingers found in such proteins as TFIIIA, however, are quite different (Miller et al., 1985; Lee et al., 1989).

Table 1 Mutations in the DNA-binding domain of the GR. Point mutations are indicated by the single-letter amino acid designation for a wild-type residue, followed by its position within the protein, and then the single letter code for the mutant protein. Insertion mutants have the letters In followed by the position after which the amino acids are inserted, followed by the amino acids inserted. Δ481–484 has the amino acids R A S S P replacing amino acids 481–484 inclusive. The numbering system used throughout the table refers to the mouse GR. Conversion to the rat GR numbering system can be affected by adding 12 to each number, and to the human numbering system by deleting 7 from each number. Trans, the ability to activate transcription; all values refer to mammalian cells (m) unless followed by y which indicates yeast. DNA, the ability to bind to DNA; the assays used can be found in the relevant references. +, at least 25%, +/−, less than 25%, −, less than 5% of wild-type values. cs, cold sensitive. pc, bind DNA, do not trans-activate in yeast (y) or in mammalian cells (m). References: (1) Hollenberg and Evans (1988); (2) Schena et al. (1989); (3) Danielsen et al. (1986); (4) Severne et al. (1988); (5) Rusconi and Yamamoto (1987); (6) Giguere et al. (1986); (7) Danielsen et al. (1989); (8) M. Danielsen et al. (unpubl. res.); (9) Umesono and Evans (1989).

	Phenotype					Phenotype			
Mutation	Trans	DNA	Other	Ref.	Mutation	Trans	DNA	Other	Ref.
P 425 S	+			8	R 467 G	+	+		1
K 426 I	+			8	D 469 G	+	+		1
K 426 N	+			8	C 470 G	−	−		1
L 427 P	+ (y)			2	C 470 R	− (y)	−		2
C 428 G	−	−		1	C 470 Y	− (ym)	−		2
C 428 R	− (ym)	−		2	C 470 H	+			4
C 428 Y	− (y)			2	C 470 S	−			4
V 430 G	−	−		1	I 472 G	−	−		1
C 431 G	−	−		1	K 474 G	+	+/−		1
S 432 P	− (ym)	−		2	R 476 Q	− (y)	+	pcy	2
D 433 G	+	+		1		+ (m)			
G 437 V	+			3	R 477 G	−	−		1
C 438 G	+	+/−		1	R 477 K	+/− (y)	+/−	cs	2
C 438 S	+			4		+ (m)			
C 438 Y	+			7	K 478 G	+/−	+/−		1
H 439 G	+/−	+		1	N 479 S	+ (y)	+	pcm	2

3. THE GLUCOCORTICOID RECEPTOR

Mutation	Phenotype Trans	DNA	Other	Ref.	Mutation	Phenotype Trans	DNA	Other	Ref.
H 439 Y	+			4		− (m)			
H 439 Q	+			8	C 480 G	−	−		1
Y 440 G	−	−		1	C 480 R	− (ym)	−		2
G 441 E	+ (ym)			2	C 480 S	−			4
V 442 A	+ (y)			2	P 481 G	+	+		1
T 444 G	+/−	+/−		1	C 483 G	−	−		1
C 445 G	−	−		1	C 483 Y	− (y)	−		2
C 445 Y	− (y)	−		2	C 483 S	+			4
G 446 E	+			9	R 484 G	−	−		1
S 447 T	+			8	R 484 H	−	−		3
C 448 G	−	−		1	Y 485 L	+			7
C 448 Y	− (ym)	−		2	R 486 G	+	+		1
C 448 S	−			4	K 487 G	+	+		1
C 448 A	−			4	C 488 G	−	−		1
K 449 G	−	+	pcm?	1	C 488 R	− (y)	−		2
V 450 A	+/−			7	C 488 Y	+			8
F 451 G	−	−		1	C 488 Y	− (y)			2
F 451 S	− (ym)	−		2	C 488 S	+			4
F 452 G	−	−		1	C 488 A	+			4
K 453 G	+	+/−		1	L 489 P	− (y)	−		2
R 454 G	−	−		1	G 492 R	+ (y)			2
R 454 G	− (y)	−		2	G 492 E	+ (y)			2
R 454 K	+/− (m)	−		2	M 493 G	+/−	−		1
	− (y)				M 493 I	− (y)	−		2
E 457 G	+	+		1	M 493 T	− (y)	−		2
E 457 K	+ (y)			2	I 507 T	+ (y)			2
H 460 G	+	+		1	P 481 R + A 482 S	−	+	pcm	2
Y 462 G	+	+/−		1	In429–GSV	−			6
Y 462 H	+ (y)			2	In435–RIRA	−			6
L 463 P	+ (y)			2	In447–ADPR	−			6
C 464 G	−	−		1	In495–GSV	+/−			6
C 464 A	−			4	In497–RIRA	+/−			6
C 464 H	+/−			4	In522–RIR	+			6
R 467 K	+ (y)			2	Δ481–484	−	−		5

The DNA-binding domain of the GR contains nine cysteines which are conserved in all steroid receptors, one more than is required to tetrahedrally co-ordinate two molecules of zinc. This has led to the development of two different co-ordination schemes for the formation of the second finger. Cys_{480} and Cys_{483} could be used to produce a finger containing nine amino acids as shown in Fig. 2, or Cys_{483} and Cys_{488} could be used to form a finger with 12 amino acids in the loop. Severne et al. (1988) have shown that Cys_{488}, but not Cys_{480} and Cys_{483}, can be converted to serine without compromising the transcriptional activity of the receptor. Since serine would

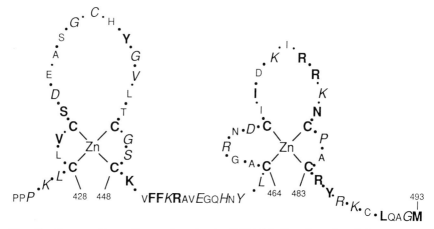

Fig. 2 Compendium of mutations in the DNA-binding domain of the GR. The DNA-binding domain is drawn as two zinc fingers using the co-ordination scheme suggested by Severne et al. (1988). References to each mutation are given in Table 1. Mutations that have at least 25% of wild-type transcriptional activity are in italics, and less than 5% activity in bold. The mouse GR numbering system is used throughout.

not be expected to interact with zinc (Vallee and Auld, 1990), these authors propose the co-ordination scheme shown in Fig. 2. Consistent with this proposal is the finding that replacing either Cys_{480} or Cys_{483} individually with histidines (that can also co-ordinate zinc) results in receptors that can still activate transcription, albeit less well than wild type receptor. This co-ordination scheme is also supported by the structural data obtained from 2D NMR (Härd et al., 1990).

The DNA-binding domain of the GR has been subjected to extensive mutagenesis. Perhaps the most surprising result has been the realization that mutagenesis of amino acids that are conserved across species lines and in all receptor types does not necessarily result in inactive receptors (summarized in Table 1). For instance, a His at position 439 of the mGR is present at the equivalent position of all steroid/thyroid hormone receptors cloned, yet mutation to Tyr results in a receptor with apparently normal transcriptional activity (Severne et al., 1988). A possible explanation for this dichotomy is that these mutations lead to small alterations in receptor activity that are not detected in the assays used. In the intact organism such alterations could lead to subtle changes in the expression of various genes, thereby conferring a selective disadvantage on organisms harbouring such a mutation. Alternatively one could argue that portions of the DNA-binding domain are involved in additional uncharacterized activities that are not measured in the relatively simplified transcriptional assays used to characterize these mutations.

Schena et al. (1989) have used selection in yeast to obtain a large number of point mutations in the DNA-binding domain of the GR. Although most mutations resulted in loss of both DNA-binding activity and transcriptional activity three were obtained that although transcriptionally inactive were capable of binding to DNA (Table 1). These positive control (pc) mutations were closely linked in the second zinc finger and they may define a region of the domain that is involved in protein–protein interactions. These mutations showed strongly divergent properties in yeast and mammalian cells. For instance R476Q (Table 1) binds to DNA normally and activates transcription in mammalian cells, yet is inactive in yeast. This would imply that protein–protein interactions are required for transactivation and that these interactions differ in yeast and mammalian cells. Another mutation R477K is cold sensitive in yeast, again indicating a role for this region in protein–protein interactions.

3.3.1 DNA-Binding Specificity

Green and Chambon (1987) have clearly demonstrated that the region encoding the two zinc fingers of the oestrogen receptor is involved in the recognition of specific DNA sequences (see also Chapter 2.4). They substituted the GR DNA-binding domain for the ER DNA-binding domain to generate a hybrid receptor that activates transcription through a glucocorticoid response element (GRE) rather than an oestrogen response element (ERE) in response to oestrogens. This type of experiment has been termed a 'finger swap' and has been used effectively to study the ligand binding properties of unidentified cloned receptor sequences (Giguere et al., 1987; Petkovich et al., 1987; Benbrook et al., 1988; Brand et al., 1988).

Some of the amino acids involved in determining the specificity of DNA binding have been established (Fig. 3). Changing the Gly–Ser sequence between the distal pair of cysteines in the first zinc finger of the GR to Glu–Gly (GE9, Fig. 3), residues that are found in the oestrogen receptor, leads to activation of transcription through an oestrogen response element (ERE) rather than through a glucocorticoid response element (GRE) (Danielsen et al., 1989; Umesono and Evans, 1989; reviewed in Berg, 1989). The converse of these experiments has been performed by Mader et al. (1989) starting with the oestrogen receptor. Changing the Glu–Gly of the oestrogen receptor to Gly–Ser of the glucocorticoid receptor results in a receptor that no longer activates transcription through an ERE, but can partially activate transcription through a GRE. Making an additional Ala to Val change in the region immediately adjacent to the first finger increases the activity of this hybrid receptor on a GRE (see Chapter 2).

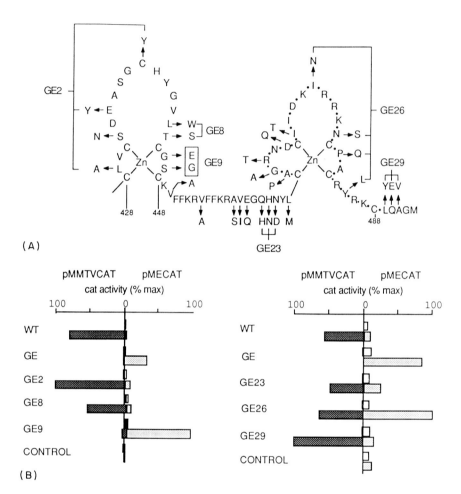

Fig. 3 Change of specificity mutations in the DNA-binding domain of the mouse GR. (A) Structure of the DNA-binding domain of the GR. Differences between the sequence of the GR and the ER are indicated by arrows. The boxed amino acids indicate those that convert the GR's ability to activate transcription through a GRE to an ERE. (B) COS-7 cells were transfected with the indicated receptor constructions, together with the GRE reporter pMMTVCAT or the ERE reporter pMECAT. Cells were re-fed the next day with either normal growth medium or with medium containing 10^{-6} M dexamethasone (dex). Cells were harvested for CAT assays two days later. All transfections were performed in triplicate, the values shown are the mean values obtained in five transfections. Open and closed bars, no dex treatment; light and heavy stippled bars, dex treated. GE, both fingers changed to the ER sequence; WT, wild-type GR sequence; control, pMMTVCAT or pMECAT alone. A more complete set of mutations and complete details of the experiment can be found in Danielsen et al. (1989).

An ERE and a GRE have very similar sequences. Martinez et al. (1987) have shown that the near-consensus GRE, AGGACActgTGTCCT, can be converted into an ERE by changing just one base pair in each half of the palindrome yielding AGGTCActgTGACCT (see Chapter 6). It is conceivable that the amino acids between the distal cysteines of the first finger interact directly with this base pair. Alternatively one could envision that the amino acids in the knuckle either direct or allow the formation of a specific structure that permits interaction with either a GRE or an ERE. Berg (1989) first suggested and structural analysis now confirms (Härd et al., 1990) that the presence of a glycine between the cysteines in both the ER and GR allows the formation of a strongly amphipathic α helix involving the 10 amino acids immediately following the second set of cysteines of the first finger. This would be functionally analogous to the helix found in Cys_2His_2 fingers that may be involved in recognizing specific DNA sequences (Lee et al., 1989, and references therein). Cadmium can be exchanged for zinc in a GR DNA-binding domain polypeptide synthesized in E. coli and the resulting complex retains its ability to bind specifically to a GRE (Freedman et al., 1988). Since cadmium and zinc have different co-ordination spheres the S–S separation could increase by as much as 0.03 nm. This indicates that whatever the constraints are that determine DNA binding they appear not to require the exact spatial configuration of the co-ordinated cysteines and thus of the intervening amino acids. It would be of interest to determine whether specificity of GRE versus ERE recognition is altered by replacement of zinc with cadmium.

The results described above demonstrate that the ability of the GR to discriminate between a GRE and an ERE resides in the first zinc finger. However, the second finger is also required for DNA binding since deletions (Hollenberg et al., 1987; Danielsen et al., 1987; Miesfeld et al., 1987), amino acid insertions, and point mutations into this region give rise to inactive receptors (Danielsen et al., 1986; Hollenberg and Evans, 1988). As expected, hybrid receptors in which parts of the second finger or interfinger regions are replaced by the corresponding oestrogen receptor sequence retain their ability to recognize a GRE (Danielsen et al., 1989). However, some of these hybrid receptors also gain the ability to activate transcription through an ERE (GE26, Fig. 3). Such promiscuous activity does not map to a single region of the second finger since a number of different hybrids, involving various amino acid substitutions, show GR and ER activity. Conceivably, substitutions of ER amino acids in the GR second finger (or inter-finger region) disrupt the structure of the protein in such a way as to reduce the ability of the receptor to distinguish between a GRE and an ERE.

The thyroid hormone receptor recognizes the sequence AGGTCATGACCT which differs from an ERE AGGTCACTGTGACCT only by the

Table 2 Characteristics of selected monoclonal and polyclonal antibodies that recognize the GR or associated proteins. M, monoclonal, P, polyclonal antibodies. CR, cross-reactivity. N, recognition. N-, no recognition of native protein. W, recognizes protein on Western blots. Species designators are: m, mouse; r, rat; h, human. Binding of 8G11-C6 to GR is not inhibited by hormone and is therefore probably not an anti-idiotypic antibody against the hormone-binding site. It was probably obtained due to an autoimmune production of antibody against the B chain of insulin which cross-reacts with receptor. An anti-idiotypic antibody has been raised against the closely related mineralocorticoid receptor (MR) (Lombes et al., 1990). Binding of this antibody to the MR is inhibited by aldosterone but not by RU486. This antibody does not recognize the GR.

Name	Immunogen	Type	Recognition sequence	Notes	References
GR49	Purified rGR	M (IgG)	120–160 rGR (mGR 108–148)	N, W, CR: mGR	Westphal et al. (1982, 1984)
BUGR 1&2	Purified rGR	M (IgG2)	407–423 rGR (mGR 395–411)	N, W, CR: mGR, rabbit GR	Gametchu and Harrison (1984), Eisen et al. (1985), Rafestin-Oblin et al. (1986), Rusconi and Yamamoto (1987)
PGR1	Recombinant gal K fusion to mGR aa 453–783	P	Multiple	N, W	A. van der Straten, M. Danielsen and G. M. Ringold (unpubl. res.)
aP1	Recombinant rGR 440–795 (428–683, mGR)	P	510–560 rGR (major) (498–548 mGR)	N, W, CR: mGR, hGR	Hoeck et al. (1989), Hoeck and Groner (1990)
mab 7	rGR	M (IgG2a)	N-terminal domain	N, W, mGR, hGR	Okret et al. (1984), Wikström et al. (1987)
mab 250	rGR	M	N-terminal domain rGR 119–273 (mGR 107–261)	N, W	Okret et al. (1984), Rusconi and Yamamoto (1987)
AP64	DNA-binding domain—basic region peptide	P	mGR 488–505	N, W, 4 S receptor only, inhibits DNA binding	Urda et al. (1989)

³A6	Unactivated rGR	M (IgM)		N, W. Inhibits steroid and DNA binding. CR: mGR, hGR, pig GR	Robertson et al. (1987)
RAP2	GR DNA-binding domain peptide	P	(mGR 471–482)	N-, W. CR: cPR, hGR, hPR	D. F. Smith et al. (1988)
266	Chick PR DNA-binding domain peptide	P	cPR 465–481 (mGR 472–488)	N, W. 4 S receptor only. CR: hGR, hPR. Inhibits DNA binding of PR	D. F. Smith et al. (1988)
268	Chick PR/GR DNA-binding domain peptide	P	cPR 450–464 identical to mGR 457–471	N, W. CR: hGR	D. F. Smith et al. (1988)
269	Chick PR DNA-binding domain peptide	P	cPR 450–481 (mGR 457–488)	N-, W. CR: hGR, hPR	D. F. Smith et al. (1988)
anti-GR 8G11-C6	hGR	M		W	Marchetti et al. (1989)
	Auto-anti-idiotypic (see legend) with TA-thyroglobulin as the original immunogen.	M (IgM)	DNA-binding domain LCLVCSD mGR 427–433	N, W. Inhibits DNA binding. CR: B-chain of human insulin amino acids 15–21 (LYLVCGE)	Cayanis et al. (1986, 1989)
αGR135	Peptide 144–172 hGR, (mGR 153–181)	P	N-terminal domain	N, W	Hollenberg et al. (1987)
anti-GR	Purified rGR	M (IgM)		N, W. CR: rPR, mGR, not hGR	Teasdale et al. (1986)
AC88	*Achlya ambisexualis* (water mould) Hsp90	M (IgG1)	Hsp90	N, W. CR: Hsp90 of mouse, human, rat, pig steroid receptors	Riehl et al. (1985), Schuh et al. (1985), Dalman et al. (1989)

(continued)

Name	Immunogen	Type	Recognition sequence	Notes	References
8D3	Partially purified mouse Ah (dioxin) receptor	M (IgM)	Hsp90	N, CR: rat, human, but not *Achlya ambisexualis* Hsp90. 9S mER, mGR	Perdew (1988), Dalman et al. (1989)
anti-Hsp90	Rat Hsp90 dissociated from unactivated rGR	P	Hsp90	W, CR: rabbit Hsp90	Denis et al. (1987), Denis (1988)
anti-Hsp90	Gel isolated mouse Hsp90	P	Hsp90	W, CR: 63 and 67-kD proteins	Rexin et al. (1988a)
D7a	Chick Hsp90	M	Hsp90	N, W	Brugge et al. (1983), Kost et al. (1989)
7D11	Chick Hsp90	M (IgG1)	Hsp90	N, W. Only recognizes free Hsp90	Sullivan et al. (1985), Kost et al. (1989), Schuh et al. (1985)
BF4	Chick PR	M (IgG2b)	Hsp90	N, W. CR: 9 S receptor —not human	Joab et al. (1984), Radanyi et al. (1983, 1989)
Kn 382/ EC1	Unactivated rabbit PR	M (IgG1)	56/59-kD receptor associated protein	N, W. CR: GR, ER, PR, AR from rabbit; hGR; not murine receptors	Tai et al. (1986), Nakao et al. (1985), Bresnick et al. (1990)

spacing between the two halves of the palindrome (see Chapter 6). For this reason it is not surprising that to convert the GR into a receptor that will recognize a TRE the amino acids Gly–Ser between the second pair of cysteines have to be changed to the TR (and ER) equivalent Glu–Gly (Umesono and Evans, 1989). In addition to these changes the amino acids between the first pair of cysteines of the second finger have to be converted to the TR equivalent in order to obtain transcriptional activity from a TRE. These authors suggest that the second finger changes are required in order to accommodate the spacing differences between the two halves of either an ERE or a TRE palindrome. An alternative explanation is that second finger changes merely increase the promiscuous activity of these receptors. This seems likely since TR/GR hybrid receptors activate transcription from both a TRE and an ERE (Umesono and Evans, 1989) whereas the authentic thyroid hormone receptor (Glass *et al.*, 1988) activates transcription from a TRE only.

Härd *et al.* (1990) have developed a model of GR binding to DNA which fits with both the structural and biochemical data described in detail above. In this model, one GR binds to each half of the GRE palindrome. The so called recognition α-helix of each receptor (mGR 447–457) binds in the major groove and interacts directly with GRE-specific bases. The two halves of the receptor dimer interact in the region of the first pair of cysteines of the second finger. The amino acids of the second finger that appear important for protein–protein interaction face away from the DNA so that they could conceivably interact with the transcriptional machinery. This model can now be tested by mutational analysis, which in turn should allow refinement of the model.

3.3.2 Role of the Basic Region in Nuclear Localization

Immediately adjacent to the second zinc finger is a region enriched in basic amino acids (mGR 498–505, Fig. 1). A basic region is present in all steroid receptors at approximately the equivalent position, but the actual sequence is not conserved. It has been suggested that this region is important for nuclear localization of the GR, since its attachment to other proteins confers nuclear localization (Picard and Yamamoto, 1987). Although this basic region confers hormone-dependent nuclear localization on hybrid proteins, if it is used in the GR itself it would probably only function in hormone-activated receptors, since a monoclonal antibody to this sequence recognizes the activated but not the non-activated forms of the receptor (Table 2; Urda *et al.*, 1989). There is another nuclear localization sequence present in the hormone-binding domain of the GR that has been shown to be hormone dependent (Picard and Yamamoto, 1987).

C-terminal deletions that extend into the basic region yield transcriptionally inactive receptors (Danielsen et al., 1987; Hollenberg et al., 1987). This has led to the suggestion that this region of the receptor interacts non-specifically with DNA. In support of this Freedman et al. (1988) have found that the basic region is necessary for DNA binding of receptor fragments produced in E. coli, and Urda et al. (1989) have found that an antibody against this region inhibits DNA binding. Schena et al. (1989), using chemical mutagenesis of a fragment of the rat GR (residues 414–556, mGR 402–544) and selection for transcription-defective receptors in yeast, did not obtain any point mutations in this basic region. This implies that single amino acid changes in the region do not yield a selectable phenotype. In contrast to these results, Hollenberg et al. (1987) have found that deletion of amino acids 492–515 (mGR 499–522), which encompasses the basic region, yields a transcriptionally active receptor. This construction is not a simple deletion, however, since a number of basic amino acids were introduced as a consequence of the cloning procedure. Whether this explains the disparity in the results of the above studies remains to be determined.

3.4 The hormone-binding domain

Mild trypsin cleavage of the GR gives rise to a 27- to 30-kD C-terminal fragment which contains the hormone-binding site. Extensive mapping of this domain by proteolysis, affinity labelling and *in vitro* mutagenesis has revealed that the whole of the C-terminal domain, approximately 262 amino acids, is required to form the hormone-binding site. In addition, this domain contains the binding site for Hsp90 and possibly other receptor-associated proteins (Section 3.6). Receptor lacking the C-terminal hormone-binding domain activates transcription, suggesting that in the intact receptor this domain represses activity in the absence of hormone, possibly due to the interaction with Hsp90, and that this repression is relieved by hormone binding. In addition, the hormone-binding domain also contains a hormone-dependent transactivation domain (Webster et al., 1988) and a nuclear localization sequence (Picard and Yamamoto, 1987), both of which are activated upon hormone binding.

3.4.1 Limits of the Hormone-Binding Site

Rusconi and Yamamoto (1987) have shown that an N-terminal deletion to amino acid 497 of the rat GR (mGR 485) results in receptor with near wild-type affinity for dexamethasone, while deletion to amino acid 547 (mGR 535) gives a receptor that has a K_d over 11 000-fold higher than wild-type receptor. Additional studies of deletion mutants, including an hGR that lacked residues 490–515 (mGR 497–522) but retained activity (Hollenberg et

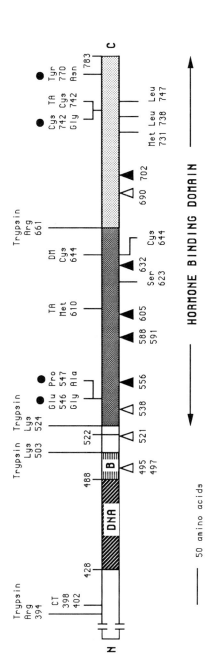

Fig. 4 Structure of the hormone-binding domain of the GR. In the upper part of the figure point mutations that reduce hormone binding of the GR are shown above the receptor (filled circles). The wild-type amino acid is followed by the residue number (mouse GR), which in turn is followed by the amino acid mutation. GR amino acids which are covalently labelled by affinity reagents are shown: Met_{610} and Cys_{742} are photoaffinity labelled by triamcinolone acetonide (TA), Cys_{644} is covalently modified by dexamethasone mesylate (DM). The major trypsin and chymotrypsin cleavage sites are shown: Lys_{524} and Arg_{661} are the cleavage sites that give rise to the 16-kD core hormone binding unit (shaded). In the lower part of the figure the triangles represent three or four amino acid insertions that reduce (filled) or do not affect (open) hormone binding. Ser_{623} is equivalent to hER Val_{400} which, when mutated to Gly, reduces oestrogen binding. Cys_{644} is equivalent to Gly_{340} of the hTRβ gene which, when mutated to Arg, results in loss of thyroid hormone binding. Met_{731} is equivalent to hAR Val_{866} which, when mutated to Met, reduces androgen binding. Leu_{747} is equivalent to hER Cys_{530} which is affinity labelled by ketononestrol aziridine and tamoxifen aziridine (Harlow et al., 1989). Met_{731} and Leu_{738} are equivalent to mER Ile_{518} and Gly_{525}, respectively, which, when mutated to Arg, yields oestrogen-deficient receptors. References are given in the text. The computer program Gap was used to align sequences (Devereux et al., 1984).

al., 1987; Oro et al., 1988a; Pratt et al., 1988) suggests that the N-terminal boundary of the domain lies somewhere between amino acid residues 522 and 535 of the mouse GR.

The C-terminal limit of the hormone-binding site is near the C-terminal end of the receptor since removal of the last five amino acids from the rat GR results in a receptor with a K_d 30-fold higher than wild-type receptor, and more extensive deletions resulted in receptors with reduced affinity (Rusconi and Yamamoto, 1987). Furthermore, Hollenberg et al. (1985) have isolated a human GR cDNA clone termed hGRβ, in which the last 50 amino acids had been replaced by 15 unrelated amino acids, that is unable to bind hormone.

The hormone-binding domain has also been mapped by inserting three or four amino acids at random throughout the human GR in a linker scanning approach (Giguere et al., 1986; Fig. 4). Insertions at or before amino acid 532 (mGR 538) give rise to receptors that can bind hormone, while those at or following amino acid 550 (mGR 556), in general, do not. This places the N-terminal boundary of the domain to between amino acids 538 and 556

Fig. 5 (Opposite) Region of homology in the hormone-binding domain of members of the steroid/thyroid hormone receptor family. Dashes represent identity with the mouse GR (mGR) sequence, a space is a gap used to align the sequences (used for EAR2 only). The same receptor types from different species are shown only if the sequences differ in this region, those with identity to the species shown are cited immediately after the sequence of identity. Prefixes: h, human; m, mouse; r, rat; c, chick; d, Drosophila. References: GR, mouse (Danielsen et al., 1986), human (Hollenberg et al., 1985), rat (Miesfeld et al., 1986); androgen receptor, AR, human and rat (Chang et al., 1988); progesterone receptor, PR, human (Misrahi et al., 1987), rabbit (Loosfelt et al., 1986), chick (Gronemeyer et al., 1987); mineralocorticoid receptor, MR, human (Arriza et al., 1987), rat (Patel et al., 1989); oestrogen receptor related receptor, human 1 and 2, ERR1 and ERR2 (Giguere et al., 1988), rainbow trout ERR (Pakdel et al., 1989); oestrogen receptor, ER, human (Green et al., 1986), chick (Krust et al., 1986), Xenopus (Weiler et al., 1987), rat (Koike et al., 1987), mouse (White et al., 1987); H2RIIBP, mouse (Hamada et al., 1989); vitamin D receptor, VDR, human (Baker et al., 1988), rat (Burmester et al., 1988); E75, Drosophila (Feigl et al., 1989); NUR/77, mouse (Hazel et al., 1988) which is the same protein as N10 (Ryseck et al., 1989), rat (called NGFI-β) (Watson and Milbrandt, 1989); retinoic acid receptor, RAR, human α (Giguere et al., 1987; Petkovich et al., 1987), β (The et al., 1987; Benbrook et al., 1988), γ (Krust et al., 1989), mouse, α,β,γ (Zelent et al., 1989), newt α,δ (Ragsdale et al., 1989); SVP (seven-up), Drosophila, (Mlodzik et al., 1990); EAR2, human (Miyajima et al., 1988); COUP, may be chicken homologue of SVP (Wang et al., 1989), also called EAR3 (Miyajima et al., 1988); thyroid hormone receptor, TRα, human (Miyajima et al., 1989), rat (Thompson et al., 1987), Xenopus laevis (Z. Mohran and M. Danielsen, unpubl. res.), TRβ, human (Weinberger et al., 1986), rat (Murray et al., 1988); Reverse TR, revTR, human (Miyajima et al., 1989), rat (Lazar et al., 1989); TR2, (Chang and Kokontis, 1988). Three members of the family from Drosophila, Knirps (Nauber et al., 1988), knirps-related (Oro et al., 1988a,b), and egon (Rothe et al., 1989) show no homology to this sequence.

	start	Trp	Ala	Lys	Ala	Ile	Pro	Gly	Phe	Arg	Asn	Leu	His	Leu	Asp	Gln	Met	Thr	Leu	end
mGR	583	Trp	---	---	---	---	---	---	---	---	---	---	---	---	---	---	---	---	Leu	602
rGR	595	---	---	---	---	---	---	---	Leu	---	---	---	---	---	---	---	---	---	---	614
hAR	717	---	---	---	---	Leu	---	---	---	---	---	---	---	Val	---	---	---	Ala	Ile	736
hPR	732	---	Ser	---	Ser	Leu	---	---	---	---	---	---	---	Ile	---	---	Ile	Val	Ile	751
cPR	585	---	Ser	---	Leu	Leu	---	---	---	---	---	---	---	Ile	---	---	Ile	---	Ile	604
hMR	783	---	---	---	Val	Leu	---	---	---	---	---	---	---	Ile	---	---	Ile	---	Ile	802
hERR1	339	Phe	---	---	Ser	---	---	---	---	Lys	---	---	Pro	Ser	Glu	---	---	Ser	---	358
hERR2	257	---	---	---	His	---	---	---	---	Ser	Ser	---	Ser	---	Ser	---	---	Ser	Val	276
hER	360	---	---	---	Arg	Val	---	---	---	Val	Asp	---	Thr	---	Gly	---	Val	His	---	379
rER	365	---	---	---	Arg	Val	---	---	---	Gly	Asp	---	Thr	---	His	---	Val	His	---	384
H2RIIBP	266	---	---	---	Arg	---	---	His	---	Ser	Ser	---	Pro	---	His	---	Val	Ile	---	285
hVDR	244	Phe	---	---	Met	---	---	---	---	---	Asp	---	Thr	Ser	Glu	---	Ile	Val	---	263
rVDR	240	Phe	---	---	Met	---	---	---	---	---	Asp	---	Thr	Ser	---	---	Ile	Val	---	259
dE75B	631	Phe	Gly	Met	---	---	---	---	---	Gln	Leu	---	Thr	Gln	---	Lys	Phe	---	---	650
NUR77	423	---	Glu	---	Lys	---	---	---	---	Ile	Leu	---	Cys	Pro	Gly	---	Asp	Leu	---	442
hRARα	242	Phe	---	---	Gln	Leu	---	---	---	Thr	Thr	---	Thr	Ile	Ala	---	Ile	---	---	261
hRARβ	235	Phe	---	---	Arg	Leu	---	---	---	Thr	Thr	---	Thr	Ile	Ala	---	Ile	---	---	254
hRARγ	244	Phe	---	---	Arg	Leu	---	---	---	Thr	Gly	---	Ser	Ile	Ala	---	Ile	---	---	263
dSVP	355	---	---	---	Asn	---	---	Phe	---	Pro	Glu	---	Gln	Val	Thr	---	Val	Ala	---	374
hEAR2	216	---	Arg	---	His	Gly	Phe	---	---	Pro	Glu	---	Pro	Val	Ala	---	Val	Ala	---	234
COUP	---	---	Arg	---	Asn	---	Phe	---	---	Pro	Asp	---	Gln	Ile	Thr	---	Val	Ser	---	---
hTRα	232	Phe	---	---	Lys	Leu	Met	---	---	Ser	Glu	---	Pro	Cys	Glu	---	Ile	Ile	---	251
hTRβ	281	Phe	---	---	Lys	Leu	Met	---	---	Cys	Glu	---	Pro	Cys	Glu	---	Ile	Ile	---	300
revTR	453	---	---	---	His	---	---	---	---	---	Asp	---	Ser	Gln	His	---	Val	Ser	---	472
hTR2	407	---	Leu	---	Ser	---	Ser	---	---	Gln	Ala	---	Gly	Gln	Glu	Asn	Ile	---	Val	426

of the mGR, which agrees well with the deletion data described above. Interestingly, the hormone-binding domain of the GR contains a sequence that is conserved among all members of the nuclear hormone receptor family (Danielsen et al., 1986; Fig. 5). Certain insertions within this region disrupt hormone binding (Giguere et al., 1986). In addition to a possible role in hormone binding, this part of the receptor may also be required for interaction with Hsp 90 (Section 3.6).

Another approach to mapping this domain has been to use protease digestion to create fragments of the receptor which can still bind steroid and then to map the location of these fragments. Simons et al. (1989) have carried out such a study on the rat GR and have been able to obtain a 16-kD fragment that binds dexamethasone with an affinity only 23-fold lower than wild-type receptor. Mapping of this fragment using SDS polyacrylamide gel electrophoresis, comparison with other protease digestion products, and computer generation of possible tryptic peptides based on the protein sequence, located it at residues 537–673 (mGR 525–661). This fragment encompasses amino acid residues 546 and 547, which are important for hormone binding and residues Cys_{644} and Met_{610}, which have been identified by affinity labelling (see below). The authors claim that this is the core hormone-binding unit of the receptor. It is difficult to reconcile these results with C-terminal deletion studies which have found that deletion of only 29 amino acids results in a receptor with a K_d 1000-fold less than full-length receptor (Rusconi and Yamamoto, 1987). In addition, this fragment does not encompass Cys_{742} which is affinity labelled by triamcinolone acetonide and thus near to the hormone-binding site, nor does it encompass Tyr_{770} whose mutation to Asn results in a modest increase in the K_d for dexamethasone.

It is tempting to speculate that the difference in these results is due to the intrinsic differences in the design of these experiments. Experiments using protease treatment start with a molecule that is already in the correct conformation for hormone binding. Protease is then used to cut at exposed basic amino acids leaving a hormone-binding core. Hormone binding by deleted receptors, on the other hand, requires not only the presence of a hormone-binding core, but also the ability to fold into a productive hormone-binding domain. This raises the possibility that the most C-terminal amino acids are required for the correct folding of the hormone binding domain, but are less important for tight steroid binding. Interestingly, a four amino acid insertion in the human GR at 684 (mGR 690), which is just C-terminal of the tryptic cleavage site (mGR 661) at the C-terminus of the 16-kD fragment, does not effect hormone binding (Giguere et al., 1986). These data suggest that the region from amino acid 661 to 690 is on the surface of the protein accessible to proteolytic cleavage and in a position where amino acid insertions would be expected to have the least structural effect.

3.4.2 Affinity Labelling Studies to Characterize the Hormone-Binding Site

It is possible to map amino acids in the vicinity of the hormone-binding site by using steroids containing an activated side group such as dexamethasone 21-mesylate (DM) or triamcinolone acetonide. The labelled amino acids can be identified following receptor purification by generating peptides which can be purified and sequenced. Simons et al. (1987) found that Cys_{656} in the rat GR (mGR Cys_{644}) is covalently modified by [^{3}H]DM using a combination of receptor purification, protease digestion, purification and sequencing of the ^{3}H-labelled peptides. Equivalent results have also been obtained by two other groups working on either the rat (Carlstedt-Duke et al., 1988) or mouse GR (L. I. Smith et al., 1988). Since the labelling of Cys_{644} by [^{3}H]DM is competed by excess dexamethasone, this amino acid must be located near the hormone binding site. The results do not necessarily mean that the steroid normally interacts with this cysteine since DM has a great preference for cysteines (Simons, 1987) and will thus presumably react with any nearby cysteine. The mesylate group is positioned at the 20α position of the steroid so that Cys_{644} must be located near the side chain of the D ring of the steroid.

Carlstedt-Duke et al. (1988) have used triamcinolone acetonide (TA) to photoaffinity label the rat receptor and found that two amino acids became covalently modified, Met_{622} and Cys_{754} (mGR Met_{610}, Cys_{742}; Fig. 4). TA contains an A ring with two double bonds and a 3-keto group. Presumably photoactivation results in activation of this 3-keto group which then becomes cross-linked to receptor. Thus Met_{610} and Cys_{742} are in close proximity in the receptor protein. Cys_{644} was not cross-linked by TA indicating that, as expected, there is stereo-specific binding of the steroid to the receptor. The 3-keto group is approximately 1.5 nm from the 21-mesylate group so that Cys_{644} must be approximately 1.5 nm from Met_{610} and Cys_{742}. That both the 3-keto group and the 21-mesylate group interact with the receptor is not surprising since the 3-keto-Δ^4 structure and the 20-keto, 21-hydroxy structure are important functions that, together with the 11β-hydroxyl group, determine the specificity of the interaction of the steroid with the GR. These cross-linking studies indicate that the hormone-binding site is made up by the complex folding of a large domain since three amino acids located over a 132 amino acid region of the receptor are brought to within 1.5 nm of each other.

3.4.3 Characterization of Hormone-Binding Deficient Receptors

It has been possible to generate GR mutants in mouse lymphoma and human leukaemia cells because their growth is inhibited by glucocorticoids and therefore hormone-resistant variants can be isolated. The mouse lymphoma-derived cell line S49nt- contains two mutant forms of the GR (Danielsen et al., 1986). One form binds to hormone with reduced affinity and cannot bind to DNA. The hormone-binding and DNA-binding lesions in this receptor are separable: the DNA-binding lesion is due to an Arg_{484} to His change located in the DNA-binding domain of the receptor. The hormone-binding lesion mapped to a Tyr_{770} to Asn change and was found to result in about a three-fold decrease in the affinity of the receptor for dexamethasone. The other GR present in these cells is unable to bind hormone and can only be detected in these cells by antibody (Northrop et al., 1985). This receptor contains a single point mutation Glu_{546} to Gly that gives rise to this phenotype. Thus the two mutations that affect hormone binding are located at opposite ends of the C-terminal domain, 224 amino acids apart. More recently, defective receptors have been cloned and sequenced from the mouse lymphoma-derived cell line W7.320b that is resistant to 10^{-6} M dexamethasone (Byravan et al., 1990). These cells, which only respond to concentrations of dexamethasone greater than 10^{-5} M (Rabindran et al., 1987), contain two mutant GR alleles. One allele has a Pro_{547} to Ala change and hormone-binding studies indicate that the K_d of this receptor for dexamethasone is more than 100-fold higher than wild-type receptor. The other allele contains a Cys_{742} to Gly change that results in a receptor that does not bind hormone appreciably. These two mutations are in regions that have already been shown to be important for hormone binding. The Pro_{547} to Ala mutation is adjacent to the previously described Glu_{546} to Gly mutation that prevents hormone binding (Danielsen et al., 1986), and the Cys_{742} to Gly change is a mutation of the cysteine that is covalently modified by photoaffinity labelling with triamcinolone acetonide (Carlstedt-Duke et al., 1988). Although this could be a coincidence, it is tempting to speculate that only a limited number of sites are present in the hormone-binding domain where a point mutation can sufficiently decrease steroid binding to allow selection of a mutant fully resistant to 10^{-6} M dexamethasone. That the region around Glu_{546} is important for steroid binding is underscored by the fact that all the mapping studies to date place this amino acid in a crucial part of the hormone-binding domain.

Mutations have also been characterized in the hormone-binding domain of other steroid receptors. For instance, in a human androgen receptor isolated from a family with complete androgen insensitivity, a point mutation Val_{866} to Met (Met_{731} in mGR) reduces the affinity of the receptor

3. THE GLUCOCORTICOID RECEPTOR

for dihydrotestosterone three-fold while increasing the affinity for progesterone 2.5-fold (Lubahn et al., 1989). Interestingly, the human progesterone receptor, like the GR, has a methionine at this position. However, this mutation may have more than just hormone-binding effects since in vivo there is complete androgen insensitivity. In the mouse oestrogen receptor Fawell et al. (1990) have identified a heptad repeat of hydrophobic residues that is conserved in all steroid receptors (see Chapter 16.3). This region extends from mER 505 to 539 (equivalent to mGR 718–752) and point mutations within this region of the mER, Ile_{518} to Arg (mGR Met_{731}) and Gly_{525} to Arg (mGR Leu_{738}), yield hormone-binding deficient receptors. The human syndrome of generalized resistance to thyroid hormone has been shown to be associated with a point mutation in the thyroid hormone receptor β gene (Sakurai et al., 1989). A Gly_{340} to Arg change results in a receptor that cannot bind thyroid hormone. Sequence comparisons reveal that Gly_{340} is equivalent to Cys_{644} of the mGR which is covalently modified by dexamethasone mesylate. A point mutation has also been characterized in the human oestrogen receptor that decreases its affinity for oestrogen at 25°C but not at 4°C (Tora et al., 1989). This mutation was probably due to a cloning artifact and is an hER Gly_{400} to Val change (equivalent to mGR Ser_{623}).

3.4.4 Role of the Hormone-Binding Region in Transcriptional Activation

In addition to the major transcriptional activation domain in the N-terminal region of the GR an activation function has been identified in the hormone-binding region. A fusion protein consisting of the DNA-binding domain of the yeast GAL4 transcription factor and the hormone-binding region of the hGR (amino acids 500–777 (mGR 507–783)) was able to stimulate transcription 16-fold from a GAL4 operator (Webster et al., 1988). The activity of deletion mutants suggested the presence of a transactivation domain between residues 515 and 550 (mGR 522–556) (Hollenberg and Evans, 1988). Moreover, an acidic region between residues 526 and 556 (mGR 532–562) was able to mediate transcriptional activation, albeit weakly, when it was subsituted for the N-terminal transactivation domain. This region τ2 is distinct from τ1, although both regions have an acidic character. For this reason Hollenberg and Evans (1988) have suggested that the activator regions in the GR may function by mechanisms similar to those of yeast GAL4 and GCN4, which depend on negatively charged residues on the surface of the DNA-bound protein (Sigler, 1988). In contrast, it is unlikely that transcriptional activation by the ER is mediated by acidic regions (see Chapter 2.10).

3.5 Phosphorylation

Although phosphorylation of GR has been studied for over 10 years, the role of receptor phosphorylation has not been determined. Phosphorylation is not required for DNA binding since production of the DNA-binding domain of the GR as a recombinant protein in *Escherichia coli* yields protein that can bind to both specific and non-specific DNA sequences with similar specificity as receptor isolated from mammalian cells (Freedman *et al.*, 1988; Dahlman *et al.*, 1989). Similarly, hormone binding does not appear to require receptor phosphorylation since few if any phosphorylation sites are present in the hormone-binding domain, and GR expressed in *E. coli* can bind hormone (Nemoto *et al.*, 1990), though it does so with reduced affinity. In addition, receptors lacking the N-terminal domain (the domain which contains most of the phosphate, and which contains all steroid-induced phosphorylation sites) associate with Hsp90 and other components of the untransformed GR complex to form complexes that bind hormone with wild-type affinity and are activated to a DNA-binding form. It is possible that phosphorylation modifies the ability of the receptor to activate transcription. Removal of the N-terminal domain results in a receptor with only 5% of the activity of wild-type receptor (Danielsen *et al.*, 1987). Whether this loss of transcriptional activity is at least partly due to removal of phosphorylation sites is unknown.

3.5.1 Sites of Phosphorylation

Dalman *et al.* (1988) metabolically labelled mouse L-cells with [^{32}P]orthophosphate in the absence of glucocorticoids and characterized the GR by immunoprecipitation of intact and partially proteolysed forms of the receptor. This technique not only allows the phosphorylation state of the individual proteolytic fragments to be analysed, but also allows the separation of the receptor from possible contaminants, such as the phosphoprotein Hsp90, which comigrates with receptor on an SDS polyacrylamide gel. They concluded that the receptor is exclusively phosphorylated on serine, that three phosphorylation sites lie between mouse amino acids 313 and 369 in the N-terminal domain of the receptor, and that one site is positioned between amino acids 398 and 447 in or just N-terminal of the DNA-binding domain. Using similar techniques Smith *et al.* (1989) have found that mouse WEHI-7 receptors are phosphorylated exclusively on serine and that there are two phosphoserines in the N-terminal domain, one phosphoserine in the hormone-binding domain and none in the DNA-binding domain. Thus both groups agree that most of the phosphorylation occurs in the N-terminal domain but disagree as to whether the hormone-binding or the DNA-binding domain is also phosphorylated. Whether these differing results are due to cell type specific phosphorylation of the GR or to different interpretations of protease digestion patterns has yet to be determined. Hoeck and Groner (1990) have extended these analyses by employing V8 protease, cyanogen bromide and hydroxylamine cleavage in addition to the trypsin and chymotrypsin used by the other two groups. They found that GR from rat hepatoma cells is phosphorylated mainly in the N-terminal domain with an additional phosphorylation site either just N-terminal to or in the DNA-binding domain. Hoeck and Groner (1990) have also transfected monkey kidney-derived CV1 cells with expression vectors encoding truncated receptors. Since these cells are GR negative they were able to measure phosphate incorporation into the transfected receptors without background incorporation into endogenous receptors. They found that a C-terminally truncated rat receptor containing amino acids 3–556 (mGR 3–544) incorporates phosphate when expressed in CV1 cells, whereas an N-terminally truncated receptor containing amino acids 407–795 (mGR 395–783) did not. Taken together these results indicate that the majority of the phosphorylated amino acids are located in the N-terminal domain of the GR, and that this general phosphorylation pattern is conserved across species lines.

3.5.2 Phosphorylation and Activation

The observation that phosphatase inhibitors stabilize non-activated receptor complexes and prevents activation led to the suggestion that dephosphorylation of receptor is required for activation. Molybdate, for example, is a potent inhibitor of transformation both *in vitro* (Leach *et al.*, 1979; Toft and Nishigori, 1979) and *in vivo* (Raaka *et al.*, 1985) and it has been used by many groups to stabilize the untransformed GR during purification. Other phosphatase inhibitors such as sodium or aluminium fluoride (Reker *et al.*, 1987; Housely, 1990) and glucose-1-phosphate (Reker *et al.*, 1987) are effective in inhibiting transformation. This idea was supported by the finding that treatment of non-activated GR by alkaline phosphatase led to at least partial activation of the GR (Barnett *et al.*, 1980; Reker *et al.*, 1987) and that treatment of endogenously inactivated receptor with molybdate and ATP restored hormone-binding activity (Sando *et al.*, 1979). However, in none of the original studies was the phosphorylation state of the receptor determined. More recent work employing immunoprecipitation of the receptor and careful separation of receptor from Hsp90 have found no evidence for decreased (or increased) phosphorylation of the GR itself on activation *in vitro* (Tienrungroj *et al.*, 1987; Housely, 1990). Indeed, receptors activated *in vitro* have the same charge distribution as measured by two-dimensional electrophoresis (Smith *et al.*, 1986) as non-activated receptors. Some studies have detected a reduction in phosphate content of the receptor complex upon activation. This has been shown to be due to dissociation of Hsp90, which is itself a phosphoprotein. Orti *et al.* (1989a) have shown that the Hsp90 associated with the GR has the same net phosphorylation state as Hsp90 free in cytosol and that upon dissociation from the receptor Hsp90 does not undergo dephosphorylation.

3.5.3 Hormone-induced Phosphorylation *in vivo*

Although, as discussed above, the process of activation does not require phosphorylation or dephosphorylation of the GR, treatment of cells with glucocorticoids does lead to an increase in the phosphate content of receptors compared to untreated cells. For instance Orti *et al.* (1989c) found that GR in hormone-treated mouse WEHI-7 lymphoma cells had 70% more phosphate than the GR in untreated cells. Similar results have been obtained by Hoeck and his collaborators (Hoeck *et al.*, 1989; Hoeck and Groner, 1990) where a three- to four-fold increase in phosphate content of the GR was observed after hormone treatment of mouse NIH 3T3, rat FTO 2B hepatoma, and human HeLa cells. The increase in phosphate content is not specific for non-activated or activated receptors since activated and non-activated receptors isolated from the same cells have the same degree of phosphorylation (Mendel *et al.*, 1987). The phosphorylation of GR in response to glucocorticoids is very rapid with a $t_{1/2}$ of less than 10 min (Orti *et al.*, 1989b), indicating that the effect does not require *de novo* protein synthesis. This effect is specific for glucocorticoid agonists such as cortisol, triamcinolone acetonide and dexamethasone, but not the antagonist RU486.

Upon lysis of glucocorticoid-treated cells, GR can be detected in both cytosolic and nuclear fractions. Most of the nuclear GR can be extracted with high-salt or by nuclease digestion. However, approximately 10-20% of nuclear GR cannot be extracted unless the nuclei are boiled in SDS. Interestingly, this is the only fraction of GR that does not show increased phosphorylation upon hormone treatment (Orti *et al.*, 1989c). The authors suggest that this is due to dephosphorylation of hyperphosphorylated GR and that such receptors provide evidence for a cycling of receptors between hyperphosphorylated and phosphorylated states. There is evidence for such a cycle in ATP-depleted cells where GR binds to the nucleus in the absence of hormone. These receptors cannot bind to glucocorticoids and are hypophosphorylated compared to cytosolic receptors (Orti *et al.*, 1989b, and references therein). These receptors regain normal phosphate levels and the ability to bind hormone when ATP levels are restored. However, there is no evidence that nuclear unextractable receptors in hormone-treated cells are equivalent to tightly bound nuclear receptors in ATP-depleted cells.

3.6 Role of heat-shock proteins in glucocorticoid receptor function

Cytosolic extracts of hypotonic cell lysates contain glucocorticoid receptor complexes that sediment at 8–9 S on density gradients and have an estimated molecular weight of approximately 300 000 (reviewed in Pratt *et al.*, 1989). This form of the receptor is stabilized by low-salt, transition metal oxyanions such as molybdate and low temperatures. The 8–9 S complexes can bind glucocorticoids and will undergo activation to a 4 S (100-kD) form that can bind to DNA on warming to 25°C (Sanchez *et al.*, 1987b; Denis *et al.*, 1988b). Affinity chromatography of the non-activated form of the receptor on deoxycorticosterone agarose columns or immunoprecipitation or immunoadsorption with receptor-specific antibodies has identified a number of proteins that are present in the 8–9 S complex. These include the 100-kD receptor itself which can be distinguished by being the only protein that is affinity labelled with the specific affinity labelling reagent dexamethasone mesylate, a 90-kD protein now known to be Hsp90, a 56/59-kD protein, possibly Hsp70, and a number of low molecular weight proteins and RNA obtained in variable yield and not well characterized. All steroid receptors can be isolated as an 8–9 S form from cytosolic extracts and these high molecular weight complexes have a similar structure to the non-activated GR complexes. These data have led to the hypothesis that *in vivo* hormone binding leads to subunit dissociation which activates receptor to its DNA-binding form, and a necessary corollary of this is that the GR-associated proteins must keep the receptor in its inactive state.

3.6.1 Hsp90

The evidence to support the proposal that GR is associated with Hsp90 in cytosol can be derived from a number of observations:

(1) steroid affinity chromatography of molybdate-stabilized non-activated receptor, but not of activated receptor, gives rise to the copurification of a 90-kD protein (Housely *et al.*, 1985);
(2) immunoadsorption of non-activated receptors from cytosol with monoclonal antibodies against the hormone-binding protein copurify Hsp90 (Housely *et al.*, 1985; Sanchez *et al.*, 1985, 1987b; Mendel *et al.*, 1986; Howard and Distelhorst, 1988);
(3) antibodies against Hsp90 increase the sedimentation velocity of the 9 S complex (Joab *et al.*, 1984; Lefebvre *et al.*, 1989);

(4) Hsp90 specific antibodies immunoadsorb/immunoprecipitate a complex containing Hsp90 and GR (Denis *et al.*, 1988a; Dalman *et al.*, 1989);
(5) triamcinolone acetonide labelled 9 S complexes separated by sucrose gradient centrifugation are immunoadsorbed by anti-Hsp 90 antibody (Sanchez *et al.*, 1987a).

Hsp90 is also associated with other steroid receptors in cytosolic extracts, including the oestrogen, progesterone, androgen, and mineralocorticoid receptors (Joab *et al.*, 1984; Catelli *et al.*, 1985; Schuh *et al.*, 1985; Rafestin-Oblin *et al.*, 1989), and the aryl hydrocarbon receptor (Perdew, 1988; Wilhelmsson *et al.*, 1990), and in each case these receptors are found associated with Hsp90 in the non-activated but not the activated states.

Hsp90 is undoubtedly associated with the GR in cell extracts, but is it associated with GR in the intact cell? The most compelling evidence that it is, is that Hsp90 can be cross-linked to the GR in whole cells (Rexin *et al.*, 1988a,b). Whole cells were treated with the bifunctional reagent dithiobis-(succinimidyl propionate), the reaction quenched with excess lysine, and receptor complexes isolated. The receptor was found to be present in a salt-stable complex of 328 kD which yielded a 116-kD hormone-binding protein on reductive cleavage with mercaptoethanol. Immunoaffinity chromatography of cross-linked complexes, using the GR-specific monoclonal antibody GR49, yielded complexes containing Hsp90, as shown by Western blotting with an Hsp90 specific antibody after reductive cleavage.

The release of a dimer of Hsp90 from immunoadsorbed receptor on activation, and size determinations of the 8–9 S complex has led to the suggestion that the non-activated GR complex contains two molecules of Hsp90 and one 100-kD hormone-binding subunit (Denis *et al.*, 1987). Rexin *et al.* (1988b) have treated the non-activated receptor with cross-linking agents and analysed the products on SDS polyacrylamide gels. By comparing the results with wild-type S49 cells which contain a 100-kD receptor, and S49nt^i cells, which contain a mutated receptor that lacks the N-terminal domain and is therefore only 50 kD, they were able to show that one molecule of hormone-binding protein (105 kD) is associated with two 90-kD molecules which are presumably Hsp90. Perhaps the most convincing demonstration of the 1:2 receptor:Hsp90 stoichiometry comes from the metabolic labelling studies of Mendel and Orti (1988). They labelled cells with [^{35}S]methionine to steady state and characterized non-activated complexes on SDS polyacrylamide gels. Since the number of methionines in each subunit is known from their protein sequences, they were able to determine subunit stoichiometry from the amount of radioactivity incorporated into each protein. More recently Bresnik *et al.* (1990), using [^{35}S]methionine to label cells to steady state, immunopurification of the receptor complex, and quantitative immunoblotting, have found that between two and six molecules of Hsp90 are associated with the receptor and that this figure depends on the stringency with which antibody–receptor complexes are washed. Whether these higher order structures are due to non-specific adsorption of free Hsp90 to the core untransformed receptor (2 Hsp90:1 GR) or whether they are of physiological significance is open to question.

The GR associates with 2 isoforms of Hsp90 in murine cells as determined by SDS polyacrylamide gel electrophoresis (Mendel and Orti, 1988; Bresnick et al., 1990). There are at least two genes encoding members of the Hsp90 family of heat-shock proteins in murine cells, Hsp86 and Hsp84, both of which have been cloned (Moore et al., 1989). However, it has not been determined whether the two forms of Hsp90 associated with the GR are Hsp86 and Hsp84 or whether they represent differential modification of one of these gene products. Although it is known that Hsp90 is present as a dimer in cells (Lefebvre et al., 1989; Radanyi et al., 1989) it is not known whether it forms homo- or hetero-dimers, or which of these forms interacts with receptor.

3.6.2 Hsp90 and Receptor Activation

Truncated receptors that do not contain a hormone-binding domain are constitutively active (Danielsen et al., 1987; Godowski et al., 1987; Hollenberg et al., 1987). This indicates that binding of hormone does not induce a conformational change in the DNA-binding domain which allows the receptor to bind to DNA. Rather, DNA binding is repressed in the absence of hormone and derepressed on hormone binding (or truncation of the hormone-binding domain). The DNA-binding site appears to be buried in the non-activated receptor since antibodies against the DNA-binding domain only recognize the activated receptor (Table 2; Urda et al., 1989; D. F. Smith et al., 1988). The masking of the DNA-binding site in the non-activated receptor may be envisioned in two ways. One possibility is that the hormone-binding domain itself interacts with the DNA-binding domain, thereby preventing DNA binding. There is little, if any, evidence in favour of this hypothesis. Indeed, since the hormone-binding domain is position independent (Picard et al., 1988), and can function to control unrelated DNA-binding domains in various fusion proteins (Godowski et al., 1988; Webster et al., 1988), this possibility seems very unlikely. A more attractive possibility is that masking/unmasking of the DNA-binding site is mediated by a receptor-associated protein. The role of the ligand would be to bring about dissociation of this protein from the receptor, thereby exposing the

DNA-binding domain. This would explain why purified GR (i.e. dissociated from other proteins in the non-activated complex) binds to DNA in the absence of hormone (Willman and Beato, 1986), whereas *in vivo* hormone is required for DNA binding (Becker *et al.*, 1986). A good candidate for a masking protein is Hsp90 since it is associated with the non-DNA-binding but not the DNA-binding form of the receptor.

If Hsp90 functions to repress the activity of the GR in the absence of hormone, then C-terminally truncated receptors, which are constitutively active, should not associate with Hsp90. Pratt *et al.* (1988) have analysed a series of C-terminal truncations of the human GR and have found a correlation between constitutive activity and lack of association with Hsp90. Within this domain of the GR is a region that is conserved amongst receptor types (Fig. 5). Deletion of these sequences (mGR 574–632) results in a receptor that has some constitutive transcriptional activity, associates only weakly with Hsp90, and is susceptible to proteolytic cleavage *in vivo* (Housely *et al.*, 1990). This has led to the hypothesis that the GR contains two binding sites for Hsp90. Similar studies with the rat GR have indicated that amino acids between residues 509 and 616 (mGR 497–604) are essential for the formation of a stable receptor–Hsp90 complex (Howard *et al.*, 1990).

If Hsp90 is an important modulator of steroid receptor activity in general, then it should be present in the same intracellular compartments as steroid receptors. Hsp90 is a cytoplasmic protein and this correlates with the partially cytoplasmic location of the glucocorticoid receptor (Gustafsson *et al.*, 1987; Gasc *et al.*, 1989). Other steroid receptors, such as the progesterone (Gasc *et al.*, 1989) and oestrogen (King and Greene, 1984) receptors are thought to be located in the nucleus in the absence of hormone, so that some Hsp90 should also be present in the nucleus associated with these receptors. Indeed, there is some indirect evidence from work on the dioxin receptor that Hsp90 may be nuclear (Wilhelmsson *et al.*, 1990).

3.6.3 Hsp90 and Hormone Binding

Activation of the GR results in a decrease in affinity of the receptor for glucocorticoids by at least 100-fold (Bresnick *et al.*, 1989; Nemoto *et al.*, 1990). These data have led to the suggestion that Hsp90 is required to keep the receptor in a conformation capable of binding hormone with high affinity. This is supported by the recent finding that GR produced in *E. coli* does not form a 9 S receptor complex, implying that it does not associate with the *E. coli* equivalent of Hsp90, and that it binds hormone with approximately the same affinity as activated receptor (Nemoto *et al.*, 1990). GR synthesized in a reticulocyte lysate *in vitro* translation system associates with Hsp90 and binds hormone with high affinity (Dalman *et al.*, 1989; Denis and Gustafsson, 1989). However, GR translated in wheat germ lysates is incapable of binding steroid even though the protein synthesized is identical to receptor made in a reticulocyte lysate when analysed by SDS gel electrophoresis (Dalman et al., 1990). This may be due to a deficiency of Hsp90 in these wheat germ extracts since they do not contain any immunologically (antibody AC88, Table 1) recognizable Hsp90.

3.6.4 p56/p59

Nakao et al. (1985) raised monoclonal antibodies against partially purified rabbit progesterone receptor and obtained one monoclonal KN 382/EC1 (Table 1) that recognized only the molybdate-stabilized untransformed receptor. They went on to show that this antibody recognized a 59-kD protein that was present in untransformed rabbit progesterone, oestrogen, androgen, and glucocorticoid receptor complexes (Tai et al., 1986). More recently Sanchez et al. (1990) have found that this antibody recognizes the untransformed human GR but not the transformed receptor. The human protein recognized by the EC1 antibody is a 56-kD protein (p56) that is a unique protein as determined by N-terminal sequencing. The protein is of moderate abundance; less abundant than Hsp90 but much more abundant than the GR in human IM9 cells. From the amounts of p56, Hsp90 and GR immunoadsorbed by the EC1 antibody, the authors calculate that p56 and Hsp90 must be associated with each other in cytosolic extracts even when they are not associated with the receptor. The stoichiometric excess of Hsp90 over p56 in vivo would mean that only a small fraction of the Hsp90 present in a cell would be associated with p56. Since the EC1 antibody shifts all of the 9S receptor to a higher form on glycerol gradients, all receptor complexes must contain at least one molecule of p56. For these reasons, p56 is as likely to be a modulator of steroid receptor function as is Hsp90.

Rexin et al. (1989b) found that mouse untransformed GR could be cross-linked not only to Hsp90, but also to a 50-kD protein both in vitro and in vivo. They calculated that the untransformed receptor complex contains one hormone-binding protein, two molecules of Hsp90 and one 50-kD protein (p50). It is tempting to speculate that p50 is the mouse homologue of p56. Unfortunately the EC1 antibody does not recognize any proteins in either mouse or rat cytosols (Bresnick et al., 1990), so that a quick test of this hypothesis is not possible. Since Rexin et al. (1988b) were able to show that p50 can be cross-linked directly to both the receptor and to Hsp90, all three proteins must be in close proximity in the untransformed GR complex.

Acknowledgements

I would like to thank my colleagues for helpful discussions, with special thanks to all those who generously shared their data with me before publication. The author's research is supported by a grant DK 42552 from the National Institutes of Health.

References

Arriza, J. L., Weinberger, C., Cerelli, G., Glaser, T. M., Handelin, B. L., Housman, D. E. and Evans, R. M. (1987). Cloning of human mineralocorticoid receptor cDNA: structural and functional kinship with the glucocorticoid receptor. *Science* **237**, 268–275.
Baker, A. R., McDonnell, D. P., Hughes, M., Crisp, T. M., Mangelsdorf, D. J., Haussler, M. R., Picke, J. W., Shine, J. and O'Malley, B. W. (1988). Cloning and expression of full length cDNA encoding human vitamin D receptor. *Proc. Natl. Acad. Sci. USA* **85**, 3294–3298.
Barnett, C. A., Schmidt, T. J. and Litwack, G. (1980). Effects of calf intestinal alkaline phosphatase, phosphatase inhibitors, and phosphorylated compounds on the rate of activation of glucocorticoid receptor complexes. *Biochemistry* **19**, 5446–5455.
Becker, P. B., Gloss, B., Schmid, W., Strahler, U. and Schutz, G. (1986). *In vivo* protein-DNA interactions in a glucocorticoid response element require the presence of hormone. *Nature* **324**, 686–688.
Benbrook, D., Lernhardt, E. and Pfahl, M. (1988). A new retinoic acid receptor identified from a hepatocellular carcinoma. *Nature* **33**, 669–672.
Berg, J. M. (1989). DNA binding specificity of steroid receptors. *Cell* **57**, 1065–1068.
Brand, N., Petkovich, M., Krust, A., Chambon, P., de The, H., Marchio, A., Tiollais, P. and Dejean, A. (1988). Identification of a second human retinoic acid receptor. *Nature* **332**, 850–853.
Bresnick, E. H., Dalman, F. C., Sanchez, E. R. and Pratt, W. B. (1989). Evidence that the 90-kDa heat shock protein is necessary for the steroid binding conformation of the L-cell glucocorticoid receptor. *J. Biol. Chem.* **264**, 4992–4997.
Bresnick, E. H., Dalman, F. C. and Pratt, W. B. (1990). Direct stoichiometric evidence that the untransformed M_r 3 000 000, 9 S, glucocorticoid receptor is a core unit derived from a larger heteromeric complex. *Biochemistry* **29**, 520–527.
Brugge, J., Yonemoto, W. and Darrow, D. (1983). Interaction between Rous sarcoma virus transforming protein and two cellular phosphoproteins: analysis of the turnover and distribution of this complex. *Mol. Cell. Biol.* **3**, 9–19.
Burmester, J. K., Wiese, R. J., Maeda, N. and DeLuca, H. F. (1988). Structure and regulation of the rat 1,25-dihydroxyvitamin D_3 receptor. *Proc. Natl. Acad. Sci. USA* **85**, 9499–9502.
Byravan, S., Milhon, J., Rabindran, S. K., Olinger, B., Danielsen, M. and Stallcup, M. R. (1990). Two point mutations that dramatically reduce the hormone binding affinity of the mouse glucocorticoid receptor. Submitted.
Carlstedt-Duke, J., Stromstedt, P.-E., Persson, B., Cederlund, E., Gustafsson, J.-A. and Jornvall, H. (1988). Identification of hormone-interacting amino acid residues within the steroid-binding domain of the glucocorticoid receptor in relation to other steroid hormone receptors. *J. Biol.Chem.* **263**, 6842–6846.

Catelli, M. G., Binart, N., Jung-Testas, I., Renoir, J. M., Baulieu, E. E., Feramisco, J. R. and Welch, W. J. (1985). The common 90-kD protein component of nontransformed 8 S steroid receptors is a heat shock protein. *EMBO J.* **4**, 3131–3135.

Cayanis, E., Rajagopalan, R., Cleveland, W. L., Edelman, I. S. and Erlanger, B. F. (1986). Generation of an auto-anti-idiotypic antibody that binds to glucocorticoid receptor. *J. Biol. Chem.* **261**, 5094–5103.

Cayanis, E., Sarangarajan, R., Lombes, M., Nahon, E., Edelman, I. S. and Erlanger, B. F. (1989). Identification of an epitope shared by the DNA-binding domain of glucocorticoid receptor and the B chain of insulin. *Proc. Natl. Acad. Sci. USA* **86**, 2138–2142.

Chang, C. and Kokontis, J. (1988). Identification of a new member of the steroid receptor super-family by cloning and sequencing analysis. *Biochem. Biophys. Res. Commun.* **155**, 971–977.

Chang, C., Kokontis, J. and Liao, S. (1988). Structural analysis of complementary DNA and amino acid sequences of human and rat androgen receptors. *Proc. Natl. Acad. Sci. USA* **85**, 7211–7215.

Dahlman, K., Stromstedt, P.-E., Rae, C., Jornvall, J. I., Carlstedt-Duke, J. and Gustafsson, J.-A. (1989). High level expression in *Escherichia coli* of the DNA-binding domain of the glucocorticoid receptor in a functional form utilizing domain-specific cleavage of a fusion protein. *J. Biol. Chem.* **264**, 804–809.

Dalman, F. C., Sanchez, E. R., Lin, A. L.-Y., Perini, F. and Pratt, W. B. (1988). Localization of phosphorylation sites with respect to the functional domains of the mouse L cell glucocorticoid receptor. *J. Biol. Chem.* **263**, 12 259–12 267.

Dalman, F. C., Bresnick, E. H., Patel, P. D., Perdew, G. H., Watson, S. J. and Pratt, W. B. (1989). Direct evidence that the glucocorticoid receptor binds to Hsp90 at or near the termination of receptor translation *in vitro*. *J. Biol. Chem.* **264**, 19 815–19 821.

Dalman, F. C., Koenig, R. J., Perdew, G. H., Massa, E. and Pratt, W. B. (1990). In contrast to the glucocorticoid receptor, the thyroid hormone receptor is translated in the DNA binding state and is not associated with Hsp90. *J. Biol. Chem.* **265**, 3615–3618.

Danielsen, M., Northrop, J. P. and Ringold, G. M. (1986). The mouse glucocorticoid receptor: mapping of functional domains by cloning, sequencing and expression of wild-type and mutant receptor proteins. *EMBO J.* **5**, 2513–2522.

Danielsen, M., Northrop, J. P., Jonklaas, J. and Ringold, G. M. (1987). Domains of the glucocorticoid receptor involved in specific and nonspecific deoxyribonucleic acid binding, hormone activation, and transcriptional enhancement. *Mol. Endocrinol.* **1**, 816–822.

Danielsen, M., Hinck, L. and Ringold, G. M. (1989). Two amino acids within the knuckle of the first zinc finger specify DNA response element activation by the glucocorticoid receptor. *Cell* **57**, 1131–1138.

Denis, M. (1988). Two-step purification and N-terminal amino acid analysis of the rat M_r = 90 000 heat shock protein. *Anal. Biochem.* **173**, 405–411.

Denis, M. and Gustafsson, J.-A. (1989). Translation of glucocorticoid receptor mRNA *in vitro* yields a nonactivated protein. *J. Biol. Chem.* **264**, 6005–6008.

Denis, M., Wikström, A. C. and Gustafsson, J. A. (1987). The molybdate-stabilized nonactivated glucocorticoid receptor contains a dimer of M_r 90 000 non-hormone-binding protein. *J. Biol. Chem.* **262**, 11 803–11 806.

Denis, M., Gustafsson, J.-A. and Wikström, A.-C. (1988a). Interaction of the M_r = 90 000 heat shock protein with the steroid-binding domain of the glucocorticoid receptor. *J. Biol. Chem.* **263**, 18 520–18 523.

Denis, M., Poellinger, L., Wikström, A.-C. and Gustafsson, J.-A. (1988b). Requirement of hormone for thermal conversion of the glucocorticoid receptor to a DNA binding state. *Nature* **333**, 686–688.

Devereux, J., Haeberli, P. and Smithies, O. (1984). A comprehensive set of sequence analysis programs for the vax. *Nucl. Acids Res.* **12**, 387–395.

Eisen, L. P., Reichman, M. E., Thompson, E. B., Gametchu, B., Harrison, R. W. and Eisen, H. J. (1985). Monoclonal antibody to the rat glucocorticoid receptor. *J. Biol. Chem.* **260**, 11 805–11 810.

Eriksson, P. and Wrange, O. (1990). Protein–protein contacts in the glucocorticoid receptor homodimer influence its DNA binding properties. *J. Biol. Chem.* **265**, 3535–3542.

Evans, R. M. (1988). The steroid and thyroid hormone receptor superfamily. *Science* **240**, 889–895.

Fawell, S. E., Lees, J. A. and Parker, M. G. (1989). A proposed consensus steroid binding sequence—a reply. *Mol. Endocrinol.* **3**, 1002–1005.

Fawell, S. E., Lees, J. A., White, R. and Parker, M. G. (1990). Characterization and localization of steroid binding and dimerization activities in the mouse estrogen receptor. *Cell* **60**, 953–962.

Feigl, G., Gram, M. and Pongs, O. A. (1989). A member of the steroid hormone receptor gene family is expressed in the 20-OH-ecdysone inducible puff 75B in *Drosophila melanogaster*. *Nucl. Acids Res.* **17**, 7167–7178.

Freedman, L. P., Luisi, B. F., Korszun, Z. R., Basavappa, R., Sigler, P. B. and Yamamoto, K. R. (1988). The function and structure of the metal coordination sites within the glucocorticoid receptor DNA binding domain. *Nature* **334**, 543–546.

Gametchu, B. and Harrison, R. W. (1984). Characterization of a monoclonal antibody to the rat liver glucocorticoid receptor. *Endocrinology* **114**, 274–279.

Gasc, J. M., Delahaye, F. and Baulieu, E. E. (1989). Compared intracellular localization of the glucocorticosteroid and progesterone receptors: an immunocytochemical study. *Exp. Cell Res.* **181**, 492–504.

Gehring, U. and Arndt, H. (1985). Heteromeric nature of glucocorticoid receptors. *FEBS Lett.* **179**, 138–142.

Giguere, V., Hollenberg, S. M., Rosenfeld, M. G. and Evans, R. M. (1986). Functional domains of the human glucocorticoid receptor. *Cell* **46**, 645–652.

Giguere, V., Ong, E. S., Segui, P. and Evans, R. M. (1987). Identification of a receptor for the morphogen retinoic acid. *Nature* **330**, 624–629.

Giguere, V., Yang, N., Segui, P. and Evans, R. M. (1988). Identification of a new class of steroid hormone receptors. *Nature* **331**, 91–94.

Glass, C. K., Holloway, J. M., Devary, O. V. and Rosenfeld, M. G. (1988). The thyroid hormone receptor binds with opposite transcriptional effects to a common sequence motif in thyroid hormone and estrogen elements. *Cell* **54**, 313–323.

Godowski, P. J., Rusconi, S., Miesfeld, R. and Yamamoto, K. R. (1987). Glucocorticoid receptor mutants that are constitutive activators of transcriptional enhancement. *Nature* **325**, 365–368.

Godowski, P. J., Didier, P. and Yamamoto, K. R. (1988). Signal transduction and transcriptional regulation by glucocorticoid receptor-LexA fusion proteins. *Science* **241**, 812–816.

Green, S. and Chambon, P. (1987). Oestradiol induction of a glucocorticoid-responsive gene by a chimeric receptor. *Nature* **325**, 75–78.
Green, S., Walter, P., Kumar, V., Krust, A., Bornert, J. M., Argos, P. and Chambon, P. (1986). Human oestrogen receptor cDNA: sequence, expression and homology to v-*erb*-A. *Nature* **320**, 134–139.
Gronemeyer, H., Turcotte, B., Quirin-Stricker, C., Bocquel, M.T., Meyer, M. E., Krozowski, Z., Jeltsch, J. M., Lerouge, T., Garnier, J. M. and Chambon, P. (1987). The chicken progesterone receptor: sequence, expression and functional analysis. *EMBO J.* **6**, 3985–3994.
Gustafsson, J.-A., Carlstedt-Duke, J., Poellinger, L., Okret, S., Wikström, A.-C., Bronnegard, M., Gillner, M., Dong, Y., Fuxe, K., Cintra, A., Harfstrand, A. and Agnati, L. (1987). Biochemistry, molecular biology, and physiology of the glucocorticoid receptor. *Endocr. Rev.* **8**, 185–455.
Ham, J. and Parker, M. G. (1989). Regulation of gene expression by nuclear hormone receptors. *Curr. Opin. Cell Biol.* **1**, 503–511.
Hamada, K., Gleason, S. L., Levi, B.-Z., Hirschfeld, S., Appella, E. and Ozato, K. (1989). H-2BIIBP, a member of the nuclear hormone receptor superfamily that binds to both the regulatory element of major histocompatibility class I genes and the estrogen response element. *Proc. Natl. Acad. Sci. USA* **86**, 8289–8293.
Härd, T., Kellenbach, E., Boelens, R., Maler, B. A., Dahlman, K., Freedman, L. P., Carlstedt-Duke, J., Yamamoto, K. R., Gustafsson, J.-A. and Kaptein, R. (1990). Solution structure of the glucocorticoid receptor DNA-binding domain. *Science* **249**, 157–160.
Harlow, K. W., Smith, D. N., Katzenellenbogen, J. A., Greene, G. L. and Katzenellenbogen, B. S. (1989). Identification of cysteine 530 as the covalent attachment site of an affinity-labelling estrogen (ketononestrol aziridine) and antiestrogen (tamoxifen aziridine) in the human estrogen receptor. *J. Biol. Chem.* **264**, 17 476–17 485.
Hazel, T. G., Nathans, D. and Lau, L. F. (1988). A gene inducible by serum growth factors encodes a member of the steroid and thyroid hormone receptor superfamily. *Proc. Natl. Acad. Sci. USA* **85**, 8444–8448.
Hoeck, W. and Groner, B. (1990). Hormone-dependent phosphorylation of the glucocorticoid receptor occurs mainly in the amino terminal transactivation domain t1/enh2. *J. Biol. Chem.* **265**, 5403–5408.
Hoeck, W., Rusconi, S. and Groner, B. (1989). Down-regulation and phosphorylation of glucocorticoid receptors in cultured cells. *J. Biol. Chem.* **264**, 14 396–14 402.
Hollenberg, S. M., Weinberger, C., Ong, E. S., Carelli, G., Oro, A., Lebo, R., Thompson, E. B. and Evans, R. M. (1985). Primary structure and expression of a functional human glucocorticoid receptor. *Nature* **318**, 635–641.
Hollenberg, S. M., Giguere, V., Sequi, P. and Evans, R. M. (1987). Colocalization of DNA binding and transcriptional activation functions in the human glucocorticoid receptor. *Cell* **49**, 39–46.
Hollenberg, S. M. and Evans, R. M. (1988). Multiple and cooperative transactivation domains of the human glucocorticoid receptor. *Cell* **55**, 899–906.
Housely, P. R. (1990). Aluminum fluoride inhibition of glucocorticoid receptor inactivation and transformation. *Biochem.* **29**, 3578–3585.
Housely, P. R., Sanchez, E. R., Westphal, H. M., Beato, M. and Pratt, W. B. (1985). The molybdate-stabilized L-cell glucocorticoid receptor isolated by affinity chromatography or with a monoclonal antibody is associated with a 90–92 kDa nonsteroid binding phosphoprotein. *J. Biol. Chem.* **260**, 13 810–13 817.

Housely, P. R., Sanchez, E. R., Danielsen, M., Ringold, G. M. and Pratt, W. B. (1990). Evidence that the conserved region in the steroid binding domain of the glucocorticoid receptor is required for both optimal binding of Hsp90 and protection from proteolytic cleavage: a two-site model for Hsp90 binding to the steroid binding domain. *J. Biol. Chem.* **265**, 12 778–12 781.

Howard, K. J. and Distelhorst, C. W. (1988). Evidence for intracellular association of the glucocorticoid receptor with the 90-kDa heat shock protein. *J. Biol. Chem.* **263**, 3474–3481.

Howard, K. J., Holley, S. J., Yamamoto, K. R. and Disselhorst, C. W. (1990). Mapping the Hsp90 binding region of the glucocorticoid receptor. *J. Biol. Chem.* **265**, 11 928–11 935.

Joab, I., Radanyi, C., Renoir, J. M., Buchou, T., Catelli, M. G., Binart, N., Mester, J. and Baulieu, E. E. (1984). Common non-hormone binding component of nontransformed chick oviduct receptors of four steroid hormones. *Nature* **308**, 850–853.

King, W. J. and Greene, G. L. (1984). Monoclonal antibodies localize oestrogen receptor in the nuclei of target cells. *Nature* **307**, 745–747.

Koike, S., Sakai, M. and Muramatsu, M. (1987). Molecular cloning and characterization of rat estrogen receptor cDNA. *Nucl. Acids Res.* **15**, 2499–2513.

Kost, S. L., Smith, D. F., Sullivan, W. P., Welch, W. J. and Toft, D. O. (1989). Binding of heat shock proteins to the avian progesterone receptor. *Mol. Cell. Biol.* **9**, 3829–3838.

Krust, A., Green, S., Argos, P., Kumar, V., Walter, P., Bornert, J.-M. and Chambon, P. (1986). The chicken oestrogen receptor sequence: homology with v-*erb*A and the human oestrogen and glucocorticoid receptors. *EMBO J.* **5**, 891–897.

Krust, A., Kastner, P. H., Petkovich, M., Zelent, A. and Chambon, P. (1989). A third retinoic acid receptor, hRAR-γ. *Proc. Natl. Acad. Sci. USA* **86**, 5310–5314.

Kumar, V., Green, S., Staub, A. and Chambon, P. (1986). Localization of the oestradiol-binding and putative DNA-binding domains of the human estrogen receptor. *EMBO J.* **5**, 2231–2236.

Kumar, V., Green, S., Stack, G., Berry, M., Jin, J.-R. and Chambon, P. (1987). Functional domains of the human estrogen receptor. *Cell* **51**, 941–951.

Lazar, M. A., Hodin, R. A., Darling, D. S. and Chin, W. W. (1989). A novel member of the thyroid/steroid hormone receptor family is encoded by the opposite strand of the rat c-*erb*Aα transcriptional unit. *Mol. Cell. Biol.* **9**, 1128–1136.

Leach, K. L., Dahmer, M. K., Hammond, N. D., Sando, J. J. and Pratt, W. B. (1979). Molybdate inhibition of glucocorticoid receptor inactivation and transformation. *J. Biol. Chem.* **254**, 11 884–11 890.

Lee, M. S., Gippert, G. P., Soman, K. V., Case, D. A. and Wright, P. E. (1989). Three-dimensional solution structure of a single zinc finger DNA-binding domain. *Science* **245**, 635–637.

Lefebvre, P., Sablonniere, B., Tbarka, N., Formstecher, P. and Dautrevaux, M. (1989). Study of the heteromeric structure of the untransformed receptor using chemical cross-linking and monoclonal antibodies against the 90K heat-shock protein. *Biochem. Biophys. Res. Commun.* **159**, 677–686.

Lombes, M., Farman, N., Oblin, M. E., Baulieu, E. E., Bonvalet, J. P., Erlanger, B. F. and Gasc, J. M. (1990). Immunohistochemical localization of renal mineralocorticoid receptor by using an anti-idiotypic antibody that is an internal image of aldosterone. *Proc. Natl. Acad. Sci. USA* **87**, 1086–1088.

3. THE GLUCOCORTICOID RECEPTOR

Loosfelt, H., Atger, M., Misrahi, M., Guiochon-Mantelo, A., Meriel, C., Logeat, F., Benarous, R. and Milgrom, E. (1986). Cloning and sequence analysis of rabbit progesterone-receptor cDNA. *Proc. Natl. Acad. Sci. USA* **83**, 9045–9049.

Lubahn, D. B., Brown, T. R., Simental, J. A., Higgs, H. N., Migeon, C. J., Wilson, E. M. and French, F. S. (1989). Sequence of the intron/exon junctions of the coding regions of the human androgen receptor gene and identification of a point mutation in a family with complete androgen insensitivity. *Proc. Natl. Acad. Sci. USA* **86**, 9534–9538.

Mader, S., Kumar, V., de Verneuil, H. and Chambon, P. (1989). Three amino acids of oestrogen receptor are essential to its ability to distinguish an oestrogen from a glucocorticoid-responsive element. *Nature* **338**, 271–274.

McDonnell, D. P., Scott, R. A., Kerner, S. A., O'Malley, B. W. and Pike, J. W. (1989). Functional domains of the human vitamin D_3 receptor regulate osteocalcin gene expression. *Mol. Endocrinol.* **3**, 635–644.

Marchetti, D., Van, N. T., Gametchu, B., Thompson, E. B., Kobayashi, Y., Watanabe, F. and Barlogie, B. (1989). Flow cytometric analysis of glucocorticoid receptor using monoclonal antibody and fluoresceinated ligand probes. *Cancer Res.* **49**, 863–869.

Martinez, E., Givel, F. and Wahli, W. (1987). The estrogen-responsive element as an inducible enhancer: DNA sequence requirements and conversion to a glucocorticoid-responsive element. *EMBO J.* **6**, 3719–3727.

Mendel, D. B. and Orti, E. (1988). Isoform composition and stoichiometry of the ∼90-kDa heat shock protein associated with glucocorticoid receptors. *J. Biol. Chem.* **263**, 6695–6702.

Mendel, D. B., Bodwell, J. E., Gametchu, B., Harrison, R. W. and Munck, A. (1986). Molybdate-stabilized nonactivated glucocorticoid-receptor complexes contain a 90-kDa non-steroid-binding phosphoprotein that is lost on activation. *J. Biol. Chem.* **261**, 3758–3763.

Mendel, D. B., Bodwell, J. E. and Munck, A. (1987). Activation of cytosolic glucocorticoid–receptor complexes in intact WEHI-7 cells does not dephosphorylate the steroid-binding protein. *J. Biol. Chem.* **262**, 5644–5648.

Miesfeld, R., Rusconi, S., Godowski, P. J., Maler, B. A., Okret, S., Wikström, A.-C., Gustafsson, J.-A. and Yamamoto, K. R. (1986). Genetic complementation of glucocorticoid receptor deficiency by expression of cloned receptor cDNA. *Cell* **46**, 389–399.

Miesfeld, R., Godowski, P. J., Maler, B. A. and Yamamoto, K. R. (1987). Glucocorticoid receptor mutants that define a small region sufficient for enhancer activation. *Science* **236**, 423–427.

Miller, J., McLachlan A. D. and Klug, A. (1985). Repetitive zinc-binding domains in the protein transcription factor IIIA from *Xenopus* oocytes. *EMBO J.* **4**, 1609–1614.

Misrahi, M., Atger, M., D'Auriol, L., Loosfelt, H., Meriel, C., Fridlansky, F., Guiochon-Mantel, A., Galibert, F. and Milgrom, E. (1987). Complete amino acid sequence of the human progesterone receptor deduced from cloned cDNA. *Biochem. Biophys. Res. Commun.* **143**, 740–748.

Miyajima, N., Kadowaki, Y., Fukushige, S., Shimizu, S., Semba, K., Yamanashi, Y., Matsubara, K., Toyoshima, K. and Yamamoto, T. (1988). Identification of two novel members of *erb*A superfamily by molecular cloning: the gene products of the two are highly related to each other. *Nucl. Acids Res.* **16**, 11 057–11 074.

Miyajima, N., Horiuchi, R., Shibuja, Y., Fukushige, S., Matsubara, K., Toyoshima, K. and Yamamoto, T. (1989). Two *erb*A homologs encoding proteins with different T3 binding capacities are transcribed from opposite DNA strands of the same genetic locus. *Cell* **57**, 31–39.

Mlodzik, M., Hiromi, Y., Weber, U., Goodman, C. S. and Rubin, G. M. (1990). The *Drosophila seven-up* gene, a member of the steroid receptor gene superfamily, controls photoreceptor cell fates. *Cell* **60**, 211–224.

Moore, S. K., Kozak, C., Robinson, E. A., Ullrich, S. J. and Appella, E. (1989). Murine 86- and 84-kDa heat shock proteins, cDNA sequences, chromosome assignments, and evolutionary origins. *J. Biol. Chem.* **264**, 5343–5351.

Murray, M. B., Zilz, N. D., McCreary, N. L., MacDonald, M. J. and Towle, H. C. (1988). Identification and characterization of rat cDNA clones for two distinct thyroid hormone receptors. *J. Biol. Chem.* **263**, 12 770–12 777.

Nakao, K., Myers, J. E. and Faber, L. E. (1985). Development of a monoclonal antibody to the rabbit 8.5S uterine progestin receptor. *Can. J. Biochem. Cell Biol.* **63**, 33–40.

Nauber, U., Pankratz, M. J., Kienlin, A., Seifert, E., Klemm, U. and Jackle, H. (1988). Abdominal segmentation of the *Drosophila* embryo requires a hormone receptor-like protein encoded by the gap gene knirps. *Nature* **336**, 489–492.

Nemoto, T., Ohara-Nemoto, Y., Denis, M. and Gustafsson, J.-A. (1990). The transformed glucocorticoid receptor has a lower steroid-binding affinity than the nontransformed receptor. *Biochemistry* **29**, 1880–1886.

Northrop, J. P., Gametchu, B., Harrison, R. W. and Ringold, G. M. (1985). Characterization of wild type and mutant glucocorticoid receptors from rat hepatoma and mouse lymphoma cell lines. *J. Biol. Chem.* **260**, 6398–6403.

Okret, S., Wikström, A.-C., Wrange, O., Andersson, B. and Gustafsson, J.-A. (1984) Monoclonal antibodies against the rat liver glucocorticoid receptor. *Proc. Natl. Acad. Sci. USA* **81**, 1609–1613.

Oro, A. E., Hollenberg, S. M. and Evans, R. M. (1988a). Transcriptional inhibition by a glucocorticoid receptor-β-galactosidase fusion protein. *Cell* **55**, 1109–1114.

Oro, A. E., Ong, E. S., Margolis, J. S., Posakony, J. W., McKeown, M. and Evans, R. M. (1988b). The *Drosophila* gene *knirps-related* is a member of the steroid-receptor gene superfamily. *Nature* **336**, 493–496.

Orti, E., Mendel, D. B. and Munck, A. (1989a). Phosphorylation of glucocorticoid receptor-associated and free forms of the $\sim$90-kDa heat shock protein before and after receptor activation. *J. Biol. Chem.* **264**, 231–237.

Orti, E., Mendel, D. B., Smith, L. I., Bodwell, J. E. and Munck, A. (1989b). A dynamic model of glucocorticoid receptor phosphorylation and cycling in intact cells. *J. Steroid Biochem.* **34**, 85–96.

Orti, E., Mendel, D. B., Smith, L. I. and Munck, A. (1989c). Agonist-dependent phosphorylation of glucocorticoid receptors in intact cells. *J. Biol. Chem.* **264**, 9728–9731.

Pakdel, F., Guellec, C. L., Vaillant, C., Roux, M. G. L. and Valotaire, Y. (1989). Identification and estrogen induction of two estrogen receptors (ER) messenger ribonucleic acids in the rainbow trout liver: sequence homology with other ERs. *Mol. Endocrinol.* **3**, 44–51.

Patel, P. D., Sherman, T. G., Goldman, D. J. and Watson, S. J. (1989). Molecular cloning of a mineralocorticoid (type I) receptor complementary DNA from rat hippocampus. *Mol. Endocrinol.* **3**, 1877–1885.

Payvar, F. and Wrange, O. (1984). Relative selectivities and efficiencies of DNA binding by purified intact and protease cleaved glucocorticoid receptor. In "Steroid Hormone Receptors: Structure and Function (eds H. Eriksson, J. A. Gustafsson and B. Hogberg), pp. 267–282. Elsevier/North Holland Biomedical Press, Amsterdam.

Perdew, G. H. (1988). Association of the Ah receptor with the 90-kDa heat shock protein. *J. Biol. Chem.* **263**, 13 802–13 805.

Petkovich, M., Brand, N. J., Krust, A. and Chambon, P. (1987). A human retinoic acid receptor which belongs to the family of nuclear receptors. *Nature* **330**, 444–450.

Picard, D. and Yamamoto, K. R. (1987). Two signals mediate hormone-dependent nuclear localization of the glucocorticoid receptor. *EMBO J.* **6**, 3333–3340.

Picard, D., Salser, S. J. and Yamamoto, K. R. (1988). A movable and regulable inactivation function within the steroid binding domain of the glucocorticoid receptor. *Cell* **54**, 1073–1080.

Pratt, W. B., Sanchez, E. R., Bresnick, E. H., Meshinchi, S., Scherrer, L. C., Dalman, F. C. and Welsh, M. J. (1989). Interaction of the glucocorticoid receptor with the M_r 90 000 heat shock protein: an evolving model of ligand-mediated receptor transformation and translocation. *Cancer Res.* **49**, 2222s–2229s.

Pratt, W. B., Jolly, D. J., Pratt, D. V., Hollenberg, S. M., Giguere, V., Cadepond, F. M., Schweizer-Groyer, G., Catelli, M.-G., Evans, R. M. and Baulieu, E.-E. (1988). A region in the steroid binding domain determines formation of the non-DNA-binding, 9 S glucocorticoid receptor complex. *J. Biol. Chem.* **263**, 267–273.

Raaka, B., Finnerty, M., Sun, E. and Samuels, H. H. (1985). Effects of molybdate on steroid receptors in intact GH_1 cells. *J. Biol. Chem.* **260**, 14 009–14 015.

Rabindran, S. K., Danielsen, M., Firestone, G. L. and Stallcup, M. R. (1987). Glucocorticoid-dependent maturation of viral proteins in mouse lymphoma cells: Isolation of defective and hormone-independent cell variants. *Somatic Cell Mol. Genet.* **13**(2), 131–143.

Radanyi, C., Joab, I., Renoir, J.-M., Richard-Foy, H. and Baulieu, E.-E. (1983). Monoclonal antibody to chicken oviduct progesterone receptor. *Proc. Natl. Acad. Sci. USA* **80**, 2854–2858.

Radanyi, C., Renoir, J.-M., Sabbah, M. and Baulieu, E.-E. (1989). Chick heat shock protein of $M_r = 90 000$, free or released from progesterone receptor, is in a dimeric form. *J. Biol. Chem.* **264**, 2568–2573.

Ragsdale, C. W., Petkovich, M., Gates, P. B., Chambon, P. and Brockes, J. P. (1989). Identification of a novel retinoic acid receptor in regenerative tissues of the newt. *Nature* **341**, 654–657.

Rafestin-Oblin, M.-E., Lombes, M., Harrison, R., Blanchardie, P. and Clarie, M. (1986). Cross-reactivity of a monoclonal antiglucocorticoid receptor antibody BuGR1 with glucocorticoid and mineralocorticoid receptors of various species. *J. Steroid Biochem.* **24**, 259–262.

Rafestin-Oblin, M.-E., Couette, B., Radanyi, C., Lombes, M. and Baulieu, E.-E. (1989). Mineralocorticoid receptor of the chick intestine. *J. Biol. Chem.* **264**, 9304–9309.

Reker, C. E., LaPointe, M. C., Kovacic-Milivojevic, B., Chiou, W. J. H. and Vedeckis, W. V. (1987). A possible role for dephosphorylation in glucocorticoid receptor transformation. *J. Steroid Biochem.* **26**, 653–665.

Rexin, M., Busch, W. and Gehring, U. (1988a). Chemical cross-linking of heteromeric glucocorticoid receptors. *Biochemistry* **27**, 5593–5601.

Rexin, M., Busch, W. Segnitz, B. and Gehring, U. (1988b). Tetrameric structure of the nonactivated glucocorticoid receptor in cell extracts and intact cells. *FEBS Lett.* **241**, 234–238.

Riehl, R. M., Sullivan, W. P., Vroman, B. J., Bauer, V. J., Pearson, G. R. and Toft, D. O. (1985). Immunological evidence that the nonhormone binding component of avian steroid receptors exists in a wide range of tissues and species. *Biochemistry* **24**, 6586–6591.

Robertson, N. M., Fusmik, W. F., Grove, B. F., Miller-Diener, A., Webb, M. L. and Litwack, G. (1987). Characterization of a monoclonal antibody that probes the functional domains of the glucocorticoid receptor. *Biochem. J.* **246**, 55–65.

Rothe, M., Nauber, U. and Jackle, H. (1989). Three hormone receptor-like *Drosophila* genes encode an identical DNA binding finger. *EMBO J.* **8**, 3087–3094.

Rusconi, S. and Yamamoto, K.R. (1987). Functional dissection of the hormone and DNA binding activities of the glucocorticoid receptor. *EMBO J.* **6**, 1309–1315.

Ryseck, R. P., Macdonald-Bravo, H., Mattei, M.-G., Ruppert, S. and Bravo, R. (1989). Structure, mapping, and expression of a growth factor inducible gene encoding a putative nuclear hormonal binding receptor. *EMBO J.* **8**, 3327–3335.

Sakurai, A., Takeda, K., Ain, K., Ceccarelli, P., Nakai, A., Seino, S., Bell, G. I., Refetoff, S. and DeGroot, L. J. (1989). Generalized resistance to thyroid hormone associated with a mutation in the ligand binding domain of the human thyroid hormone receptor β. *Proc. Natl. Acad. Sci. USA* **86**, 8977–8981.

Sanchez, E. R., Toft, D. O., Schlesinger, M. J. and Pratt, W. B. (1985). Evidence that the 90 kDa phosphoprotein associated with the untransformed L-cell glucocorticoid receptor is a murine heat shock protein. *J. Biol. Chem.* **260**, 12 398–12 401.

Sanchez, E. R., Meshinchi, S., Schlesinger, M. J. and Pratt, W. B. (1987a). Demonstration that the 90-kilodalton heat shock protein is bound to the glucocorticoid receptor in its 9S nondeoxynucleic acid binding form. *Mol. Endocrinol.* **1**, 908–912.

Sanchez, E. R., Meshinchi, S., Tienrungroj, W., Schlesinger, M. J., Toft, D. O. and Pratt, W. B. (1987b). Relationship of the 90-kDa murine heat shock protein to the untransformed and transformed states of the L cell glucocorticoid receptor. *J. Biol. Chem.* **262**, 6986–6991.

Sanchez, E. R., Faber, L. E., Henzel, W. J. and Pratt, W. B. (1990). The 56 kDa protein in the untransformed glucocorticoid receptor complex is a unique protein that exists in cytosol in a complex with both the 70 kDa and 90 kDa heat shock proteins. *Biochem.* **29**, 5145–5152.

Sando, J. J., Hammond, N. D., Stratford, C. A. and Pratt, W. B. (1979). Activation of thymocyte glucocorticoid receptors to the steroid binding form. *J. Biol. Chem.* **254**, 4779–4789.

Schena, M., Freedman, L. P. and Yamamoto, K. R. (1989). Mutations in the glucocorticoid receptor zinc finger region that distinguish interdigitated DNA binding and transcriptional enhancement activities. *Genes Dev.* **3**, 1590–1601.

Schuh, S. W., Yonemoto, J., Brugge, J., Bauer, R., Riehl, R. M., Sullivan, W. P. and Toft, D. P. (1985). A 90 000 dalton binding protein common to both steroid receptors and the Rous sarcoma virus transforming protein, pp60^v-src. *J. Biol. Chem.* **260**, 14 292–14 296.

Severne, Y., Wieland, S., Schaffner, W. and Rusconi, S. (1988). Metal binding 'finger' structures in the glucocorticoid receptor defined by site-directed mutagenesis. *EMBO J.* **7**, 2503–2508.

Sigler, P. B. (1988). Acid blobs and negative noodles. *Nature* **333**, 210–212.

Simons, S. S. (1987). Selective covalent labeling of cysteins in bovine serum albumin and in hepatoma tissue culture cell glucocorticoid receptors by dexamethasone 21-mesylate. *J. Biol. Chem.* **262**, 9669–9675.

Simons, S. S., Pumphrey, J. G., Rudikoff, S. and Eisen, H. I. (1987). Identification of cysteine 656 as the amino acid of hepatoma tissue culture cell glucocorticoid receptors that is covalently labeled by dexamethasone 21-mesylate. *J. Biol. Chem.* **262**, 9676–9680.

Simons, S. S., Sistare, F. D. and Chakraborti, P. K. (1989). Steroid binding activity is retained in a 16-kDa fragment of the steroid binding domain of rat glucocorticoid receptors. *J. Biol. Chem.* **264**, 14 493–14 497.

Smith, A. C., Elsasser, M. S. and Harmon, J. M. (1986). Analysis of glucocorticoid receptor activtion by high resolution two-dimensional electrophoresis of affinity-labeled receptor. *J. Biol. Chem.* **261**, 13 285–13 292.

Smith, D. F., Lubahn, D. B., McCormick, D. J., Wilson, E. M. and Toft, D. O. (1988). The production of antibodies against the conserved cysteine region of steroid receptors and their use in characterizing the avian progesterone receptor. *Endocrinology* **122**, 2816–2825.

Smith, L. I., Bodwell, J. E., Mendel, D. B., Ciardelli, T., North, W. G. and Munck, A. (1988). Identification of cysteine-644 as the covalent site of attachment of dexamethasone 21-mesylate to murine glucocorticoid receptors in WEHI-7 cells. *Biochemistry* **27**, 3747–3753.

Smith, L. I., Mendel, D. B., Bodwell, J. E. and Munck, A. (1989). Phosphorylated sites within the functional domains of the ~100-kDa steroid-binding subunit of glucocorticoid receptors. *Biochemistry* **28**, 4490–4498.

Sullivan, W. P., Vroman, B. T., Bauer, V. J., Puri, R. J., Riehl, G. R., Pearson, G. R. and Toft, D. O. (1985). Isolation of steroid receptor binding protein from chick oviduct and production of monoclonal antibodies. *Biochemistry* **24**, 4214–4222.

Tai, P.-K. K., Maeda, Y., Nakao, K., Wakim, N. G., Duhring, J. L. and Faber, L. E. (1986). A 59-Kilodalton protein associated with progestin, estrogen, androgen, and glucocorticoid receptors. *Biochemistry* **25**, 5269–5275.

Teasdale, J., Lewis, F. A., Barrett, I. D., Abbott, A. C., Wharton, J. and Bird, C. C. (1986). Immunocytochemical application of monoclonal antibodies to rat liver glucocorticoid receptor. *J. Pathol.* **150**, 227–237.

The H. de-, Marchio, A., Tiollais, P. and Dejean, A. (1987). A novel steroid thyroid hormone receptor-related gene inappropriately expressed in human hepatocellular carcinoma. *Nature* **330**, 667–670.

Thompson, C. C., Weinberger, C., Lebo, R. and Evans, R. M. (1987). Identification of a novel thyroid hormone receptor expressed in the mammalian central nervous system. *Science* **237**, 1610–1614.

Tienrungroj, W., Sanchez, E. R., Housely, P. R., Harrison, R. W. and Pratt, W. B. (1987). Glucocorticoid receptor phosphorylation, transformation, and DNA binding. *J. Biol. Chem.* **262**, 17 342–17 349.

Toft, D. O. and Nishigori, H. (1979). Stabilization of the avian progesterone receptor by inhibitors. *J. Steroid Biochem.* **11**, 413–416.

Tora, L., Gronemeyer, H., Turcotte, B., Gaub, M.-P. and Chambon, P. (1988). The N-terminal region of the chicken progesterone receptor specifies target gene activation. *Nature* **333**, 185–188.

Tora, L., Mullick, A., Metzger, D., Ponglikitmongkol, M., Park, I. and Chambon, P. (1989). The cloned human oestrogen receptor contains a mutation which alters its hormone binding properties. *EMBO J.* **8**, 1981–1986.

Umesono, K. and Evans, R. M. (1989). Determinants of target gene specificity for steroid/thyroid hormone receptors. *Cell* **57**, 1139–1146.
Urda, L. A., Yen, P. M., Simons, S. S. and Harmon, J. M. (1989). Region-specific antiglucocorticoid receptor antibodies selectively recognize the activated form of the ligand-occupied receptor and inhibit the binding of activated complexes to deoxyribonucleic acid. *Mol. Endocrinol.* **3**, 251–260.
Vallee, B. L. and Auld, D. S. (1990). Active site ligands and activated H_2O of zinc enzymes. *Proc. Natl. Acad. Sci. USA* **87**, 220–224.
Wang, L. H., Tsai, S. Y., Cook, R. G., Beattie, W. G., Tsai, M.-J. and O'Malley, B. W. (1989). COUP transcription factor is a member of the steroid receptor superfamily. *Nature* **340**, 163–166.
Watson, M. A. and Milbrandt, J. (1989). The NGFI-B gene, a transcriptionally inducible member of the steroid receptor gene superfamily: genomic structure and expression in rat brain after seizure induction. *Mol. Cell. Biol.* **9**, 4213–4219.
Webster, N. J. G., Green, S., Jin, J. R. and Chambon, P. (1988). The hormone binding domains of the estrogen and glucocorticoid receptors contain an inducible transcription activation function. *Cell* **54**, 199–207.
Weiler, I. J., Lew, D. and Shapiro, D. J. (1987). The *Xenopus laevis* estrogen receptor: sequence homology with human and avian receptor and identification of multiple estrogen messenger ribonucleic acids. *Mol. Endocrinol.* **1**, 355–362.
Weinberger, C., Thompson, C. C., Ong, E. S., Lebo, R., Gruol, D. J. and Evans, R. M. (1986). The c-*erb*-A gene encodes a thyroid hormone receptor. *Nature* **324**, 641–646.
Westphal, H. M., Moldenhauer, G. and Beato, M. (1982). Monoclonal antibodies to rat liver glucocorticoid receptors. *EMBO J.* **1**, 1467–1471.
Westphal, H. M., Mugele, M., Beato, M. and Gehring, U. (1984). Immunochemical characterization of wild type and variant glucocorticoid receptors by monoclonal antibodies. *EMBO J.* **3**, 1493–1498.
White, R., Lees, J. A., Needham, M., Ham, J. and Parker, M. (1987). Structural organization and expression of the mouse estrogen receptor. *Mol. Endocrinol.* **1**, 735–744.
Wikström, A.-C., Bakke, O., Okret, S., Bronnegard, M. and Gustafsson, J.-A. (1987). Intracellular localization of the glucocorticoid receptor: evidence for cytoplasmic and nuclear localization. *Endocrinology* **120**, 1232–1242.
Wilhelmsson, A., Cuthill, S., Denis, M., Wikström, A.-C., Gustafsson, J.-A. and Poellinger, L. (1990). The specific DNA binding activity of the dioxin receptor is modulated by the 90 kD heat shock protein. *EMBO J.* **9**, 69–76.
Willmann, T. and Beato, M. (1986). Steroid free glucocorticoid receptor binds specifically to mouse mammary tumour virus DNA. *Nature* **324**, 688–691.
Zelent, A., Krust, A., Petkovich, M., Kastner, P. and Chambon, P. (1989). Cloning of murine α and β retinoic acid receptors and a novel receptor γ predominantly expressed in skin. *Nature* **339**, 714–717.

4
Nuclear Thyroid Hormone Receptors

W. W. CHIN
Division of Genetics, Department of Medicine, Brigham and Women's Hospital, Howard Hughes Medical Institute and Harvard Medical School, Boston, MA 02115, USA

4.1 Introduction

Thyroid hormones (TH) are biologically active iodothyronines that are synthesized and secreted by the thyroid glands of most vertebrates. The major TH (Fig. 1) include 3,5,3'5'-tetraiodo-L-thyronine (T_4) and 3,5,3'-triiodo-L-thyronine (T_3) (the most biologically active form) and have effects

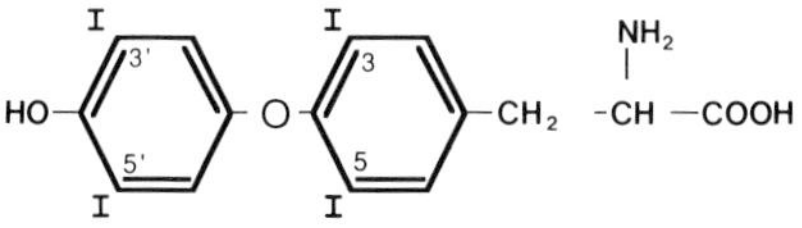

3,5,3',5'-tetraiodo-L-thyronine (L-thyroxine, T_4)

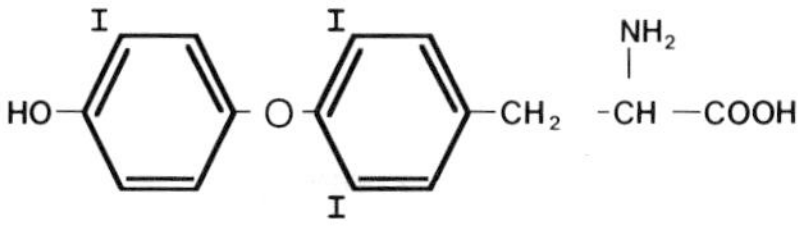

3,5,3'-triiodo-L-thyronine (T_3)

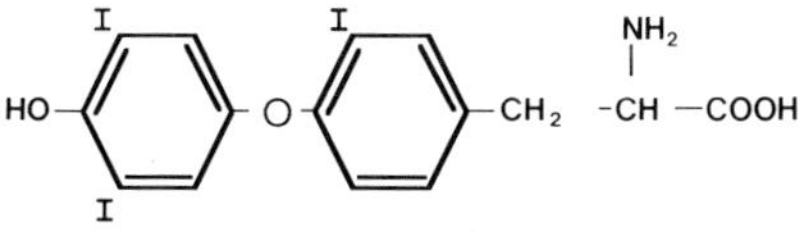

3,3',5'-triiodo-L-thyronine (reverse T_3)

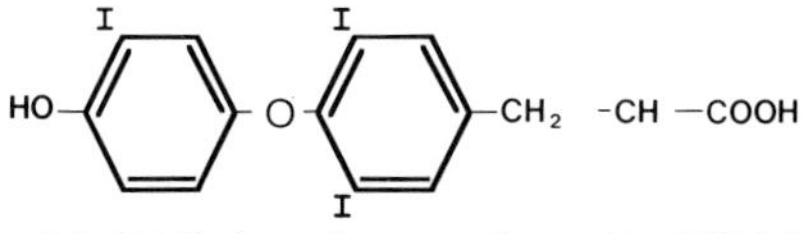

3,5,3'-triiodo-L-thyroacetic acid (TRIAC)

Fig. 1 Major thyroid hormones.

on nearly every cell, influencing general metabolism, growth and development as well as specific gene expression (Oppenheimer, 1983; Oppenheimer et al., 1987). As major examples of their role in development, TH play a pivotal action in frog tadpole metamorphosis (Baker and Tata, 1990), and dendrite formation and myelination in the maturation of the neonatal rat brain (Schwartz, 1983). TH are also involved in a complex regulatory feedback circuit involving the pituitary and thyroid glands. Specifically, the thyroid gland produces TH which, in turn, act on the thyrotrope cell of the pituitary gland to decrease production of thyrotropin (TSH), a hormone that stimulates the thyroid gland to produce more TH. In addition, TH may affect neuroendocrine factors in the central nervous system. Thus, TH are involved in a classic endocrine negative feedback loop (Shupnik et al., 1989). TH exert their effects largely by influencing gene expression. They enter a target cell via passive mechanisms and may proceed to the nucleus where they interact with a nuclear protein known as the thyroid hormone receptor (TR). The TH/TR complex then may rapidly trigger specific thyroid hormone-responsive genes to initiate their biological actions. Thus, an appreciation of the nature of TRs in mammalian cells and their interactions with TH, as well as target genes, will allow a greater understanding of the molecular mechanisms of thyroid hormone action.

4.2 Biochemistry of thyroid hormone receptors

Over the past two decades, major insights into the nature of the nuclear TR have been achieved (Oppenheimer, 1983). It is known that the TR is a nuclear protein that binds TH with high affinity ($K_d = 10^{-10} - 10^{-11}$ M) and specificity, and is tightly associated with chromatin. It is present in low abundance, and is generally unstable to general biochemical isolation procedures. As a result, the biochemistry of the TRs is not well understood. There are approximately 2000–10 000 thyroid hormone receptors per cell in most tissues, although fewer TRs have been observed in the spleen and testes as determined by studies of the nuclear binding of labelled T_3 and T_4 (Oppenheimer et al., 1974). It is also known that TRs bind TH and their analogues (Fig. 1) in a fixed hierarchy of relative affinity: TRIAC > T_3 > T_4 ≫ rT_3. Equilibrium sedimentation and photoaffinity labelling studies have provided an initial glimpse into the molecular nature of the TRs. These studies by Samuels, Baxter and their co-workers (Latham et al., 1976; Casanova et al., 1984; Samuels et al., 1988) suggested that TRs are heterogeneous in nature, having a number of forms with molecular sizes ranging from 47 to 57 kD with the variation in size thought to be due to protein degradation or post-translational modifications. Furthermore, tissue localization studies were performed revealing that TRs are present in most mammalian tissues. The best studied tissues are the pituitary gland with focus on a number of clonal pituitary cell lines, liver and heart. Importantly, the biological action of TH has been well correlated with TH occupancy of nuclear TR and binding affinity of various TH analogues. Also, TH-unresponsive tissues have decreased nuclear TH-binding activity (Oppenheimer, 1983; Samuels, 1983).

4.3 Thyroid hormone receptors are encoded by the proto-oncogene, c-*erb*A

A breakthrough in our understanding of the nature of TRs was achieved in 1986 when Sap *et al.* (1986) and Weinberger *et al.* (1986) reported the molecular cloning of cDNAs encoding putative TRs. These studies revealed that the proto-oncogene, c-*erb*A, the cellular homologue of v-*erb*A (derived from the avian erythroblastosis virus, as discussed in Chapter 15 by Berg and Vennström) (Graf and Beug, 1978; Vennström and Bishop, 1982), encodes a protein that can bind TH and thus represents a putative TR. Weinberger *et al.* (1986) isolated a cDNA encoding a TR from human placenta, whereas Sap *et al.* (1986) isolated a cDNA encoding a TR derived from chicken embryo. The deduced amino acid sequences of these cDNAs revealed proteins that are members of the steroid hormone receptor superfamily (Green and Chambon, 1986; Evans, 1988). Near the N-terminal portion of the molecule there is a region spanning approximately 70 amino acids that can form a potential dual zinc finger structure reminiscent of the DNA-binding domains in the steroid hormone receptors. Also, there is a carboxy-terminal region that corresponds to a ligand-binding domain. Indeed, *in vitro* synthesized proteins directed by these cDNAs bind radiolabelled TH and their analogues in a hierarchy of affinity that is consistent with data observed *in vivo*. Furthermore, tissue surveys showed that TR mRNAs are widely distributed. Thus, c-*erb*A likely encodes the TR, although the ability of these TRs to activate thyroid hormone-responsive genes was not examined in these early reports.

A surprising observation in the comparison of the chicken versus human TR forms was the existence of similar but not identical TRs, suggesting the possibility of multiple TRs and corresponding genes in a given species. It has been shown that c-*erb*A is localized to two human chromosomes, chromosome 3 (3p22–3p24.1) (Weinberger *et al.*, 1986; Drabkin *et al.*, 1988) and chromosome 17 (17q11.2–17q21) (Jansson *et al.*, 1983; Dayton *et al.*, 1984).

Hence, the chromosomal and cDNA data together suggest the likelihood of at least two genes encoding TRs. Indeed, the chicken embryo and human placental TRs are designated α and β, respectively.

4.4 A bevy of thyroid hormone receptors

Soon after these initial observations, Thompson *et al.* (1987) isolated a cDNA encoding an homologue of the chicken embryo TRα from rat brain and called c-*erb*Aα-1 or TRα-1 (Figs 2–4). The major TRα-1 mRNA observed in brain was apparently 2.6 kb in size and was widely distributed, being present in most other tissues. Of note, they emphasized that this αTR form was most abundant in rat brain, suggesting the presence of a 'neuronal' TR type in the species. This initial assignment proved to be erroneous, as discussed below. Benbrook and Pfahl (1987) and Nakai (1988a) then described human cDNAs from testes and kidney, respectively, encoding a form of TRα that was similar to the rat TRα-1 reported by Thompson *et al.* (1987) but differed in the deduced amino acid sequence at the carboxy-terminal end. Benbrook and Pfahl (1987) suggested that the detection of this TR isoform, called *erb*-A-T-1 or c-*erb*Aα-2, confirmed the heterogeneous nature of the TRs. Furthermore, they claimed that this form bound TH but with lower affinity than the human TRβ or rat TRα-1. Similar results were described by Nakai *et al.* (1988a). As we shall observe when considering the rat counterpart and recent work by Lazar *et al.* (1989a) and Schueler *et al.* (1990), it is clear that the initial observation of the ability of human c-*erb*Aα-2 to bind thyroid hormones was due to the detection of background levels of TH binding in unprogrammed reticulocyte lysates, and that c-*erb*Aα-2 does *not* bind TH. Thus, c-*erb*Aα-2 is not a TR inasmuch as it fails the key test: high affinity binding of TH. Nakai *et al.* (1988b) also isolated the human homologue of the rat TRα-1 clone, derived from kidney, described by Thompson *et al.* (1987).

Further work by Murray *et al.* (1988) and Koenig *et al.* (1988) resulted in the identification and characterization of cDNAs encoding rat TRα-1 and TRβ-1. The rat TRα-1 cDNA, from liver, described by Murray *et al.* (1988), is essentially similar to that described by Thompson (1987). The TRβ-1 cDNAs (Koenig *et al.*, 1988; Murray *et al.*, 1988), from liver and pituitary GH$_3$ cells, are homologues of the form originally described by Weinberger *et al.* (1986) derived from human placenta. Importantly, Koenig *et al.* (1988) demonstrated the ability of the rat TRβ-1 to transactivate a target gene containing the putative thyroid hormone-response element (TRE) found in the rat growth hormone gene, utilizing cotransfection paradigms. These data

4. NUCLEAR THYROID HORMONE RECEPTORS

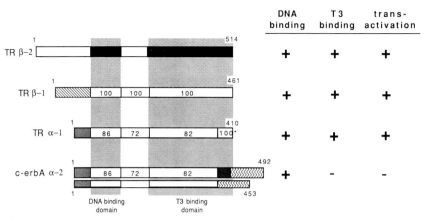

Fig. 2 Major rat hormone receptors and related isoforms: key properties of DNA binding, T_3 binding, and trans-activation are noted. Homology of different domains of the TRs are compared (numbers in each box refer to percentage similarity at the amino acid level compared to TRβ-2). The numbers at the N- and C-termini of the proteins refer to the amino residue in each protein. The variant shown below c-erbAα-2 has been described by Mitsuhashi and Nikodem [variant II] (1988). Domains with variant structures are depicted with different shading.

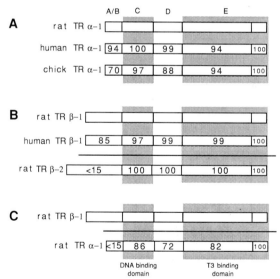

Fig. 3 Comparison of homologous domains among the thyroid hormone receptors. The C and E regions represent the DNA- and ligand-binding domains, respectively, while A/B and D regions depict the NH_2-terminal and hinge domains, respectively. The numbers in each box refer to the percentage similarity of each comparable domain to the top TR form in each group. (A) Comparison of TRα-1s. (B) Comparison of TRβs. (C) Comparison of rat TRβ-1 with TRα-1.

	RAT		HUMAN		CHICKEN	
	SIZE (AAs)	mRNA(kb)	SIZE (AAs)	mRNA(kb)	SIZE (AAs)	mRNA(kb)
TRβ–1	461	6.2	456	6.0 (2.0)		
TRβ–2	514	6.2				
TRα–1	410	6.0 (5.0)	490	6.0	408	4.5/3.0
c-erbA α–2 v II	492 453	2.6	410	2.5/2.0		
Rev-erbA α	503	3.0	614	3.0		

Fig. 4 Thyroid hormone receptors and isoforms in rat, human and chicken: protein and mRNA sizes.

showed that the TRs encoded by c-erbA were indeed *trans*-acting factors, and confirmed their identity as bona fide TRs.

Another interesting aspect in the story of heterogeneous TRs was revealed in work of Lazar and Chin (1988), Lazar et al. (1988), Izumo and Mahdari (1988) and Mitsuhashi et al. (1988), in which they independently describe cDNAs isolated from GH_3 cells, rat brain, and rat liver, respectively, encoding a form of the rat c-erbAα which was identical to the rat TRα-1 form described previously from amino acid 1 to 370. Beginning with amino acid 371 there was complete digression from the sequence in rat TRα-1. This suggested that a c-erbAα-2 form, similar to the human c-erbAα-2 form (Benbrook and Pfahl, 1987; Nakai et al., 1988a), is also present in the rat. The most curious feature of rat c-erbAα-2 is that it does not bind thyroid hormone but yet can interact, if less effectively, with putative TREs (Lazar et al., 1988). Furthermore, Izumo and Mahdavi (1988) indicated there were multiple splicing variants of the c-erbAα-2 cDNA (as many as nine forms), although the major form was the c-erbAα-2. In addition, Mitsuhashi et al. (1988) described an additional minor alternate splice form (variant II) which differs from rat c-erbAα-2 (variant I) by a deletion of 39 amino acids at the beginning of the divergent carboxy-terminal region. Finally, mouse TRα-1 and c-erbAα-2 cDNA homologues have been cloned from the heart (Prost et al., 1988). In a later section of this Chapter, a proposed function for c-erbAα-2 is described.

Hodin et al. (1989b) described a cDNA, derived from GH_3 cells, that encodes a bonafide TR by the criteria of TH and TRE binding and TRE transactivation that is homologous to rat TRβ-1 and called TRβ-2. Interestingly, this protein is identical to TRβ-1, from a region one codon N-terminal from the putative DNA-binding domain until the end of the molecule. However, the N-terminal regions are divergent. Another intriguing aspect is that this rat TRβ-2 form is expressed only in the pituitary gland. Furthermore, thyroid hormones strongly negatively regulate TRβ-2 mRNA in GH_3 pituitary cell lines. The negative regulation is similar to that observed for the TRα-1 and TRα-2 mRNAs, but to a much greater extent. This contrasts with the modest positive (two-fold) regulation of TRβ-1 mRNA in these same cells.

4.5 Thyroid hormone response elements

Our understanding of the DNA targets of the TRs has improved over the last several years with the identification and characterization of putative TREs among the rat growth hormone (Glass *et al.*, 1987; Koenig *et al.*, 1987; West *et al.*, 1987; Wight *et al.*, 1987; Ye and Samuels, 1987; Ye *et al.*, 1988), TSHα (Burnside *et al.*, 1989; Chatterjee *et al.*, 1989), TSHβ (Carr *et al.*, 1989; Darling *et al.*, 1989; Wondisford *et al.*, 1989; Wood *et al.*, 1989), malic enzyme (Petty *et al.*, 1989) and α-myosin heavy chain (Izumo and Mahdari, 1988) genes. The net result of these efforts is a consensus TRE half-site, represented by the sequence: AGGT(C/A)A (Brent *et al.*, 1989a) (Fig. 5). In the best studied TRE, derived from the 5'-flanking region of the rat growth hormone gene, there are three half-sites located betweeen −190 and −164 (Glass *et al.*, 1987; Brent *et al.*, 1989a,b). Brent *et al.* (1989b) have shown that each is required for full expression of the TH-regulated response. A

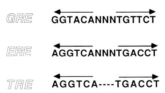

Fig. 5 Hormone-response elements. Consensus glucocorticoid (GRE), oestrogen (ERE), and thyroid hormone response elements (TRE) are shown. N = A, C, G or T. Arrows depict palindromic sequences.

similar tripartite TRE arrangement is observed in the promoter of the rat α-myosin heavy chain gene (Izumo and Mahdavi, 1988). Interestingly, Glass *et al.* (1988) noted that the oestrogen-response element (ERE) and TRE contain identical consensus half-binding sites but are different only in terms of spacing between the palindromic sequences: a three nucleotide spacer appears in the ERE (Fig. 5). It is noteworthy that TRs may bind to an ERE without evident activation, suggesting that DNA binding, although a necessary condition, is not sufficient for transactivation of associated genes. Indeed, such interaction can inhibit oestrogen-dependent transactivation. In addition, it has been shown that thyroid hormone receptors may also interact with related retinoic acid receptor (RAR) response elements and vice versa (Umesono *et al.*, 1988; Glass *et al.*, 1989; Bedo *et al.*, 1989; Brent *et al.*, 1989c; Graupner *et al.*, 1989). Of special interest is the observation that TRs can interact with RARs to form heterodimers which possess different regulatory potential depending on the target TRE and host cell (Glass *et al.*, 1989).

4.6 Functional domains of thyroid hormone receptors

Thompson and Evans (1989) have used domain swaps of the human TRβ and glucocorticoid receptors to provide further insight into the domain structure of the TR (Fig. 6). They provide evidence supporting the presence of DNA- and ligand-binding domains as well as weak transactivation domains within these regions. In other structure–function analyses, Horowitz *et al.* (1989) and others have confirmed these general domain features. Of note, is the absence of a strong transactivation domain in the N-terminal region (A/B in Fig. 6) as observed in other receptors in the steroid hormone receptor family such as the glucocorticoid receptor (Hollenberg *et al.*, 1987; Hollenberg and Evans, 1988; see Chapter 3.2). The N-terminal domains of the TRs are generally much shorter than those of the steroid receptors but may yet play an undetermined role in specific gene transactivation in different tissues (Tora *et al.*, 1988). In addition, the significance of multiple in-frame translational start-sites in the 5′-untranslated regions of most TR mRNAs is not clear (Weinberger *et al.*, 1986; Sap *et al.*, 1986).

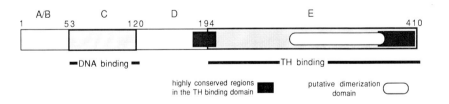

Fig. 6 Functional domains of the thyroid hormone receptor.

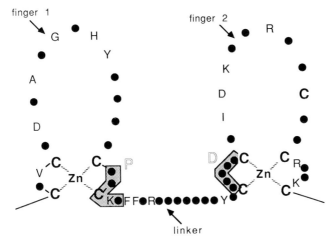

Fig. 7 The DNA binding domain of the thyroid/steroid hormone receptor family. The two zinc fingers are depicted; invariant amino acid residues are shown in standard single letter code, and variable residues with solid dots. P refers to the proximal element in the 'knuckle' of the first finger, and D to the distal element in the 'knuckle' of the second finger.

Umesono and Evans (1989) and Danielson *et al.* (1989) have performed detailed structure–function mapping of the DNA-binding domain of the thyroid/steroid hormone receptors (Fig. 7). They have shown that a short amino acid sequence in the 'knuckle' of the first putative zinc finger (P box in Fig. 7) serves to group subfamilies in this superfamily of *trans*-acting proteins. Thus, EGCK (A/G) is found in the P box of TR, oestrogen (ER), retinoic acid (RAR) and vitamin D receptors while GSCKV is seen in glucocorticoid, mineralocorticoid, progesterone and androgen receptors. In the second zinc finger, another short sequence located again in a 'knuckle' (D box in Fig. 7) serves to differentiate among members within a subfamily. The molecular basis of specific response element interactions involving these subtle changes in the DNA-binding domain is not yet elucidated. Also, heptad repeats with a leucine zipper motif (Forman *et al.*, 1989) are located in the central and C-terminal regions of the ligand binding domain that may be responsible for dimerization (Kumar and Chambon, 1988; Forman *et al.*, 1989; Glass *et al.*, 1989). A similar domain has been described in the mouse oestrogen receptor (Fawell *et al.*, 1990). Furthermore, nuclear protein(s) may interact with specific TR domains to increase TR binding of TREs, and may be important for gene activation by TR (Murray and Towle, 1989; Burnside *et al.*, 1990). These factors are present in a wide variety of tissues and may be more abundant than TRs. Analogous apparent nuclear protein/receptor interactions have been described (Feavers *et al.*, 1987; Tsai *et al.*, 1987; Edwards *et al.*, 1989). TRs do not appear to bind Hsp90, a cytosolic protein as observed with the glucocorticoid and other receptors (Dalman *et al.*, 1990; see Chapter 3.6). Last, another functional domain may be one responsible for nuclear localization (Beato, 1989).

4.7 Function of thyroid hormone receptors

The major rat TRs are TRα-1, TRβ-1 and TRβ-2 (Fig. 2). These results were confirmed by their ability to transactivate TREs derived from rat growth hormone (Koenig *et al.*, 1988; Glass *et al.*, 1989; Horowitz *et al.*, 1989), and α-myosin heavy chain genes (Izumo and Mahdavi, 1988). In addition, these receptors have been shown to regulate negatively the human α glycoprotein hormone gene (Chatterjee *et al.*, 1989). A major question concerns the role(s) of multiple TRs in a cell. Do they each have equivalent function, or does each TR possess discrete activities that are possibly gene target or tissue specific? Unfortunately, no definite answer is available. From limited early studies, there appears no large differential ability of the TRs to transactivate specific genes. Thompson and Evans (1989) have shown that rat TRα-1 can achieve greater transactivation of the 5′-flanking TRE of the rat growth hormone gene than the human TRβ. However, it is difficult to assess whether equal amounts of 'active' TR protein are expressed with each transfected cDNA in these experiments. Other subtle differences between the TRs have also been noted. First, it is well known that nuclear TR, partially purified from TH-responsive cells, bind to TRIAC with slightly greater affinity than T_3 (Oppenheimer, 1983). Similar results have been generally seen with rTRα-1, rTRβ-1, hTRβ and rTRβ-2 (Weinberger *et al.*, 1986; Murray *et al.*, 1988; Koenig *et al.*, 1988; Hodin *et al.*, 1989b). Thompson and Evans (1989), in contrast, noted that TRIAC bound equally well, compared to T_3, to the rat TRα-1. Schueler *et al.* (1990) compared the TH and analogue binding of a number of *in vitro* translated TRs. They concluded that the TRIAC/T_3 affinity ratio was 2.2-fold greater between TRβ and TRα. Furthermore, they noted that TRβ-1 had greater affinity for T_3 than TRα-1. Second, Koenig *et al.* (1989) showed that c-*erb*Aα-2 competed the activity of rat TRβ-1 more completely than rat TRα-1 for the rat growth hormone gene TRE in a cotransfection paradigm. Thus, further studies on other TREs in different tissues will be critical in defining potential differences in the functions of the diverse TRs.

Last, the variable relative levels of the TRα-1, TRβ-1, and TRβ-2 mRNAs in different tissues allow for potential differential regulation at the TR isoform level (Murray *et al.*, 1988; Hodin *et al.*, 1989a,b; Mitsuhashi and Nikodem, 1989; Sakurai *et al.*, 1989a).

4.8 Negative regulation by thyroid hormone receptors and related forms

Using a cotransfection paradigm, Koenig et al. (1989) demonstrated that rat c-*erb*Aα-2 may serve as a negative regulator of thyroid hormone action. This is of particular note since c-*erb*Aα-2 is a naturally occurring alternate splice product of the α gene that can bind TREs (it posseses the same DNA-binding region found in rTRα-1) but *not* TH (Lazar et al., 1988; Izumo and Mahdavi, 1988; Mitsuhashi et al., 1988). Interestingly, c-*erb*Aα-2 has a lower affinity for TREs than the TRs, suggesting that the variant C-terminal region may alter DNA binding by allosteric or other conformational effects. Thus, within a given cell, the expression of α gene may yield mutually antagonistic proteins. The mechanism by which c-*erb*Aα-2 negatively regulates or competitively inhibits thyroid hormone action is still unclear, however. The most likely explanations involve formation of inactive heterodimers and competition at the DNA-binding levels (Brent et al., 1989c; Damm et al., 1989; Forman et al., 1989; Glass et al., 1989;). It is important to note that v-*erb*A may act in a similar manner inasmuch as it is also defective in TH binding but can bind TREs, albeit with slightly lower affinity than bona fide TRs (Damm et al., 1989; Graupner et al., 1989; Sap et al., 1989).

Another critical development in our understanding of TH action was recently described by Brent et al. (1989c), Damm et al. (1989), Graupner et al. (1989), and Sap et al., (1989). These studies involve the concept of dominant negative regulation of gene expression by thyroid hormone receptors. Specifically, it has been shown that *unliganded* authentic thyroid hormone receptors can bind TREs and may constitutively negatively regulate genes containing these elements. Furthermore, TH presumably act by binding to the ligand-binding domain and hence altering the TR conformation to favour transactivation. The strength of this constitutive repression or ligand-mediated induction is strictly dependent on the strength of DNA–protein interactions (Brent et al., 1989c; Damm et al., 1989). Several non-mutually exclusive mechanisms have been proposed for this type of regulation as noted above for the action of c-*erb*Aα-2. One involves competition at the DNA-binding level, another invokes the formation of inactive heterodimers, and yet another suggests competition for critical transcriptional factors (squelching) (Adler et al., 1988; Gill and Ptashne, 1988; Meyer et al., 1989). Damm et al. (1989), Graupner et al. (1989) and Brent et al. (1989c) suggest that this negative regulation by the unliganded TR (apoTR) is dependent on the presence of an intact DNA-binding region. However, Forman et al. (1989) provide strong evidence that only the putative dimerization domain may be necessary for this effect. This contrasts with the work of Damm et al. (1989) who shows that a point mutation in the DNA-binding domain can result in inactivation of this dominant negative effect. However, one concern is that such a mutation in the DNA-binding domain may yield ineffective nuclear transport of the molecule, a result that has not been excluded. This regulation contrasts with the regulation by ligands of other members of the steroid hormone receptor family in which ligand is essential for biological activity of their respective receptors (Evans, 1988; Dalman et al., 1990). Furthermore, since TRs may interact with other receptors such as the retinoic acid and vitamin D receptors to form heterodimers (Umesono et al., 1988; Glass et al., 1989), the constitutive regulatory effect of unliganded TR may have far-reaching consequences.

4.9 Heterodimer formation involving thyroid hormone receptors

Forman et al. (1989) and Glass et al. (1989) have provided data supporting the interaction of different TRs with other nuclear receptors. Interestingly, Glass et al. (1989) show that the interaction of TR and RAR can result in divergent regulation that depends largely on the TRE and host cell being examined. For instance, TR can augment RAR effects on a palindromic TRE but antagonize them in the rat α-myosin heavy chain gene TRE studied by cotransfection in CV-1 cells. The involvement of heterodimers in the regulation of TH and retinoic acid responsive genes is reminiscent of the interactions of c-Fos and c-Jun (Kouzarides and Ziff, 1988; Turner and Tjian, 1989). That TRs can form homodimers has not yet been directly demonstrated. However, such oligomerization is considered likely, in analogy with other nuclear receptors such as the oestrogen (Kumar and Chambon, 1988; Fawell et al., 1990) and progesterone (Tsai et al., 1988) receptors.

4.10 Other aspects of thyroid hormone receptor biology

The generation of TR isoforms by alternate splicing is well documented. It appears that there is a single α and a single β gene in the mammalian genome. Each one is involved in alternative splicing to yield multiple mRNAs encoding proteins with different functions and tissue specificity. Thus, the initial TR α gene tanscript is alternatively spliced (donor choice in the penultimate 3' exon of the c-erbAα gene) to yield either the TRα-1 or c-erbAα-2 mRNAs (Izumo and Mahdavi, 1988; Mitsuhashi et al., 1988; Lazar *et al.*, 1989a,b) (Fig. 8). Whether alternative polyadenylation site choice is involved in this process is not known. Also, the β gene utilizes either alternative promoter choice and/or alternative RNA splicing in order to generate the TRβ-1 and TRβ-2 mRNAs in the rat pituitary (Hodin *et al.*, 1989b). The absence of TRβ-2 in non-pituitary rat tissues raises interesting hypotheses concerning the mechanisms of selective TRβ-2 mRNA production in the pituitary.

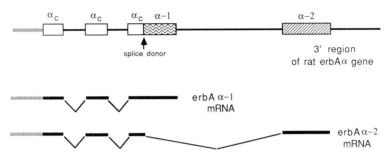

Fig. 8 Alternative splicing of the TRα gene in the rat to yield either *erb*Aα-1 (TRα-1) or c-*erb*Aα-2 mRNAs. Only the 3' region of the rat c-*erb*Aα (TRα) gene is shown.

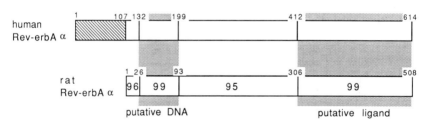

Fig. 9 Rev-erbAα in man and rat: schematic of cDNA structures. Numbers above structures refer to amino acid residue number in each respective protein. The percentage similarity of comparable domains, and putative DNA- and ligand-binding domains are noted.

Also present on the TRα gene in rat and man is a transcript on the opposite strand that encodes yet another member of the thyroid/steroid hormone receptor superfamily but is still uncharacterized with respect to ligand or DNA target. This gene is referred to as Rev-*erb*Aα in rat by Lazar *et al.* (1989b) and *ear-1* in human by Miyajima *et al.* (1989) (Fig. 9). This interesting 'orphan' nuclear receptor (Evans, 1988) possesses both putative DNA- and ligand-binding domains by homology (Figs 4 and 9). Lazar *et al.* (1989b) indicate that, in the rat, this form does *not* bind THs, whereas the Japanese group (Miyajima *et al.*, 1989) indicates that there may be weak interactions of the human form with these hormones. Lazar *et al.* (1990) have failed to confirm the findings of Miyajima et al. (1989) for human Rev-erbAα. In addition, we have data indicating that the same molecule may be expressed in other species including mouse, suggesting the conservation of a biologically important molecule in evolution.

Of special note is the genomic location of rat and human Rev-erbAα (Fig. 10). This transcript contains a region (269 bp) of an exon that is transcribed in common with c-erbAα-2; both regions involve structural coding sequences (Lazar et al., 1989b, 1990a; Miyajima et al., 1989). The presence of such an overlapping transcript on the reverse strand suggests possible interesting regulatory mechanisms in addition to its presence as an orphan member of the nuclear receptor superfamily. One involves regulation of transcription of the TRα gene by the expression of Rev-erbAα. Another involves possible effects on alternative rat c-erbAα RNA splicing. Lazar et al. (1990b) have preliminary data in support of this latter mechanism. Yet another involves translational control by anti-sense mRNAs.

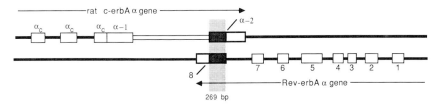

Fig. 10 Genomic organization of the c-erbAα and Rev-erbAα gene locus in the rat. REV-erbAα is transcribed on the opposite strand that encodes TRα-1 and c-erbAα-2. Note the overlap of 269 bp in the terminal exons of the c-erbAα-2 and REV-erbAα transcripts. The entire Rev-erbAα gene and 3'- end of the c-erbAα gene are shown. The small long box between the last two exons of the c-erbAα gene depicts uncertainty regarding the end of the α_c/α-1 exon.

In addition, the various TR isoforms are expressed in a tissue-specific fashion (Weinberger et al., 1986; Thompson et al., 1987; Freake et al., 1988; Lazar and Chin, 1988; Lazar et al., 1988; Murray et al., 1988; Santos et al., 1988; Hodin et al., 1989a,b, 1990; Mitsuhashi and Nikodem, 1989; Sakurai et al., 1989a) (Fig. 11). The TRα-1 mRNA is expressed in nearly all tissues and predominantly in brain, brown fat and other adipose tissue, and cardiac and other muscle. c-erbAα-2 mRNA is also expressed in a wide range of tissues including the testes, where it is the predominant TR-related isoform. The TRβ-1 mRNA is expressed in a number of tissues but is most abundant in the liver and kidney. In contrast, as discussed previously, the TRβ-2 mRNA is expressed only in the pituitary gland and may have a special function in that tissue (Hodin et al., 1989b). In the brain, TRα-1, TRβ-1, c-erbAα-2 and Rev-erbAα are expressed in most regions as determined by hybridization histochemistry using radiolabelled synthetic oligonucleotides specific for each isoform (Bradley et al., 1989). TRα-1 and

4. NUCLEAR THYROID HORMONE RECEPTORS

c-*erb*Aα-2 mRNAs are most widely distributed, with c-*erb*Aα-2 mRNA being highly abundant. Regions with the highest TRα-1 and c-*erb*Aα-2 mRNAs include the olfactory bulb, hippocampus and the granular layer of the cerebellar cortex. The REV-*erb*Aα and TRβ-1 mRNAs are limited in distribution with peak levels in the neocortex and the parvocellular pararenticular nucleus of the hypothalamus, respectively. The latter observation suggests a role of the TRβ-1 in the regulation of the hypothalamic–pituitary–thyroid endocrine axis inasmuch as the major hypothalamic regulation of thyrotropin production, thyrotropin-releasing-hormone (TRH), is secreted by the paraventricular nucleus.

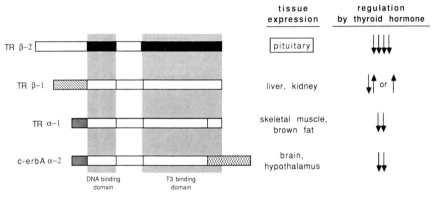

Fig. 11 Tissue-specific expression and thyroid hormone regulation of the mRNAs encoding the rat TRs and related isoforms. Note that TRβ-2 is expressed and TRβ-1 is stimulated by TH only in the pituitary gland.

Also, it has long been observed that the adult brain is modestly responsive to TH as measured by oxygen consumption and glucose uptake (Oppenheimer, 1983) in the face of abundant nuclear TH-binding capacity. The large amount of c-*erb*Aα-2 mRNA, a potential negative regulator of TR function, in many brain regions, provides a possible mechanism to explain the conundrum of ample TH binding but low TH activity. Unfortunately, preliminary studies in the brain of the neonatal rat, known for vigorous TH responses, show no difference in the c-*erb*Aα-2/TR mRNA ratios in these two states (Mitsuhashi and Nikodem, 1989).

The hormonal regulation of thyroid hormone receptor may be important in thyroid hormone action. In addition, it provides a glimpse into the regulation of *trans*-acting proteins by exogenous factors. As already alluded to above, thyroid hormones can variably regulate TR isoform mRNAs in specific tissues (Fig. 11). In general, rat TRα-1 and c-*erb*Aα-2 mRNAs are negatively regulated in most tissues (Lazar and Chin, 1988; Mitsuhashi and Nikodem, 1989; Hodin *et al.*, 1990). In contrast, TRβ-1 mRNA is not well regulated at all in most tissues except for the pituitary where, in fact, it is stimulated several-fold (Lazar and Chin, 1988; Hodin et al., 1989b, 1990). Finally, the pituitary-specific TRβ-2 mRNA is profoundly negatively regulated by TH in the pituitary gland. These data are consistent with the regulation of TR by TH, as assessed by labeled TH nuclear binding in cultured pituitary cells (Samuels et al., 1977).

4.11 Clinical syndromes

A number of rare syndromes involving thyroid hormone receptor abnormalities have been described. These include the syndromes of peripheral and pituitary resistance to thyroid hormones (Gershengorn and Weintraub, 1975; Refetoff, 1982). Clinically, these patients have chemical abnormalities manifested by hyper, eu- or hypothyroidism in the presence of elevated levels of biologically active thyroid hormones and thyrotropin or TSH. Hence, a situation where there is inappropriate secretion of TSH is observed. In some cases, there is apparent lack of sensitivity of various organs including peripheral and pituitary tissues to ambient levels of thyroid hormone often with documented altered TH affinity of lymphocyte or fibroblast nuclear TRs (Menezes-Ferreira et al., 1984). These heterogeneous disorders are inherited generally in an *autosomal dominant* pattern. Although only a few families have been described in the literature, the recognition of this syndrome is critical because of the importance of early recognition in order to reverse symptoms or to prevent incorrect medical management. Two groups recently have described the molecular defects in several of these families. Usala et al. (1988, 1990) have examined three kindreds with this syndrome. Of interest, individuals in pedigree A were often short in stature, had lower intelligence quotients (IQs), and displayed frequent hyperactivity. Such hyperactivity, although in milder states, has been noted in pedigrees B and C in this study. Such somatic and neurological alterations may be due to the TRβ abnormality during development. Using restriction enzyme length polymorphic analyses, they have shown that the disorder is tightly linked to the c-*erb*Aβ gene on chromosome 3 (Usala et al., 1988). In fact a logarithm of the odds ratio (lod) score of 5.77 in these three kindreds confirm this tight linkage. Furthermore, they proceed to show that a point mutation in codon 448, resulting in a change of proline to histidine (CCT to CAT) in the ligand-binding domain of a single TRβ gene, is the likely cause of this disorder (Usala et al., 1990). This alteration occurs in the invariant 37–40 amino acids at the extreme C-terminus of all TRs (Fig. 7). Although direct TH-binding studies were not performed on this mutant TRβ, the 448 (Pro to His) mutation was not observed in 92 random individuals, a finding strongly suggesting the aetiological importance of this genetic change and the lower likelihood of its being a random polymorphism. In independent work, Sakurai *et al.* (1989b) have studied a pedigree with an affected father and son who have inherited the disorder as heterozygotes. These individuals have elevated T_4, T_3 and free T_4 indices, indicating the presence of increased bioavailable thyroid hormones along with detectable TSH. The presence of TSH in the face of increased thyroid hormone levels in the blood is inappropriate and suggests relative insensitivity of the pituitary to elevated levels of thyroid hormone. This group, using molecular genetic analyses, showed that codon 340 has a point mutation (GGT to CGT; glycine to arginine). This point mutation results in a change in the ligand-binding domain of the β thyroid hormone receptor (in the putative dimerization domain (Fig. 7)) resulting in lack of TH binding; this has been proven using *in vitro* analyses.

These data convincingly show that several pedigrees with the syndrome of peripheral resistance to thyroid hormone acquired disease due to a point mutation, which may be different in different pedigrees, of the ligand-binding domain resulting in deficient thyroid hormone binding. The molecular defects present in other pedigrees remain to be studied. It is likely that the mutations will be heterogeneous in nature, as has been demonstrated in other clinical disorders due to nuclear receptor defects (Brown *et al.*, 1988; Hughes *et al.*, 1988; see Chapter 14). Another interesting observation is the autosomal-dominant inheritance pattern of this disorder. Several explanations may be offered to explain this finding. Importantly, it should be noted that the abnormal TRβs are functionally equivalent to c-*erb*Aα-2 or v-*erb*A and hence could act as dominant negative regulators of TH action as discussed above. The defective β product from the single abnormal allele could result in inactive heterodimers in addition to contributing to diminished active TR homo- or heterodimers.

4.12 Summary

Classic biochemical and physiological studies suggested the presence of a single TR involved in TH action. The recent explosion of information on TR structure and function has revealed a number of important details. First, TR is encoded by the proto-oncogene, c-*erb*A. Second, there are multiple members of the c-*erb*A family in rat, mouse, man and chicken. These represent bona fide receptors that are expressed in most tissues, although in varying relative amounts. Little is presently known about the possible gene- and tissue-specific roles of these heterogeneous TR forms, although the existence of different functions is postulated for each form. A major example of tissue-specific expression is the detection of TRβ-2 only in the pituitary gland. Third, alternative RNA splicing primary TRα and TRβ transcripts appears to result in TR isoforms that have biological function. c-erbAα-2 seems to act as a constitutive, non-ligand-dependent suppressor of TH action by interfering or competing with the activities of bona fide TRs. Differential splicing of the TRβ gene results in the pituitary-specific TRβ-2. Fourth, opposite-strand transcription of a small overlapping transcriptional region corresponding to the c-erbAα-2-specific exon leads to the production of a protein that is an 'orphan' member of the steroid hormone superfamily. Its role in physiology is probably important, given its high structural conservation in rodents and man, and may include regulation of relative amounts of biologically active TRα-1 and a TR-inhibitor, c-erbAα-2. Fifth, studies in the TR gene family have revealed functional aspects that are distinct from those operative in steroid hormone receptors. They lack cytosolic proteins that interact with the 'non-transformed' form as in the glucocorticoid receptor, and may be active albeit *negative* in the unliganded state. Sixth, the apparent promiscuous association with retinoic acid and possibly other nuclear receptors opens the door for even more complex and wide-reaching effects on growth, differentiation, and development. Seventh, other nuclear proteins may be important in collaboration with the TRs to effect TH action, as has been observed with other nuclear receptors. Finally, studies of the TR genes in man have provided insight into the molecular defects in a number of heterogeneous, rare disease states involving TH resistance. Clearly, we are only in our infancy with regard to our understanding of how TRs function, but our horizon has been expanded greatly and we will no doubt be treated to more exciting news in the near future.

References

Adler, S., Waterman, M. L., He, X. and Rosenfeld, M. G. (1988). Steroid receptor-mediated inhibition of rat prolactin gene expression does not require the receptor DNA-binding domain. *Cell* **52**, 685–695.

Baker, B. S. and Tata, J. R. (1990). Accumulation of proto-oncogene c-*erb*-A related transcripts during *Xenopus* development: association with early acquisition of response to thyroid hormone and estrogen. *EMBO J.* **9**, 879–885.

Beato, M. (1989). Gene regulation by steroid hormones. *Cell* **56**, 335–344.

Bedo, G., Santisteban, P. and Aranda, A. (1989). Retinoic acid regulates growth hormone gene expression. *Nature* **339**, 231–234.

Benbrook, D. and Pfahl, M. (1987). A novel thyroid hormone receptor encoded by a cDNA clone from a human testis library. *Science* **238**, 788–791.

Berg, J. M. (1989). DNA binding specificity of steroid receptors. *Cell* **57**, 1065–1068.

Bradley, D. J., Young, W. S. III and Weinberger, C. (1989). Differential expression of alpha and beta thyroid hormone receptor genes in rat brain and pituitary. *Proc. Natl. Acad. Sci. USA* **86**, 7250–7254.

Brent, G. A., Harney, J. W., Chen, Y., Warne, R. L., Moore, D. D. and Larsen, P. R. (1989a). Mutations of the rat growth hormone promoter which increase and decrease response to thyroid hormone define a consensus thyroid hormone response element. *Mol. Endocrinol.* **3**, 1996–2004.

Brent, G. A., Larsen, P. R., Harney, J. W., Koenig, R. J. and Moore, D. D. (1989b). Functional characterization of the rat growth hormone promoter elements required for induction by thyroid hormone with and without a co-transfected beta-type thyroid hormone receptor. *J. Biol. Chem.* **264**, 178–182.

Brent, G. A., Dunn, M. K., Harney, J. W., Gulick, T., Larsen, P. R. and Moore, D. D. (1989c). Thyroid hormone aporeceptor represses T3-inducible promoters and blocks activity of the retinoic acid receptor. *The New Biologist* **1**, 329–336.

Brown, T. R., Lubahn, D. B., Wilson, E. M., Joseph, D. R., French, F. S. and Migeon, C. J. (1988). Deletion of the steroid-binding domain of the human androgen receptor gene in one family with complete androgen insensitivity syndrome. Evidence for further genetic heterogeneity in this syndrome. *Proc. Natl. Acad. Sci. USA* **85**, 8151–8155.

Burnside, J., Darling, D. S., Carr, F. E. and Chin, W. W. (1989). Thyroid hormone regulation of the rat glycoprotein hormone alpha-subunit gene promoter activity. *J. Biol. Chem.* **264**, 6886–6891.

Burnside, J., Darling, D. S. and Chin, W. W. (1990). A nuclear factor that enhances binding of thyroid hormone receptors to thyroid hormone response elements. *J. Biol. Chem.* **265**, 2500–2504.

Carr, F. E., Burnside, J. and Chin, W. W. (1989). Thyroid hormones regulate rat thyrotropin-β gene promoter activity expressed in GH3 cells. *Mol. Endocrinol.* **3**, 709–716.

Casanova, J., Horowitz, S. D., Copp, R. P., McIntyre, W. R., Pascual, A. and Samuels, H. H. (1984). Photoaffinity labeling of thyroid hormone receptors. *J. Biol. Chem.* **259**, 12 084–12 091.

Chatterjee, V. K. K., Lee, J.-K., Rentoumis, A. and Jameson, J. L. (1989). Negative regulation of the thyroid-stimulating hormone alpha gene by thyroid hormone: Receptor interaction adjacent to the TATA box. *Proc. Natl. Acad. Sci. USA* **86**, 9114–9118.

Danielsen, M., Hinck, L. and Ringold, G. M. (1989). Two amino acids within the knuckle of the first zinc finger specify DNA response element activation by the glucocorticoid receptor. *Cell* **57**, 1131–1138.

Dalman, F. C., Koenig, R. J., Perdew, G. H., Massa, E. and Pratt, W. B. (1990). In contrast to the glucocorticoid receptor, the thyroid hormone receptor is translated in the DNA binding state and is not associated with hsp90. *J. Biol. Chem.* **265**, 3615–3618.

Damm, K., Thompson, C. C. and Evans, R. M. (1989). Protein encoded by v-*erb*A functions as a thyroid-hormone receptor antagonist. *Nature* **339**, 593–597.

Darling, D. S., Burnside, J. and Chin, W. W. (1989). Binding of thyroid hormone receptors to the rat thyrotropin-β gene. *Mol. Endocrinol.* **3**, 1359–1368.

Dayton, A. I., Selden, J. R., Laws, G., Dorney, D. J., Finan, J., Tripputi, P., Emanuel, B. S., Roveera, G., Nowell, P. C. and Croce, C. (1984). A human c-*erb*A oncogene homologue is closely proximal to the chromosome 17 breakpoint in acute promyelocytic leukemia. *Proc. Natl. Acad. Sci. USA* **81**, 4495–4499.

Drabkin, H., Kao, F.-T., Hartz, J., Hart, I., Gazdar, A., Weinberger, C., Evans, R. and Gerber, M. (1988). Localization of human ERBA2 to the 3p22–3p24.1 region of chromosome 3 and variable deletion in small cell lung cancer. *Proc. Natl. Acad. Sci. USA* **85**, 9258–9262.

Edwards, D. P., Kuhnel, B., Estes, P. A. and Nordeen, S. K. (1989). Human progesterone receptor binding to mouse mammary tumor virus deoxyribonucleic acid: dependence on hormone and non-receptor nuclear factors. *Mol. Endocrinol.* **3**, 381–391.

Evans, R. M. (1988). The steroid and thyroid hormone receptor superfamily. *Science* **240**, 889–895.

Fawell, S. E., Lees, J. A., White, R. and Parker, M. G. (1990). Characterization and colocalization of steroid binding and dimerization activities in the mouse estrogen receptor. *Cell* **60**, 953–962.

Feavers, I. M., Jiricny, J., Moncharmont, B., Saluz, H. P. and Jost, J. P. (1987). Interactions of two non-histone proteins with the estradiol response element of the avian vitellogenin gene modulates the binding of the estrogen-receptor complex. *Proc. Natl. Acad. Sci. USA* **84**, 7453–7457.

Forman, B.M., Yang, C.-R., Au, M., Casanova, J., Ghysdael, J. and Samuels, H. H. (1989). A domain containing a leucine-zipper-like motif mediates novel *in vivo* interactions between the thyroid hormone and retinoic acid receptors. *Mol. Endocrinol.* **3**, 1610–1620.

Freake, H. C., Santos, A., Goldberg, Y., Ghysdael, J. and Oppenheimer, J. H. (1988). Differences in antibody recognition of the triiodothyronine nuclear receptor and c-*erb*A products. *Mol. Endocrinol.* **2**, 986–991.

Gershengorn, M. C. and Weintraub, B. D. (1975). Thyrotropin-induced hyperthyroidism caused by selective pituitary resistance to thyroid hormone. A new syndrome of "inappropriate secretion of TSH". *J. Clin. Invest.* **56**, 633–643.

Gill, G. and Ptashne, M. (1988). Negative effect of the transcriptional activator GAL4. *Nature* **334**, 721–724.

Glass, C. K., Franco, R., Weinberger, C., Albert, V. R., Evans, R. M. and Rosenfeld, M. G. (1987). A c-*erb*A binding site in rat growth hormone gene mediates trans-activation by thyroid hormone. *Nature* **329**, 738–741.

Glass, C. K., Holloway, J. M., Devary, O. V. and Rosenfeld, M. G. (1988). The thyroid hormone receptor binds with opposite transcriptional effects to a common sequence motif in thyroid hormone and estrogen response elements. *Cell* **54**, 313–323.

Glass, C. K., Lipkin, S. M., Devary, O. V. and Rosenfeld, M. G. (1989). Positive and negative regulation of gene transcription by a retinoic acid-thyroid hormone receptor heterodimer. *Cell* **59**, 697–708.

Graf, T. and Beug, H. (1983). Role of the v-*erb*A and v-*erb*B oncogenes of avian erythroblastosis virus in erythroid cell transformation. *Cell* **34**, 7–9.

Graupner, G., Wills, K. N., Tzukerman, M., Zhang, X. K. and Pfahl, M. (1989). Dual regulatory role for thyroid-hormone receptors allows control of retinoic-acid receptor activity. *Nature* **340**, 653–656.

Green, S. and Chambon, P. (1986). A superfamily of potentially oncogenic hormone receptors. *Nature* **32**, 617–619.

Hodin, R. A., Lazar, M. A. and Chin, W. W. (1989a). The pituitary-specific form of rat c-*erb*A is a biologically active thyroid hormone receptor. *Curr. Surg.* **46**, 298–301.

Hodin, R. A., Lazar, M. A., Wintman, B. I., Darling, D. S., Koenig, R. J., Larsen, P. R., Moore, D. D. and Chin, W. W. (1989b). Identification of a thyroid hormone receptor that is pituitary-specific. *Science* **244**, 76–79.

Hodin, R. A., Lazar, M. A. and Chin, W. W. (1990). Differential and tissue-specific regulation of multiple rat c-*erb*A messenger RNA species by thyroid hormone. *J. Clin. Invest.* **85**, 101–105.

Hollenberg, S. M. and Evans R. M. (1988). Multiple and cooperative transactivation domains of the human glucocorticoid receptor. *Cell* **65**, 899–906.

Hollenberg, S. M., Giguere, V., Segui, P. and Evans R. M. (1987). Co-localization of DNA-binding and transcriptional activation functions in the human glucocorticoid receptor. *Cell* **49**, 39–46.

Horowitz, Z. D., Yang, C. R., Forman, B. M., Casanova, J. and Samuels, H. H. (1989). Characterization of the domain structure of chick c-*erb*A by deletion mutation: *in vitro* translation and cell transfection studies. *Mol. Endocrinol.* **3**, 148–156.

Hughes, M. R., Malloy, P. J., Keiback, D. G., Kesterson, R. A., Pike, J. W., Feldman, D. and O'Malley, B. W. (1988). Point mutations in the human vitamin D receptor gene associated with hypocalcemic rickets. *Science (Wash. DC)* **242**, 1702–1705.

Izumo, S. and Mahdavi, V. (1988). Thyroid hormone receptor α isoforms generated by alternative splicing differentially activate myosin HC gene transcription. *Nature* **334**, 539–542.

Jansson, M., Philipson, L. and Vennström, B. (1983). Isolation and characterization of multiple human genes homologous to the oncogenes of avian erythroblastosis virus. *EMBO J.* **2**, 561–565.

Koenig, R. J., Brent, G. A., Warne, R. L., Larsen, P. R. and Moore, D. D. (1987). Thyroid hormone receptor binds to a site in the rat growth hormone promoter required for induction by thyroid hormone. *Proc. Natl. Acad. Sci. USA* **84**, 5670–5674.

Koenig, R. J., Warne, R. L., Brent, G. A., Harne, J. W., Larsen, P. R. and Moore, D. D. (1988). Isolation of a cDNA clone encoding a biologically active thyroid hormone receptor. *Proc. Natl. Acad. Sci. USA* **85**, 5031–5035.

Koenig, R. J., Lazar, M. A., Hodin, R. A., Brent, G. A., Larsen, P. R., Chin, W. W. and Moore, D. D. (1989). Inhibition of thyroid hormone action by a nonhormone binding c-*erb*A protein generated by alternative splicing. *Nature* **337**, 659–661.

Kouzarides, T. and Ziff, E. (1988). The role of the leucine zipper in the *fos-jun* interaction. *Nature* **336**, 646–651.

Kumar, V. and Chambon, P. (1988). The estrogen receptor binds tightly to its responsive element as a ligand-induced homodimer. *Cell* **55**, 145–156.

Latham, K. R., Ring, J. C. and Baxter, J. D. (1976). Solubilized nuclear 'receptors' for thyroid hormones. Physical characteristics and binding properties; evidence for multiple forms. *J. Biol. Chem.* **251**, 7387–7388.

Lazar, M. A. and Chin, W. W. (1988). Regulation of two c-*erb*A messenger ribonucleic acids in rat GH_3. *Mol. Endocrinol.* **2**, 479–484.

Lazar, M. A., Hodin, R. A., Darling, D. S. and Chin, W. W. (1988). Identification of a rat c-*erb*-Aα-related protein which binds DNA acid but does not bind thyroid hormone. *Mol. Endocrinol.* **2**, 893–901.

Lazar, M. A., Hodin, R. A. and Chin, W. W. (1989a). Human carboxyl-terminal variant of α-type c-*erb*A inhibits trans-activation by thyroid hormone receptors without binding thyroid hormone. *Proc. Natl. Acad. Sci. USA* **86**, 7771–7774.

Lazar, M. A., Hodin, R. A., Darling, D. S. and Chin, W. W. (1989b). A novel member of the thyroid/steroid hormone receptor family is encoded by the opposite strand of the rat c-*erb*A-alpha transcriptional unit. *Mol. Cell. Biol.* **9**, 1128–1136.
Lazar, M. A., Jones, K. E. and Chin, W. W. (1990a). Isolation of a cDNA encoding human Rev-ErbAα: Transcription from the non-coding DNA stand of a thyroid hormone receptor gene results in a related protein which does not bind thyroid hormone. *DNA Cell Biol.* **9**, 77–83.
Lazar, M. A., Hodin, R. A., Cardona, G. and Chin, W. W. (1990b). Gene expression from the c-*erb*Aα/Rev-*Erb*Aα genomic locus: Potential regulation of alternative splicing by opposite strand transcription. *J. Biol. Chem.* **265**, 12 859–12 863.
Menezes-Ferreira, M. M., Eil, C., Wortsman, J. and Weintraub, B. D. (1984). Decreased nuclear uptake of ^{125}I triiodothyronine in fibroblasts from patients with peripheral thyroid hormone resistance. *J. Clin. Endocrinol. Metab.* **59**, 1081–1087.
Meyer, M.-E., Gronemeyer, H., Turcotte, B., Bocquel, M.-T., Tasset, D. and Chambon, P. (1989). Steroid hormone receptors compete for factors that mediate their enhancer function. *Cell* **57**, 433–442.
Mitsuhashi, T., Tennyson, G. E. and Nikodem, V. M. (1988). Alternative splicing generates messages encoding rat c-*erb*A proteins that do not bind thyroid hormone. *Proc. Natl. Acad. Sci. USA* **85**, 5804–5808.
Mitsuhashi, T. and Nikodem, V. M. (1989). Regulation of expression of the alternative mRNAs of the rat alpha-thyroid hormone receptor gene. *J. Biol. Chem.* **264**, 8900–8904.
Miyajima, N., Horiuchi, R., Shibuya, Y., Fukushige, S.-I., Matsubara, K.-I., Toyoshima, K. and Yamamoto, T. (1989). Two *erb*A homologs encoding proteins with different T_3 binding capacities are transcribed from opposite DNA strands of the same genetic locus. *Cell* **57**, 31–39.
Murray, M. B. and Towle, H. C. (1989). Identification of nuclear factors that enhance binding of the thyroid hormone receptor to a thyroid hormone response element. *Mol. Endocrinol.* **3**, 1434–1442.
Murray, M. B., Zilz, N. D., McCreary, N. L., MacDonald, M. J. and Towle, H. C. (1988). Isolation and characterization of rat cDNA clones for two distinct thyroid hormone receptors. *J. Biol. Chem.* **263**, 12 770–12 777.
Nakai, A., Seino, S., Sakurai, A., Szilak, I., Bell, G. I. and DeGroot, L. J. (1988a). Characterization of a thyroid hormone receptor expressed in human kidney and other tissues. *Proc. Natl. Acad. Sci. USA* **85**, 2781–2785.
Nakai, A., Sakurai, A., Bell, G. I. and DeGroot, L. J. (1988b). Characterization of a third human thyroid hormone receptor coexpressed with other thyroid hormone receptors in several tissues. *Mol. Endocrinol.* **2**, 1087–1092.
Oppenheimer, J. H. (1983). The nuclear receptor-triiodothyronine complex: relationship to thyroid hormone distribution. *In* "Molecular Basis of Thyroid Hormone Action" (eds J. H. Oppenheimer and J. H. Samuels), pp. 1–35. Academic Press, New York.
Oppenheimer, J. H., Schwartz, H. L. and Surks, M. I. (1974). Tissue differences in the concentration of triiodothyronine nuclear binding sites in the rat: liver, kidney, pituitary, heart, brain, spleen and testis. *Endocrinology* **95**, 897–903.
Oppenheimer, J. H., Schwartz, H. L., Mariash, C. N., Kinlaw, W. B., Wong, N. C. W. and Freake, H. C. (1987). Advances in our understanding of thyroid hormone action at the cellular level. *Endocr. Rev.* **8**, 288–308.

Petty, K. J., Morioka, H., Mitsuhashi, T. and Nikodem, V. M. (1989). Thyroid hormone regulation of transcription factors involved in malic enzyme gene expression. *J. Biol. Chem.* **264**, 11 483–11 490.

Prost, E., Koenig, R. J., Moore, D. D., Larsen, P. R. and Whalen, R. G. (1988). Multiple sequences encoding potential thyroid hormone receptors isolated from mouse skeletal muscle libraries. *Nucl. Acids Res.* **16**, 6248.

Refetoff, S. (1982). Syndromes of thyroid hormone resistance. *Am. J. Physiol.* **243**, E88–E98.

Sakurai, A., Nakai, A. and DeGroot, L. J. (1989a). Expression of three forms of thyroid hormone receptor in human tissues. *Mol. Endocrinol.* **3**, 392–399.

Sakurai, A., Takeda, K., Ain, K., Ceccarelli, P., Nakai, A., Seino, S., Bell, G. I., Refetoff, S. and DeGroot, L. J. (1989b). Generalized resistance to thyroid hormone associated with a mutation in the ligand-binding domain of the human thyroid hormone receptor beta. *Proc. Natl. Acad. Sci. USA* **86**, 8977–8981.

Samuels, H. H. (1983). Identification and characterization of thyroid hormone receptors and action using cell culture techniques. *In* "Molecular Basis of Thyroid Hormone Action" (eds J. H. Oppenheimer and J. H. Samuels), pp. 36–66. Academic Press, New York.

Samuels, H. H., Stanley, F. and Shapiro, L. E. (1977). Modulation of thyroid hormone nuclear receptor levels by 3,5,3'-triiodothyronine in GH1 cells. *J. Biol. Chem.* **252**, 6052–6060.

Samuels, H. H., Forman, B. M., Horowitz, Z. D. and Ye, Z.-S (1988). Regulation of gene expression by thyroid hormone. *J. Clin. Invest.* **81**, 957–967.

Santos, A., Freake, H. D., Rosenberg, M. E., Schwartz, H. L. and Oppenheimer, J. H. (1988). Triiodothyronine nuclear binding capacity in rat tissues correlates with a 6.0 kilobase (kb) and not a 2.6 kb messenger ribonucleic acid hybridization signal generated by a human c-*erb*A probe. *Mol. Endocrinol.* **2**, 992–998.

Sap, J., Munoz, A., Damm, K., Goldberg, Y., Ghysdael, J., Leutz, A., Beug, H. and Vennström, B. (1986). The c-*erb*-A protein is a high-affinity receptor for thyroid hormone. *Nature* **324**, 635–640.

Sap, J., Munoz, A., Schmitt, J., Stunnenberg, H. and Vennström, B. (1989). Repression of transcription mediated at a thyroid hormone response element by the v-*erb*A oncogene product. *Nature* **340**, 242–244.

Schueler, P. A., Schwartz, H. L., Strait, K. A., Mariash, C. N. and Oppenheimer, J. H. (1990). Binding of 3,4,3'-triiodothyronine (3) and its analogs to the *in vitro* translational products of c-*erb*A protooncogenes: Differences in the affinity of the α- and β-forms for the acetic acid analog and failure of the human testis and kidney products to bind T3. *Endocrinology* **4**, 277–234.

Schwartz, H. L. (1983). Effect of thyroid hormone on growth and development. *In* "Molecular Basis of Thyroid Hormone Action" (eds J. H. Oppenheimer and J. H. Samuels), pp. 413–444. Academic Press, New York.

Shupnik, M. A., Ridgway, E. C. and Chin, W. W. (1989). Molecular biology of thyrotropin. *Endocr. Rev.* **10**, 459–475.

Thompson, C. C. and Evans, R. M. (1989). Trans-activation by thyroid hormone receptors: Functional parallels with steroid hormone receptors. *Proc. Natl. Acad. Sci. USA* **86**, 3494–3498.

Thompson, C. C., Weinberger, C., Lebo, R. and Evans, R. M. (1987). Identification of a novel thyroid hormone receptor expressed in the mammalian central nervous system. *Science* **237**, 1610–1614.

Tora, L., Gronemeyer, H., Turcotte, B., Gaub, M.-P. and Chambon, P. (1988). The N-terminal region of the chicken progesterone receptor specifies target gene activation. *Nature* **333**, 185–188.

Tsai, S. Y., Sagami, I., Wang, H., Tsai, M. J. and O'Malley, B. W. (1987). Interactions between a DNA-binding transcription factor (COUP) and a non-DNA binding factor (S300-II). *Cell* **50**, 701–709

Tsai, S. Y., Carlstedt-Duke, J., Weigel, N. L., Dahlman, K., Gustafsson, J., Tsai, M. and O'Malley, B. W. (1988). Molecular interactions of steroid hormone receptor with its enhancer element: evidence for receptor dimer formation. *Cell* **55**, 361–369.

Turner, R. and Tjian, R. (1989). Leucine repeats and an adjacent DNA binding domain mediate the formation of functional *Fos*-c*Jun* heterodimers. *Science* **243**, 1689–1694.

Umesono, K. and Evans, R. M. (1989). Determinants of target gene specificity for steroid/thyroid hormone receptors. *Cell* **57**, 1139–1146.

Umesono, K., Giguere, V., Glass, C. K., Rosenfeld, M. G. and Evans, R. M. (1988). Retinoic acid and thyroid hormone induce gene expression through a common responsive element. *Nature* **336**, 262–265.

Usala, S. J., Bale, A. E., Gesundheit, N., Weinberger, C., Lash, R. W., Wondisford, F. E., McBride, O. W. and Weintraub, B. D. (1988). Tight linkage between the syndrome of generalized thyroid hormone resistance and the human c-*erb*Aβ gene. *Mol. Endocrinol.* **2**, 1217–1220.

Usala, S. J., Tennyson, G. E., Bale, A. E., Lash, R. W., Gesundheit, N., Wondisford, F. E., Accilli, D., Hauser, P. and Weintraub, B. D. (1990). A base mutation of the c-*erb*Aβ thyroid hormone receptor in a kindred with generalized thyroid hormone resistance. Molecular heterogeneity in two other kindreds. *J. Clin. Invest.* **85**, 93–100.

Vennström, B. and Bishop, J. M. (1982). Isolation and characterization of chicken DNA homologous to the two putative oncogenes of avian erythroblastosis virus. *Cell* **28**, 135–143.

Weinberger, C., Thompson, C. P., Ong, E. S., Lebo, R., Gruol, D. J. and Evans, J. M. (1986). The c-*erb*-A gene encodes a thyroid hormone receptor. *Nature* **324**, 641–646.

West, B L., Catanzaro, D. F., Mellon, S. H., Cattini, P. A., Baxter, J. D. and Reudelhuber, T. L. (1987). Interaction of a tissue-specific factor with an essential rat growth hormone gene promoter element. *Mol. Cell Biol.* **7**, 1193–1197.

Wight, P. A., Crew, M. D. and Spindler, S. R. (1987). Discrete positive and negative thyroid hormone-responsive transcription regulatory elements of the rat growth hormone gene. *J. Biol. Chem.* **262**, 5659–5663.

Wondisford, F. E., Farr, E. A., Radovick, S., Steinfelder, J. H., Moates, J. M., McClaskey, J. H. and Weintraub, B. D. (1989). Thyroid hormone inhibition of human thyrotropin beta-subunit gene expression is mediated by a *cis*-acting element located in the first exon. *J. Biol. Chem.* **264**, 14 601–14 604.

Wood, W. M., Kao, M. Y., Gordon, D. F. and Ridgway, E. C. (1989). Thyroid hormone regulates the mouse thyrotropin β-subunit-gene promoter in transfected primary thyrotropes. *J. Biol. Chem.* **264**, 14 840–14 847.

Ye, Z.-S. and Samuels, H. H. (1987). Cell- and sequence-specific binding of nuclear proteins to 5'-flanking DNA of the rat growth hormone gene. *J. Biol. Chem.* **262**, 6313–6317.

Ye, Z.-S., Forman, B. M., Aranda, A., Pascual, A., Park, H.-Y., Casanova, J. and Samuels, H. H. (1988). Rat growth hormone gene expression. Both cell-specific and thyroid hormone response elements are required for thyroid hormone regulation. *J. Biol. Chem.* **263**, 7821–7829.

5
The Steroid Receptor Superfamily: Transactivators of Gene Expression

S. Y. TSAI, M.-J. TSAI and B. W. O'MALLEY

Department of Cell Biology, Baylor College of Medicine, Houston, TX 77030, USA

5.1 Introduction

The regulatory regions of eukaryotic genes are complex and consist of multiple distinct elements assembled into a functional unit (Dynan and Tjian, 1985; McKnight and Tjian, 1986; Maniatis *et al.*, 1987; Jones *et al.*, 1988). Individual *cis*-elements may function independently or in co-operation with other sequences to achieve the requisite overall promoter strength and regulatory characteristics for a particular gene. Each *cis*-element is recognized by a *trans*-acting factor which can be unique or part of a family of specific binding protein(s). The regulatory regions of steroid hormone responsive genes are no exception and consist of a combination of specific steroid hormone response elements (SREs) and multiple general or tissue-specific *cis*-elements (Yamamoto, 1985; Beato, 1989). Only now are we unravelling the mechanism by which these *cis* elements and their cognate transactivation factors act in a combinatorial fashion to effect regulation of gene expression. These studies allow us to conclude that steroid hormone receptors interact with their cognate response elements and function as transcription factors to confer hormone inducibility to target genes.

The objective of this review is to summarize our current understanding of the molecular mechanism of steroid receptor-mediated stimulation of target gene expression. We will discuss first our studies on the chicken ovalbumin upstream promoter transcription factor (COUP-TF), a promoter binding protein which is essential for efficient transcription of the ovalbumin gene. The recent cloning of the COUP-TF revealed that it is a member of the steroid hormone receptor superfamily. Next, we will discuss our evidence obtained from *in vitro* transcription studies which shows that steroid receptors act directly at the DNA level to effect expression of SRE-containing genes. Steroid hormone receptors enhance transcription by facilitating the formation of a stable pre-initiation complex of *trans*-acting factors and RNA polymerase, thus augmenting the initiation of target gene expression. Finally, we will discuss the role of hormone in transformation of receptor to an active conformation.

5.2 Identification and characterization of chicken ovalbumin upstream promoter transcription factor

The chicken ovalbumin gene is a good model system for studying hormonal regulation of target gene expression. Synthesis of ovalbumin mRNA in chicken oviducts is controlled by steroid hormones primarily at the level of transcription (O'Malley and Means, 1974; O'Malley *et al.*, 1979). Overall expression of the ovalbumin gene is regulated also by factors other than steroid receptors. For instance, we have defined a *cis*-acting element which is essential for the efficient transcription of the ovalbumin gene (Pastorcic *et al.*, 1986; Sagami *et al.*, 1986). This upstream promoter element, which includes the sequence GTGTCAAAGGTCAAA, is designated as the chicken ovalbumin upstream promoter (COUP) element. Exonuclease III and DNase I footprinting experiments revealed the presence of a *trans*-acting factor in oviduct extract which binds specifically to the COUP sequence (Elbrecht *et al.*, 1985; Sagami *et al.*, 1986; Tsai *et al.*, 1987; Wang *et al.*, 1987). Subsequently we demonstrated that this factor was required for efficient transcription of the ovalbumin gene in a cell-free system (Pastorcic *et al.*, 1986; Sagami *et al.*, 1986; Bagchi *et al.*, 1987; Tsai *et al.*, 1987; Wang *et al.*, 1987). This COUP-TF was found to be present in HeLa, chicken oviduct, HIT and many other cell types (Sagami *et al.*, 1986; Bagchi *et al.*, 1987; Wang *et al.*, 1987; Hwung *et al.*, 1988a,b); it was purified from HeLa cells using a combination of conventional and sequence-specific DNA affinity column chromatography (Wang *et al.*, 1987, 1989). A purified preparation exhibited a few distinct polypeptides when analysed by SDS polyacrylamide gel electrophoresis. These polypeptides were in the molecular size range 43–53 kD and bound specifically to the COUP sequence. In addition, the 43–45-kD proteins were shown to enhance transcription of the ovalbumin promoter (Wang *et al.*, 1987). It is likely that all polypeptides purified by this procedure are functional in transcriptional activation. Since S300 column chromatography, glycerol gradient centrifugation analyses, and UV cross-linking experiments indicated that the native binding component has a molecular weight of 90 000–100 000 (Sagami *et al.*, 1986), we concluded that the functional molecular species exists as a dimer composed of two 43–53-kD subunits.

Amino acid sequence analyses of the 53-kD and 45-kD polypeptides indicate that these two polypeptides correspond to the 'orphan' receptors, ERRI and COUP-TF, respectively. Therefore, it is unlikely that these multiple polypeptides represent a family of distinct but closely related proteins capable of binding to the same recognition sequence (Wang et al., 1987, 1989). In this context, all members of the *Jun* family, e.g. AP_1 (c-*Jun*), *Jun*B and *Jun*D, recognize the AP_1 binding sites (Bohmann et al., 1987; Curran and Franza, 1988; Halazonetis, 1988; Nakabeppu et al., 1988; Bohmann and Tjian, 1989); the Oct_1 (O'Neill et al., 1988; Sturm et al., 1988), Oct_2 (Clerc et al., 1988; Scheidereit et al., 1988), Pit-1 (Mangalam et al., 1989) and Unc-85 proteins (Finney et al., 1988) recognize an AT-rich sequence (Bodner and Karin, 1987); and the *Myo*D (Lassar et al., 1989), Myogenin (Wright et al., 1989) and *Myb* (Pinney et al., 1988) factors all recognize muscle-specific elements, even though different members may possess different affinities for a specific recognition sequence.

5.3 Chicken ovalbumin upstream promoter transcription factor binds specifically to apparently dissimilar promoter sequences

COUP-TF recognition sequences are also present in the 5' flanking sequence of the apolipoprotein genes, apo-VLDL and apo-CIII, and are functionally important for the expression of these genes (Wijnholds et al., 1988; Leff et al., 1989). Surprisingly, the most purified COUP-TF also binds to a promoter region of the rat insulin II gene which has a rather dissimilar recognition sequence (Hwung et al., 1988a,b). The binding site in the ovalbumin promoter is AT rich (37.5% GC) while in the insulin promoter it is GC rich (73% GC). Also, the DNase I protected regions are quite different in these two promoter elements, but methylation interference and mutational analyses indicate that certain nucleotides that are important for the binding of COUP-TF are conserved. Computer-aided graphics illustrate that the purine contacts of COUP-TF on both sequences are located on one side of the DNA double helix. In contrast, mapping of the phosphate contacts showed that COUP-TF wraps around the ovalbumin promoter while it binds to only one face of the DNA in the case of the insulin promoter. This raises a question as to how a single protein can bind differently to two elements. It is likely that the topology of the binding sites may be similar enough to be recognized initially by COUP-TF and the protein then demonstrates a flexible mode of binding to accommodate the divergent linear sequences of the two promoters.

5.4 The chicken ovalbumin upstream promoter transcription factor belongs to the steroid hormone receptor superfamily

COUP-TF was cloned by screening a cDNA expression library from HeLa cells using catenated COUP recognition sequences and COUP-TF antiserum. The amino acid sequence deduced from the COUP-TF cDNA revealed a remarkable homology to members of the steroid hormone receptor superfamily of genes (Evans, 1988). The DNA binding domain of COUP-TF was identified as a 66 amino acid zinc finger motif in which all 20 invariant amino acids of the receptor superfamily are conserved and 11 out of 12 conserved residues are identical (Wang et al., 1989). The comparison of amino acid sequence of COUP-TF with the human steroid hormone receptor superfamily is shown schematically in Fig. 1. In addition to the zinc finger motif (region I), there are two regions (II and III) residing in the ligand-binding domain which share significant similarity among members of this superfamily. Although the significance of these two newly defined regions is not yet clear, their strong amino acid sequence conservation indicates that they are functionally important for transactivation. Comparison with different members of the steroid receptor superfamily show that COUP-TF is most homologous to hERR-1 and hERR-2, two 'orphan receptors' whose functions and ligands have yet to be defined.

The hallmark of a steroid receptor is the ability to bind its characteristic hormonal ligand. To date we have not identified a ligand for COUP-TF. Nevertheless, the conservation of regions II and III between COUP-TF and other members of the steroid receptor superfamily suggests that a ligand may exist and be required to activate COUP-TF. When introduced into animal cells, COUP-TF appears to be constitutively active. However, mutational analysis of the ligand-binding region (C-terminus) of COUP-TF reveals a behaviour indistinguishable from the receptors for classical steroid hormones. For instance, small deletions within the C-terminus lead to loss of gene regulatory capacity but removal of the entire C-terminus, excluding the DNA-binding region, regains the constitutive function. Our operational hypothesis is that the ligand for COUP-TF, and perhaps other orphan receptors, is generally present within eukaryotic cells.

Recently, a *Drosophila* homologue of COUP-TF, the seven-up gene (SVP), was identified by Mlodzik et al. (1990). The seven-up gene product is required for photoreceptor cell formation during eye development in the fruit fly. Mutation of SVP leads to two defects in ommatidial structure: the apparent transformation of the outer photoreceptor cell types R_1, R_3, R_4 and R_6 towards the central photoreceptor R_7 cell type, and the presence of extra photoreceptor cells. These phenotypic defects indicate that this gene

5. THE STEROID RECEPTOR SUPERFAMILY

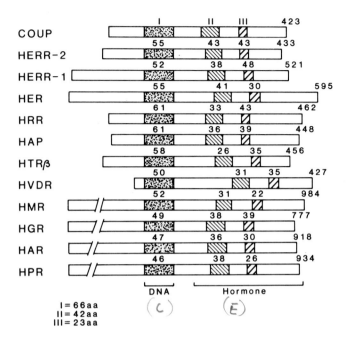

Fig. 1 Sequence homology of the steroid receptor superfamily. Alignment of amino acid sequences on the basis of maximum similarity identified three regions shown as I, II and III. The numbers above each box indicate the percentage of similarity in each region as compared to the corresponding region of COUP-TF. The numbers at the C-terminal of individual receptors represent the total amino acid residues. Specific references are: COUP, COUP transcription factor (Wang et al., 1989); HERR-1 and HERR-2, oestrogen receptor-related cDNAs (Giguere et al., 1988); HER, oestrogen receptor (Green et al., 1986); HRR, retinoic acid receptor (Giguere et al., 1987; Petkovich et al., 1987; Robertson, 1987); HAP, another retinoic acid receptor (deThe et al., 1987); HTRβ, thyroid hormone receptor (Weinberger et al., 1986) HDVR, vitamin D receptor (Baker et al., 1988); HMR, mineralocorticoid receptor (Arriza et al., 1987); HGR, glucocorticoid receptor (Hollenberg et al., 1985); HAR, androgen receptor (Lubahn et al., 1988; Chang et al., 1988); and HRP, progesterone receptor (Misrahi et al., 1987).

plays a central role in cellular fate and communication. Comparison of *Drosophila* SVP with human COUP-TF revealed that SVP protein is 109 amino acids longer than COUP-TF in the region N-terminal to the DNA-binding domain, containing little sequence conservation. In contrast, the DNA-binding and the putative C-terminal ligand-binding regions are virtually identical. This striking C-terminal conservation (93% identity) suggests that *Drosophila* and human COUP-TF may be regulated by the same ligand. It also raises an interesting possibility that COUP-TF may be important in human developmental processes.

5.5 Chicken ovalbumin upstream promoter transcription factor requires S300-II for its function

We and others have shown that COUP-TF and other members of the steroid receptor family are bona fide transcription factors (Pastorcic *et al.*, 1986; Kumar *et al.*, 1987; Hollenberg and Evans, 1988; see Chapters 2 and 3). They contain both DNA-binding domains and 'transactivation domains'. However, upon examining the amino acid sequences of these 'activation domains', no obvious negative charge regions (Hope and Struhl, 1986; Ma and Ptashne, 1987) or regions rich in glutamine or serine/threonine (Courey and Tjian, 1988) are found. This raises a question as to what 'transactivation domains' mediate the activation of target gene transcription. It is possible that these receptors may contain regions which serve as binding sites for adaptor (activator) proteins, such as VP16 which transactivates viral immediate early gene expression by interacting with the octa nucleotide binding protein (Oct_1) (O'Hare and Goding, 1988; Triezenberg *et al.*, 1988). Such an adaptor protein could provide the final activation function to COUP-TF or other members of the steroid receptor family, particularly the smaller receptors. A candidate adaptor protein for certain members of this family is S300-II transcription factor (Tsai *et al.*, 1987). We have demonstrated previously that in addition to COUP-TF, the non-DNA-binding protein S300-II is required for transcription of the ovalbumin gene *in vitro*. In addition, S300-II stabilized the binding of COUP-TF to the COUP sequence (Tsai *et al.*, 1987). Therefore, it is likely that S300-II functions by interacting with a DNA-binding protein such as COUP-TF to activate promoter function is illustrated in Fig. 2. It will be of interest to determine whether SVP requires an adaptor protein such as S300-II for its function. Conceivably, the extra 109 amino acid residues at the N-terminal domain of SVP may obviate the need for S300-II by providing a functional activation domain which interacts with other *trans*-acting factors to facilitate the initiation of RNA synthesis.

5.6 Chicken ovalbumin upstream promoter transcription factor and orphan receptors

Indeed, one of the most fascinating observations to evolve from the cloning of cDNAs for steroid receptors is the unexpected large size of the steroid receptor superfamily of related genes. After elucidation of the receptors for the more traditional members of this family (glucomineralocorticoids, sex steroids, thyroid hormone, vitamin D_3, and retinoic acid), a large number of cloned receptoroids have been discovered (Evans, 1988). These molecules can be considered to be 'orphan receptors' in search of a function and a ligand. Since they were cloned by cDNA cross-hybridization screening using cDNA probes, we have little clue as to their cellular physiology. A function is implicit, however, since they are expressed in cells as fully processed cytoplasmic mRNAs.

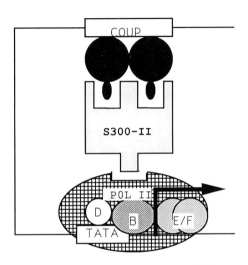

Fig. 2 Model for the function of S300-II factor. It acts as a bridge protein with other *trans*-acting factors, e.g. those at the TATA box.

The first report of two such molecules was by the Evans laboratory; they were termed ERR-1 (oestrogen-receptor related) and ERR-2 (Giguere *et al.*, 1988). Their function remains unknown to date. A number of 'orphan receptor' sequences have been published to date (deThe *et al.*, 1987; Nauber *et al.*, 1988; Oro *et al.*, 1988; Lazar *et al.*, 1989). They are recognized easily by their extensive homology in the DNA-binding region and their absolute conservation of type II zinc fingers. We estimate that more than 20 additional related molecules may have been cloned in the combined laboratories in our field. Six of these molecules have been cloned from *Drosophila*. Until recently, all of these putative receptors were without designated functions.

The assignment of COUP-TF to the steroid receptor superfamily of proteins has a number of apparent implications. For the first time, it designates promoter regulatory proteins as a legitimate subtype in this family. The family had been thought previously to include only enhancer regulatory proteins. Second, it provides information which is useful for understanding the evolution of this family of transcriptional regulators. Third, it raises the question as to whether COUP-TF, and other promoter activators for that matter, may be ligand-activated gene regulators. This latter query remains to be answered experimentally. Deductive reasoning should permit us to conclude that if one of these 'orphan receptors' has been adopted for a function, then in time, others will follow suit.

Although admittedly speculatory, the question as to whether the 'orphan receptors' have endogenous ligands is clearly the most exciting to be considered. If they do have ligands, we predict that the ligands are indigenous to the cells of origin since transfection into most cultured cells shows them to be active in an apparent constitutive manner. Our best guess is that many of these putative ligands will be hydrophobic in nature and some may be nutritionally or metabolically derived. In fact, this could be the tip of a new intracrine iceberg wherein a series of yet to be described hormones are discovered to be indigenous to the specific target cells expressing these 'orphan receptors'. If true, the elucidation of a new intracrine regulatory system will inject more than a little interest and excitement into the fields of hormone action and molecular endocrinology.

5.7 Progesterone receptor: its role in transactivation of target gene expression

As mentioned earlier, COUP-TF, a member of the steroid receptor superfamily, functions as a distal promoter binding protein in enhancing transcription of the ovalbumin promoter *in vitro* (Pastorcic *et al.*, 1986). However, our efforts to demonstrate transcriptional regulation of target gene expression by 'authentic' steroid receptors, such as oestrogen receptor, glucocorticoid receptor or progesterone receptor, in an *in vitro* system, were not successful until recently. Corthesy *et al.* (1988) demonstrated hormone-dependent transcription of the vitellogenin gene, in crude extracts of *Xenopus* nuclei. Subsequently, it was demonstrated that bacterially-expressed truncated glucocorticoid receptor fragment (Freedman *et al.*, 1989) and native progesterone receptor (Kalff *et al.*, 1990; Klein-Hitpass *et al.*, 1990) are capable of enhancing RNA synthesis in an *in vitro* transcription assay (see Chapter 9.8). So, after many years of experimentation, we are finally able to demonstrate progesterone receptor-mediated regulation of target gene transcription directly in a cell-free system. As expected, enhancement of transcription depends on the presence of progesterone response elements (PREs) in the template (Klein-Hitpass *et al.*, 1990) and is inhibited by addition of competitor DNA containing authentic PREs. Test constructs containing two PREs yield 30-fold higher levels of transcription in the presence of chicken progesterone receptor. The transcription of test genes lacking PREs is unaffected by addition of receptor.

There are two forms (A and B) of progesterone receptor in chicken and human cells. These two forms arise from differential initiation of translation on a single mRNA species (Conneely *et al.*, 1989a). Both A and B forms of progesterone receptor activate *in vitro* transcription, suggesting that both forms of receptor interact with the response element and the transcriptional machinery in a similar manner to achieve transcriptional enhancement. Similarly, both forms of receptor interact equally well with an MMTV promoter to achieve high levels of expression in transient transfection assays (Conneely *et al.*, 1989b). Interestingly, progesterone receptor form A transactivates the chicken ovalbumin promoter more efficiently than form B, suggesting that the N-terminal region of the receptor plays an important role in the differential activation of target promoters (Tora *et al.*, 1988; Conneely *et al.*, 1989a).

It is well documented that multiple general transcription factors, TFIID, IIA, IIB, IIE, IIF, and polymerase II are required for regulation of the initiation of transcription (Hawley and Roeder, 1985; Reinberg *et al.*, 1987; Buratowski *et al.*, 1989). Hawley and Roeder (1985) proposed that the formation of a stable initiation complex for transcription can be biochemically dissected in the multiple steps:

(1) a template committed step during which transcription factors bind DNA;
(2) a step in which a rapid-start complex is formed which poises factors and RNA polymerase for RNA synthesis; and
(3) the initiation of transcription.

Each step requires interactions among core promoter factors, polymerase and the template DNA. Recently, Van Dyke *et al.* (1989) demonstrated that TFIID alone is sufficient to confer template commitment to a minimal promoter containing only a TATA box but lacking distal promoter elements. Sawadogo and Roeder (1985) demonstrated that major late transcription factor (MLTF), together with the TATA binding protein, are sufficient to confer template commitment to a construct harbouring the MLTF recognition sequence in its distal promoter. Hai *et al.* (1988) showed that TFIID, IIB, polymerase and ATF (a transcription factor required for cellular and adenovirus transcription) are all required for the formation of the pre-initiation complex on the E_4 promoter and that ATF interacts with TFIID to facilitate the formation of a pre-initiation complex (Horikoshi *et al.*, 1988).

To elucidate the mechanism by which steroid hormone receptors interact with core promoter factors to enhance the initiation of transcription, we examined whether the progesterone receptor is essential for the formation of a template-committed pre-initiation complex at target gene promoters. Our results indicate that in the presence of general transcription factors provided by the HeLa cell extracts progesterone receptor directs the formation of a stable pre-initiation complex (Klein-Hitpass et al., 1990). Further analyses of the requirements for general transcription factors indicate that TFIID, and TFIIE/F together with progesterone receptor, are sufficient to confer template commitment to PRE containing promoters. This result suggests that progesterone receptor bound to a PRE interacts with TFIID and TFIIE/F to facilitate the enhancement of gene transcription (Fig. 3). Our findings suggest that different enhancer or promoter binding proteins may interact with one or a subset of general transcription factors to promote and stabilize the formation of pre-initiation complexes and thus enhance the initiation of transcription. Since TFIIE/F have been thought to be associated with RNA polymerase II, it will be of great interest to investigate the biochemical mechanism by which progesterone receptor interacts with these factors in the transcription initiation process. Our preliminary investigations using other steroid receptors (e.g. glucocorticoid and oestrogen) indicate that this may be a general mechanism by which all receptors act to regulate the expression.

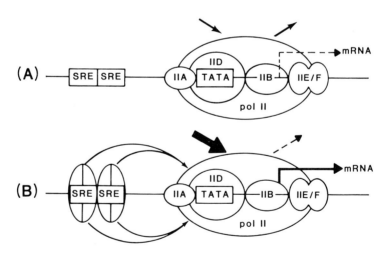

Fig. 3 Steroid receptor stabilizes the pre-initiation complex for transcription. As a mechanism of action steroid receptor enhances recruitment and/or stabilizes *trans*-acting factors within the promoter.

5.8 Multiple hormone response elements mediate synergistic transcriptional activation of target genes

Multiple steroid hormone response elements (SREs) are often found in the 5'-flanking regions of hormone responsive genes (Renkawitz *et al.*, 1984; Klein-Hitpass *et al.*, 1986; Jantzen *et al.*, 1987; Chalepakis *et al.*, 1988; Glass *et al.*, 1988; see Chapter 6). Transient transfection studies demonstrate that deletion or mutation of one of two SREs leads to a dramatic decrease in the inducibility of a target gene, suggesting that the SREs co-operate with one another to confer synergistic induction. Two mechanistic models, which may contribute to this synergism, have been proposed by Ptashne (1988) and are depicted in Fig. 4. First, co-operative binding of *trans*-acting factors to two recognition sequences may induce synergism through protein–protein interaction, as in the case of λ-bacteriophage repressor (Ptashne *et al.*, 1980). Second, *trans*-acting factors which bind independently to their recognition sequences but interact simultaneously with other target protein(s), such as a component of the general transcriptional machinery, may also generate co-operativity.

(A) BINDING AFFINITY

(B) INTERACTION WITH TARGET SITE

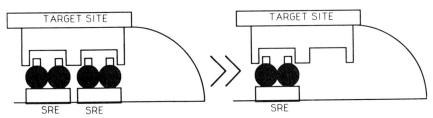

Fig. 4 Models of co-operativity. (A) Receptor multimers co-operate to form very high affinity complexes. (B) Receptors contribute to the overall surface interaction with target site (e.g. TATA box) proteins.

In vitro binding experiments with purified preparations of chicken progesterone receptor indicate that receptors bind directly to two tandemly linked PREs and that binding of a receptor dimer to the first PRE site enhances the binding affinity of a subsequent receptor dimer to the second site by 100-fold (Fig. 4A). Thus, the co-operative binding of receptors to PREs appears to contribute to the hormone-induced synergism in gene expression observed *in vivo* (Tsai *et al.*, 1989). Using the *in vitro* transcription assay to analyse levels of transcription of test genes containing varying copy numbers of PREs, we showed that while one copy of PRE enhances receptor dependent transcription four-fold, two copies enhance transcription 30-fold. These results indicate that two copies of PRE function co-operatively also under our *in vitro* transcription conditions. Similar regulation through co-operative interactions between DNA-binding proteins such as glucocorticoid receptor (Schmid *et al.*, 1989), heat-shock transcription factor (Topol *et al.*, 1985) or Oct_2 (LeBowitz *et al.*, 1989; Poellinger *et al.*, 1989) have been observed.

In addition to co-operativity induced by two similar response elements, synergism has also been demonstrated with heterologous response elements such as a GRE with NF-1 in the LTR of MMTV (Schule *et al.*, 1988a; Buetti *et al.*, 1989), a GRE with a CACA element, and others (Schule *et al.*, 1988a,b). We have carried out studies of *in vitro* transcription with complex regulatory regions containing steroid response elements, upstream promoters and TATA elements. Using transcription-competition assays, we observed that when glucocorticoid receptor, isolated from a bacculovirus expression system (Srinivasan and Thompson, 1990) is bound to GREs, the receptor co-operates with NF-1 to confer high levels of expression to the MMTV promoter; glucocorticoid receptor or NF-1 alone provide only low levels of basal expression (see Chapters 9.7 and 10.3). Interestingly, progesterone receptors purified either from chicken oviduct or from T47D cells failed to enhance high levels of expression from an MMTV promoter, resembling the results obtained *in vivo* by Nordeen *et al.* (1989) and Gowland and Buetti (1989). In contrast, both progesterone and glucocorticoid receptors were equally efficient in conferring inducibility to test genes containing both COUP and GRE/PRE sequences in their promoter *in vitro*. Based on these observations, we suggest that glucocorticoid and progesterone receptors possess qualitatively different types of activation domains. Consequently, the activation surfaces of glucocorticoid receptors can interact well with NF-1, but those of progesterone receptor are less effective. On the other hand, both activation domains can function equally well in conjunction when COUP-TF is bound to the upstream promoter. Whether the observed synergism is partially primarily dependent on co-operative binding or on overall conformation is not yet known.

High levels of induction can also be achieved by placing a combination of heterologous elements, such as oestrogen and progesterone response elements, upstream of a minimal promoter. This synergistic induction is not due to co-operative binding, but due to combinatorial interactions between different activation domains in the heterologous receptor complex (Fig. 4B). Taken together, these results demonstrate that complex regulation of eukaryotic gene expression can be achieved by assembling unique subsets of cis-elements. Through either co-operative binding or co-operative interactions of specific activation domains with target sites (transcription factors), expression of a given specific gene can be regulated over a wide range.

5.9 The role of ligand in steroid receptor function

It is well documented that steroid receptor-mediated induction of target gene expression *in vivo* is dependent on the presence and concentration of steroid hormone (O'Malley *et al.*, 1979; Yamamoto, 1985; Beato, 1989). Consistent with this model, specific hypersensitive sites, which correlate well with the state of expression of hormone responsive genes, are detected in and around the SREs in the presence of the cognate hormone (Fritton *et al.*, 1984; Kaye *et al.*, 1984; Jantzen, K. *et al.*, 1987). A nuclear genomic footprint, demonstrating receptor binding to the PRE/GRE element of the tyrosine aminotransferase gene, is only observed in the presence of hormone (Becker *et al.*, 1986). Furthermore, oestrogen (Kumar and Chambon, 1988) or progesterone (Bagchi *et al.*, 1988) receptors in crude nuclear extracts bind to their respective SREs only when the extracts were prepared from hormone-treated cells. However, *in vitro* binding studies using 'purified' chicken and rabbit progesterone receptors (Bailey *et al.*, 1986; Rodriguez *et al.*, 1989), rat liver glucocorticoid (Willmann and Beato, 1986) and oestrogen receptors (Klein-Hitpass *et al.*, 1989; Lees *et al.*, 1989) illustrate that receptors bind specifically to SREs in a hormone-independent manner (see Chapter 9.3). This observed discrepancy in hormone-dependent binding could be explained by the hypothesis that in crude extracts the receptor is associated with inhibitory proteins which prevent specific binding to SREs. These proteins could be released during hormonal treatment or during purification (Denis *et al.*, 1988; Pratt *et al.*, 1988; Kost *et al.*, 1989). Once a receptor was freed from associated inhibitory proteins, such as Hsp90 or Hsp70, it could then bind to its respective SRE with high affinity and confer hormone inducibility (Fig. 5). A similar mechanism of activation has been proposed for another transcription factor NF-KB (Baeuerle and Baltimore, 1988a,b, 1989). In T cells, NF-KB is associated with the inhibitor IFKB. Once IFKB is phosphorylated, NFKB is released from IFKB, translocated to the nucleus where it binds to its specific recognition site, and functions in enhancing immunoglobulin K chain synthesis.

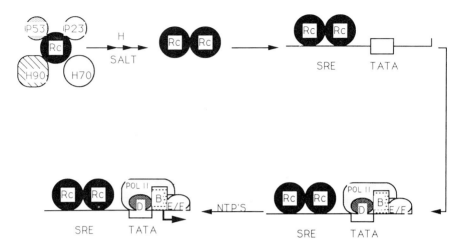

Fig. 5 Model of steroid receptor action. Rc: receptor; SRE: steroid response elements; B, D, E/F, and Pol II denote the transcription factors TFIIB, TFIID, TFIIE/F and RNA polymerase II, respectively.

The results of our *in vitro* transcription experiments demonstrate that purified progesterone receptors bind and function to enhance the expression of PRE-containing target genes in a hormone-independent manner. In contrast, crude nuclear extracts isolated from T47D cells contain ligand-free progesterone receptors which do not bind to PRE sites; these receptors also fail to enhance transcription of PRE-containing test genes. Upon treatment with progesterone ligand, either *in vitro* or *in vivo*, the receptors in such extracts now bind specifically to their respective PREs and enhance transcription of PRE-containing test genes (Bagchi et al., 1990).

This transcriptional stimulation is specific for progestins and is inhibited by 70% when the anti-progestin Ru486 is added to the reaction. Surprisingly, salt extraction converts the unliganded progesterone receptor to a 4 S form, as determined by sucrose gradient centrifugation. This 4 S form of receptor does not contain Hsp90 and 59-kD protein, but is still inactive in the cell-free transcription assay. Thus, conversion of the 8 S complex to 4 S form, accompanied by the release of Hsp90 and 59-kD protein by salt treatment, is not sufficient to activate the receptor. The discovery of this novel inactive 4 S form of receptor (free of Hsp90) could only have been detected in and defined in cell-free studies such as these (Fig. 6). Our results are consistent with the observations that neither thyroid nor vitamin D receptors are associated with heat-shock protein in their native unliganded state but are still inactive. We suggest that hormone is needed to effect an additional structural alteration(s) in the receptor molecule. *In situ*, this could occur either by dissociation of additional inhibitory proteins, by inducing a specific conformation change in the receptor, or via covalent modifications such as phosphorylation. Following such an alteration, the receptor gains the capacity to bind and activate hormone-responsive genes (Fig. 5). In highly purified forms of receptor, this conformation may be achieved via the purification process itself.

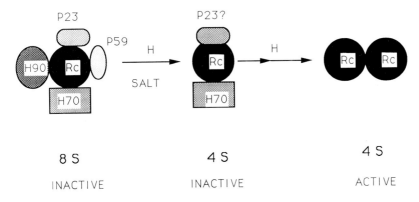

Fig. 6 Influence of hormone on receptor action. Hormone converts a 4 S inactive receptor form to a 4 S form which is active in cell-free transcription.

5.10 Conclusion and perspectives

Following cloning of the cDNAs for steroid hormone receptors, we have made progress toward an understanding of the structure–function relationships of steroid hormone receptors. Cell-free transcription studies promise to provide detailed information on their mechanism of action. There are a number of additional important questions, however, that have yet to be answered. For instance, we need more precise information as to how steroid receptors interact with distal promoter binding proteins and with general transcription factors during the formation of transcriptional initiation complex. Second, we would like to determine the mechanism by which receptor molecules distinguish positive and negative response elements and act accordingly to induce or to repress the expression of a given hormone-responsive gene. Third, we must define the role that chromatin structure plays in the enhancement of overall levels of expression of hormonal target genes. Fourth, we would like to know how authentic hormones activate and anti-hormones inhibit transcription of a hormone-responsive gene. Fifth, we would like to determine what chemical transduction signals render 'orphan receptors' functional in transactivation. Finally, we would like to know whether there is a requisite adaptor or activator protein for receptors and understand its role in target gene regulation by receptor family members. For the most part, these questions can be answered by cell-free transcription using reconstituted transcription factors and partially purified, highly purified and mutated steroid receptors. Such information should enable us then to understand the functional role of steroid hormones in more complex processes such as the development and differentiation of their target tissues.

References

Arriza, J. L., Weinberger, C., Cerelli, G., Glasser, T. M., Handelin, B. L., Housmann, D. E. and Evans, R. M. (1987). Cloning of human mineralo-corticoid receptor cDNA: structural and functional kinship with the glucocorticoid receptor. *Science* **237**, 268–275.

Baeuerle, P. A. and Baltimore, D. (1988a). Activation of DNA-binding activity in an apparently cytoplasmic precursor of the NF-kB transcription factor. *Cell* **53**, 211–217.

Baeuerle, P. A. and Baltimore, D. (1988b). IkB: A specific inhibitor of the NF-kB transcription factor. *Science* **242**, 540–546.

Baeuerle, P. A. and Baltimore, D. (1989). A 65-kD subunit of active NF-kB is required for inhibition of NF-kB by IkB. *Gene and Dev.* **3**, 1689–1698.

Bagchi, M. K., Tsai, S. Y., Tsai, M.-J. and O'Malley, B. W. (1987). Purification and characterization of chicken ovalbumin gene upstream promoter transcription factor from homologous oviduct cells. *Mol. Cell. Biol.* **12**, 4151–4158.

Bagchi, M. K., Elliston, J. F., Tsai, S. Y., Edwards, D. P., Tsai, M.-J. and O'Malley, B. W. (1988). Steroid hormone-dependent interaction of human progesterone receptor with its target enhancer element. *Mol. Endocrinol.* **2**, 1221–1229.

Bagchi, M. K., Tsai, S. Y., Tsai, M.-J. and O'Malley, B. W. (1990). Progesterone-dependent cell free transcription: identification of a functional intermediate in receptor activation. *Nature* **345**, 547–550.

Bailey, M., LePage, C., Rauch, M. and Milgrom, E. (1986). Sequence-specific DNA binding of the progesterone receptor to the uteroglobin gene: effect of hormone, antihormone and receptor phosphorylation. *EMBO J.* **5**, 3235–3241.

Baker, A. R., McDonnell, D. P., Hughes, M., Crisp, T. M., Mangelsdorf, D. J., Haussler, M. R., Pike, J. W., Shine, J. and O'Malley, B. W. (1988). Cloning and expression of full-length cDNA encoding human vitamin D receptor. *Proc. Natl. Acad. Sci. USA* **85**, 3294–3298.

Beato, M. (1989). Gene regulation by steroid hormones. *Cell* **56**, 335–344.

Becker, P. B., Gloss, B., Schmid, W., Strahle, V. and Schutz, G. (1986). In vivo protein–DNA interactions in a glucocorticoid response element require the presence of hormone. *Nature* **324**, 686–688.

Bodner, M. and Karin, M. (1987). A pituitary-specific transacting factor can stimulate transcription from the growth hormone promoter in extracts of non-expressing cells. *Cell* **50**, 267–275.

Bohmann, D. and Tjian, R. (1989). Biochemical analysis of transcriptional activation by Jun: differential activity of c and v Jun. *Cell* **59**, 709–717.

Bohmann, D., Bos, T. J., Admon, A., Nishimura, T., Vogt, P. K. and Tjian, R. (1987). Human protooncogene c-*Jun* encodes a DNA-binding protein with structural and functional properties of transcription factor Ap-1. *Science* **238**, 1386–1392.

Buetti, E., Kuknel, B. and Diggelmann, H. (1989). Dual function of a nuclear factor 1 binding site in MMTV transcription regulation. *Nucl. Acids Res.* **17**, 3065–3078.
Buratowski, S., Hahn, S., Guarente, L. and Sharp, P. A. (1989). Five intermediate complexes in transcription initiation by RNA polymerase II. *Cell* **56**, 549–561.
Chalepakis, G., Annemann, J., Slater, E., Bruller, H.-J., Gross, B. and Beato, M. (1988). Differential gene activation by glucocorticoids and progestins through the hormone regulatory element of mouse mammary tumor virus. *Cell* **53**, 371–382.
Chang, C., Kokonitis, J. and Liao, S. (1988). Molecular cloning of the human and rat cDNA encoding androgen receptors. *Science* **240**, 324–326.
Clerc, R. G., Corcoran, L. M., LeBowitz, H. H., Baltimore, D. D. and Sharp, P. A. (1988). The β-cell specific Oct-2 protein contains POU-box and homeo-box type domains. *Genes Dev.* **2**, 1570–1581.
Conneely, O. M., Kettleberger, D. M., Tsai, M.-J. and O'Malley, B. W. (1989a). Promoter-specific activating domains of the chicken progesterone receptor. In "Gene Regulation by Steroid Hormone IV" (eds A. K. Roy and J. Clark), pp. 220–233. Springer-Verlag, New York.
Conneely, O. M., Kettleberger, D. M., Tsai, M-J., Schrader, W. T. and O'Malley, B. W. (1989b). The chicken progesterone receptor A and B isoforms are products of an alternate translation event. *J. Biol. Chem.* **264**, 14 062–14 064.
Corthesy, B., Hipskind, R., Theulaz, I. and Wahli, W. (1988). Estrogen dependent *in vitro* transcription from the vitellogenin promoter in liver nuclear extracts. *Science* **239**, 1137–1139.
Courey, A. J. and Tjian, R. (1988). Analysis of SP_1 *in vivo* reveals multiple transcriptional domains, including a novel glutamine-rich activation motif. *Cell* **55**, 887–898.
Curran, T. and Franza, B. R. (1988). *Fos* and *Jun*: The AP-1 connection. *Cell* **55**, 395–397.
Denis, M., Poellinger, L., Wikstom, A.-C. and Gustafsson, J.-A. (1988). Requirement of hormone for thermal activation of the glucocorticoid receptor to a DNA-binding state. *Nature* **333**, 686–688.
deThe, H., Marchio, A., Tiollois, P. and Dejean, A. (1987). A novel steroid thyroid hormone receptor-related gene inappropriately expressed in human hepatocellular carcinoma. *Nature* **330**, 667–670.
Dynan, W. S. and Tjian, R. (1985). Control of eukaryotic messenger RNA synthesis by sequence-specific DNA-binding proteins. *Nature* **316**, 774–778.
Elbrecht, A., Tsai, S. Y., Tsai, M.-J. and O'Malley, B. W. (1985). Identification of exonuclease footprinting of a distal promoter-binding protein from HeLa cell extracts. *DNA* **4**, 233–240.
Evans, R. M. (1988). The steroid and thyroid hormone receptor superfamily. *Science* **242**, 889–895.
Finney, M., Ruvkin, G. and Horvitz, H. R. (1988). The *C. elegans* cell lineage and differentiation gene UNC-86 encodes a protein containing a homeodomain and extended sequence similarity to mammalian transcription factors. *Cell* **56**, 757–769.
Freedman, L. P. Yoshinaga, S. K., Vanderbilt, J. N. and Yamamoto, K. R. (1989). *In vitro* transcription enhancement by purified derivatives of the glucocorticoid receptor. *Science* **245**, 298–301.
Fritton, H. P., Igo-Kemenes, T. I., Nowock, J., Strech-Jurk, U., Theisen, M. and Sippel, A. E. (1984). Alternative sets of DNaseI-hypersensitive sites characterize the various functional states of the chicken lysozyme gene. *Nature* **311**, 163–165.

Glass, C. K., Holloway, J. M., Devary, O. V. and Rosenfeld, M. G. (1988). The thyroid hormone receptor binds with opposite transcriptional effects to a common sequence motif in the thyroid hormone and estrogen response elements. *Cell* **54**, 313–323.
Giguere, V., Ong, E. S., Sequi, P. and Evans, R. M. (1987). Identification of a receptor for the morphogen retinoic acid. *Nature* **330**, 624–629.
Giguere, V., Yang, N., Segui, P. and Evans, R. M. (1988). Identification of a new class of steroid hormone receptors. *Nature* **331**, 91–94.
Gowland, P. and Buetti, E. (1989). Mutations in the hormone regulatory element of mouse mammary tumor virus differentially affect the response to progestins, androgen and glucocorticoids. *Mol. Cell. Biol.* **9**, 3999–4008.
Green, S., Walter, P., Kumar, V., Krust, A., Bornet, J. M., Argos, P. and Chambon, P. (1986). Human estrogen receptor cDNA: sequence, expression and homology to v-*erb*A. *Nature* **320**, 134–139.
Hai, T., Horikoshi, M., Roeder, R. G. and Green, M. R. (1988). Analysis of the role of the transcription factor ATF in the assembly of a functional preinitiation complex. *Cell* **54**, 1043–1051.
Halazonetis, T. D., Georgopoulos, K., Greenberg, M. E. and Leder, P. (1988). C-*Jun* dimerizes with itself and with c-*fos*, forming complexes of different DNA binding affinities. *Cell* **55**, 917–924.
Hawley, D. K. and Roeder, R. G. (1985). Separation and partial characterization of three functional steps in transcription initiation by human RNA polymerase II. *J. Biol. Chem.* **260**, 8163–8172.
Hollenberg, S. M., Weinberger, C., Ong, E. S., Cerelli, G., Oro, A., Lebo, R., Thompson, E. B., Rosenfeld, M. G. and Evans, R. M. (1985). Primary structure and expression of a functional human glucocorticoid receptor cDNA. *Nature* **318**, 635–641.
Hollenberg, S. M. and Evans, R. M. (1988). Multiple and cooperative transactivation domains of the human glucocorticoid receptor. *Cell* **55**, 899–906.
Hope, I. A. and Struhl, K. (1986). Functional dissection of a eukaryotic transcriptional activator protein GCN4 of yeast. *Cell* **46**, 885–894.
Horikoshi, M., Hai, T., Lin, Y. S., Green, M. R. and Roeder, R. G. (1988). Transcription factor ATF interacts with the TATA factor to facilitate establishment of a preinitiation complex. *Cell* **54**, 1033–1042.
Hwung, Y. P., Crowe, D. T., Wang, L.-H., Tsai, S. Y. and Tsai, M.-J. (1988). The COUP transcription factor binds to an upstream promoter element of the rat insulin II gene. *Mol. Cell. Biol.* **8**, 2070–2077.
Hwung, Y.-P., Wang, L.-H., Tsai, S. Y. and Tsai, M.-J. (1988). Differential binding of the chicken ovalbumin upstream promoter (COUP) transcription factor to two different promoters. *J. Biol. Chem.* **263**, 13 470–13 474.
Jantzen, K., Fritton, H. P., Igo-Kemenes, T., Espel, E., Janich, S., Cato, A. C. B., Mugele, K. and Beato, M. (1987). Partial overlapping of binding sequences for steroid hormone receptors and DNase hypersensitive sites in the rabbit uteroblogin gene region. *Nucl. Acids. Res.* **15**, 4535–4552.
Jantzen, H. M., Strahle, U., Gloss, B., Stewart, F., Schmid, W., Boshart, M., Miksicek, R. and Schutz, G. (1987). Cooperativity of glucocorticoid response elements located far upstream of the tyrosine aminotransferase gene. *Cell* **49**, 29–38.
Jones, N. C., Rigby, P. W. J. and Ziff, E. B. (1988). Trans-acting protein factors and the regulation of eukaryotic transcription: lessons from studies on DNA tumor virus. *Genes Dev.* **2**, 267–281.

Kalff, M., Gross, B. and Beato, M. (1990). Progesterone receptor stimulates transcription of mouse mammary tumour virus in a cell free system. *Nature* **344**, 360–362.
Kaye, J. S., Bellard, M., Dretzen, G., Bellard, F. and Chambon, P. (1984). A close association between sites of DNase I hypersensitivity and sites of enhanced cleavage by micrococcal nuclease in the 5′-flanking region of the actively transcribed ovalbumin gene. *EMBO J.* **3**, 1137–1144.
Klein-Hitpass, L., Schorpp, M., Wagner, U. and Ryffel, G. U. (1986). An estrogen-responsive element derived from the 5′ flanking region of the *Xenopus* vitellogenin A2 gene function in transfected human cells. *Cell* **46**, 1053–1061.
Klein-Hitpass, L., Tsai, S. Y., Greene, G., Clark, M.-J. and O'Malley, B. W. (1989). Specific binding of estrogen receptor to the estrogen response element. *Mol. Cell. Biol.* **9**, 43–49.
Klein-Hitpass, L., Tsai, S. Y., Weigel, N. L., Allan, G. F., Riley, D., Rodriguez, R., Schrader, W. T., Tsai, M.-J. and O'Malley, B. W. (1990). The progesterone receptor stimulates cell-free transcription by enhancing the formation of a stable preinitiation complex. *Cell* **60**, 247–257.
Kost, S. L., Smith, D., Sullivan, W., Welch, W. J. and Toft, D. O. (1989). Binding of heat shock proteins to the avian progesterone receptor. *Mol. Cell Biol.* **9**, 3829–3838.
Kumar, V. and Chambon, P. (1988). The estrogen receptor binds tightly to its responsive element as a ligand-induced homodimer. *Cell* **55**, 145–156.
Kumar, V., Green, S., Stack, G., Berry, M., Jin, J.-R. and Chambon, P. (1987). Functional domains of the human estrogen receptor. *Cell* **51**, 941–951.
Lassar, A. B., Buskin, J. E., Lockshon, D., Davis, R. L., Apone, S., Hauschka, S. D. and Weintraub, H. (1989). *Myo*D is a sequence-specific DNA binding protein requiring a region of myc homology to bind to the muscle creatine kinase enhancer. *Cell* **58**, 823–831.
Lazar, M. A., Hodin, R. A., Darling, D. S. and Chin, W. W. (1989). A novel member of the thyroid/steroid hormone receptor family is encoded by the opposite strand of the rat c-*erb*Aα transcription unit. *Mol. Cell. Biol.* **9**, 1128–1136.
LeBowitz, J. H., Clerc, R. G., Brenowitz, M. and Sharp, P. A. (1989). The Oct-2 protein binds cooperatively to adjacent octamer sites. *Genes Dev.* **3**, 1625–1638.
Lees, J. A., Farrell, S. E. and Parker, M. G. (1989). Identification of two transactivation domains in the mouse oestrogen receptors. *Nucl. Acids Res.* **17**, 5477–5488.
Leff, T., Reue, K., Melian, A., Culver, H. and Breslow, J. (1989). A regulatory element in the Apo CIII promoter that directs hepatic specific transcription binds to proteins in expressing and nonexpressing cell types. *J. Biol. Chem.* **264**, 16132–16137.
Lubahn, D. B., Joseph, D. R., Sullivan, P. M., Willard, H. F., French, F. S. and Wilson, E. M. (1988). Cloning of the human androgen receptor cDNA and localization to the X-chromosome. *Science* **240**, 327–330.
Ma, J. and Ptashne, M. (1987). Deletion analysis of Gal4 defines two transcriptional activating segments. *Cell* **48**, 847–853.
Mangalam, H. J., Albert, V. R., Ingraham, H. A., Kapiloff, M., Wilson, L., Nelson, C., Elsholtz, H. and Rosenfeld, M. G. (1989). A pituitary POU domain protein, Pit-1 activates both growth hormone and prolactin promoters transcriptionally. *Genes Dev.* **3**, 946–958.
Maniatis, T., Goodbourn, S. and Fischer, J. A. (1987). Regulation of inducible and tissue-specific gene expression. *Science* **236**, 1237–1245.

McKnight, S. and Tjian, R. (1986). Transcription selectivity of viral genes in mammalian cells. *Cell* **46**, 795–805.
Misrahi, M., Atger, M., D'Auriol, L. Loosfeit, H., Meriel, C., Friedlansky, F., Guichon-Mantel, A., Galibert, F. and Milgrom, E. (1987). Complete amino acid sequence of the human progesterone receptor deduced from clone cDNA. *Biochem. Biophys. Res. Commun.* **143**, 740–748.
Mlodzik, M., Hiromi, Y., Weber, U., Goodman, C. S. and Rubin, G. M. (1990). The *Drosophila* seven-up gene, a member of the steroid receptor gene superfamily, controls photoreceptor cell fates. *Cell* **60**, 211–224.
Nakabeppu, Y., Ryder, K. and Nathans, D. (1988). DNA binding activities of three murine *Jun* proteins: stimulation by Fos. *Cell* **55**, 907–915.
Nauber, U., Pankratz, M. J., Kienlin, A., Seifert, E., Klemm, U. and Jackle, H. (1988). Abdominal segmentation of the *Drosophila* embryo requires a hormone receptor-like protein encoded by the gap gene knirps. *Nature* **336**, 489–492.
Nordeen, S. K., Kunhnel, B., Lawler-Heavner, J., Barber, D. A. and Edwards, D. P. (1989). A quantitative comparison of dual control of a hormone response element by progestins and glucocorticoids in the same cell line. *Mol. Endocrinol.* **3**, 1270–1278.
O'Hare, P. and Goding, C. R. (1988). Herpes simplex virus regulatory elements and immunoglobulin octamer domain bind a common factor and are both targets for virion transactivation. *Cell* **52**, 435–445.
O'Malley, B. W. and Means, A. R. (1974). Female steroid hormone and target cell nuclei. *Science* **183**, 610–620.
O'Malley, B. W., Roop, D. R., Lai, E. C., Nordstrom, J. L., Catterall, J. F., Swaneck, G. E., Colbert, D. A., Tsai, M.-J., Dugaiczyk, A. and Woo, S. L. C. (1979). The ovalbumin gene: organization, structure, transcription and regulation. *Rec. Progr. Hormone Res.* **35**, 1–42.
O'Neill, E. A., Fletcher, C., Burrow, C. R., Heintz, N., Roeder, R. G. and Kelley, T. J. (1988). Transcription factor OTF-1 is functionally identical to the DNA replication factor NFIII. *Science* **241**, 1210–1213.
Oro, A. E., Org, E. S., Margolis, J. S., Posakony, J. W., McKeown, M. and Evans, R. M. (1988). The *Drosophila* gene knirps-related is a member of the steroid-receptor gene superfamily. *Nature* **336**, 493–496.
Pastorcic, M., Wang, H., Elbrecht, A., Tsai, S. Y., Tsai, M.-J. and O'Malley, B. W. (1986). Control of transcription initiation *in vitro* requires binding of a transcription factor to the distal promoter of the ovalbumin gene. *Mol. Cell. Biol.* **6**, 2784–2791.
Petkovich, M., Brand, N. J., Krust, A. and Chambon, P. (1987). A human retinoic acid receptor belongs to the family of nuclear receptors. *Nature* **330**, 444–450.
Pinney, D. F., Pearson-White, S. H., Konieczny, S. F., Latham, K. E. and Emerson, C. P. (1988). Myogenic lineage determination and differentiation: evidence for a regulatory gene pathway. *Cell* **53**, 781–793.
Poellinger, L., Yoza, B. K. and Roeder, R. G. (1989). Functional cooperativity between protein molecules bound at two distinct sequence elements of the immunoglobulin heavy chain promoter. *Nature* **337**, 573–576.
Pratt, W. B., Jolly, D. J., Pratt, D. V., Hollenberg, S. M., Giguere, V., Cadepond, F. M., Schweizer-Groyer, G., Catelli, M.-G., Evans, R. M. and Baulieu, E.-E. (1988). A region in the steroid binding domain determines formation of the non-DNA binding 9S glucocorticoid receptor complex. *J. Biol. Chem.* **263**, 267–273.

Ptashne, M. (1988). How eukaryotic trancriptional activators work. *Nature* **335**, 683–689.
Ptashne, M., Jeffrey, A., Johnson, A. D., Maurer, R., Meyer, B. J., Pabo, C. O., Roberts, T. M. and Sauer, R. T. (1980). How the λ repressor and CRO work. *Cell* **19**, 1–11.
Reinberg, D., Horikoshi, M., and Roeder, R. G. (1987). Factors involved in specific transcription in mammalian RNA polymerase II. Functional analysis of initiation factors IIA and IID and identification of a new factor operating at sequences downstream of the initiation site. *J. Biol. Chem.* **262**, 3322–3330.
Renkawitz, R., Schutz, G., von der Ahe, D. and Beato, M. (1984). Sequences in the promoter region of the chicken lysozyme gene required for steroid regulation and receptor binding. *Cell* **37**, 503–510.
Robertson, M. (1987). Retinoic acid receptor: Towards a biochemistry of morphogenesis. *Nature* **330**, 420–421.
Rodriguez, R., Carson, M. A., Weigel, N. L., O'Malley, B. W. and Schrader, W. T. (1989). Hormone-induced changes in the *in vitro* DNA-binding activity of the chicken progesterone receptor. *Mol. Endocrinol.* **3**, 356–362.
Sagami, I., Tsai, S. Y., Wang, H., Tsai, M.-J. and O'Malley, B. W. (1986). Identification of two factors required for transcription of the ovalbumin gene. *Mol. Cell. Biol.* **6**, 4259–4267.
Sawadogo, M. and Roeder, R. G. (1985). Interaction of a gene-specific transcription factor with the adenovirus major late promoter upstream of the TATA box region. *Cell* **43**, 165–175.
Scheidereit, C., Cromlish, J. A., Gerster, T., Kawakami, K., Balmaceda, C.-G., Currie, R. A. and Roeder, R. G. (1988). A human lymphoid-specific transcription factor that activates immunoglobulin gene is a homeo-box protein. *Nature* **336**, 551–557.
Schmid, W., Strahle, U., Schutz, G., Schmitt, J. and Stunnenberg, H. (1989). Glucocorticoid receptor binds cooperatively to adjacent recognition sites. *EMBO J.* **8**, 2257–2263.
Schule, R., Muller, M., Kaltschmidt, C. and Renkawitz, R. (1988a). Many transcription factors interact synergistically with steroid receptors. *Science* **242**, 1418–1420.
Schule, R., Muller, M., Otsuka-Murakami, H. and Renkawitz, R. (1988). Cooperativity of the glucocorticoid receptor and the caccc-box binding factor. *Nature* **332**, 87–90.
Srinivasan, G. and Thompson, E. B. (1990). Overexpression of full-length human glucocorticoid receptor in spodoptera frugiperda cells using the baculovirus expression vector system. *Mol. Endocrinol.* **4**, 209–216.
Sturm, R. A., Das, G. and Herr, W. (1988). The ubiquitous octamer binding protein Oct-1 contains a Pou domain with a homeobox subdomain. *Genes Dev.* **2**, 1582–1599.
Topol, J., Ruden, D. M. and Parker, C. S. (1985). Sequences required for *in vitro* transcriptional activation of a *Drosophila* Hsp70 gene. *Cell* **42**, 527–537.
Tora, L., Gronemeyer, H., Turcotte, B., Gaub, M.-P. and Chambon, P. (1988). The N-terminal region of the chicken progesterone receptor specifies target gene activation. *Nature* **333**, 185–188.
Triezenberg, S. J., LaMarco, K. L. and McKnight, S. L. (1988). Evidence of DNA: protein interactions that mediate HSV-1 immediate early gene activation by VP16. *Genes Dev.* **2**, 730–742.

Tsai, S. Y., Sagami, I., Wang, H., Tsai, M.-J. and O'Malley, B. W. (1987). Interaction between a DNA-binding transcription factor (COUP) and a non-DNA binding factor (S300-II). *Cell* **50**, 701–709.

Tsai, S. Y., Carlstedt-Duke, J.-A., Weigel, N. L., Dahlman, K., Gustafsson, J.-A., Tsai, M.-J. and O'Malley, B. W. (1988). Molecular interactions of steroid hormone receptor with its enhancer element: evidence for receptor dimer formation. *Cell.* **55**, 361–369.

Tsai, S. Y., Tsai, M.-J. and O'Malley, B. W. (1989a). Cooperative binding of steroid hormone receptors contribute to transcriptional synergism at target enhancer elements. *Cell* **57**, 443–448.

Van Dyke, M. W., Sawadogo, M. and Roeder, R. G. (1989). Stability of transcription complexes on class II genes. *Mol. Cell. Biol.* **9**, 342–344.

Wang, L.-H., Tsai, S. Y., Sagami, I., Tsai, M.-J. and O'Malley, B. W. (1987). Purification and characterization of chicken ovalbumin upstream promoter transcription factor from HeLa cells. *J. Biol. Chem.* **262**, 16 080–16 086.

Wang, L.-H., Tsai, S. Y., Cook, R. G., Beattie, W. G., Tsai, M.-J. and O'Malley, B. W. (1989). COUP transcription factor is a member of the steroid receptor superfamily. *Nature* **340**, 163–166.

Weinberger, C., Thompson, C. C., Ong, E. S., Lebo, R., Gruol, D. J. and Evans, R. M. (1986). The c-*erb*-A gene encodes a thyroid hormone receptor. *Nature* **324**, 641–646.

Willmann, T. and Beato, M. (1986). Steroid-free glucocorticoid receptor binds specifically to mouse mammary tumor virus DNA. *Nature* **324**, 688–691.

Wijnholds, J., Philipsen, J. N. J. and Ab, G. (1988). Tissue-specific and steroid-dependent interaction of transcription factors with the oestrogen-inducible apoVLDLII promoter *in vivo*. *EMBO J.* **7**, 2757–2763.

Wright, W. E., Sassoon, D. A. and Lin, V. K. (1989). Myogenin, a factor regulating myogenesis has a domain homologous to *MyoD*. *Cell* **56**, 607–617.

Yamamoto, K. R. (1985). Steroid receptor-regulated transcription of specific genes and gene networks. *Annu. Rev. Genet.* **19**, 209–252.

6
Characterization of Hormone Response Elements

E. MARTINEZ and W. WAHLI

Institut de Biologie Animale, Université de Lausanne, Bâtiment de Biologie, CH-1015 Lausanne, Switzerland

6.1 Introduction

Steroid and thyroid hormones regulate the transcription of specific genes in target cells containing the corresponding intracellular receptors. These hormone receptors belong to a family of structurally and functionally related ligand-dependent transcription factors including, in addition to the steroid and thyroid hormone receptors, the vitamin D_3 and the retinoic acid receptors (see Chapters 1 to 5 and references therein). Although steroid/thyroid hormones can both activate and repress (see Chapter 8) transcription of specific protein-coding genes, it is the mechanisms of transcription activation by this class of hormones that have been the most extensively studied. In particular, functional experiments using chimaeric genes transferred into cultured cells have led to the identification of small DNA sequence motifs in the vicinity of positively hormone-regulated genes, which are required for hormone-dependent activation of transcription. These regulatory DNA sequences have been termed hormone response elements (HREs) and are the specific high affinity DNA binding sites for the activated hormone–receptor complexes.

This chapter reviews the structural and functional characteristics of the HREs identified thus far. The finding of a high structural similarity between all the consensus HRE sequences suggests that they belong to a family of transcription regulatory DNA sequences that have evolved by duplication of a common ancestor DNA sequence motif and co-evolved with their cognate receptors.

6.2 Hormone response elements are usually but not always palindromic DNA sequences

The identification of hormone response elements (HREs) has been the result of (1) gene transfer experiments into cultured cells, often using chimaeric

Table 1 Compilation of HREs found in different genes. In the column headed 'Experimental data' filled circles symbolize functionally important HREs tested for induction of transcription; open circles mean that only receptor-DNA binding data are available; open squares stand for DNA elements that are bound by the corresponding receptors but do not seem to be functionally important; and a question mark (?) indicates that no functional data are available. The co-ordinates of the elements are relative to the initiation of transcription (+1) except for the co-ordinates of the Mo MLV TRE, which are relative to the start of the U3 region of the LTR (Sap *et al.*, 1989 and references therein).

HREs	Gene	Localization	Experimental data	References
1. GREs	Mouse mammary tumour virus (MMTV)	LTR/5'-flanking: $-184/-170$ $-128/-114$ $-107/-93$ $-92/-78$	● ● ● ●	1 to 16
	Moloney murine sarcoma virus (MoMSV)	LTR/5'-flanking: $-362/-326$ $-270/-250$ $-197/-176$	□ ● ●	17, 18
	Human metallothionein IIA (hMTIIA)	5'-flanking: $-262/-248$	●	19
	Chicken lysozyme (chLYS)	5'-flanking: ~ -2125 ~ -2090 ~ -2050 ~ -1980 ~ -1930 ~ -1900 $-202/-172$ $-80/-49$	○ ○ ○ ○ ○ ○ ● ○	20–23
	Chicken vitellogenin II (chVITII)	5'-flanking: $-605/-591$	●	24, 25
	Human growth hormone (hGH)	1st intron: $+93/+112$	●	26, 27
	Rabbit uteroglobin (rbUTG)	5'-flanking: $-2709/-2684$ $-2674/-2652$ $-2643/-2619$ $-2394/-2376$	○ ○ ○ ○	28, 29
	Rat tyrosine aminotransferase (rTAT)	5'-flanking: $-5450/-5436$ $-2614/-2599$ $-2509/-2495$ $-2440/-2426$	● □ ● ●	30, 79

Table 1 Continued

HREs	Gene	Localization	Experimental data	References
	Rat tryptophan oxygenase (rTO)	5'-flanking: $-1195/-1157$ $-433/-394$	● ●	31
	Rat T-kininogen (rT-KIN)	5'-flanking: $-167/-100$	●	32
	Rat phosphoenol-pyruvate carboxykinase (rPEPCK)	5'-flanking: $-1264/-1111$ $-468/-420$	● ●	78
	Drosophila pSC7 (dSC7)	5'-flanking: $-338/-316$	○	24
	Mouse ribosomal RNA (mRIB)	Start site: $-5/+21$	○	24
2. PREs	MMTV	LTR/5'-flanking: $-184/-170$ $-128/-114$ $-107/-93$ $-92/-78$	● ● ● ●	14, 21, 33, 34, 35
	hMTIIA	5'-flanking: $-268/-245$	●	36
	chLYS	5'-flanking: ~ -2125 ~ -2090 ~ -2050 ~ -1980 ~ -1930 ~ -1900 $-202/-160$ $-80/-49$	□ □ □ ● ● ● ● ○	20–23
	rbUTG	5'-flanking: $-2709/-2690$ $-2680/-2660$ $-2642/-2620$ $-2486/-2471$ $-2426/-2402$ $-2394/-2376$	○ ○ ○ ○ ○ ○	29, 37
	rTAT	5'-flanking: $-2509/-2495$	●	38
3. AREs	MMTV	LTR/5'-flanking: $-184/-170$ $-128/-114$ $-107/-93$ $-92/-78$	● ● ● ●	14, 34, 39, 40
4. MREs	MMTV	LTR/5'-flanking: $-184/-78$	●	41, 42

Table 1 Continued

HREs	Gene	Localization		Experimental data	References
5. EREs	*Xenopus* vitellogenin (xVIT):				
	xVITA1	5'-flanking:	−335/−323	●	43–50
	xVITA2	5'-flanking:	−331/−319	●	
			−311/−299	●	
	xVITB1	5'-flanking:	−555/−543	□	
			−334/−322	●	
			−314/−302	●	
	xVITB2	5'-flanking:	−335/−323	●	
			−315/−303	●	
	chVITII	5'-flanking:	−627/−615	●	25, 51
			−335/−323	●	
			−299/−286	●	
	Chicken apo-VLDLII (ch apo-VLDLII)	5'-flanking:	−221/−209	○	52
			−177/−165	○	
	Chicken ovalbumin (chOVA)	5'-flanking:	−47/−42	●	53
	rT-KIN	5'-flanking:	−167/−100	●	32
	Rat prolactin (rPRL)	5'-flanking:	−1580/−1568	●	54
	Rat luteinizing hormone β (rLHβ)	5'-flanking:	−1631/−1078	●	57
	Human MCF-7 cells pS2	5'-flanking:	−405/−393	●	55, 56
6. TREs	Moloney murine leukaemia virus (MoMLV)	LTR:	+331/+351	●	80
	Rat growth hormone (rGH)	5'-flanking:	−189/−179 ⎫ −178/−167 ⎭	●	58–62
			−162/−154 ⎫ −153/−142 ⎭	●	
			−61/−53 ⎫ −52/−41 ⎭	●	
			−23/−12	●	
		3rd intron:	+1342/+1390	●	81
	Synthetic palindromic TRE oligonucleotide:		G17-2	●	63
	rPRL	5'-flanking:	−1555/−1544	●	64

Table 1 Continued

HREs	Gene	Localization	Experimental data	References
	Rat α-myosin heavy chain (r α-Mhc)	5′-flanking: −1700/−972 −612/−374 −161/−71	● ● ●	65–67
7. RREs	rGH	5′-flanking: −178/−167	●	68, 69
	r α-Mhc	5′-flanking: −163/−81	●	70
	Synthetic palindromic TRE oligonucleotide:	G17-2	●	68
	Mouse laminin B1 (mLB1)	5′-flanking: −477/−432	●	77
	Human retinoic acid receptor β (hRAR β)	5′-flanking: −59/−33	●	82
8. VDRE	Human osteocalcin (hOS)	5′-flanking: −513/−493	●	71, 72, 73
9. EcdREs	*Drosophila* Hsp27 (dHsp27)	5′-flanking: −551/−529	●	74
	dHsp23	5′-flanking: −202/−180 −239/−217	● ●	75
	dHsp22	5′-flanking: −194/−134	?	76

References: 1, Buetti and Diggelmann (1981); 2, Huang et al. (1981); 3, Lee et al. (1981); 4, Payvar et al. (1981); 5, Geisse et al. (1982); 6, Chandler et al. (1983); 7, Payvar et al. (1983); 8, Scheidereit et al. (1983); 9, Pfahl et al. (1983); 10, Hynes et al. (1983); 11, Scheidereit and Beato (1984); 12, Kühnel et al. (1986); 13, Buetti and Kühnel (1986); 14, Ham et al. (1988); 15, Chalepakis et al. (1988a); 16, Buetti et al. (1989); 17, Miksicek et al. (1986); 18, DeFranco and Yamamoto (1986); 19, Karin et al. (1984); 20, Renkawitz et al. (1984); 21, von der Ahe et al. (1985); 22, von der Ahe et al. (1986); 23, Hecht et al. (1988); 24, Scheidereit et al. (1986); 25, Ankenbauer et al. (1988); 26, Moore et al. (1985); 27, Slater et al. (1985); 28, Cato et al. (1984); 29, K. Jantzen et al. (1987); 30, H. M. Jantzen et al. (1987); 31, Danesch et al. (1987); 32, Anderson and Lingrel (1989); 33, Cato et al. (1986); 34, Otten et al. (1988); 35, Chalepakis et al. (1988b); 36, Slater et al. (1988); 37, Bailly et al. (1986); 38, Strähle et al. (1987); 39, Cato et al. (1987); 40, Darbre et al. (1986); 41, Arriza et al. (1987); 42, Cato and Weinmann (1988); 43, Klein-Hitpass et al. (1986); 44, Seiler-Tuyns et al. (1986); 45, Martinez et al. (1987); 46, ten Heggeler-Bordier et al. (1987); 47, Klein-Hitpass et al. (1988a); 48, Klein-Hitpass et al. (1988b); 49, Klein-Hitpass et al. (1989); 50, Martinez and Wahli (1989); 51, Burch et al. (1988); 52, Wijnholds et al. (1988); 53, Tora et al. (1988); 54, Maurer and Notides (1987); 55, Berry et al. (1989); 56, Kumar and Chambon (1988); 57, Shupnik et al. (1989); 58, Glass et al. (1987); 59, Koenig et al. (1987); 60, Wight et al. (1988); 61, Brent et al. (1989); 62, Norman et al. (1989); 63, Glass et al. (1988); 64, Day and Maurer (1989); 65, Markham et al. (1987); 66, Gustafson et al. (1987); 67, Izumo and Mahdavi (1988); 68, Umesono et al. (1988); 69, Bedo et al. (1989); 70, Graupner et al. (1989); 71, McDonnell et al. (1989); 72, Kerner et al. (1989); 73, Morrison et al. (1989); 74, Riddihough and Pelham (1987); 75, Mestril et al. (1986); 76, Klemenz and Gehring (1986); 77, Vasios et al. (1989); 78, Petersen et al. (1988); 79, Grange et al. (1989); 80, Sap et al. (1989); 81, Sap et al. (1990); 82, de Thé et al. (1990).

expression vectors composed of regulatory DNA sequences fused to hormone-insensitive reporter genes, and (2) *in vitro* protein–DNA interaction analyses using partially purified hormone receptors.

HREs were first identified in the promoter region of viral and cellular genes whose transcription is induced by glucocorticoids. The glucocorticoid response elements (GREs) are now well characterized in different genes (Table 1) and a consensus sequence has been derived from all the GREs identified thus far (Fig. 1A). This consensus GRE is a 15 base pair (bp) imperfect palindromic DNA sequence (Scheidereit *et al.*, 1986; H. M. Jantzen *et al.*, 1987; Fig. 1A). Note that in order to increase sequence similarity between HREs, we have always used the GRE consensus sequence 5'-AG(A/G)ACANNNTGTACC-3', corresponding to the DNA strand which is complementary to the consensus usually shown, i.e. 5'-GGTACANNNTGT(T/C)CT-3'. Surprisingly, in all the genes tested, the GREs always coincide with DNA sequences which are also required for progesterone, androgen and mineralocorticoid regulation of transcription (Table 1 and references therein). Thus, the nucleotide sequence for the progesterone response element (PRE), the androgen response element (ARE) and the mineralocorticoid response element (MRE) cannot be discriminated from the GRE consensus sequence (Fig. 1A).

Similarly, oestrogen response elements (EREs), first identified in the 5'-flanking region of the *Xenopus* vitellogenin genes (Klein-Hitpass *et al.*, 1986; Seiler-Tuyns *et al.*, 1986; Martinez *et al.*, 1987), have also been found in the promoter region of chicken, rat and human genes (Table 1). The ERE consensus sequence derived from functional analyses (Fig. 1B) matches perfectly the proposed ERE consensus sequence that was obtained by DNA sequence comparison of the promoter region of six oestrogen-regulated genes (Walker *et al.*, 1984). The minimal functional ERE consensus sequence is a 13 bp perfect palindromic DNA element (Fig. 1B) which is structurally related to the GRE but functionally distinct from it (see Section 6.5).

More recently, thyroid hormone response elements (TREs) and retinoic acid response elements (RREs) have been identified in several different genes (Table 1 and references therein). However, since only few TREs and RREs have been identified thus far, it is difficult to define a consensus sequence for these elements. Furthermore, it appears that TREs as well as RREs can be formed of either inverted or direct repeats. A compilation of all known TRE sequences is presented in Fig. 1C. From these data two possible consensus TRE sequences can be proposed. The first is a 12 bp palindromic DNA element very similar to the ERE but missing the 3 bp spacer between the two arms of the palindrome present in the latter. However, the 3' arm of this palindromic sequence is not well conserved among different TREs (Fig. 1C).

The second is a direct repeat of the more conserved arm of the palindromic TRE described above (see Fig. 1C). At the moment, RREs seem also to function as TREs, except for that in the murine laminin B1 (mLB1) gene (Table 1 and references therein). Thus, two consensus RREs are also possible that are indistinguishable from the two tentative consensus TREs proposed above (see Fig. 1D). The fact that no perfect palindromic natural TRE nor RRE has been found yet, together with the observation that the natural responsive sequences for thyroid hormones and retinoic acid characterized thus far are often formed by reiterated copies of one arm of the above-proposed consensus palindromic TRE and RRE (Fig. 1C and D), suggests that co-operative interactions between receptors and other factors could play a major role in the discriminating mechanism that allows the thyroid hormone and the retinoic acid receptors to regulate their specific target genes (see Section 6.5).

Functional analyses of the 5'-flanking region of the human osteocalcin (hOS) gene have allowed the identification of the first vitamin D_3 response element (VDRE) which contains a core sequence forming a double palindromic DNA element (Morrison et al., 1989). The arms of the palindromic VDRE core sequence are separated by one bp spacer and show striking similarity with those of the consensus ERE (Fig. 1E; see also Fig. 4).

The ecdysteroid moulting hormones regulate gene expression in insects and crustacea and functional ecdysteroid response elements (EcdREs) have been identified in the promoter region of the *Drosophila* Hsp27 (dHsp27) and Hsp23 (dHsp23) genes. In addition, possible ecdysone-responsive DNA sequences have also been pointed out in the *Drosophila* Hsp22 (dHsp22) gene promoter (Table 1 and references therein). Although these different ecdysone regulatory DNA sequences are poorly conserved, a consensus EcdRE core sequence formed by 11 to 13 bp is obtained that shows imperfect palindromicity (Fig. 1F). Furthermore, in Section 6.5 we will present new results that demonstrate an increased ecdysone responsiveness of a synthetic EcdRE having a perfect palindromic core sequence compared to the wild-type dHsp27 imperfect palindromic EcdRE.

The fact that most of the consensus HREs that have been characterized so far present a palindromic structure is in agreement with the finding that oestrogen and glucocorticoid receptors bind as dimers to their respective response elements (Kumar and Chambon, 1988; Tsai et al., 1988; Fawell et al., 1990) and suggests that this could also be the case for the other hormone receptors of the steroid/thyroid hormone receptor superfamily. Interestingly, a recent report presents evidence that the human thyroid hormone receptor can form a heterodimer with the human retinoic acid receptor that binds to a subset of TREs/RREs and confers either positive or negative effect on transcription, depending on the target HRE. These results suggest that a

A.

Consensus GRE   5'- A G $\begin{smallmatrix}A\\G\end{smallmatrix}$ A C A N N N T G T A C C - 3'
(PRE, ARE, MRE)

B.

Consensus ERE   5'- N_A G G T C A N_C N N_G T G A C C N_T - 3'

C.

```
TREs  MoMLV      -349   G G G T C A T T T C A G     -338
      rGH       +1373   A G G T C A C T A G C T    +1384
       "         -189   A G G T A A G A T C A .     -179
       "         -178   G G G A C G T G A C C G     -167
       "         -162   A G A A G C A G T G . . .   -154
       "         -153   G G G A C G C G A T G T     -142
       "          -61   G G A G C C T T G . . .      -53
       "          -52   G G G T C G A G G A A A      -41
       "          -23   A G G G C A T G C A A G      -12
      Synth. G17-2     A G G T C A T G A C C T
      rPRL      -1555   G G G T C A G A A G A G    -1544
      r α-Mhc    -594   G G G A A A C T . . . .     -587
       "         -586   G A G T C A T G C A C C     -575
       "         -137   T G G A G G T G A C A G     -148

Consensus TRE  5'-  G G G T C A T G A C A G  -3'
                    A       A T     T A C T
               or 5'- ←——————————— -3'  in direct repeats
```

D.

```
RREs  hRARβ      -52   G G T T C A C C G A A .     -43
       "         -42   A G T T C A C T C G C A     -31
      mLB1      -476   C A G A C A G G T . . .    -468
       "        -467   . . . . . T G A C C C      -462
       "        -453   A G G G C T T A A C C T    -442
       "        -441   . . A G C T C A C C T      -433
      rGH       -178   G G G A C G T G A C C G    -167
      r α-Mhc   -137   T G G A G G T G A C A G    -148
      Synth. G17-2    A G G T C A T G A C C T

Consensus RRE  5'-  A G G A C A T G A C C T  -3'
                    G     T               G
               or 5'- ←——————————— -3'  in direct repeats
```

E.

VDRE   hOS   -513   T T G G T G A C T C A C C G G G T G A A C   -493

F.

```
EcdREs  dHsp27  -551   G A C A A G G G T T C A A T G C A C T T G T C   -529
        dHsp23  -239   G G A G A T T G C T C A T T T T C C A T A G C   -217
          "     -202   C A A T G G C A G A T A A T G C G T A A T T G   -180
      ? ┌dHsp22 -176   C G C T A A A G A A C A G T G A A C A A C C C   -154
        └  "   -157   A C C C C T A A C T A A A T G C C A T T G C C   -135
```

Consensus "core EcdRE"   5'- N G N T C A A T G C $\begin{smallmatrix}C\\A\end{smallmatrix}$ C $\begin{smallmatrix}A\\T\end{smallmatrix}$ -3' more refined control of hormone-dependent transcription can be achieved through HREs interacting with different receptor combinations (Glass *et al.*, 1989; see Chapter 16.3).

6.3 Hormone response elements as hormone-inducible enhancers

The fact that HREs are naturally found in both possible orientations, as well as at different positions relative to the transcription initiation site (see Table 1), suggests that they function as hormone-inducible transcriptional enhancers. This was directly demonstrated in gene transfer experiments using chimaeric genes, by showing that HRE-containing DNA fragments confer hormone-dependent transcriptional activation to heterologous hormone insensitive promoters in a relatively position- and orientation-independent manner. For instance, such studies are particularly well described: (1) for GRE/PRE-containing DNA fragments derived from viral LTRs (Chandler *et al.*, 1983; Majors and Varmus, 1983; Ponta *et al.*, 1985; Cato *et al.*, 1986; DeFranco and Yamamoto, 1986; Miksicek *et al.*, 1986; Ham *et al.*, 1988), and from cellular genes (Karin *et al.*, 1984; Slater *et al.*, 1985; Danesch *et al.*, 1987; H. M. Jantzen *et al.*, 1987; Hecht *et al.*, 1988); and (2) for ERE-containing promoter DNA fragments from cellular genes (Klein-Hitpass *et al.*, 1986; Seiler-Tuyns *et al.*, 1986; Martinez *et al.*, 1987; Maurer and Notides, 1987; Burch *et al.*, 1988; Klein-Hitpass *et al.*, 1988a; Berry *et al.*, 1989; Shupnik *et al.*, 1989).

Strong enhancers, characterized by their capacity to activate transcription when located at long distances from promoters, are known to be modular units composed of several regulatory elements which are DNA-binding sites for transcription factors. Moreover, their strength has been shown to depend on the number of regulatory elements that they are composed of (Herr and Gluzman, 1985; Zenke *et al.*, 1986; Maniatis *et al.*, 1987; and references

---

Fig. 1 (Opposite) Consensus HRE sequences are mostly palindromic DNA structures. Abbreviations are as in Table 1. Inverted arrows show the palindromes. The black dots mark the centres of symmetry. N means that any nucleotide is found in the corresponding position. In the consensus ERE (panel B), $N_{A,G,T \text{ or } C}$ indicate that the nucleotide in the subscript is the one which is the most represented at that position (found in about 50% of EREs) but it can be replaced by any other nucleotide without any apparent functional difference. Points are gaps. Negative numbers are the natural co-ordinates of the sequences relative to the transcription start site ($+1$), taken from the corresponding references listed in Table 1. Two possible consensus TRE and RRE sequences (i.e. a palindromic and a direct repeat sequence) are given in panels C and D respectively (see text). The question mark (?, in panel F) indicates that the two sequences from the dHsp22 gene have not been tested functionally.

therein). This is also true for strong hormone-inducible enhancers, also called hormone-responsive units (HRUs; Klein-Hitpass et al., 1988b), which are formed by several HREs or minimally by one HRE associated with one or several other flanking regulatory DNA elements, since only HRUs, but not isolated single HREs, have been shown to significantly activate transcription over long distances (see below).

6.4 Synergistic action of hormone response elements

In natural genes, HREs are often found clustered, forming hormone-responsive units (HRUs) that behave as strong inducible enhancers. Interestingly, in many hormone-regulated genes, highly active HRUs are formed by a combination of several imperfect HREs that are poorly active when tested separately. These HREs are rarely perfect consensus responsive elements, but generally correspond to imperfect versions deviating from the consensus HRE sequences. Analyses of the ability of different single natural HREs to confer hormone-dependent activation to closely adjacent heterologous promoters have shown that poorly conserved HREs have a low or no activity at all, but that HREs which are more conserved, or which match perfectly with their corresponding consensus sequences, are by themselves highly active. Moreover, it has been shown that the transcription stimulatory activity of the HREs tested so far correlates with the *in vitro* binding affinity of the hormone receptors for these regulatory DNA sequences. These types of experiments have demonstrated that HREs can act in synergy to confer high hormone-dependent transcription activation to the regulated target genes. For instance, the HRU of the MMTV LTR contains four GREs that activate transcription synergistically (Buetti and Kühnel, 1986; Cato et al., 1988a; Chalepakis et al., 1988b; Buetti et al., 1989) and two GREs located 2.5 kb upstream of the tyrosine aminotransferase gene also act in synergy (H. M. Jantzen et al., 1987). Similarly, the *Xenopus* vitellogenin B1 gene HRU is formed by two imperfect EREs that co-operate only when they are closely adjacent (Martinez et al., 1987; Klein-Hitpass et al., 1988b; Martinez and Wahli, 1989; see also Chapter 2.9) and in the chicken lysozyme gene three GREs/PREs located far upstream of the transcription start site work in a co-operative manner (Hecht et al., 1988). Furthermore, different associated HREs, acting through the binding of their respective hormone receptors, also activate transcription synergistically. This has been shown for the distal chicken vitellogenin II (chVITII) gene HRU that is formed by an ERE closely adjacent to a GRE/PRE, which confers synergistic activation of transcription to chimaeric genes, in response to the combination of both

6. HORMONE RESPONSE ELEMENTS 135

hormones, oestrogen and glucocorticoid (Ankenbauer et al., 1988) or oestrogen and progesterone (Cato et al., 1988b). However, in contrast to the ERE there is no evidence yet that this GRE/PRE is functionally important for the *in vivo* regulation of the chVITII gene in the liver. Moreover, it has been reported that many regulatory DNA sequences, recognized by specific *trans*-acting factors, including cell-specific transcription factors, can also act in synergy in combination with closely adjacent HREs (Schüle et al., 1988; Strähle et al., 1988; Tora et al., 1988). In some cases, these functional data have been complemented by *in vitro* receptor-DNA binding studies. For instance, it has been demonstrated that a pair of GREs/PREs that function synergistically *in vivo*, are bound *in vitro* by the progesterone receptor (Tsai et al., 1989) or the glucocorticoid receptor (Schmid et al., 1989) in a co-operative manner. Similarly, it has been shown that the *in vivo* synergistic action of the two imperfect EREs of the vitellogenin B1 gene oestrogen-responsive unit correlates with the co-operative binding *in vitro* of oestrogen receptor molecules to these elements (Martinez and Wahli, 1989). At the moment, it is not known whether co-operative DNA binding of receptors with other transcription factors also contributes to the observed functional synergism between HREs and different closely adjacent regulatory DNA elements (see Chapters 7.3–7.5).

Interestingly, some strong HREs which are active when present as a single copy close to a basal promoter (basal transcription elements + TATA box + transcription start site) or to a core promoter (TATA box + transcription start site) can no longer activate transcription, or have a much lower activity, when they are placed at a much longer distance from the transcription initiation site. However, at such a distant position, two adjacent HREs, or one HRE closely associated with a different transcription factor binding site, can function synergistically and constitute an HRU which functions as a strongly active inducible enhancer (Strähle et al., 1988). Moreover, the maximal enhancer activity of such HRUs, when present at a long distance from the transcription initiation site, is highly dependent on the presence of basal transcription elements present just upstream of the TATA box. For instance, the NF-I binding site present in the basal promoter of the *Xenopus* vitellogenin B1 gene (Corthésy et al., 1989) acts in synergy with the oestrogen-responsive unit located 220 bp upstream of it, to confer maximal transcription activity in the presence of hormone *in vitro* as well as *in vivo* (Wahli et al., 1989; and references therein). Similar observations have also been made for the glucocorticoid regulation of MMTV transcription (Buetti and Kühnel, 1986; Miksicek et al., 1987; Cato et al., 1988a; Buetti et al., 1989; and references therein).

In summary, two types of hormone-dependent synergistic transcription activation mechanisms can be considered. The first is the synergistic interactions between hormone receptors or hormone receptors and other transcription factors at a hormone-responsive unit (HRU) functioning as an inducible enhancer. In some cases, this type of synergy has been shown to result from co-operative DNA binding of hormone receptors and to depend on the distance separating the adjacent HREs forming the HRU, suggesting that protein–protein contacts are involved. This co-operative interaction leads to an increased stability of the receptor–enhancer DNA complexes (Martinez and Wahli, 1989; Schmid et al., 1989; Tsai et al., 1989). Alternatively, non-DNA-binding protein(s) may interact co-operatively with the receptor–DNA complex to stabilize it.

The second type of mechanism is involved in the synergy between a distant HRU and basal transcription DNA elements located close to the core promoter. In this case, synergism may result from protein–protein interactions between receptors, or from enhancer binding factors and basal transcription factors, bringing the receptors or enhancer factors at the proximity of the core promoter through looping-out of the intervening DNA (Ptashne, 1988), thus allowing a more efficient formation of the transcription initiation complex. This is also consistent with the observation that, at a distance from the promoter, several adjacent HREs are needed to have maximal enhancer activity (see above), possibly as a result of an increased stability of the interactions between the receptors and basal transcription factors (see Chapters 7.3–7.5).

It is quite likely that *in vivo* both types of mechanisms combine to achieve maximal synergistic induction of gene expression. Furthermore, the synergistic action of hormone receptors with tissue-specific transcription factors may also contribute to the targeting of hormone-dependent specific gene activation in defined cell types (Tora et al., 1988).

6.5 HREs constitute a family of closely related regulatory DNA sequences: the problem of specificity

As described under Section 6.2, most of the consensus HREs identified thus far have a palindromic structure consistent with the fact that some of them (i.e. the GRE and the ERE) have been shown to be the binding sites for receptor dimers. Moreover, all these hormone-responsive consensus sequences present a high degree of sequence similarity and sometimes are even indistinguishable. For instance, a GRE acts also as a PRE, as an ARE and an MRE (see Section 6.2, Fig. 1A and Chapter 9.4). Similarly, TREs and RREs are in most cases indistinguishable (see Section 6.2, Fig. 1C and D and Chapter 9.4). Interestingly, the consensus ERE sequence shows striking

6. HORMONE RESPONSE ELEMENTS

similarities to both the GRE and the TRE consensus sequences (Fig. 1A, B and C). Furthermore, mutagenesis experiments have allowed definition of a strongly active TRE as a consensus perfect palindromic ERE lacking the 3 bp spacer between the arms of the palindrome. This perfect palindromic TRE (G17-2 in Fig. 1C) has a three- to four-fold higher affinity for the thyroid hormone receptor than the wild-type rat growth hormone gene TRE that is located at position −178 bp from the transcription start site (Glass et al., 1988). In addition, it has been shown that the thyroid hormone receptor binds to the consensus ERE with a high affinity. However, a single TRE does not confer oestrogen regulation and reciprocally a consensus ERE does not confer thyroid hormone regulation to a heterologous promoter (Glass et al., 1988; see Chapter 4.5). It has also been demonstrated that, in spite of the similarity between the consensus GRE and ERE sequences, each of these elements responds specifically to glucocorticoid and oestrogen, respectively. This specific response correlates with the specific binding of the corresponding receptor to these sites (Kumar and Chambon, 1988; Klein-Hitpass et al., 1989; Martinez and Wahli, 1989; and references therein). However, two base pair substitutions at symmetric positions in a perfect palindromic ERE are sufficient to convert this element into a functional GRE (Fig. 2; Klock et al., 1987; Martinez et al., 1987).

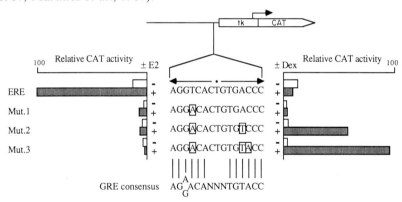

Fig. 2 Two base pair substitutions convert an ERE into a functional GRE. Human adenocarcinoma MCF-7 cells have been transiently transfected with chimaeric tk-CAT expression vectors having the indicated HREs inserted upstream of the tk promoter (tk is the HSV thymidine kinase promoter, positions −105 to +51; CAT is the bacterial chloramphenicol acetyltransferase gene coding region). The relative CAT activities present in extracts from transfected cells cultured in the presence (+) or in the absence (−) of hormone (E2 is oestradiol and Dex is dexamethasone) are shown for each transfected construct by hatched and open bars, respectively (the CAT activity conferred by the ERE in the presence of E2 was taken arbitrarily as 100). Mutated nucleotides relative to the ERE sequence are boxed in the mutants (Mut.1, Mut.2 and Mut.3). Mut.3 matches perfectly the consensus GRE sequence. Drawn from results of Martinez et al. (1987).

Surprisingly, an ecdysone response element (EcdRE) that has been identified in the 5'-flanking region of the *Drosophila* Hsp27 gene (Riddihough and Pelham, 1987), presents DNA sequence similarities with the vertebrate HREs and more particularly with the ERE (see Fig. 1F and Fig. 3). In order to test whether the EcdRE is functionally related to the ERE, the *Drosophila* Hsp27 EcdRE has been mutated to increase its similarity with the ERE. Wild-type and mutated EcdREs were then cloned upstream of a chimaeric tk-CAT reporter gene and subsequently transferred into ecdysone-responsive *Drosophila* Schneider cells to test if they confer ecdysone responsiveness to the heterologous thymidine kinase (tk) promoter. Alternatively, oestrogen responsiveness of the same constructs was analysed by cotransfecting a human oestrogen receptor expression vector (HE0: Green *et al.*, 1986). The results (Fig. 3; Martinez, 1989; Martinez *et al.*, 1991) show that the wild-type *Drosophila* EcdRE induces two- to three-fold the tk-CAT gene expression in response to ecdysone, but does not stimulate transcription after oestrogen administration. Substitution of three base pairs in the core imperfect palindromic Hsp27 EcdRE (mt.3 in Fig. 3), allows the formation of a perfect palindromic core motif that is identical to a consensus ERE, but has only one base pair, instead of three, in the centre of the palindrome. Interestingly, these mutations increase the ecdysone response about three-fold (Fig. 3, mt.3) compared to the wild-type Hsp27 EcdRE. This mutant which confers a 6.5-fold induction of tk-CAT gene expression in response to ecdysone, does not respond to oestrogen. However, the insertion of two extra nucleotides in the middle of the element abolishes ecdysone response and converts it into an active ERE (Fig. 3, mt.3 + Δ). These experiments demonstrate that ecdysteroid regulation of gene expression in insects is achieved through EcdREs that are highly similar to the vertebrate ERE, but are distinct by the fact that the core EcdRE has only one base pair separating the two arms of the palindrome instead of three as for the ERE.

The chicken ovalbumin upstream promoter (COUP) element is a transcription regulatory DNA sequence that is also found in the 5'-flanking region of the rat insulin gene and is the DNA-binding site for the ubiquitous COUP transcription factor (COUP-TF: Tsai *et al.*, 1987; Wang *et al.*, 1987; Hwung *et al.*, 1988a,b). The COUP-TF cDNA has been cloned and its deduced amino acid sequence has revealed that it is a member of the nuclear receptor superfamily for which the ligand, if any, has not yet been identified. The region of highest similarity with the other members of the family is in the putative DNA-binding domain (Wang *et al.*, 1989). Consistent with this finding, the DNA sequence of the COUP element also shows strong similarities with the ERE consensus sequence. However, the COUP elements that have been identified (Hwung *et al.*, 1988b; and references therein) do not have a palindromic structure, but rather are formed by a tandem

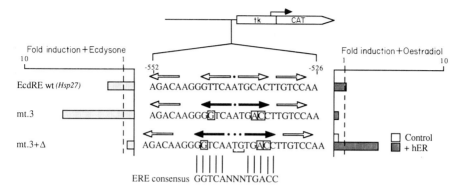

Fig. 3 The insect EcdRE is related to the vertebrate ERE. Chimaeric tk-CAT expression vectors having either the *Drosophila* Hsp27 EcdRE (EcdRE wt) or mutants of it (mt.3 and mt.3 + Δ) inserted upstream of the tk promoter (see Fig.2), have been transiently transfected into the ecdysone-responsive *Drosophila* Schneider cells SL-3. The cells were then cultured in the presence or in the absence of either ecdysone or oestradiol. The human oestrogen receptor (hER) cDNA expression vector (HE0, Green *et al.*, 1986) or a control vector (pKCR2, Breathnach and Harris, 1983) have been cotransfected with the reporter genes in cells that were tested for oestradiol induction. The induction factors were calculated from the CAT activities present in extracts of transfected cells (CAT activity in the presence of hormone divided by CAT activity in the absence of hormone) and are represented by hatched bars. The open bar shows the oestradiol induction factor conferred by the mt.3 + Δ construct in SL-3 in the absence of oestrogen receptor (Control). Negative numbers above the EcdRE wt sequence are the natural co-ordinates of this element in the 5′-flanking region of the Hsp27 gene. Arrows indicate the arms of palindromic sequence elements similar to those of the ERE. Open arrows are imperfect ERE arms and black arrows are arms identical to those of the ERE consensus sequence (ERE consensus). Mutated bases relative to the EcdRE wt sequence are boxed. In mt.3 + Δ, the two inserted bases are underlined with a bracket. Drawn from results of Martinez (1989).

duplication of a half-ERE (Fig. 4) (see Chapter 5.2). It has been reported that the regulatory region II of the major histocompatibility complex class I genes (MHCI-RII) is specifically bound by a protein (i.e. H-2RIIBP) that belongs to the superfamily of nuclear hormone receptors (Hamada *et al.*, 1989). For H-2RIIBP, as for the COUP-TF (see above), no ligand has yet been identified. Interestingly, the MHCI-RII enhancer element bound by this protein contains a perfect half-palindromic ERE sequence (see MHCI-RII in Fig. 4). In addition, H-2RIIBP also binds to a perfect palindromic ERE (Hamada *et al.*, 1989), suggesting that this protein could have an important transcription regulatory function not only for MHC class I genes, but also for oestrogen-regulated genes.

The alignment of all the functional palindromic consensus HRE sequences that have been characterized, along with the COUP and MHCI-RII elements, clearly shows that all these regulatory DNA sequences are related (Fig. 4). This is consistent with the observation that the DNA-binding domain of the members belonging to the nuclear receptor superfamily is highly conserved (Evans, 1988; Green and Chambon, 1988). However, from the functional studies described above, some important differences exist that allow the receptors to discriminate between different HRE sequences. Considering a common consensus hormone response element, matching with most of the consensus HREs described in Fig. 4, that has the following palindromic structure: 5'-AGGN$_a$CA N$_{(x)}$ TGN$_b$CCT-3', it becomes first apparent that HREs can differ at a defined symmetric nucleotide position (N$_a$ and N$_b$) within the arms of the palindrome. For instance, in HREs where 3 bp separate the arms of the palindrome, if N$_a$ is an A and N$_b$ is a T, the response element is a GRE, thus most likely also a PRE, an ARE and an MRE. However, if it is the inverse, i.e. N$_a$ is a T and N$_b$ is an A, the element functions as an ERE (Figs 2 and 4). The second main difference between HREs is in the spacing between the arms of the palindromic structure (N$_{(x)}$). This allows a functional discrimination by the different receptors of HREs having the same recognition sequences in the stem but different loop sizes, as for instance the ERE (x is 3), the EcdRE (x is 1) and the TRE (x is 0; see Fig. 4). This also suggests, that nuclear receptors use at least two different mechanisms to recognize their specific HREs and activate transcription (see below).

From all these observations it is possible to classify HREs into two main subfamilies according to their nucleotide sequences in the arms, i.e. the GRE subfamily (N$_a$ is an A) and the ERE/COUP subfamily (N$_a$ is a T; see above and consensus half-sites in Fig. 4). Detailed functional analyses of the DNA-binding domain of different nuclear receptors have allowed the identification of three amino acids present in the first 'zinc-finger' (Fig. 4, CI, amino acids labelled 1, 2 and 3) that determine specific target HRE sequence recognition (Danielsen *et al.*, 1989; Mader *et al.*, 1989; Umesono and Evans, 1989; see Chapters 2.4 and 3.3) and of five amino acids localized between the two N-terminal cysteines of the second 'zinc-finger' (CII) that are involved in half-HRE spacing functional discrimination (Umesono and Evans, 1989). Other amino acids present in the second 'zinc-finger' (CII) have also been shown to be important for preventing promiscuous HRE recognition (Danielsen *et al.*, 1989). The fact that receptor dimers interact with palindromic HREs (Kumar and Chambon, 1988; Tsai *et al.*, 1988; Fawell *et al.*, 1990), together with the above observations that the first 'zinc-finger' determines target HRE specificity (see also Green *et al.*, 1988), suggest that the first finger (CI) in each of the two molecules forming the dimer directly contacts one half-HRE. Interestingly, and consistent with this idea, the comparison, in the first finger (CI) of all the receptors, of the first two amino

6. HORMONE RESPONSE ELEMENTS

acids that are involved in half-HRE sequence discrimination (Fig. 4, amino acids labelled 1 and 2), reveals that they are highly conserved within two subfamilies of receptors, i.e. (1) the glucocorticoid receptor (GR) subfamily, including the progesterone receptor (PR), the androgen receptor (AR) and the mineralocorticoid receptor (MR); and (2) the oestrogen receptor (ER)/COUP-TF subfamily including the thyroid hormone receptor (TR), the retinoic acid receptor (RAR), the vitamin D_3 receptor (VD_3R) and the H-2RII binding protein (H-2RIIBP). All members of the GR subfamily have a glycine and a serine as the two first discriminating amino acids (G and S respectively, in Fig. 4), while all the receptors of the ER/COUP-TF subfamily have a glutamate and a glycine at these same positions (E and G respectively, in Fig. 4). Thus, these two subfamilies of receptors correlate with the two subfamilies of HREs that they recognize, i.e. the GRE subfamily and the ERE/COUP subfamily, suggesting, as a first HRE recognition mechanism, a direct interaction between the two conserved amino acids involved in half-HRE sequence discrimination and the single base pair that differentiates the two HRE subfamilies (see consensus half-sites in Fig. 4). If this reasoning is correct, one can predict that in the ecdysteroid receptor (EcdR), whose cloning has not been reported yet, the two first amino acids involved in half-HRE sequence discrimination are a glutamate and a glycine (E and G in positions 1 and 2 respectively, Fig. 4), since the EcdRE belongs to the ERE/COUP subfamily (Fig. 4). The third sequence-discriminating amino acid (number 3, Fig. 4), which is conserved within the GR subfamily but not within the ER/COUP-TF subfamily, is however also necessary for a maximal HRE-specific response (Mader et al., 1989; Umesomo and Evans, 1989).

A second mechanism of specific HRE recognition by the receptors allows the functional discrimination of half-HRE spacing. For instance, the ERE and the TRE that have the same nucleotide sequence in the arms of their palindromic structure differ in the spacing of their arms (Fig. 4) and confer activation of gene transcription only in response to oestrogen and thyroid hormone, respectively. However, the thyroid hormone receptor has been shown to bind with high affinity to the ERE (Glass et al., 1988). This implies that receptor binding to a HRE is not sufficient for transactivation and that it is likely that the conformation of the receptor dimer once bound to the HRE is also important. This receptor conformation might be at least partly determined by the spacing between the arms of the palindrome. The half-HRE spacing-dependent transactivation mechanism has been shown to be somehow controlled by the second finger (CII) of the receptor DNA-binding domain, which is also thought to be involved in the dimerization of the receptor. Interestingly, the five amino acids present between the two N-terminal cysteines of the second finger of the thyroid hormone receptor

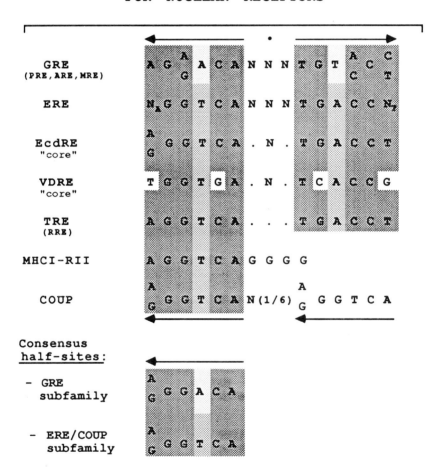

Fig. 4 (See caption on page 144.)

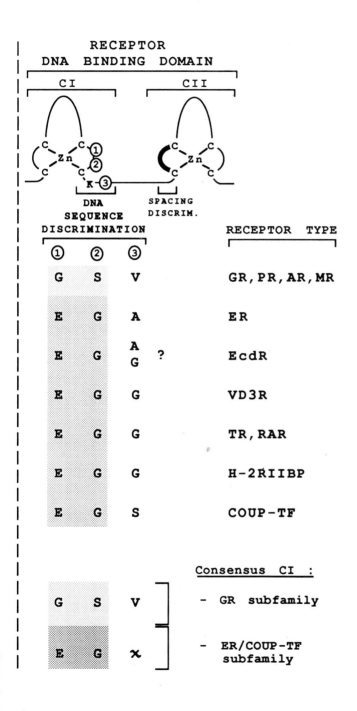

(TR), have been shown to be involved in the relief of the half-HRE spacing constraint as follows: a receptor recognizing specifically a HRE (ERE or GRE) having a 3 bp spacer between the arms of the palindrome, can also functionally recognize it in the absence of these 3 bp, when the five amino acids localized between both N-terminal cysteines of the second finger (CII) are replaced by the five corresponding amino acids of the thyroid hormone receptor whose DNA recognition site lacks any spacer sequence (Umesono and Evans, 1989).

The two mechanisms of specific HRE functional recognition by the nuclear receptors, presented above, i.e. sequence and half-palindrome spacing discrimination, are not sufficient to account for the differential *in vivo* transactivation activities conferred by the various receptors of the GR subfamily through similar or even identical HREs in different cell types (Cato *et al.*, 1986, 1988a; Chalepakis *et al.*, 1988b; Gowland and Buetti, 1989; and references therein). Thus, in addition to the fact that receptors of the GR subfamily bind to various GRE/PRE sequences with different affinities and interact differentially with their DNA sequences (von der Ahe *et al.*, 1985, 1986; K. Jantzen *et al.*, 1987; Chalepakis *et al.*, 1988b; Ham *et al.*, 1988), there is also evidence that differential co-operative interactions with other transcription factors are also required for specific target gene regulation by the members of this subfamily (Buetti and Kühnel, 1986;

Fig. 4   HREs are closely related and form two subfamilies of response elements that correlate with two subfamilies of nuclear receptors. On the left-hand side are represented the consensus functional responsive elements for nuclear receptors. The functional GRE consensus sequence shown takes account of the results presented in Fig. 2 (Martinez *et al.*, 1987) and of those of Klock *et al.* (1987). For the VDRE sequence only one functional element has been described so far (see Table 1 and Fig. 1). N is any nucleotide and points represent gaps. The DNA sequence of the MHCI-RII element and the amino acid sequence of the H-2RIIBP putative DNA-binding domain (see text) are from Hamada *et al.* (1989). In the COUP element, N(1/6) means that either 1 or 6 nucleotides are present in the chicken ovalbumin and in the rat insulin genes, respectively (Hwung *et al.*, 1988b). The conserved nucleotide positions are shaded. Inverted repeated sequences are shown by inverted arrows. The consensus half-site HREs characterizing the two element subfamilies are also shown (see also text for more details). On the right-hand side, upper level, a schematic drawing shows the two 'zinc fingers' (putative structures: CI and CII) present in the DNA-binding domain of the nuclear receptors. The two amino acids involved in half-HRE sequence discrimination, situated between the two cysteines (C) of the C-terminal part of the first finger (CI), are numbered 1 and 2. The third half-HRE sequence discriminating amino acid (numbered 3) is also shown (see text for references). The N-terminal region of the CII finger involved in half-HRE spacing discrimination is draw in bold and is indicated with a bracket (see text). Below the drawing, the amino acids present at positions 1, 2 and 3 in different nuclear receptors are listed (for references, see Baker *et al.*, 1988; Chang *et al.*, 1988; Evans, 1988; McDonnell *et al.*, 1989) as well as the derived consensus amino acids at those positions for the two receptor subfamilies (Consensus CI). The x is for any amino acid. The question mark (?) indicates our predicted amino acid sequence of the EcdR that is not known yet. See text for more details.

Miksicek *et al.*, 1987; Cato *et al.*, 1988a; Buetti *et al.*, 1989; Gowland and Buetti, 1989). Similarly, TREs and RREs are highly related if not identical DNA sequences (see Fig. 1C and D) and most of the RREs tested so far also function as TREs (see Section 6.2). However, the RRE(s)-containing mouse laminin B1 gene promoter is not regulated by thyroid hormones (Vasios *et al.*, 1989). This also suggests that in a natural promoter context receptor–receptor interactions and/or receptor interactions with other factors play a major role in hormone-specific gene transcription activation. Finally, other mechanisms of specific functional HRE recognition by the right nuclear receptor certainly take place *in vivo*. These could possibly involve receptor-specific derepression of target gene promoters, cell-specific expression and/or post-translation modifications of receptors and differential receptor-HRE interactions in genes having a particular chromatin organization (see Chapters 9.6 and 10).

6.6 Conclusions

The striking conservation of all the characterized HRE sequences from insects to man, and the high similarity between the DNA-binding domains of the nuclear receptors, suggest that the different HREs are evolutionary related and that they have co-evolved with the receptor proteins that they bind. From the comparison of all the HRE consensus sequences presented in Fig. 4, it is tempting to propose that all the HREs have evolved from a common ancestor regulatory DNa sequence motif having the following consensus sequence: 5'-(A/G)GGTCA-3', which corresponds to the site recognized by one receptor molecule. If this were the case, tandem and inverted duplications of this motif would have given rise to COUP-like elements and to the different palindromic elements of the ERE subfamily, respectively (Fig. 4). Subsequently, the members of the GRE subfamily would have further evolved through point mutations in the arms of the ERE.

The HREs that have been reviewed in this chapter correspond to positively-acting hormone regulatory elements. Although it is well known that HREs act through the binding of their respective nuclear receptors, which contain specific transactivation domains (see Chapter 2.10), very little is yet known about the receptor-mediated transactivation mechanisms. In addition, a few negatively-acting HREs (nHREs) have also been identified. These elements usually do not share extensive similarity with the positively-acting consensus HREs that have been described here, but nevertheless act through the binding of their cognate receptors (see Chapter 8; and references therein). Thus, it will be interesting to see whether different nHREs acting through the binding of nuclear hormone receptors also share a high degree of similarity and what structurally and functionally differentiates them from positively-acting HREs.

Acknowledgements

We thank Drs G. Krey and N. Mermod for critically reading the manuscript and F. Givel for technical assistance. The work done in the authors' laboratory was supported by the Etat de Vaud and the Swiss National Science Foundation.

References

Anderson, K. P. and Lingrel, J. B. (1989). Glucocorticoid and estrogen regulation of a rat T-kininogen gene. *Nucl. Acids Res.* **17**, 2835–2848.
Ankenbauer, W., Strähle, U. and Schütz, G. (1988). Synergistic action of glucocorticoid and estradiol responsive elements. *Proc. Natl. Acad. Sci. USA* **85**, 7526–7530.
Arriza, J. L., Weinberger, C., Cerdli, G., Glaser, T. M., Handelin, B. L., Housman, D. E. and Evans, R. M. (1987). Cloning of human mineralocorticoid receptor complementary DNA: structural and functional kinship with the glucocorticoid receptor. *Science* **237**, 268–275.
Bailly, A., Le Page, C., Rauch, M. and Milgrom, E. (1986). Sequence-specific DNA binding of the progesterone receptor to the uteroglobin gene: effects of hormone, antihormone and receptor phosphorylation. *EMBO J.* **5**, 3235–3241.
Baker, A. R., McDonnell, D. P., Hughes, M., Crisp, T. M., Mangelsdorf, D. J., Haussler, M. R., Pike, J. W., Shine, J. and O'Malley, B. W. (1988). Cloning and expression of full-length cDNA encoding human vitamin D receptor. *Proc. Natl. Acad. Sci. USA* **85**, 3294–3298.
Bedo, G., Santisteban, P. and Aranda, A. (1989). Retinoic acid regulates growth hormone gene expression. *Nature* **339**, 231–234.
Berry, M., Nunez, A.-M. and Chambon, P. (1989). Estrogen-responsive element of the human pS2 gene is an imperfectly palindromic sequence. *Proc. Natl. Acad. Sci. USA* **86**, 1218–1222.
Breathnach, R. and Harris, B. A. (1983). Plasmids for the cloning and expression of full-length double-stranded cDNAs under control of the SV40 early or late gene promoter. *Nucl. Acids Res.* **11**, 7119–7136.
Brent, G. A., Larsen, P. R., Harney, J. W., Koenig, R. J. and Moore, D. D. (1989). Functional characterization of the rat growth hormone receptor elements required for induction by thyroid hormone with and without a co-transfected beta type thyroid hormone receptor. *J. Biol. Chem.* **264**, 178–182.
Buetti, E. and Diggelmann, H. (1981). Cloned mouse mammary tumor virus DNA is biologically active in transfected mouse cells and its expression is stimulated by glucocorticoid hormones. *Cell* **23**, 335–345.
Buetti, E. and Kühnel, B. (1986). Distinct sequence elements involved in the glucocorticoid regulation of the mouse mammary tumor virus promoter identified by linker scanning mutagenesis. *J. Mol. Biol.* **190**, 379–389.

Buetti, E., Kühnel, B. and Diggelmann, H. (1989). Dual function of nuclear factor I binding site in MMTV transcription regulation. *Nucl. Acids Res.* **17**, 3065–3078.
Burch, J. B. E., Evans, M. I., Friedman, T. M. and O'Malley, P. J. (1988). Two functional estrogen response elements are located upstream of the major chicken vitellogenin gene. *Mol. Cell. Biol.* **8**, 1123–1131.
Cato, A. C. B. and Weinmann, K. (1988). Mineralocorticoid regulation of transfected mouse mammary tumour virus DNA in cultured kidney cells. *J. Cell. Biol.* **106**, 2119–2125.
Cato, A. C. B., Geisse, S., Wenz, M., Westphal, H. M. and Beato, M. (1984). The nucleotide sequences recognized by the glucocorticoid receptor in the rabbit uteroglobin gene region are located far upstream from the initiation of transcription. *EMBO J.* **3**, 2771–2778.
Cato, A. C. B., Miksicek, R., Schütz, G., Arnemann, J. and Beato, M. (1986). The hormone regulatory element of mouse mammary tumour virus mediates progesterone induction. *EMBO J.* **5**, 2237–2240.
Cato, A. C. B., Henderson, D. and Ponta, H. (1987). The hormone response element of the mouse mammary tumour virus DNA mediates the progestin and androgen induction of transcription in the proviral long terminal repeat region. *EMBO J.* **6**, 363–368.
Cato, A. C. B., Skroch, P., Weinmann, J., Butkeraitis, P. and Ponta, H. (1988a). DNA sequences outside the receptor-binding sites differentially modulate the responsiveness of the mouse mammary tumour virus promoter to various steroid hormones. *EMBO J.* **7**, 1403–1410.
Cato, A. C. B., Heitlinger, E., Ponta, H., Klein-Hitpass, L., Ryffel, G. U., Bailly, A., Rauch, C. and Milgrom, E. (1988b). Estrogen and progesterone receptor-binding sites on the chicken vitellogenin II gene: synergism of steroid hormone action. *Mol. Cell. Biol.* **8**, 5323–5330.
Chalepakis, G., Postma, J. P. M. and Beato, M. (1988a). A model for hormone receptor binding to the mouse mammary tumour virus regulatory element based on hydroxyl radical footprinting. *Nucl. Acids Res.* **16**, 10 237–10 247.
Chalepakis, G., Arnemann, J., Slater, E., Brüller, H.-J., Gross, B. and Beato, M. (1988b). Differential gene activation by glucocorticoids and progestins through the hormone regulatory element of mouse mammary tumor virus. *Cell* **53**, 371–382.
Chandler, V. L., Maler, B. A. and Yamamoto, K. R. (1983). DNA sequences bound specifically by glucocorticoid receptor *in vitro* render a heterologous promoter hormone responsive *in vivo*. *Cell* **33**, 489–499.
Chang, C., Kokontis, J. and Liao, S. (1988). Molecular cloning of human and rat complementary DNA encoding androgen receptors. *Science* **240**, 324–326.
Corthésy, B., Cardinaux, J.-R., Claret, F.-X. and Wahli, W. (1989). A nuclear factor I-like activity and a liver-specific repressor govern estrogen-regulated *in vitro* transcription from the *Xenopus* vitellogenin B1 promoter. *Mol. Cell. Biol.* **9**, 5548–5562.
Danesch, U., Gloss, B., Schmid, W., Schütz, G. and Renkawitz, R. (1987). Glucocorticoid induction of the rat tryptophan oxygenase gene is mediated by two widely separated glucocorticoid-responsive elements. *EMBO J.* **6**, 625–630.
Danielsen, M., Hinck, L. and Ringold, G. M. (1989). Two amino acids within the knuckle of the first zinc finger specify DNA response element activation by the glucocorticoid receptor. *Cell* **57**, 1131–1138.
Darbre, P., Page, M. and King, R. J. B. (1986). Androgen regulation by the long terminal repeat of mouse mammary tumour virus. *Mol. Cell. Biol.* **6**, 2847–2854.

Day, R. N. and Maurer, R. A. (1989). Thyroid hormone-responsive elements of the prolactin gene: evidence for both positive and negative regulation. *Mol. Endocrinol.* **3**, 931–938.

DeFranco, D. and Yamamoto, K. (1986). Two different factors act separately or together to specify functionally distinct activities at a single transcriptional enhancer. *Mol. Cell. Biol.* **6**, 993–1001.

de Thé, H., Vivanco-Ruiz, M. M., Tiollais, P., Stunnenberg, H. and Dejean, A. (1990). Identification of a retinoic acid responsive element in the retinoic acid receptor β gene. *Nature* **343**, 177–180.

Evans, R. M. (1988). The steroid and thyroid hormone receptor superfamily. *Science* **240**, 889–895.

Fawell, S. E., Lees, J. A., White, R. and Parker, M. G. (1990). Characterization and localization of steroid binding and dimerization activities in the mouse estrogen receptor. *Cell* **60**, 953–962.

Geisse, S., Scheidereit, C., Westphal, H. M., Hynes, N. E., Groner, B. and Beato, M. (1982). Glucocorticoid receptors recognize DNA sequences in and around murine mammary tumour virus DNA. *EMBO J.* **1**, 1613–1619.

Glass, C. K., Franco, R., Weinberger, C., Albert, V. R., Evans, R. M. and Rosenfeld, M. G. (1987). A c-*erb*-A binding site in rat growth hormone gene mediates trans-activation by thyroid hormone. *Nature* **329**, 738–741.

Glass, C. K., Holloway, J. M., Devary, O. V. and Rosenfeld, M. G. (1988). The thyroid hormone receptor binds with opposite transcriptional effects to a common sequence motif in the thyroid hormone and estrogen response elements. *Cell* **54**, 313–323.

Glass, C. K., Lipkin, S. M., Devary, O. V. and Rosenfeld, M. G. (1989). Positive and negative regulation of gene transcription by retinoic acid-thyroid hormone receptor heterodimer. *Cell* **59**, 697–708.

Gowland, P. L. and Buetti, E. (1989). Mutations in the hormone regulatory element of mouse mammary tumor virus differentially affect the response to progestins, androgens and glucocorticoids. *Mol. Cell. Biol.* **9**, 3999–4008.

Grange, T., Roux, J., Rigaud, G. and Pictet, R. (1989). Two remote glucocorticoid responsive units interact cooperatively to promote glucocorticoid induction of rat tyrosine aminotransferase gene expression. *Nucl. Acids Res.* **17**, 8695–8709.

Graupner, G., Wills, K. N., Tzukerman, M., Zhang, X.-K. and Pfahl, M. (1989). Dual regulatory role for thyroid-hormone receptors allows control of retinoic-acid receptor activity. *Nature* **340**, 653–656.

Green, S. and Chambon, P. (1988). Nuclear receptors enhance our understanding of transcription regulation. *Trends Genet.* **4**, 309–314.

Green, S., Walter, P., Kumar, V., Krust, A., Bonert, J. M., Argos, P. and Chambon, P. (1986). Human oestrogen receptor cDNA: sequence, expression and homology to v-*erb*A. *Nature* **320**, 134–139.

Green, S., Kumar, V., Theulaz, I., Wahli, W. and Chambon, P. (1988). The N-terminal DNA-binding "zinc finger" of the oestrogen and glucocorticoid receptors determines target gene specificity. *EMBO J.* **7**, 3037–3044.

Gustafson, T. A., Markham, B. E., Bahl, J. J. and Morkin, E. (1987). Thyroid hormone regulates expression of a transfected alpha-myosin heavy-chain fusion gene in fetal heart cells. *Proc. Natl. Acad. Sci. USA* **84**, 3122–3126.

Ham, J., Thomson, A., Needham, M., Webb, P. and Parker, M. (1988). Characterization of response elements for androgens, glucocorticoids and progestins in mouse mammary tumour virus. *Nucl. Acids Res.* **16**, 5263–5276.

Hamada, K., Gleason, S. L., Levi, B.-Z., Hirschfeld, S., Appela, E. and Ozato, K. (1989). H-2RIIBP, a member of the nuclear hormone receptor superfamily that binds to the regulatory element of major histocompatibility class I genes and the estrogen response element. *Proc. Natl. Acad. Sci. USA* **86**, 8289–8293.

Hecht, A., Berkenstam, A., Strömstedt, P.-E., Gustafsson, J.-A. and Sippel, A. E. (1988). A progesterone responsive element maps to the far upstream steroid dependent DNase hypersensitive site of chicken lysozyme chromatin. *EMBO J.* **7**, 2063–2073.

Herr, W. and Gluzman, Y. (1985). Duplications of a mutated simian virus 40 enhancer restore its activity. *Nature* **313**, 711–714.

Huang, A. L., Ostrowski, M. C., Berard, D. and Hager, G. L. (1981). Glucocorticoid regulation regulation of the Ha-MuSV p21 gene conferred by sequences from mouse mammary tumor virus. *Cell* **27**, 245–255.

Hynes, N., van Goyen, A. J. J., Kennedy, N., Herrlich, P., Ponta, H. and Groner, B. (1983). Subfragments of the large terminal repeat cause glucocorticoid-responsive expression of mouse mammary tumor virus and of an adjacent gene. *Proc. Natl. Acad. Sci. USA* **80**, 3637–3641.

Hwung, Y.-P., Crowe, D. T., Wang, L.-H., Tsai, S. Y. and Tsai, M.-J. (1988a). The COUP transcription factor binds to an upstream promoter element of the rat insulin II gene. *Mol. Cell. Biol.* **8**, 2070–2077.

Hwung, Y.-P., Wang, L.-H., Tsai, S. Y. and Tsai, M.-J. (1988b). Differential binding of the chicken ovalbumin upstream promoter (COUP) transcription factor to two different promoters. *J. Biol. Chem.* **263**, 13 470–13 474.

Izumo, S. and Mahdavi, V. (1988). Thyroid hormone receptor alpha isoforms generated by alternative splicing differentially activate myosin HC gene transcription. *Nature* **334**, 539–542.

Jantzen, K., Fritton, H. P., Igo-Kemenes, T., Espel, E., Janich, S., Cato, A. C. B., Mugele, K. and Beato, M. (1987). Partial overlapping of binding sequences for steroid hormone receptors and DNaseI hypersensitive sites in the rabbit uteroglobin gene region. *Nucl. Acids Res.* **15**, 4535–4552.

Jantzen, H. M., Strähle, U., Gloss, B., Stewart, F., Schmid, W., Boshart, M., Miksicek, R. and Schütz, G. (1987). Cooperativity of glucocorticoid response elements located far upstream of the tyrosine aminotransferase gene. *Cell* **49**, 29–38.

Karin, M., Haslinger, A., Holtgreve, A., Richards, R. I., Krauter, P., Westphal, H. M. and Beato, M. (1984). Characterization of DNA sequences through which cadmium and glucocorticoid hormones induce human metallothionein-IIA gene. *Nature* **308**, 513–519.

Kerner, S. A., Scott, R. A. and Pike, J. W. (1989). Sequence elements in the human osteocalcin gene confer basal activation and inducible response to hormonal vitamin D3. *Proc. Natl. Acad. Sci. USA* **86**, 4455–4459.

Klein-Hitpass, L., Schorpp, M., Wagner, U. and Ryffel, G. U. (1986). An estrogen-responsive element derived from the 5' flanking region of the *Xenopus* vitellogenin A2 gene functions in transfected human cells. *Cell* **46**, 1053–1061.

Klein-Hitpass, L., Ryffel, G. U., Heitlinger, E. and Cato, A. C. B. (1988a). A 13 bp palindrome is a functional estrogen responsive element and interacts specifically with estrogen receptor. *Nucl. Acids Res.* **16**, 647–663.

Klein-Hitpass, L., Kaling, M. and Ryffel, G. U. (1988b). Synergism of closely adjacent estrogen-responsive elements increases their regulatory potential. *J. Mol. Biol.* **201**, 537–544.

Klein-Hitpass, L., Tsai, S. Y., Greene, G. L., Clark, J. H., Tsai, M.-J. and O'Malley, B. W. (1989). Specific binding of estrogen receptor to the estrogen response element. *Mol. Cell. Biol.* **9**, 43–49.

Klemenz, R. and Gehring, W. J. (1986). Sequence requirement for expression of the *Drosophila melanogaster* heat shock protein hsp22 gene during heat shock and normal development. *Mol. Cell. Biol.* **6**, 2011–2019.

Klock, G., Strähle, U. and Schütz, G. (1987). Oestrogen and glucocorticoid responsive elements are closely related but distinct. *Nature* **339**, 734–736.

Koenig, R. J., Brent, G. A., Warne, R. L., Larsen, P. R. and Moore, D. D. (1987). Thyroid hormone receptor binds to a site in the rat growth hormone promoter required for induction by thyroid hormone. *Proc. Natl. Acad. Sci. USA* **84**, 5670–5674.

Kühnel, B., Buetti, E. and Diggelman, H. (1986). Functional analysis of the glucocorticoid regulatory elements present in the mouse mammary tumor virus long terminal repeat. *J. Mol. Cell. Biol.* **190**, 367–378.

Kumar, V. and Chambon, P. (1988). The estrogen receptor binds tightly to its responsive element as a ligand-induced homodimer. *Cell* **55**, 145–146.

Lee, F., Mulligan, R., Berg, P. and Ringold, G. (1981). Glucocorticoids regulate expression of dihydrofolate reductase cDNA in mouse mammary tumor virus chimaeric plasmids. *Nature* **294**, 228–232.

Mader, S., Kumar, V., de Verneuil, H. and Chambon, P. (1989). Three amino acids of the oestrogen receptor are essential to its ability to distinguish an oestrogen from a glucocorticoid-responsive element. *Nature* **338**, 271–274.

Majors, J. and Varmus, H. E. (1983). A small region of the mouse mammary tumor virus long terminal repeat confers glucocorticoid hormone regulation on a linked heterologous gene. *Proc. Natl. Acad. Sci. USA* **80**, 5866–5870.

Maniatis, T., Goodbourn, S. and Fischer, J. A. (1987). Regulation of inducible and tissue-specific gene expression. *Science* **236**, 1237–1244.

Markham, B. E., Bahl, J. J., Gustafson, T. A. and Morkin, E. (1987). Interaction of a protein factor within a thyroid hormone-sensitive region of rat alpha-myosin heavy chain gene. *J. Biol. Chem.* **262**, 12 856–12 862.

Martinez, E. (1989). Ph.D. Thesis, University of Lausanne, Switzerland.

Martinez, E. and Wahli, W. (1989). Cooperative binding of estrogen receptor to imperfect estrogen-responsive DNA elements correlates with their synergistic hormone-dependent enhancer activity. *EMBO J.* **8**, 3781–3791.

Martinez, E., Givel, F. and Wahli, W. (1987). The estrogen-responsive element as an inducible enhancer: DNA sequence requirements and conversion to a glucocorticoid-responsive element. *EMBO J.* **6**, 3719–3727.

Martinez, E., Givel, F. and Wahli, W. (1991). A common ancestor DNA motif for invertebrate and vertebrate hormone response elements. Submitted.

Maurer, R. and Notides, A. C. (1987). Indentification of an estrogen-responsive element from the 5′-flanking region of the rat prolactin gene. *Mol. Cell. Biol.* **7**, 4247–4254.

McDonnell, D. P., Scott, R. A., Kerner, S. A., O'Malley, B. W. and Pike, J. W. (1989). Functional domains of the human vitamin D_3 receptor regulate osteocalcin gene expression. *Mol. Endocrinol.* **3**, 635–644.

Mestril, R., Schiller, P., Amin, J., Klapper, H., Ananthan, J. and Voellmy, R. (1986). Heat shock and ecdysterone activation of the *Drosophila melanogaster* hsp23 gene; a sequence element implied in developmental regulation. *EMBO J.* **5**, 1667–1673.

Miksicek, R., Heber, A., Schmid, W., Danesch, U., Posseckert, G., Beato, M. and Schütz, G. (1986). Glucocorticoid responsiveness of the transcriptional enhancer of moloney murine sarcoma virus. *Cell* **46**, 283–290.

Miksicek, R., Borgmeyer, U. and Nowock, J. (1987). Interaction of the TGGCA-binding protein with upstream sequences is required for efficient transcription of mouse mammary tumor virus. *EMBO J.* **6**, 1355–1360.

Moore, D. D., Marks, A. R., Buckley, D. I., Kapler, G., Payvar, F. and Goodman, H. M. (1985). The first intron of the human growth hormone gene contains a binding site for glucocorticoid receptor. *Proc. Natl. Acad. Sci. USA* **82**, 699–702.

Morrison, N. A., Shine, J., Fragonas, J.-C., Verkest, V., McMenemy, M. L. and Eisman, J. A. (1989). 1,25-Dihydroxyvitamin D-responsive element and glucocorticoid repression in the osteocalcin gene. *Science* **246**, 1158–1161.

Norman, M. F., Lavin, T. N., Baxter, J. D. and West, B. L. (1989). The rat growth hormone gene contains multiple thyroid response elements. *J. Biol. Chem.* **264**, 12063–12073.

Otten, A. D., Sanders, M. M. and McKnight, G. S. (1988). The MMTV LTR promoter is induced by progesterone and dihydrotestosterone but not by estrogen. *Mol. Endocrinol.* **2**, 143–147.

Payvar, F., Wrange, O., Carlstedt-Duke, J., Okret, S., Gustafsson, J.-A. and Yamamoto, K. R. (1981). Purified glucocorticoid receptors bind selectively *in vitro* to a cloned DNA fragment whose transcription is regulated by glucocorticoids *in vivo*. *Proc. Natl. Acad. Sci. USA* **78**, 6628–6632.

Payvar, F., DeFranco, D., Firestone, G. L., Edgar, B., Wrange, O., Okret, S., Gustafsson, J.-A. and Yamamoto, K. R. (1983). Sequence-specific binding of glucocorticoid receptor to MTV DNA at sites within and upstream of the transcribed region. *Cell* **35**, 381–392.

Petersen, D. D., Magnuson, M. A. and Granner, D. K. (1988). Location and characterization of two widely separated glucocorticoid response elements in the phosphoenolpyruvate carboxykinase gene. *Mol. Cell. Biol.* **8**, 96–104.

Pfahl, M., McGinnis, D. and Hendricks, M. (1983). Correlation of glucocorticoid receptor binding sites on MMTV proviral DNA with hormone inducible transcription. *Science* **222**, 1341–1343.

Ponta, H., Kennedy, N., Skroch, P., Hynes, N. E. and Groner, B. (1985). Hormonal response region in the mouse mammary tumor virus long terminal repeat can be dissociated from the proviral promoter and has enhancer properties. *Proc. Natl. Acad. Sci. USA* **82**, 1020–1024.

Ptashne, M. (1988). How eukaryotic transcriptional activators work. *Nature* **335**, 683–689.

Renkawitz, R., Schütz, G., von der Ahe, D. and Beato, M. (1984). Sequences in the promoter region of the chicken lysozyme gene required for steroid regulation and receptor binding. *Cell* **37**, 503–510.

Riddihough, G. and Pelham, H. R. B. (1987). An ecdysone response element in the *Drosophila* hsp27 promoter. *EMBO J.* **6**, 3729–3734.

Sap, J., Muñoz, A., Schmitt, J., Stunnenberg, H. and Vennström, B. (1989). Repression of transcription mediated at a thyroid hormone response element by the v-*erb*A oncogene product. *Nature* **340**, 242–244.

Sap, J., de Magistris, L., Stunnenberg, H. and Vennström, B. (1990). A major thyroid hormone response element in the third intron of the rat growth hormone gene. *EMBO J.* **9**, 887–896.

Scheidereit, C. and Beato, M. (1984). Contacts between hormone receptor and DNA double helix within a glucocorticoid regulatory element of mouse mammary tumor virus. *Proc. Natl. Acad. Sci. USA* **81**, 3029–3034.

Scheidereit, C., Geisse, S., Westphal, H. M. and Beato, M. (1983). The glucocorticoid receptor binds to define nucleotide sequences near the promoter of mouse mammary tumor virus. *Nature* **304**, 749–752.

Scheidereit, C., Westphal, H. M., Carlson, C., Bosshard, H. and Beato, M. (1986). Molecular model of the interaction between the glucocorticoid receptor and the regulatory elements of inducible genes. *DNA* **5**, 383–391.

Schmid, W., Strähle, U., Schütz, G., Schmitt, J. and Stunnenberg, H. (1989). Glucocorticoid receptor binds cooperatively to adjacent recognition sites. *EMBO J.* **8**, 2257–2263.

Schüle, R., Muller, M., Kaltschmidt, C. and Renkawitz, R. (1988). Many transcription factors interact synergistically with steroid receptors. *Science* **242**, 1418–1420.

Seiler-Tuyns, A., Walker, P., Martinez, E., Mérillat, A.-M., Givel, F. and Wahli, W. (1986). Identification of estrogen-responsive DNA sequences by transient expression experiments in a human breast cancer cell line. *Nucl. Acids Res.* **14**, 8755–8770.

Shupnik, M. A., Weismann, C. M., Notides, A. C. and Chin, W. W. (1989). An upstream region of the rat luteinizing hormone beta gene binds estrogen receptor and confers estrogen responsiveness. *J. Biol. Chem.* **264**, 80–86.

Slater, E. P., Rabenau, O., Karin, M., Baxter, J. D. and Beato, M. (1985). Glucocorticoid receptor binding and activation of a heterologous promoter by dexamethasone by the first intron of the human growth hormone gene. *Mol. Cell. Biol.* **5**, 2984–2992.

Slater, E. P., Cato, A. C. B., Karin, M., Baxter, J. D. and Beato, M. (1988). Progesterone induction of metallothionein-IIA gene expression. *Mol. Endocrinol.* **2**, 485–491.

Strähle, U., Klock, G. and Schütz, G. (1987). A DNA sequence of 15 base pairs is sufficient to mediate both glucocorticoid and progesterone induction of gene expression. *Proc. Natl. Acad. Sci. USA* **84**, 7871–7875.

Strähle, U., Schmid, W. and Schütz, G. (1988). Synergistic action of the glucocorticoid receptor with transcription factors. *EMBO J.* **7**, 3389–3395.

ten Heggeler-Bordier, B., Hipskind, R., Seiler-Tuyns, A., Martinez, E., Corthésy, B. and Wahli, W. (1987). Electron microscopic visualization of protein–DNA interactions at the estrogen-responsive element and in the first intron of the *Xenopus laevis* vitellogenin gene. *EMBO J.* **6**, 1715–1720.

Tora, L., Gaub, M.-P., Mader, S., Dierich, A., Bellard, M. and Chambon, P. (1988). Cell-specific activity of a GGTCA half-palindromic oestrogen-responsive element in the chicken ovalbumin gene promoter. *EMBO J.* **7**, 3771–3778.

Tsai, S. Y., Sagami, I., Wang, H., Tsai, M.-J. and O'Malley, B. W. (1987). Interactions between a DNA-binding transcription factor (COUP) and a non-DNA binding factor (S300-II). *Cell* **50**, 701–709.

Tsai, S. Y., Carlstedt-Duke, J., Weigel, N. L., Dahlman, K., Gustafsson, J.-A., Tsai, M.-J. and O'Malley, B. W. (1988). Molecular interactions of steroid hormone receptor with its enhancer element: evidence for receptor dimer formation. *Cell* **55**, 361–369.

Tsai, S. Y., Tsai, M.-J. and O'Malley, B. W. (1989). Cooperative binding of steroid hormone receptors contributes to transcriptional synergism at target enhancer elements. *Cell* **57**, 443–448.

Umesono, K., Giguere, V., Glass, C. K., Rosenfeld, M. G. and Evans, R. M. (1988). Retinoic acid and thyroid hormone induce gene expression through a common responsive element. *Nature* **336**, 262–265.

Umesono, K. and Evans, R. M. (1989). Determinants of target gene specificity for steroid/thyroid hormone receptors. *Cell* **57**, 1139–1146.

Vasios, G. W., Gold, J. D., Petkovich, M., Chambon, P. and Gudas, L. J. (1989). A retinoic acid-responsive element is present in the 5' flanking region of the laminin B1 gene. *Proc. Natl. Acad. Sci. USA* **86**, 9099–9103.

von der Ahe, D., Janich, S., Scheidereit, C., Renkawitz, R., Schütz, G. and Beato, M. (1985). Glucocorticoid and progesterone receptors bind to the same sites in two hormonally regulated promoters. *Nature* **313**, 706–709.

von der Ahe, D., Renoir, J.-M., Buchou, T., Baulieu, E.-E. and Beato, M. (1986). Receptor for glucocorticosteroid and progesterone recognize distinct features of a DNA regulatory element. *Proc. Natl. Acad. Sci. USA* **83**, 2817–2821.

Wahli, W., Martinez, E., Corthésy, B. and Cardinaux, J.-R. (1989). Cis- and trans-acting elements of the estrogen-regulated vitellogenin gene B1 of *Xenopus laevis*. *J. Steroid Biochem.* **34**, 17–32.

Walker, P., Germond, J.-E., Brown-Luedi, N., Givel, F. and Wahli, W. (1984). Sequence homologies in the region preceding the transcription initiation site of the liver estrogen-responsive vitellogenin and apo-VLDLII genes. *Nucl. Acids Res.* **12**, 8611–8626.

Wang, L.-H., Tsai, S. Y., Sagami, I., Tsai, M.-J. and O'Malley, B. W. (1987). Purification and characterization of chicken ovalbumin upstream promoter transcription factor from HeLa cells. *J. Biol. Chem.* **262**, 16080–16086.

Wang, L.-H., Tsai, S. Y., Cook, R. G., Beattie, W. G., Tsai, M.-J. and O'Malley, B. W. (1989). COUP transcription factor is a member of the steroid receptor superfamily. *Nature* **340**, 163–166.

Wight, P. A., Crew, M. D. and Spindler, S. R. (1988). Sequences essential for activity of the thyroid hormone responsive transcription stimulatory element of the rat growth hormone gene. *Mol. Endocrinol.* **2**, 536–542.

Wijnholds, J., Philipsen, J. N. J. and Ab, G. (1988). Tissue-specific and steroid-dependent interaction of transcription factors with the oestrogen-inducible apo-VLDLII promoter *in vivo*. *EMBO J.* **7**, 2757–2763.

Zenke, M., Grundström, T., Matthes, H., Wintzerith, M., Schatz, C., Wildeman, A. and Chambon, P. (1986). Multiple sequence motifs are involved in SV40 enhancer function. *EMBO J.* **5**, 387–397.

7
Co-operative Transactivation of Steroid Receptors

M. MULLER, C. BANIAHMAD, C. KALTSCHMIDT,
R. SCHÜLE* and R. RENKAWITZ

Max-Plank Institut für Biochemie, Genzentrum, D-8033 Martinsried, FRG

* Present address: Salk Institute for Biological Studies, La Jolla, CA 92037, USA.

7.1 Introduction

Regulation of gene transcription by steroid hormone receptors is mediated by specific DNA sequences located in the vicinity of the regulated gene (for recent reviews see Beato, 1989; Ham and Parker, 1989; see Chapter 6). Such transcriptional control regions have been identified in a variety of genes and the comparison of these sequences has allowed the identification of a consensus for steroid receptor binding sites. The binding of steroid receptors to these consensus sequences has been confirmed by DNase I and DMS methylation protection experiments using purified hormone loaded receptor (Payvar *et al.*, 1983; Scheidereit *et al.*, 1983; Cato *et al.*, 1984; Karin *et al.*, 1984; Renkawitz *et al.*, 1984; Slater *et al.*, 1985; von der Ahe *et al.*, 1985; Becker *et al.*, 1986; Danesch *et al.*, 1987; Jantzen *et al.*, 1987; Chalepakis *et al.*, 1988). Finally, synthetic oligonucleotides cloned in front of the herpes simplex thymidine kinase (tk) promoter have been found to confer responsiveness to glucocorticoids and progestins (Strähle *et al.*, 1987), oestrogens (Klock *et al.*, 1987) or androgens (Ham *et al.*, 1988) to this otherwise unresponsive transcription unit. The so-defined steroid hormone response elements (HRE) consist of palindromic sequences of about 20 bp, which have been shown to bind a dimer of the corresponding receptor (Kumar and Chambon, 1988; Tsai *et al.*, 1988; Fawell *et al.*, 1990).

The prototype glucocorticoid receptor binding site is the palindromic sequence GGTACANNNTGTTCT, which can also mediate the action of progesterone, androgen and mineralocorticoid receptors (Cato *et al.*, 1986; Arriza *et al.*, 1987; Strähle *et al.*, 1987; Ham *et al.*, 1988). The oestrogen receptor, on the other hand, specifically binds a different, although related sequence AGGTCANNNTGACCT (Klock *et al.*, 1987). These sequences conferred inducibility by the corresponding steroid to a reporter gene when cloned close to the TATA box (within 100 bp), but not when cloned at a position further upstream (Strähle *et al.*, 1988). This suggests that additional components of the transcription machinery are involved in regulation by those HREs which are naturally located at large distances from the promoter which they control.

Steroid hormones have in some cases been shown to negatively regulate gene expression by a process involving competition in binding to DNA between the receptor and another transcription factor whose binding site overlaps the HRE (Akerblom *et al.*, 1988; Oro *et al.*, 1988; Sakai *et al.*, 1988; see Chapter 8.3).

In summary, HREs appear to function very similarly to other transcriptional regulatory sequences which modulate transcription via interaction with protein factors, some of which (e.g. NF-1, CP1, Sp1 or OTF1) are present in a large variety of cell types.

7.2 Co-operation between hormone response elements

A detailed analysis of the long terminal repeat (LTR) of the mouse mammary tumour virus (MMTV) revealed the presence of several glucocorticoid receptor binding sites (Pfahl, 1982; Buetti and Diggelman, 1983; Chandler *et al.*, 1983; Hynes *et al.*, 1983; Majors and Varmus, 1983; Payvar *et al.*, 1983; Scheidereit *et al.*, 1983; Lee *et al.*, 1984; Ponta *et al.*, 1985; see Chapter 9.2), the progressive deletion of which resulted in a gradual loss of inducibility (Hynes *et al.*, 1983; Kühnel *et al.*, 1986). Similar observations were made during the analysis of other glucocorticoid regulated genes including the chicken lysozyme gene (Renkawitz *et al.*, 1984; von der Ahe *et al.*, 1985), the rabbit uteroglobin gene (Cato *et al.*, 1984), the rat tyrosine aminotransferase (TAT) gene (Jantzen *et al.*, 1987) and the rat tryptophan oxygenase (TO) gene (Danesch *et al.*, 1987). In the case of the TAT gene, induction mediated by two GREs could only be explained by co-operative effects. Similarly, oestrogen regulation of the *Xenopus* vitellogenin B1 gene has been found to require the presence of at least two copies of a non-palindromic ERE upstream of the start site (Martinez *et al.*, 1987; Klein-Hitpass *et al.*, 1988; see Chapter 6.4). The presence of different HREs in the chicken vitellogenin II gene allowed the demonstration of a synergism in induction in the presence of both steroids (Cato *et al.*, 1988a; Ankenbauer *et al.*, 1988). All these observations indicate that several hormone receptor binding sites can co-operate to yield the strong induction observed in natural genes. In order to study the synergism in a more defined environment, synthetic regulatory units consisting of several receptor binding sites were constructed. In each case, the combination of two palindromic GREs showed a synergistic effect (Schüle et al., 1988a; Strähle et al., 1988). Further addition of one or two GREs, however, did not lead to a further increase in induction (Strähle et al., 1988; Tsai et al., 1989).

7.3 Functional co-operativity between hormone response elements and other transcription factor binding sites

7.3.1 Natural Genes

During the study of sequences responsible for glucocorticoid regulation of the rat tryptophan oxygenase (TO) gene, a puzzling observation was made (Danesch et al., 1987). Footprinting experiments revealed a receptor binding site located from positions -440 to -387 relative to the transcriptional start site (see Fig. 1). Nevertheless, a 5'-deletion mutant ($\Delta-444$) containing the intact receptor binding site failed to show inducibility to glucocorticoids. A construct containing additional 5' sequences ($\Delta-472$) was, however, regulated by the steroid. These observations suggested that a sequence adjacent to the GRE contributes to hormone induction of the TO gene. Footprinting experiments in this region using nuclear extracts revealed the presence of a binding site containing a CACCC-sequence (see Fig. 1) identical to that found in the human β-globin gene (Dierks et al., 1983). Another example of such a functional co-operation between receptor and other transcription factors was seen in the MMTV promoter. Deletion of an NF-1 binding site in the vicinity of a GRE almost completely abolished the inducibility by glucocorticoids (Buetti and Kühnel, 1986; Miksicek et al., 1987). Mutations at this site, or in some other non-receptor-binding sequences, impeded the function of the regulatory element (Cato et al., 1988b). Interestingly, different mutations had varying effects on the inducibility depending on the type of receptor involved, indicating that requirements for neighbouring sequences are receptor specific (Chalepakis et al., 1988). In a different approach, Cordingley et al. (1987) showed that in vivo protection against exonuclease III digestion through the NF-1 binding site could only be seen in chromatin isolated from hormone-induced cells, although the factor is also present in uninduced cells. This suggests that the NF-1 binding site is inaccessible in uninduced cells. Clearly this is one of several possible mechanisms for functional co-operativity, one factor helps the other in binding by repositioning nucleosomes (Richard-Foy and Hager, 1987; see Chapters 9.6 and 10.3).

Fig. 1 5′ Deletion analysis of the rat tryptophan oxygenase gene. 5′ deletions of the TO gene upstream region were cloned in front of the tk promoter/chloramphenicol acetyl transferase coding sequences. The constructs were transfected into mouse fibroblasts and tested for glucocorticoid inducibility (Danesch et al., 1987). The bars indicate the regions protected from DNaseI digestion by purified glucocorticoid receptor (thin lines; the arrow heads indicate close contact points) or by crude nuclear extract (thick lines; sequences identical to the β-globin CACCC box are boxed).

In vivo analysis of the TAT gene revealed steroid-dependent changes in the pattern of specific sites protected from DMS methylation. In addition to the glucocorticoid receptor binding sites, other sequence elements were protected upon glucocorticoid induction (Becker *et al.*, 1986). One of these regions is a CAAT box immediately adjacent to a GRE, which was shown in transfection experiments to be essential for inducibility in its natural environment (Strähle *et al.*, 1988).

Since the detailed analysis of MMTV and the TO and TAT genes showed the importance of non-receptor binding sites for steroid induction, a computer search for known transcription factor binding sites in the neighbourhood of glucocorticoid response elements was carried out (Fig. 2). For several other genes potential binding sequences were detected in the vicinity of receptor binding sites.

Transcription Factor Binding Sites in the Vicinity of GREs

	GRE distance		Binding sequence		GRE distance
CACCC			GAGCCACACCC		
TO	13	←	GAGCCACACCC		
TAT			TAACCACACCC	→	31
ch lys			CTCCCACCCCC	→	20
h metIIa			CACCCAGCCCC	→	17
octamer			ATTTGCAT		
TO			ATTGGCAT	→	31
TAT	17	←	ATTTGCAA	→	45
NF1			TGGCNNNNTGCCA		
h GH			TGGCCTGCGGCCA	→	35
MMTV			TGGATAGATTCCA	→	25
Mo-MuLV			CGGCTCAGGGCCA	→	8
Mo-MuLV	10	←	TGAATATGGGCCA	→	53
Mo-MuLV	55	←	CGGCTCAGGGCCA		
SV40 core			GTGGAAA		
rab UG	10	←	GTGGAAA	→	16
rab UG	50	←	TTGGAAA		
Mo-MuLV			GTGGTAT	→	34
Mo-MuLV	35	←	GTGGTAA	→	34
ch lys			TGGGAAA	→	33

Fig. 2 Potential transcription factor binding sites in the vicinity of glucocorticoid responsive elements. A computer search was performed for known transcription factor binding sites in well characterized hormone responsive units. The sequences searched for are the CACCC box, the SV40 enhancer core motif, the immunoglobulin heavy chain octamer motif and an NF1 binding sequence. The HRUs tested are from the rat tryptophan oxygenase (TO) gene, the rat tyrosine aminotransferase (TAT) gene, the chicken lysozyme (ch lys) gene, the human metallothionein IIa (h metIIa) gene, the human growth hormone (hGH) gene, the MMTV LTR (MMTV), the Moloney murine sarcoma virus (Mo-MuLV) and the rabbit uteroglobin (rab UG) gene. The numbers indicate the distance between each binding site and the consensus receptor binding sequence, the arrows indicate the relative orientation of the two motifs.

These observations supported the idea that the strength of a hormone responsive unit (HRU) (Klein-Hitpass et al., 1988) is determined by both receptor and by non-receptor binding sites.

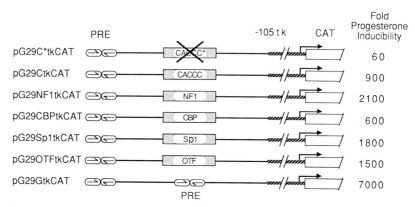

Fig. 3 Synergistic activity of the progesterone receptor with adjacent transcription factors. The MMTV progesterone responsive element (PRE) was combined with binding sites for another progesterone receptor, for OTF, Sp1, CBP, NF-1 or the CACCC box. For reference a mutagenized CACCC element (CACCC*), which cannot bind a protein, was combined with the PRE. After fusion to the tkCAT reporter gene and transfection into human T47D breast cancer cells the inducibility by progesterone was determined. Reproduced with permission from Renkawitz et al. (1989) (see also Schüle et al., 1988a).

7.3.2 Synthetic Genes

Because single interactions between pairs of bound transcription factors cannot be analysed within a complex cluster of binding sites for nuclear proteins, the effect of neighbouring sequences on the action of a GRE was studied in a more precisely controlled environment by combining single, well-characterized transcription factor binding sites with a single GRE. The use of a common vector for these synthetic hormone responsive units allowed a direct comparison of the effects of different binding sites. The constructs depicted in Fig. 3 all contain a thymidine kinase promoter/chloramphenicol acetyl transferase (CAT) gene transcription unit controlled by an upstream GRE/PRE from MMTV. In addition, different transcription factor binding sites were inserted at 29 bp from the GRE/PRE. The sequences used are a CACCC box as found in the rat TO gene, binding sites for NF-1, CBP, Sp1, OTF, a second GRE/PRE and, as a control, a mutated CACCC box which does not bind any factor (CACCC*). It is clearly apparent that the progesterone inducibility of these constructs is synergistically increased by the presence of the second transcription factor binding site when compared to the CACCC* control (Fig. 3 and Schüle et al., 1988a). The best co-operation is seen between two PREs, the other sequences co-operate to various extents. Similar results were obtained for dexamethasone inducibility by transfection in different cell lines (Schüle et al., 1988a; Strähle et al., 1988) and also with different GRE sequences. Interestingly, the effects of the different transcription factors were strongly dependent on the cell line used for the transfection experiments, probably reflecting the relative abundance of the various factors in these cell lines (Strähle et al., 1988). These observations offer a possible explanation for the variable inducibility of natural genes in different cell types, which does not always correlate with corresponding amounts of receptor in the cell (Tora et al., 1988; Bocquel et al., 1989).

Thus it appears that in many cases a hormone responsive unit is composed of one or several receptor binding sites intermingled with binding sites for other transcription factors, which act synergistically to yield the observed expression and inducibility pattern.

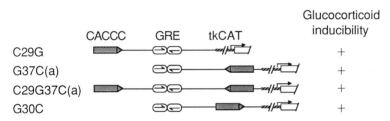

Fig. 4 Different arrangements of CACCC box and GRE are equally inducible. Several constructs fused to the tkCAT reporter gene are schematically drawn. The position of the CACCC element is indicated by the shaded box and its orientation is shown by the arrowhead. Glucocorticoid induction was measured in mouse L cell fibroblasts. Reproduced with permission from Renkawitz et al. (1989) (see also Schüle et al., 1988b).

7.4 Functional details of co-operativity

7.4.1 Arrangement of Co-operating Binding Sites

For a better understanding of synergism in transcription activation, it is important to define the requirements for this phenomenon. First of all, the influence of the relative arrangement of the binding sites was investigated by comparing constructs containing the CACCC box either downstream from a GRE, in sense or anti-sense orientation, upstream from a GRE or in two copies flanking a GRE (Fig. 4). All these constructs showed a similar induction by dexamethasone (Schüle et al., 1988b). This result indicates that the relative orientation and arrangement of GRE and CACCC box is of no importance for the functional co-operativity. Interestingly, an additional CACCC box does not further increase inducibility. This might be related to the fact discussed earlier that two GREs show a clear synergism while more binding sites do not further increase inducibility.

Thus, within several transcription factor binding sites only two are involved in synergism, and this synergism occurs independently of their relative arrangement one to the other.

7.4.2 Stereo-specific Alignment

It is feasible that functional co-operativity may involve direct interaction between transcription factors. This possibility was examined by testing the glucocorticoid or progesterone induction of constructs containing a CACCC box at various distances from a GRE/PRE (Fig. 5). In both cases a clear distance dependence of synergism is seen (Fig. 5). This distance dependence shows a cyclic pattern with a period of about 10 bp, i.e. one complete turn of the double helix. The induction maxima are different from the two receptors analysed (Schüle et al., 1988a,b; Renkawitz et al., 1989). These results suggest a requirement for a stereo-specific alignment of the two factors; insertion or deletion of 5 bp displaces the factors around one half turn of the double helix. We conclude that the synergism involves direct or indirect protein–protein interactions between the two factors. The different optima observed for the two receptors probably reflect differences in their sizes and/or binding geometry. An observation which was made during analysis of the MMTV HRU (Chalepakis et al., 1988) leads to a similar interpretation. These authors observed that individual mutations in the GRE sequences or in their vicinity, as well as insertion of 5 bp between two groups of GREs, differentially affect glucocorticoid and progesterone inducibility.

The relative responsiveness to progesterones or glucocorticoids of each of several GREs in such an HRE may then be determined by its spatial relationship to companion transcription factor binding sites.

A different result is seen by changing the spacing between a GRE and the TATA box of the tk promoter. Both strong distance dependence and an optimal spacing, but no apparent periodicity comparable to that seen within an upstream HRE were observed (Ham et al., 1988). It seems that the interaction of receptors with the TATA box factor does not require a stereo-specific alignment, an observation that has also been made with other transcription factors upstream of the TATA box as well (Chodosh et al., 1987; Wirth et al., 1987; Ruden et al., 1988).

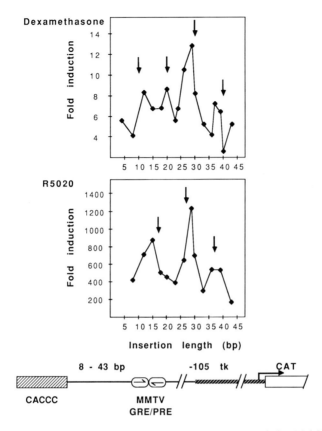

Fig. 5 Receptor-specific space requirements for the GRE/PRE and the CACCC element. DNA fragments of different length have been inserted between the receptor binding site (GRE/PRE) and the CACCC element. These sequences were cloned upstream of the tkCAT reporter gene and tested for glucocorticoid inducibility in mouse L cells, or for progesterone inducibility in T47D cells. The arrows point to a 10 bp periodicity. Reproduced with permission from Renkawitz *et al.* (1989).

7.5 Functional synergism and DNA-binding affinity

7.5.1 Two Receptor Binding Sites

The simplest explanation for the synergism between two transcription factors would be co-operative binding to two adjacent binding sites. This possibility was investigated using the gel retardation method. Tsai *et al.*

(1989) showed that incubation of increasing amounts of partially purified progesterone receptor in the presence of two tandemly arranged GRE/PREs resulted in the co-operative formation of a complex containing two receptor dimer molecules. The binding affinity was 100 times greater for a double GRE/PRE than for a single site. Similarly, co-operative binding to a double GRE by purified glucocorticoid receptor was demonstrated by Schmid et al. (1989). In addition, these authors showed a dependence of the increased binding affinity on the space between the binding sites; affinities changed in a cyclical pattern with a periodicity at 10 bp intervals. Moreover, the stability of the formed DNA–protein complex was increased for two properly spaced GREs. Similar results were obtained for double EREs (Klein-Hitpass et al. 1988) by using DNA-cellulose binding competition or gel-shift assays (Martinez and Wahli, 1989; see Chapter 6.4).

Thus it appears that the functional synergism of two HREs is due, at least in part, to the increase of binding affinity of the receptor to the regulatory element.

7.5.2 Combination of a Glucocorticoid Response Element with Another Transcription Factor

To analyse whether the synergism of non-receptor factors with a receptor is similarly mediated by increased DNA binding affinities, gel retardation experiments addressing this question were carried out (Fig. 6). DNA fragments containing a GRE combined with a NF-1 or OTF1 binding site were assayed for complex formation. Incubation with HeLa cell nuclear extracts in the presence of purified glucocorticoid receptor did not yield more shifted DNA than the sum of the complexes obtained using each protein component alone (Muller et al., 1990). The same observation was made using purified OTF1 and glucocorticoid receptor. Other experiments (not shown) using purified or in vitro translated receptor in combination with in vitro translated NF-1 or Sp1 showed no evidence for co-operative binding (Muller et al., 1990). Thus it appears that the functional synergism of a GRE with another transcription factor binding site, in contrast to that of a dimerized GRE, cannot be explained by an increase in DNA binding affinity of either of the two factors.

7.6 Synergistic domains of the receptor

Since the cloning of cDNAs coding for several steroid hormone receptor proteins, the detailed analysis of functional domains in these complex

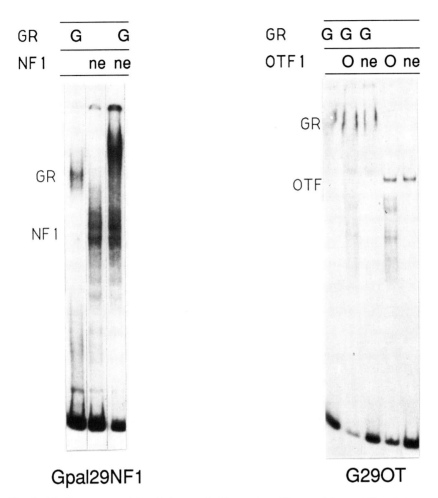

Fig. 6 Binding co-operativity of glucocorticoid receptor with several transcription factors. Gel retardation experiments with single or combined transcription factors are shown. The top line (GR) indicates the presence of the purified receptor (G); the second line indicates the source of the other transcription factor used ('ne' for unfractionated HeLa nuclear extract 'O' for purified OTF1 from HeLa cells). The bottom line shows the probe used; the names indicate the binding sequence (Gpal for a palindromic GRE, NF-1 for a NF-1 binding site, G for the MMTV GRE, OT for an OTF1 binding site, each spaced by 29 base pairs).

proteins has been possible. By expression of deletion mutants of the receptor and of chimaeric proteins, discrete functions have been attributed to precise regions of the protein (see Chapters 2 and 3). In general, a central basic region containing two potential 'zinc finger' domains is responsible for DNA binding, and a large C-terminal domain constitutes the hormone-binding region. In the oestrogen receptor a constitutive transactivation domain has been identified in the N-terminal region of the protein (Lees *et al.*, 1989), which, in the case of chicken receptor, was shown to be promoter and cell specific (Kumar *et al.*, 1987; Tora *et al.*, 1989). In addition, a hormone-dependent transactivation function is closely associated with the hormone binding domain (Green and Chambon, 1987; Kumar *et al.*, 1987; Lees *et al.*, 1989; Tora *et al.*, 1989). The chicken progesterone receptor also contains two transactivation domains centred around the DNA-binding regions (Gronemeyer *et al.*, 1987). Deletion of 127 amino acids at the N-terminus yields the naturally occurring form A receptor, which is also active, but shows very different promoter and cell specificity as compared to the full-length form B receptor (Tora *et al.*, 1988).

Functional dissection of the rat glucocorticoid receptor has led to similar results. A central DNA-binding domain, two transactivation regions and a C-terminal hormone-binding domain (Danielsen *et al.*, 1986, 1987; Godowski *et al.*, 1987; Miesfeld *et al.*, 1987; Rusconi and Yamamoto, 1987; Picard *et al.*, 1988) were identified.

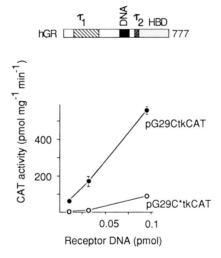

Fig. 7 Glucocorticoid receptor domains involved in co-operativity. Expression plasmid coding for wild-type glucocorticoid receptor was cotransfected with reporter construct pG29CtkCAT or pG29C*tkCAT into T47D cells. CAT activities for both of the reporter constructs were determined. In this range of receptor DNA concentrations the gene activity increases linearly with a six-fold steeper slope (co-operativity) for pG29CtkCAT (GRE plus CACCC box) as compared to pG29C*tkCAT (GRE without CACCC box). The domain structure of the receptor (see text) is depicted at the top of the figure.

A localization of the transactivation domains has been possible for the human glucocorticoid receptor (Hollenberg and Evans, 1988). One transac-

7. CO-OPERATIVE TRANSACTIVATION OF STEROID RECEPTORS

tivating domain (τ1) spans amino acids 77–262, the second domain (τ2) is located at amino acids 526–556. Both mediate their function independently and gain activity when multimerized. The relative position of these domains are depicted in Fig. 7 (top of figure).

In an attempt to localize the regions of the human glucocorticoid receptor which are responsible for the synergism with other transcription factors, different receptor mutant expression vectors were cotransfected with plasmids carrying a reporter gene in cells containing no endogenous receptor. In order to assess the co-operation function independently of effects on transactivation, nuclear transport, DNA binding, hormone binding, RNA or protein stability or dimerization, we compared the inducibility of two different reporter plasmids. pG29C*tkCAT shows the activation mediated by one isolated GRE, while pG29CtkCAT measures the synergism obtained by addition of a CACCC box. In each case several amounts of receptor expression plasmid were tested to ensure detection of effects by weak activator mutants and to compare the slopes of gene activity relative to receptor DNA concentrations as depicted for the wild-type receptor (Fig. 7). Thus co-operativity is calculated by dividing the slopes achieved with both reporter constructs. We have identified a number of domains which contribute to the synergism (M. Muller *et al.*, unpubl. res.). Some are involved in transactivation, others are not.

Table 1 Differential effects of receptor deletions on co-operativity with different transcription factors. For design of reporter constructs see Fig. 3 and for calculation of co-operativity see Fig. 7.

Reporter construct	Co-operativity measured with receptor expression plasmids[a]		
	hGR	418–777	1–488
pG29CtkCAT	+ +	+ +	+ +
pG29CBPtkCAT	+	−	+
pG29NF1tkCAT	+ +	+ +	+ +
pG29OTFtkCAT	+ +	+ +	+ +
pG29Sp1tkCAT	+	+ +	+
pG29GtkCAT	+ +	+ +	+

[a] + +, co-operativity up to eight-fold; +, co-operativity up to two-fold.

Two of the receptor mutants were analysed for differential co-operativity with different transcription factors. The results shown in Table 1 indicate that different parts of the receptor molecule co-operate differently with specific transcription factors. For example, a C-terminally deletion (mutant 1–488) displays weak or no synergism with Sp1, CBP or a second GRE. An N-terminal deletion (mutant 418–777), on the contrary, co-operates well with most of the different transcription factors, but not at all with CBP. Similar differences between functional characteristics of the amino- and carboxy-terminal transactivation domain have also been found for the human oestrogen receptor (Tora et al., 1989).

7.7 Conclusion

The transcriptional control region of a steroid hormone-regulated gene is composed of one or several receptor binding sites and binding sites for other transcription factors. Although several steroid receptors recognize identical DNA sequences, some of the steroids affect gene expression selectively depending on the gene and the cell type studied. One possible explanation for this observation is the differential expression of the respective hormone receptors in different cells. Transfection of an expression vector for the progesterone receptor conferred progestin inducibility to several genes in cells where these genes were previously unaffected (Strähle et al., 1989). Another explanation is given by the synergism described here between HREs, or between HREs and other transcription factor binding sites. Different stereo-specific requirements for each receptor, or differential expression of neighbouring transcription factors, may lead to the precise fine tuning of expression of different genes in different organs in the living organism. Although the control region of some of these genes is a complex cluster of protein-binding sites, it appears that productive interaction is only possible between two neighbouring transcription factors.

7.7.1 Mechanisms of Functional Synergism

(a) DNA Binding

DNA binding affinity studies have shown that the synergism between two identical receptor binding sites is due, at least in part, to an increased affinity for the corresponding receptor (Klein-Hitpass et al., 1988; Schmid et al., 1989; Tsai et al., 1989; Martinez and Wahli, 1989).

In contrast, co-operation between glucocorticoid receptor and other transcription factors does not seem to involve DNA binding (Muller et al., 1990).

(b) Functional Co-operation

In principle one could envisage different mechanisms for functional co-operativity (Fig. 8).

7. CO-OPERATIVE TRANSACTIVATION OF STEROID RECEPTORS

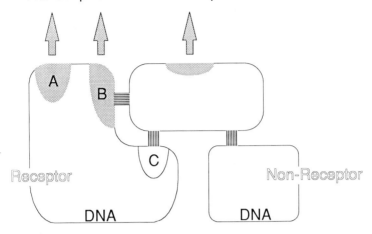

Fig. 8 Model for potential transactivating and co-operating domains. The model schematically depicts three types of transactivating (shaded areas) and co-operating domains (B and C). Domain A transactivates without showing synergism, domain B shows transactivation by itself, but the activity can be synergistically increased by an adjacent non-receptor transcription factor. Both may be connected by a transactivating bridging protein. Domain C does not transactivate by itself, requiring the presence of an adjacent transcription factor. Transactivation (arrows) may trigger the transcription initiation complex directly or may involve further intermediary factors.

The model for functional synergism presented by Ptashne (1988) predicts that a transcription factor interacts with a non-DNA-binding protein which then mediates the activation to the transcription machinery. Two neighbouring transcription factors would be recognized by the target protein on the basis of a common feature, the number and strength of these interactions would determine the resulting transcription efficiency, thus synergism would be mediated by the larger surface interacting with the non-DNA-binding protein. The existence of such a 'bridging' protein is supported by the observation that overexpression of one steroid receptor can inhibit the effects of another receptor, presumably by titrating out a functionally limiting transcription factor (Adler et al., 1988; Meyer et al., 1989). For this model, it was predicted (Ptashne, 1988) that the transactivation and synergism domains coincide (domain B of the model in Fig. 8). In contrast, a transactivation domain may directly interact with the transcription initiation complex (domain A in Fig. 8), its function would then be independent of neighbouring factors.

In addition to the co-operativity mediated by transacting domains, one could envisage the existence of synergizing domains. In such a system one would predict the presence of a protein interacting with the synergizing domain on one hand and a neighbouring factor on the other hand (domain C; Fig. 8). Only this protein would have a transactivation function similar to the regular transactivation domains. In addition, the non-receptor transcription factor may or may not contain a transactivation domain of its own.

Our results suggest that several mechanisms are involved, the glucocorticoid receptor comprises each of the domains depicted in the model (Fig. 8).

Acknowledgements

We wish to thank Ronald M. Evans and Stanley M. Hollenberg for receptor expression vectors, Ernst-L. Winnacker and Lars Rogge for the NF-1 expression vector and purified NF-1 from pig liver, Robert Tjian and Albert J. Courey for the Sp1 expression vector, Miguel Beato for purified receptor, and Peter C. van der Vliet and Ger Pruijn for purified OTF1. We thank Andrea Schweizer for synthesizing oligonucleotides, and Karin Schulz and Dagmar Wolf for technical assistance. This work was supported by grants from the Deutsche Forschungsgemeinschaft (Re 433/6-4) and from the Bundesministerium für Forschung und Technologie.

References

Adler, S., Waterman, M. L., He, X. and Rosenfeld, M. G. (1988). Steroid receptor-mediated inhibition of rat prolactin gene expression does not require the receptor DNA-binding domain. *Cell* **52**, 685–695.
Arriza, J. L., Weinberger, C., Cerelli, G., Glaser, T. M., Handelin, B. L., Housman, D. E. and Evans, R. E. (1987). Cloning of human mineralocorticoid receptor complementary DNA: structural and functional kinship with the glucocorticoid receptor. *Science* **237**, 268–275.
Akerblom, I. E., Slater, E. P., Beato, M., Baxter, J. D. and Mellon, P. L. (1988). Negative regulation by glucocorticoids through interference with a cAMP responsive enhancer. *Science* **241**, 350–353.
Ankenbauer, W., Strähle, U. and Schütz, G. (1988). Synergistic action of glucocorticoid and estradiol responsive elements. *Proc. Natl. Acad. Sci. USA* **85**, 7526–7530.
Beato, M. (1989). Gene regulation by steroid hormones. *Cell* **56**, 335–344.
Becker, P. B., Gloss, B., Schmid, W., Strähle, U. and Schütz, G. (1986). In vivo protein-DNA interactions in a glucocorticoid response element require the presence of the hormone. *Nature* **324**, 686–688.
Bocquel, M. T., Kumar, V., Stricker, C., Chambon, P. and Gronemeyer, H. (1989). The contribution of the N- and C-terminal regions of steroid receptors to activation of transcription is both receptor and cell-specific. *Nucl. Acids Res.* **17**, 2581–2595.
Buetti, E. and Diggelmann, H. (1983). Glucocorticoid regulation of mouse mammary tumor virus: identification of a short essential region. *EMBO J.* **2**, 1423–1429.

Buetti, E. and Kühnel, B. (1986). Distinct sequence elements involved in the glucocorticoid regulation of the mouse mammary tumor virus promoter identified by linker scanning mutagenesis. *J. Mol. Biol.* **190**, 379–389.

Cato, A. C. B., Geisse, S., Wenz, M., Westphal, H. M. and Beato, M. (1984). The nucleotide sequences recognized by the glucocorticoid receptor in the rabbit uteroglobin gene region are located far upstream from the initiation of transcription. *EMBO J.* **3**, 2731–2736.

Cato, A. C. B., Miksicek, R., Schütz, G., Arnemann, J. and Beato, M. (1986). The hormone regulatory element of mouse mammary tumor virus mediates progesterone inducibility. *EMBO J.* **5**, 2237–2240.

Cato, A. C. B., Heitlinger, E., Ponta, H., Klein-Hitpass, L., Ryffel, G. U., Bailly, A., Rauch, C. and Milgrom, E. (1988a). Estrogen and progesterone receptor-binding sites on the chicken vitellogenin II gene: synergism of steroid hormone action. *Mol. Cell Biol.* **8**, 5323–5330.

Cato, A. C. B., Skroch, P., Weinmann, J., Butkeraitis, P. and Ponta, H. (1988b). DNA sequences outside the receptor binding sites differentially modulate the responsiveness of the mouse mammary tumor virus promoter to various steroid hormones. *EMBO J.* **7**, 1403–1407.

Chalepakis, G., Arnemann, J., Slater, E., Brüller, H.-J., Gross, B. and Beato, M. (1988). Differential gene activation by glucocorticoids and progesterone through the hormone regulatory element of mouse mammary tumor virus. *Cell* **53**, 371–382.

Chandler, V. L., Maler, B. A. and Yamamoto, K. R. (1983). DNA sequences bound specifically by glucocorticoid receptor *in vitro* render a heterologous promoter hormone responsive *in vivo*. *Cell* **33**, 489–499.

Chodosh, L. A., Carthew, R. W., Morgan, J. G., Crabtree, G. R. and Sharp, P. A. (1987). The adenovirus major late transcription factor activates the rat γ-Fibrinogen Promoter. *Science* **238**, 684–688.

Cordingley, M. G., Riegel, A. T. and Hager, G. L. (1987). Steroid-dependent interaction of transcription factors with the inducible promoter of mouse mammary tumor virus *in vivo*. *Cell* **48**, 261–270.

Danesch, U., Gloss, B., Schmid, W., Schütz, G., Schüle, R. and Renkawitz, R. (1987). Glucocorticoid induction of the rat tryptophan oxygenase gene is mediated by two widely separated glucocorticoid-responsive elements. *EMBO J.* **6**, 625–630.

Danielsen, M., Northrop, P. and Ringold, G. M. (1986). The mouse glucocorticoid receptor: mapping of functional domains by cloning, sequencing and expression of wild-type and mutant receptor proteins. *EMBO J.* **5**, 2513–2522.

Danielsen, M., Northrop, J. P., Jonklaas, J. and Ringold, G. M. (1987). Domains of the glucocorticoid receptor involved in specific and nonspecific deoxyribonucleic acid binding, hormone activation, and transcriptional enhancement. *Mol. Endocrinol.* **1**, 816–822.

Dierks, P., van Ooyen, A., Cochran, M. D., Dobkin, C., Reiser, J. and Weissmann, C. (1983). Three regions upstream from the cap site are required for efficient and accurate transcription of the rabbit β-globin gene in mouse 3T6 cells. *Cell* **32**, 695–706.

Fawell, S. E., Lees, J. A., White, R. and Parker, M. G. (1990). Characterization and colocalization of steroid binding and dimerization activities in the mouse estrogen receptor. *Cell* **60**, 953–962.

Godowski, P. J., Rusconi, S., Miesfeld, R. and Yamamoto, K. R. (1987). Glucocorticoid receptor mutants that are constitutive activators of transcriptional enhancement. *Nature* **325**, 365–368.

Green, S. and Chambon, P. (1987). Oestradiol induction of a glucocorticoid-responsive gene by chimeric receptor. *Nature* **325**, 75–78.

Gronemeyer, H., Turcotte, B., Quirin-Stricker, C., Bocquel, M.-T., Meyer, M.-E., Krozowski, Z., Jeltsch, J.-M., Lerouge, T., Garnier, J. M. and Chambon, P. (1987). The chicken progesterone receptor: sequence, expression and functional analysis. *EMBO J.* **6**, 3985–3994.

Ham, J. and Parker, M. G. (1989). Regulation of gene expression by nuclear hormone receptors. *Curr. Opinion Cell Biol.* **1**, 503–511.

Ham, J., Thompson, A., Needham, M., Webb, P. and Parker, M. (1988). Characterization of response elements for androgens glucocorticoids and progestins in mouse mammary tumor virus. *Nucl. Acids Res.* **16**, 5263–5277.

Hollenberg, S. M. and Evans, R. M. (1988). Multiple and cooperative transactivation domains of the human glucocorticoid receptor. *Cell* **55**, 899–906.

Hynes, N., van Ooyen, A. J. J., Kennedy, N., Herrlich, P., Ponta, H. and Groner, B. (1983). Subfragments of the larger terminal repeat cause glucocorticoid-responsive expression of the mouse mammary tumor virus and of an adjacent gene. *Proc. Natl. Acad. Sci. USA* **80**, 3637–3641.

Jantzen, H. M., Strähle, U., Gloss, B., Stewart, F., Schmid, W., Boshart, M., Miksicek, R. and Schütz, G. (1987). Cooperativity of glucocorticoid response elements located far upstream of the tyrosine aminotransferase gene. *Cell* **49**, 29–38.

Karin, M., Haslinger, A., Holtgreve, A., Richards, R. I., Krauter, P., Westphal, H. M. and Beato, M. (1984). Characterization of DNA sequences through which cadmium and glucocorticoid hormones induce human metallothionein IIA. *Nature* **308**, 513–519.

Klein-Hitpass, L., Kaling, M. and Ryffel, G. U. (1988). Synergism of closely adjacent estrogen-responsive elements increases their regulatory potential. *J. Mol. Biol.* **201**, 537–544.

Klock, G., Strähle, U. and Schütz, G. (1987). Oestrogen and glucocorticoid responsive elements are closely related but distinct. *Nature* **329**, 734–736.

Krust, A., Green, S., Argos, P., Kumar, V., Walter, P., Bornert, J.-M. and Chambon, P. (1986). The chicken oestrogen receptor sequence: homology with v-*erbA* and the human oestrogen and glucocorticoid receptors. *EMBO J.* **5**, 891–897.

Kühnel, B., Buetti, E. and Diggelmann, H. (1986). Functional analysis of the glucocorticoid regulatory elements present in the mouse mammary tumor virus long terminal repeat—a synthetic distal binding site can replace the proximal binding domain. *J. Mol. Biol.* **190**, 367–378.

Kumar, V. and Chambon, P. (1988). The estrogen receptor binds tightly to its responsive element as a ligand-induced homodimer. *Cell* **55**, 145–156.

Kumar, V., Green, S., Stack, G., Berry, M., Jin, J.-R. and Chambon, P. (1987). Functional domains of the human estrogen receptor. *Cell* **51**, 941–951.

Lee, F., Hall, C. V., Ringold, G. M., Dobson, D. E., Luh, J. and Jacob, P. E. (1984). Functional analysis of the steroid hormone control region of mouse mammary tumor virus. *Nucl. Acids Res.* **12**, 4191–4206.

Lees, J. A., Fawell, S. E. and Parker, M. G. (1989). Identification of two transactivation domains in the mouse oestrogen receptor. *Nucl. Acids Res.* **17**, 5477–5488.

Majors, J. and Varmus, H. (1983). A small region of the mouse mammary tumor virus long terminal repeat confers glucocorticoid hormone regulation on a linked heterologous gene. *Proc. Natl. Acad. Sci. USA* **80**, 5866–5870.

Martinez, E. and Wahli, W. (1989). Cooperative binding of estrogen receptor to imperfect estrogen-responsive DNA elements correlates with their synergistic hormone-dependent enhancer activity. *EMBO J.* **8**, 3781–3791.

Martinez, E., Givel, F. and Wahli, W. (1987). The estrogen-responsive element as an inducible enhancer: DNA sequence requirements and conversion to a glucocorticoid-responsive element. *EMBO J.* **6**, 3719–3727.

Meyer, M.-E., Gronemeyer, H., Turcotte, B., Bocquel, M.-T., Tasset, D. and Chambon, P. (1989). Steroid hormone receptors compete for factors that mediate their enhancer function. *Cell* **57**, 433–442.

Miesfeld, R., Godowski, P. J., Maler, B. A. and Yamamoto, K. R. (1987). Glucocorticoid receptor mutants that define a small region sufficient for enhancer activation. *Science* **236**, 423–427.

Miksicek, R., Heber, A., Schmid, W., Danesch, U., Posseckert, G., Beato, M. and Schütz, G. (1986). Glucocorticoid responsiveness of the transcriptional enhancer of Moloney murine sarcoma virus. *Cell* **46**, 283–290.

Miksicek, R., Borgmeyer, U. and Nowock, J. (1987). Interaction of the TGGCA-binding protein with upstream sequences is required for efficient transcription of mouse mammary tumor virus. *EMBO J.* **6**, 1355–1360.

Miksicek, M. R. and Schütz, G. (1987). Cooperativity of glucocorticoid responsive elements located far upstream of the tyrosin-aminotransferase gene. *Cell* **49**, 29–38.

Muller, M., Baniahmad, C., Kaltschmidt, C. and Renkawitz, R. (1990). Submitted.

Oro, A. E., Hollenger, S. M. and Evans, R. M. (1988). Transcriptional inhibition by a glucocorticoid receptor-β-galactosidase fusion protein. *Cell* **55**, 1109–1114.

Payvar, F., DeFranco, D., Firestone, G. L., Edgar, B., Wrange, Ö., Okret, S., Gustafsson, J. A. and Yamamoto, K. R. (1983). Sequence-specific binding of glucocorticoid receptor to MTV-DNA at sites within and upstream of the transcribed region. *Cell* **35**, 381–392.

Pfahl, M. (1982). Specific binding of the glucocorticoid–receptor complex to the mouse mammary tumor proviral promoter region. *Cell* **31**, 475–482.

Picard, D., Salser, S. J. and Yamamoto, K. R. (1988). A movable and regulable inactivation function within the steroid binding domain of the glucocorticoid receptor. *Cell* **54**, 1073–1080.

Ponta, H., Kennedy, N., Skroch, P., Hynes, N. E. and Groner, B. (1985). Hormonal response region in the mouse mammary tumor virus long terminal repeat can be dissociated from the proviral promoter and has enhancer properties. *Proc. Natl. Acad. Sci. USA* **82**, 1020–1024.

Ptashne, M. (1988). How eukaryotic transcriptional work. *Nature* **335**, 683–689.

Renkawitz, R., Schütz, G., von der Ahe, D. and Beato, M. (1984). Sequences in the promoter region of the chicken lysozyme gene required for steroid regulation and receptor binding. *Cell* **37**, 503–510.

Renkawitz, R., Schüle, R., Kaltschmidt, C., Baniahmad, A., Altschmied, J., Steiner, Ch. and Muller, M. (1989). Clustered arrangement and interaction of steroid hormone receptors with other transcription factors. *In* "Proceedings of the 40th Colloquium Mosbach". Springer-Verlag, Heidelberg, pp. 21–28.

Richard-Foy, H. and Hager, G. L. (1987). Sequence-specific positioning of nucleosomes over the steroid-inducible MMTV promoter. *EMBO J.* **6**, 2321–2328.

Ruden, D. M., Ma, J. and Ptashne, M. (1988). No strict alignment is required between a transcriptional activator binding site and the "TATA box" of a yeast gene. *Proc. Natl. Acad. Sci. USA* **85**, 4262–4266.

Rusconi, S. and Yamamoto, K. R. (1987). Functional dissection of the hormone and DNA binding activities of the glucocorticoid receptor. *EMBO J.* **6**, 1309–1315.

Sakai, D., Helms, D. D., Helms, S., Carlstedt-Duke, J., Gustafsson, J. A., Rottman, F. M. and Yamamoto, K. R. (1988). Hormone-mediated repression of transcription: a negative glucocorticoid responsive element from the bovine prolactin gene. *Genes Dev.* **2**, 1144–1154.

Scheidereit, C., Geisse, S., Westphal, H. M. and Beato, M. (1983). The glucocorticoid receptor binds to defined nucleotide sequences near the promoter of mouse mammary tumor virus. *Nature* **304**, 749–752.

Schmid, W., Strähle, U., Schütz, G., Schmitt, J. and Stunnenberg, H. (1989). Glucocorticoid receptor binds cooperatively to adjacent recognition sites. *EMBO J.* **8**, 2257–2263.

Schüle, R., Muller, M., Kaltschmidt, C. and Renkawitz, R. (1988a). Many transcription factors interact synergistically with steroid receptors. *Science* **242**, 1418–1420.

Schüle, R., Muller, M., Otsuka-Murakami, H. and Renkawitz, R. (1988b). Cooperativity of the glucocorticoid receptor and the CACCC-box binding factor. *Nature* **332**, 87–90.

Slater, E., Rabenau, O., Karin, M., Baxter, J. D. and Beato, M. (1985). Glucocorticoid receptor binding and activation of a heterologous promoter in response to dexamethasone by the first intron of the human growth hormone gene. *Mol. Cell Biol.* **5**, 2984–2992.

Strähle, U., Klock, G. and Schütz, G. (1987). A DNA sequence of 15 base pairs is sufficient to mediate both glucocorticoid and progesterone induction of gene expression. *Proc. Natl. Acad. Sci. USA* **84**, 7871–7875.

Strähle, U., Schmid, W. and Schütz, G. (1988). Synergistic action of the glucocorticoid receptor with transcription factors. *EMBO J.* **7**, 3389–3395.

Strähle, U., Boshart, M., Klock, G., Stewart, F. and Schütz, G. (1989). Glucocorticoid- and progesterone-specific effects are determined by differential expression of the respective hormone receptors. *Nature* **339**, 629–632.

Tora, L., Gronemeyer, H., Turcotte, B., Gaub, M.-P. and Chambon, P. (1988). The N-terminal region of the chicken progesterone receptor specifies target gene activation. *Nature* **333**, 185–188.

Tora, L., White, J., Brou, C., Tasset, D., Webster, N., Scheer, E. and Chambon, P. (1989). The human estrogen receptor has two independent nonacidic transcriptional activation functions. *Cell* **59**, 477–487.

Tsai, S. Y., Carlstedt-Duke, J., Weigel, N. L., Dahlman, K., Gustafsson, J.-A., Tsai, M.-J. and O'Malley, B. W. (1988). Molecular interactions of steroid-hormone receptor with its enhancer element: evidence for receptor dimer formation. *Cell* **55**, 361–369.

Tsai, S. Y., Tsai, M.-J. and O'Malley, B. W. (1989). Cooperative binding of steroid hormone receptors contributes to transcriptional synergism at target enhancer elements. *Cell* **57**, 443–448.

von der Ahe, D., Janich, S., Scheidereit, C., Renkawitz, R., Schütz, G. and Beato, M. (1985). Glucocorticoid and progesterone receptors bind to the same sites in two hormonally regulated promoters. *Nature* **313**, 706–709.

Wirth, T., Staudt, L. and Baltimore, D. (1987). An octamer oligonucleotide upstream of a TATA motif is sufficient for lymphoid-specific promoter activity. *Nature* **329**, 174–178.

8
Repression of Gene Expression by Steroid and Thyroid Hormones

I. E. AKERBLOM and P. L. MELLON

Regulatory Biology Laboratory, The Salk Institute for Biological Studies, La Jolla, CA 92037, USA

8.1 Introduction

Negative regulation of transcription is important in the control of many aspects of differentiation, development, growth control, and homeostasis. In the bacteria *Escherichia coli*, for example, the lactose operon is silenced by a transcriptional repressor molecule unless lactose is present (Riggs and Bourgeois, 1968; Miller and Reznikoff, 1978). In the yeast *Saccharomyces cerevisiae*, the expression of mating-type genes is controlled by transcription factors which repress genes inappropriate for the specific mating phenotype (Sprague *et al.*, 1983; Nasmyth and Shore, 1987). The ontogeny of *Drosophila* occurs by controlled temporal and spatial expression of transcription factors which combine to either suppress or activate genes involved in pattern formation (Akam, 1987; Ingham, 1988; Scott and Carroll, 1987; Han *et al.*, 1989). A number of mammalian oncogenes whose proper regulation is crucial to the growth control of the cell are transcriptionally repressed after their acute induction is triggered by the presence of growth factors (Hay *et al.*, 1987; Sassone-Corsi *et al.*, 1988). Finally, in endocrine systems, negative regulation may occur through the interplay of specific protein hormone/antihormone pairs which act antagonistically such as growth hormone and somatostatin, or through negative feedback loops (Tepperman and Tepperman, 1987).

Steroid hormones exert a number of important negative regulatory functions, in particular through feedback loops involving their own production. For example, stress results in the production of glucocorticoids by the adrenal gland (Tepperman and Tepperman, 1987) which feedback to repress the synthesis of the peptide hormones responsible for signalling their production (hypothalamic corticotropin-releasing factor and pituitary adrenocorticotropin) (ACTH; a product of the pro-opiomelanocortin gene, POMC; Charron and Drouin, 1986; Drouin *et al.*, 1987). Glucocorticoids specifically repress transcription through the 5'-flanking sequences of several genes including collagen (types I and IV), stromelysin, and α-fetoprotein, in addition to POMC (Israel and Cohen, 1985; Charron and Drouin, 1986; Frisch and Ruley, 1987; Weiner *et al.*, 1987; Guertin *et al.*, 1988). Thyroid hormones, whose multiple receptors have been identified as members of the steroid hormone receptor gene superfamily (for reviews, see Evans, 1988 and Chapter 4), also decrease the transcription of genes involved in the stimulation of their synthesis (such as the thyrotropin-releasing hormone genes, TSH; Shupnik *et al.*, 1985).

Steroid hormone receptors and some of the genes with which they interact have emerged as prominent model systems for the investigation of the molecular mechanisms of transcriptional repression. That these receptors can in fact exert negative transcriptional effects appears to be a paradox since steroid hormone receptors are the best characterized class of transcriptional activator proteins (see Chapters 2 and 3; for reviews see Yamamoto, 1985; Evans, 1988; Beato, 1989). More recently, other factors involved in transcriptional activation have also been implicated in repression of transcription (Brand *et al.*, 1987; Keleher *et al.*, 1988; Mordacq and Linzer, 1989; Takimoto *et al.*, 1989).

8.2 Molecular mechanisms involved in negative regulation of transcription by steroid hormones

Several models have been proposed for transcriptional repression exerted by members of the steroid hormone receptor superfamily. One strategy is that of competition between the hormone receptor and other important transcription factors. Receptor bound to specific sites coincident with or overlapping those of positive transcription factors may cause interference with binding of the positive factor(s) and a corresponding decrease in transcriptional activation (Fig. 1a) (Guertin *et al.*, 1988; Drouin *et al.*, 1989a). Alternatively, occupation of a receptor-binding site closely apposed to a site for a transcriptional activator may mask the nearby activator protein, preventing its access to required transcriptional machinery, a model similar to one proposed for repression in the yeast mating type system (Fig. 1b) (Brent and Ptashne, 1984; Keleher *et al.*, 1988). A variation of this model proposes that receptor can interfere with interaction between an activator protein and the transcriptional machinery by itself interacting with the transcription initiation complex. These models require specific DNA binding by the receptor to sites in the target gene.

A second type of mechanism for repression involves steroid hormone

8. REPRESSION OF GENE EXPRESSION

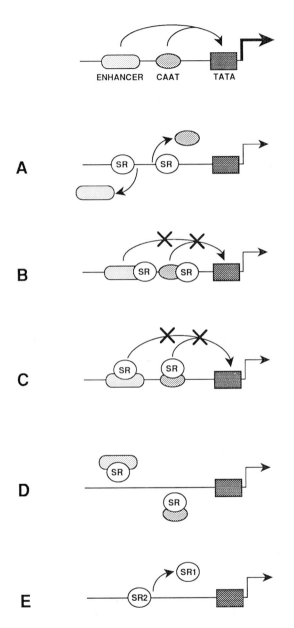

Fig. 1 Models for the molecular basis of negtive regulation by steroid and thyroid hormone receptors. SR, steroid or thyroid hormone receptor; CAAT, CAAT box binding transcription factor; TATA, TATAAA box binding transcription factor.

receptors inhibiting expression through protein–protein interactions which do not involve direct DNA binding by the receptor. This can occur either through the masking of transcription factors bound to DNA (Fig. 1c) or through sequestration of factors preventing their interaction with the DNA (Fig. 1d; Meyer et al., 1989). In at least one case, a steroid hormone receptor appears to repress transcription by in some way altering the transcription factors necessary for efficient expression of a specific gene (Mordacq and Linzer, 1989).

Members of the steroid hormone receptor superfamily can also interfere with transcription mediated by other members of the family by interacting with similar or identical DNA-binding sites; this is possible due to their highly related zinc finger-containing DNA-binding domains (Glass et al., 1988; Umesono et al., 1988; Danielsen et al., 1989; Graupner et al., 1989; Umesono and Evans, 1989). Competition between related receptors for access to a given binding site leads to compromises in induction by at least one of the receptors (Fig. 1E). In addition, different members of the superfamily can form heterodimers, which, depending on the context of the binding site, can negatively or positively affect expression (Glass et al., 1989; see Chapters 4.9 and 16.3). Consequently, the effects of hormone addition on transcription will be dependent upon the types of receptors present in the cell, their relative concentrations, and finally their relative affinities to the binding sites. In general, this type of negative regulation is more correctly defined as blocking a positive response rather than as reduction of basal level expression.

Many different parameters can contribute to the effect of a particular factor, or hormone receptor, on transcription. The context of a binding site relative to sites for other transcription factors (Han et al., 1989), the architecture of the binding site itself (Glass et al., 1988), the developmental and cellular environment (Borelli et al., 1984), and the factor's presence or absence at different stages of the cell cycle (Takimoto et al., 1989) may influence whether a factor mediates positive or negative effects. Therefore, determination of the mechanisms responsible for repression is intimately associated with knowledge of the factors contributing to a particular gene's expression (for a review on general negative regulation see Levine and Manley, 1989).

8.3 Competition between receptors and transcription factors at the level of DNA binding

Transcriptional activation of gene expression by steroid hormone receptors occurs through receptor binding to specific DNA sequences near positively regulated genes (Chandler et al., 1983; Hynes et al., 1983; Karin et al., 1984; Renkawitz et al., 1984; Miesfeld et al., 1986; see Chapter 6.2 and 9.4). In the cases of negative regulation which require the property of receptor binding to the DNA, several interesting questions arise.

(1) Are the hormone-responsive elements involved in negative regulation similar to those used by the receptor in mediating enhancement, or, are these elements qualitatively different such that they cannot support activation?
(2) How important is the context in which an element resides?
(3) Can a binding site which mediates repression in one context mediate activation in another?

8.3.1 Pro-opiomelanocortin Gene

Repression of pro-opiomelanocortin (POMC) gene expression by glucocorticoids is a classic example of negative feedback regulation and is one of the best characterized systems for repression by steroid hormones (Drouin et al., 1989a). Negative regulation of the POMC gene by glucocorticoids occurs primarily at the transcriptional level and is conferred by 5′ flanking sequences as shown in transfection experiments and in transgenic mice (Tremblay et al., 1988). A glucocorticoid receptor binding site in the promoter region is responsible for the negative effect since mutation of two nucleotides in this negative glucocorticoid response element (nGRE) results in a loss of both receptor DNA-binding and repression (Drouin et al., 1989a). When this element is removed from the context of the POMC gene, it has neither inherent negative character nor any ability to transactivate (Drouin et al., 1989b). The promoter region containing this nGRE also binds other nuclear proteins, one of which may be the chicken ovalbumin upstream promoter (COUP) transcription factor, a 'zinc finger' protein that is related to the steroid hormone receptor superfamily (Drouin et al., 1989a,b; Wang et al., 1989). The location of the POMC nGRE overlapping the binding site for this positive transcription factor strongly suggests that the mechanism responsible for repression is through direct competition for DNA binding.

8.3.2 α-Subunit Gene of the Glycoprotein Hormones

The glycoprotein hormones, luteinizing hormone (LH), follicle stimulating hormone (FSH), chorionic gonadotropin (CG), and thyroid stimulating hormone (TSH), are heterodimeric proteins composed of individual β-subunits and a common α-subunit (for reviews, see Pierce and Parsons, 1981;

Tepperman and Tepperman, 1987). The gonadotropins, LH and CG, are the two major trophic hormones involved in stimulating production of the sex steroid hormones, which include oestrogens, progestins and androgens (Fink, 1979), and are in turn negatively regulated by oestrogens and androgens (Vreeburg et al., 1984; Gharib et al., 1986). In addition, reproductive function is known to be adversely affected by stress, a condition which results in the elevation of circulating levels of glucocorticoids (Baxter and Tyrell, 1987). Some evidence points to a decrease in the synthesis or secretion of the gonadotropin hormones as being in part responsible for the negative effect of glucocorticoids on reproduction (Mann et al., 1985; Suter and Schwartz, 1985).

TSH is the major regulator of thyroid hormone synthesis in the thyroid gland (Bowers et al., 1967; Morley, 1981; Shupnik et al., 1986). Production of TSH is subject to negative feedback control by thyroid hormone (Shupnik et al., 1985). Thyroid hormone has been shown to directly decrease the transcription of the α-subunit gene in rat primary pituitary cultures (Shupnik et al., 1985). Thus, this family of hormones is negatively regulated by many of the steroid and thyroid hormones whose production they stimulate.

The α-subunit gene is negatively regulated by glucocorticoids during gene transfer into placental cells (Akerblom et al., 1988), an effect dependent upon the introduction of glucocorticoid receptor (Giguere et al., 1986; Hollenberg et al., 1987) and the addition of hormone (Akerblom et al., 1988). The same receptor mediates both positive and negative regulation depending upon the reporter gene, establishing that the identical receptor is responsible for these divergent transcriptional effects (Fig. 2; Akerblom et al., 1988). A glucocorticoid receptor mutated in the DNA-binding domain was unable to mediate either repression or activation (Akerblom et al., 1988). In fact, the entire primary amino acid sequence of the receptor outside the DNA-binding domain could be removed and repression was still observed as long as the carboxy terminus was replaced by neutral protein coding sequences (Oro et al., 1988). Thus, repression by glucocorticoid receptor of the α-subunit promoter appears to correlate with the property of receptor DNA binding.

A repression model in which DNA binding by the receptor interferes with activation by another transcription factor predicts that any receptor capable of activating or even efficiently binding to DNA should repress. Extensive analysis using mutated glucocorticoid receptor molecules to compare activation and repression confirms this prediction (Oro et al., 1988). In addition, mutant glucocorticoid receptors entirely deficient in transcriptional activation were fully capable of mediating repression, indicating that these two processes are separable (Oro et al., 1988; I. Akerblom and P. Mellon, unpubl. res.).

8. REPRESSION OF GENE EXPRESSION

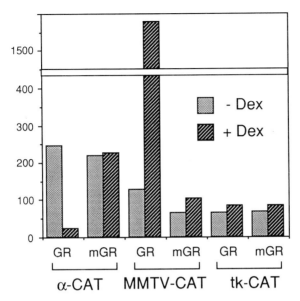

Fig. 2 Negative regulation of the α-subunit gene by glucocorticoids. CAT reporter plasmids containing either the human α-subunit gene transcriptional controlling sequences, the MMTV glucocorticoid regulatory elements, or the non-steroid responsive thymidine kinase promoter were cotransfected into placental cells along with expression vectors for either wild-type (GR) or mutant (mGR) human glucocorticoid receptor (Akerblom et al., 1988). Response to addition of the synthetic glucocorticoid dexamethasone (Dex) is expressed as acetylated [^{14}C]chloramphenicol counts per µg of cell protein per hour.

The steroid-responsive region has been defined between -152 and $+4$ of the α-subunit flanking sequences, a region which also contains the transcriptional elements involved in both tissue-specific expression and cAMP induction of this gene (Fig. 3). Two cAMP responsive elements (CREs), located between -146 and -111 (Delegeane et al., 1987; Silver et al., 1987; Deutsch et al., 1987; Jameson et al., 1987) are bound by the transcription factor CREB (Delegeane et al., 1987; Hoeffler et al., 1988; Gonzalez et al., 1989; Maekawa et al., 1989). Significantly, the basal expression of the human gene in placental cells is wholly dependent upon the CREs which act in co-operation with an adjacent trophoblast-specific element (TSE; Delegeane et al., 1987; Jameson et al., 1988) to confer placental specificity. Moreover, repression is observed only when these CRE elements are active as enhancers (Akerblom et al., 1988). However, the receptor does not affect the character of the CRE-binding protein or the TSE-binding protein since placement of their cognate elements on a heterologous promoter resulted in the loss of steroid-mediated repression (I. Akerblom and P. Mellon, unpubl. res.).

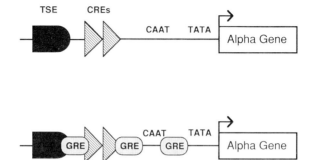

Fig. 3 Glucocorticoid receptor binding sites present in the α-subunit promoter overlap transcriptional elements important for expression in placental cells. TSE, trophoblast-specific element; CRE, cAMP-responsive element; CAAT, CAAT box; TATA, TATAAA box. The text describes the properties of the α-subunit transcription elements and their associated transcription factors (Delegeane et al., 1987). Three receptor binding sites were identified in this region using partially purified rat liver glucocorticoid receptor in DNase I and methylation protection analysis (Akerblom et al., 1988).

Three glucocorticoid receptor-binding sites have been identified *in vitro* in this region, two of which are located adjacent and overlapping the CRE elements upon which α-subunit expression in these cells is wholly dependent (Fig. 3; Akerblom et al., 1988). A third site is located overlapping the site for a more general transcription factor, the CAAT box binding protein (Delegeane et al., 1987). These sites do not have intrinsic glucocorticoid-dependent silencing activity since placement of the isolated binding sites on a heterologous promoter produced no negative effects; in fact, one of the elements was capable of mediating hormone-dependent transcriptional activation (I. Akerblom, unpubl. res.). These results suggest that it may be the context of the receptor binding sites in the α-subunit promoter which leads to negative regulation. Further, in non-placental cells, α-subunit promoter activity was increased by glucocorticoids, suggesting that the receptor can interact with α-subunit sequences in a positive manner *in vivo* when the tissue-specific factors are absent (Mellon and Akerblom, 1989). Finally, glucocorticoids negatively regulate the endogenous murine α-subunit gene in a novel α-subunit-secreting thyrotrope cell line (Akerblom et al., 1989). This was a significant observation because the experiments described above were carried out by cotransfection of the cloned glucocorticoid receptor and the human α-subunit promoter, but clearly negative regulation can also occur with endogenous glucocorticoid receptors.

Taken together, these observations lead to a model which involves binding of the receptor to the α-subunit promoter, thereby compromising the action of the transcription factors directing expression of this gene. This could be accomplished by direct competition for binding between the receptors and either the CRE or the CAAT box binding proteins (Fig. 1A). Alternatively, the location of one or all of the receptor binding sites in the α-subunit promoter may lead to a masking of the nearby transcription factors, thus preventing them from interacting with required transcriptional machinery (Fig. 1B; Keleher et al., 1988). Since expression of the α-subunit gene is entirely dependent upon CREB, which interacts with the TATA binding factor (Hai et al., 1988; Horikoshi et al., 1988), even a small loss in its ability to direct transcription would have a significant effect on the expression of the α-subunit gene.

The thyroid hormone responsiveness of both TSH subunit genes has been investigated using gene transfer into GH_3 cells (Burnside et al., 1989; Carr et al., 1989). The rat α-subunit gene is negatively regulated by thyroid hormone in these cells (Burnside et al., 1989). Through the use of promoter deletion analysis and receptor binding assays, the responsive region has been mapped to sequences which overlap the most downstream of the glucocorticoid receptor binding sites discussed above (Akerblom et al., 1988). The precise location of the thyroid hormone response element (TRE) in the α-subunit sequences has not yet been delineated, but it appears that there is no apparent homology between the thyroid hormone response sequences and the positive TRE identified in the growth hormone gene (Crew and Spindler, 1986; Larsen et al., 1986; Flug et al., 1987; Glass et al., 1987, 1988; see Chapters 4.5, 6.2, and 9.4). Therefore, there may be a fundamental difference between the binding sites of positive and negative TREs.

The TSH β-subunit gene is also negatively regulated by thyroid hormone in both rat primary pituitary cultures and in GH_3 cells (Carr et al., 1989). This effect has been mapped by deletion experiments to a region of the promoter containing 17 bp of the 5' flanking region, 27 bases of the first exon and the first 13 bases of the first intron. Within this region, there are sequences which contain some homology to the TREs of positively regulated genes; however, receptor binding to these sequences has not yet been demonstrated. If the negative effects are mediated by the receptor directly on β-subunit sequences, then perhaps the presence of the receptor in the transcriptional initiation region may interfere with efficient formation of a transcription initiation complex. It might by expected that significant differences would exist between the molecular mechanisms used for repression by thyroid hormone receptors and other steroid hormone receptors since thyroid hormone receptors are probably bound to their regulatory elements even in the absence of hormone (Damm et al., 1989; Graupner et al., 1989; Koenig et al., 1989; see Chapters 4.8 and 15.4).

8.3.3 α-Fetoprotein Gene

Negative regulation of the α-fetoprotein gene by glucocorticoids is independent of protein synthesis (Turcotte *et al.*, 1985). Repression has been mapped to a region of the promoter which contains a glucocorticoid receptor binding site defined *in vitro* (Guertin *et al.*, 1988). A fragment containing this GRE has been shown by deletion analysis to be important for both basal level activity of the promoter and for mediating repression. Thus, the GRE located in the α-fetoprotein promoter also apparently overlaps with the binding site for a positive transcription factor, similar to the situation seen in the examples cited earlier. A point mutation in the GRE consensus homology results in a loss of basal activity, as does a mutation in the gap between the two hexamer half-sites of the glucocorticoid receptor binding site, a mutation which would not be predicted to affect receptor binding. The two binding sites, one for a constitutive factor and one for the receptor, are thus coincident, indicating that competition at the level of DNA binding may occur in this case as well.

8.3.4 Prolactin Gene

Camper *et al.* (1985) have shown that the bovine prolactin promoter is repressed by glucocorticoids using gene transfer in rat pituitary GH_3 cells. Purified glucocorticoid receptor binds to seven sites, four of which appeared to be required for full repression (Sakai *et al.*, 1988). Mutagenesis of the receptor-binding sites found in the portion of the promoter responsible for repression results in loss of basal transcription and consequently hormonal regulation. Significantly, in a cell line which does not contain this constitutive enhancer activity, no repression is seen (Sakai *et al.*, 1988), indicating that no inherent silencing activity is conferred by the isolated negative GREs of the prolactin promoter. Placement of the individual putative GREs on a heterologous promoter leads to hormone-independent enhancement of basal level expression, identifying these binding sites as constitutive enhancer elements separate from their repression by hormone. Repression occurs as a compromise of this enhancement rather than as a reduction of basal level promoter activity, further suggesting that the nGREs contain no inherent negative function, a situation similar to that found for a α-fetoprotein gene. These results indicate that these nGREs are also binding sites for a cellular factor distinct from the glucocorticoid receptor.

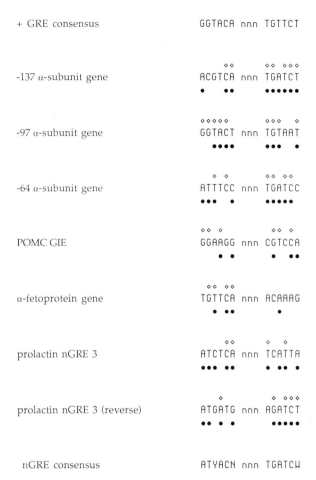

Fig. 4 Sequences of negative glucocorticoid regulatory elements. Receptor binding site in the α-subunit gene (Akerblom et al., 1988), POMC gene (Drouin et al., 1989a), the α-fetoprotein gene (Guertin et al., 1988), and the bovine prolactin gene (Sakai et al., 1988) promoters are compared with the consensus sequences for positive and negative glucocorticoid regulatory elements (Beato, 1989; Beato et al., 1987). ◇, open diamonds show identity to the positive GRE consensus; ●, closed circles show identity to the negative GRE consensus.

8.3.5 Negative Glucocorticoid Response Elements

An interesting question remains concerning the nature of the nGRE sequences. Why doesn't receptor binding to these sequences result in transcriptional activation? The GR-binding sites from the POMC, prolactin, and α-subunit genes show some homology with the consensus glucocorticoid response element (GRE) (Fig. 4). The consensus GRE was derived from a compilation of sequences which have been shown to confer transcriptional activation (Beato *et al.*, 1987; see Chapters 6.2, 9.2 and 9.4); therefore, it may not accurately reflect the consensus for DNA binding by the receptor. The lack of transcriptional activation by these less conserved binding sites may be due to a relative instability of the receptor–DNA interaction; alternatively, it may be that the particular bases found in the positive consensus sequence are required for the successful transduction of transactivation. The negative GREs discussed here are either neutral or positive when removed from the context of the repressed gene.

Glucocorticoid receptor appears to be capable of binding to less conserved sequences than the positive consensus would indicate. It is tempting to speculate that in cases of repression mediated by receptor binding, there is no selective pressure to maintain bases necessary for induction, but only those bases required for DNA binding. Thus, sequences of GREs involved in repression may show considerable drift from those of positive sites. A compilation of the sequences of GREs involved in repression apparently reflects this decreased conservation (Beato, 1989; Fig. 4).

8.4 Interactions of receptors with other transcription factors in the absence of DNA binding

Steroid receptors also appear to repress transcription through mechanisms which do not require the DNA-binding domain; these cases of negative regulation may be examples of squelching, first defined by Gill and Ptashne (1988) for the yeast transactivator GAL4. When an excess of an expression vector producing large quantities of GAL4 is cotransfected with reporter plasmids which contain a transcription unit, transcription of the reporter gene is compromised even though no DNA-binding site for the GAL4 protein is present in *cis* (Gill and Ptashne, 1988). Interestingly, this 'squelching phenomenon' was dependent upon the presence of the activation portion of the GAL4 protein, which is separable from the DNA-binding domain, suggesting that this part of the molecule may be titrating some common factor necessary for the transcription of the squelched gene.

Recently, Meyer *et al.* (1989) observed that cotransfection of the progesterone receptor and the oestrogen receptor, together with a reporter plasmid containing a hormone response element specific only for the progesterone receptor (PRE/GRE), led to an oestrogen-dependent loss of induction mediated by this PRE. This effect is not at the level of competition for binding to the DNA element as it is known that the oestrogen receptor does not recognize progesterone response elements (Meyer et al., 1989). Further, the DNA-binding domain of the oestrogen receptor was not required for repression; however, both the amino terminus and the hormone-binding domain separately were capable of mediating this negative regulation. Conversely, the glucocorticoid receptor and the progesterone receptor could also cause inhibition of an oestrogen response element, albeit at a lower level than that seen with the oestrogen receptor. Competition was also shown to occur *in vivo* using a cell line which contains both progesterone and oestrogen receptors endogenously. The authors speculate that the receptors compete for access to more general transcription factors, thus leading to a loss of activation by the receptor present in lower concentration (this effect would also be dependent upon the relative affinity of the various receptors for the relevant transcription factors).

Overall, the relative effect of competition among the various receptors on gene expression would most likely be dependent upon the receptor content of the particular cell, the concentration of the various receptors relative to each other, and the affinity of the different receptors for a particular DNA binding site. The general physiological importance of this type of competition *in vivo* awaits further investigation in these and other systems.

8.4.1 Oestrogen Regulation of the Prolactin Gene

The prolactin gene is positively regulated by oestrogen *in vivo* and in GH_3 cells after transfection. However, if the oestrogen response element (ERE) is removed, the gene is repressed by the addition of oestrogen after cotransfection with an oestrogen expression vector. In contrast to the systems described above, negative regulation of the prolactin gene by the oestrogen receptor has been characterized as independent of the DNA-binding domain (Adler et al., 1988). In addition, no steroid response element was necessary for this effect to occur (Adler et al., 1988). Moreover, the binding sites for the transcription factor pituitary factor-1 (pit 1; Nelson et al., 1988) appeared to correlate with the repressive effect (Adler et al., 1988). The authors suggest a model for inhibition which involves binding of the receptor to pit 1, preventing a productive interaction with the prolactin promoter (Fig. 1C and D).

8.4.2 Glucocorticoid Regulation of the Proliferin Gene

The proliferin gene has recently been shown to be repressed by glucocorticoids (Mordacq and Linzer, 1989). Expression of this gene is positively controlled by serum or phorbol esters (TPA), through a region of the promoter which contains binding sites for a number of transcription factors including one for activator protein-1 (AP-1) (Angel et al., 1987; Lee et al., 1987), a transcription factor which is activated by phorbol esters. Significantly, the AP-1 site has high homology with the CRE element found in the α-subunit gene and the two binding proteins display some homology, particularly in their domain structures rather than at the primary sequence level (Angel et al., 1988; Hoeffler et al., 1988; Rauscher et al., 1988; Gonzalez et al., 1989). A glucocorticoid receptor binding site has been defined in vitro just 5′ to the TPA-responsive region; however, deletion of this site does not result in a loss of negative regulation though negative regulation of this gene is mediated through the TPA-responsive region and occurs only under conditions of TPA or serum stimulation. Instead, it appears that the presence of the receptor may affect the AP-1 protein itself since a different gel-shift pattern of an AP-1 oligonucleotide was observed when extracts were prepared from glucocorticoid treated cells.

8.5 Competition between hormone receptors

A number of hormone receptors have been found to bind to overlapping sets of response elements, probably due to the close relationship amongst the DNA-binding domains of members of the steroid hormone receptor superfamily (for review see Evans, 1988; Danielsen et al., 1989; Umesono and Evans, 1989). For example, the glucocorticoid, progesterone and androgen receptors interact with highly related, if not identical, DNA-binding sites (Strähle et al., 1987; Ham et al., 1988). The activation properties of a retinoic acid receptor were first characterized through the observation that it was capable of activating transcription from a thyroid hormone response element (Umesono et al., 1988). Therefore, it seems likely that hormone receptors may in fact compete for binding to a set of permissive elements (see Chapter 6.2).

8.5.1 Retinoic Acid and Thyroid Hormone Receptors

Retinoic acid receptor ε (RARε) can induce retinoic acid-dependent transcription from a thyroid hormone response element (TRE) when the receptor and a reporter gene are cointroduced into F9 teratocarcinoma cells using gene transfer (Graupner et al., 1989). This activation can be blocked by the introduction of a thyroid hormone receptor expression plasmid (encoding

TRα or TRβ); surprisingly, this blockage occurs even in the absence of added thyroid hormone. Upon the addition of thyroid hormone, induction of the TRE can again be observed. Through the use of hybrid receptors, the DNA-binding domain of the thyroid receptor was shown to be the portion of the receptor required for this inhibitory effect. Thus, these two related receptors appear to compete for binding to at least some of the same elements. In addition, since the thyroid hormone receptor appears to bind DNA in the absence of hormone (Damm *et al.*, 1989; see Chapters 4.8 and 15.4), it is capable of blocking induction of the TRE by the activated retinoic acid receptor even when thyroid hormone is not present.

The thyroid hormone receptor and the retinoic acid receptor can also interact by formation of heterodimers which confer retinoic acid responsiveness on thyroid hormone response elements (Glass *et al.*, 1989). Surprisingly, depending on the context and specific sequence of the binding site, this response can be negative or positive.

8.5.2 Thyroid Hormone and Oestrogen Receptors

Thyroid hormone receptors have also been found to interact with oestrogen response elements (ERE; Glass *et al.*, 1988); however, they are not capable of activating an ERE with or without thyroid hormone. The half-sites of the recognition sequence for thyroid hormone receptor (TRE) are identical to those of the ERE, except that the TRE lacks the central three nucleotide spacer of the ERE dyad symmetry. Removal of this 3 bp gap in a vitellogenin ERE results in its conversion into a thyroid hormone response element *in vivo*. Interestingly, the thyroid hormone receptor also binds to the wild-type vitellogenin ERE with greater affinity than to its own response element. That the thyroid hormone receptor can bind to an ERE *in vivo* was reflected in the 75% loss of oestrogen-mediated induction of an ERE in the presence of thyroid hormone. Thyroid hormone receptor binding to an ERE *in vivo* has no inherent negative effects; repression is seen only as a loss of oestrogen-responsiveness. Interestingly, in contrast to the case presented above (Graupner *et al.*, 1989), blockage of the oestrogen-mediated response was not observed in the absence of thyroid hormone, suggesting that the thyroid hormone receptor may not bind to EREs without thyroid hormone.

8.5.3 Competition Between Multiple Forms of the Thyroid Hormone Receptor

Finally, competition may also exist between receptors which respond to the same hormone. The identification of multiple genes encoding thyroid hormone receptors (Sap et al., 1986; Weinberger et al., 1986; Thompson et al., 1987; Hodin et al., 1989), and multiple spliced forms of these receptors (Izumo and Mahdavi, 1988; Lazar et al., 1988), led to the discovery that certain forms of the receptor were capable of binding DNA but could not mediate positive transcriptional activation upon the addition of hormone. In rats, alternative splicing of thyroid hormone receptor-α results in the production of a receptor, TRα-2, which does not bind hormone and thus cannot activate transcription (Lazar et al., 1988). TRα-2 will block thyroid hormone-dependent induction of a TRE by a thyroid receptor capable of binding thyroid hormone (Koenig et al., 1989). Loss of transcriptional activation in the presence of TRα-2 may be the result of either direct binding competition between the receptors or may be due to the formation of inactive receptor heterodimers (see Chapters 4.9 and 16.3).

The thyroid hormone receptor has been identified as the cellular homologue of the oncogene v-erbA (Sap et al., 1986; Weinberger et al., 1986). Recent work on the structural and functional relationship between the wild-type and oncogenic receptor has led to the proposal that v-erbA may be a dominant negative oncogene (Damm et al., 1989). Thus, a mutant receptor compromised in its activation or hormone-binding properties, but not its DNA-binding character, may act as an antagonist of normal thyroid hormone action by preventing the wild-type receptor from interacting with its normal target sequences (see Chapter 15.4).

8.6 Summary

The molecular mechanisms responsible for negative regulation of transcription by steroid hormones vary from gene to gene. Competition with other required transcription factors is a repeating strategy by which these receptors compromise the expression of their target genes. This competition may occur at the level of direct competition for DNA binding through the presence of overlapping recognition elements for either transcription factors or other hormone receptors. Alternatively, competition for more general transcription factors may occur in an indirect manner, an effect not dependent upon the DNA-binding property of the receptor. Finally, in at least one case, the hormone receptor appears to affect a particular transcription factor directly, but not in such a way that DNA binding of the receptor is required. Steroid and thyroid receptors can also compete with each other to compromise gene expression, a phenomenon which might provide complex, higher-order levels of regulation of gene expression.

References

Adler, S., Waterman, M. L., He, X. and Rosenfeld, M. G. (1988). Steroid receptor-mediated inhibition of rat prolactin gene expression does not require the receptor DNA-binding domain. *Cell* **52**, 685–695.

Akam, M. (1987). The molecular basis for metameric pattern in the *Drosophila* embryo. *Development* **101**, 1.

Akerblom, I. E., Slater E. P., Beato, M., Baxter, J. D. and Mellon, P. L. (1988). Negative regulation by glucocorticoids through interference with a cAMP responsive enhancer. *Science* **241**, 350–353.

Akerblom, I. E., Ridgway, E. C. and Mellon, P. L. (1990). An α-subunit-secreting cell line derived from a thyrotropic tumor. *Mol. Endocrinol.* **4**, 589–596.

Angel, P., Imagawa, M., Chiu, R., Stein, B., Imbra, R. J., Ramsdorf, H. J., Jonat, C., Herrlich, P. and Karin, M. (1987). Phorbol ester-inducible genes contain a common *cis* element recognized by a TPA-modulated trans-acting factor. *Cell* **49**, 729–739.

Angel, P., Allegretto, E. A., Okino, S. T., Hattori, K., Boyle, W. J., Hunter, T. and Karin, M. (1988). Oncogene *jun* encodes a sequence-specific trans-activator similar to AP-1. *Nature* **332**, 166–171.

Baxter, J. D. and Tyrell, J. B. (1987). *In* "Endocrinology and Metabolism" (eds P. Felig, J. D. Baxter, A. E. Broadus and L. A. Frohman), p. 511. McGraw-Hill Book Company, New York, pp. 385–510.

Beato, M. (1989). Gene regulation by steroid hormones. *Cell* **56**, 335–344.

Beato, M., Arnemann, J., Chalepakis, G., Slater, E. and Willmann, T. (1987). Gene regulation by steroid hormones. *J. Steroid Biochem.* **27**, 9–14.

Borrelli, E., Hen, R. and Chambon, P. (1984). Adenovirus-2 E1A products repress enhancer-induced stimulation of transcription. *Nature* **312**, 608–612.

Bowers, C. Y., Schally, A. V., Reynolds, G. A. and Hawley, W. D. (1967). Interactions of L-thyroxine or L-triiodothyronine and thyrotropin releasing factor on the release and the synthesis of thyrotropin from the anterior pituitary gland of mice. *Endocrinology* **81**, 741–747.

Brand, A. H., Micklem, G. and Naysmyth, K. (1987). A yeast silencer contains sequences that can promote autonomous plasmid replication and transcriptional activation. *Cell* **51**, 709–719.

Brent, R. and Ptashne, M. (1984). A bacterial repressor protein or a yeast transcriptional terminator can block upstream activation of a yeast gene. *Nature* **312**, 612–615.

Burnside, J., Darling, D. S., Carr, F. E. and Chin, W. W. (1989). Thyroid hormone regulation of the rat glycoprotein hormone *a*-subunit gene promoter activity. *J. Biol. Chem.* **264**, 6886–6891.

Camper, S. A., Yao, Y. A. S. and Rottman, F. M. (1985). Hormonal regulation of the bovine prolactin promoter in rat pituitary tumor cells. *J. Biol. Chem.* **260**, 12 246–12 251.

Carr, F. E., Burnside, J. and Chin, W. W. (1989). Thyroid hormones regulate rat thyrotropin-β gene promoter activity in GH3 cells. *Mol. Endocrinol.* **3**, 709–716.

Chandler, V. L., Maler, B. A. and Yamamoto, K. R. (1983). DNA sequences bound specifically by glucocorticoid receptor *in vitro* render a heterologous promoter responsive *in vitro*. *Cell* **33**, 489–499.

Charron, J. and Drouin, J. (1986). Glucocorticoid inhibition of transcription from episomal proopiomelanocortin gene promoter. *Proc. Natl. Acad. Sci. USA* **83**, 8903–8907.

Crew, M. and Spindler, S. R. (1986). Thyroid hormone regulation of the transfected rat growth hormone promoter. *J. Biol. Chem.* **261**, 5018–5022.
Damm, K., Thompson, C. C. and Evans, R. M. (1989). Protein encoded by v-*erb*-A functions as a thyroid-hormone receptor antagonist. *Nature* **339**, 593–597.
Danielsen, M., Hinck, L. and Ringold, G. M. (1989). Two amino acids within the knuckle of the first zinc finger specify DNA response element activation by the glucocorticoid receptor. *Cell* **57**, 1131–1138.
Delegeane, A. M., Ferland, L. H. and Mellon P. L. (1987). Tissue-specific enhancer of the human glycoprotein hormone α-subunit gene: dependence on cyclic AMP-inducible elements. *Mol. Cell. Biol.* **7**, 3994–4002.
Deutsch, P. J., Jameson, J. L. and Habener, J. F. (1987). Cyclic AMP responsiveness of human gonadotropin-α gene transcription is directed by a repeated 18-base pair enhancer. *J. Biol. Chem.* **262**, 12 169–12 174.
Drouin, J., Charron, J., Gagner, J.-P., Jeannotte, L., Nemer, M., Plante, R. K. and Wrange, O. (1987). Proopiomelanocortin gene: a model for negative regulation of transcription by glucocorticoids. *J. Cell. Biochem.* **35**, 293–304.
Drouin, J., Trifiro, M. A., Plante, R. K., Nemer, M., Eriksson, P. and Wrange, O. (1989a). Glucocorticoid receptor binding to a specific DNA sequence is required for hormone-dependent repression of proopiomelanocortin gene transcription. *Mol. Cell. Biol.* **9**, 5305–5314.
Drouin, J., Sun, Y. L. and Nemer, M. (1989b). Glucocorticoid repression of pro-opiomelanocortin gene transcription. *J. Steroid Biochem.* **34**, 63–70.
Evans, R. M. (1988). The steroid and thyroid hormone receptor superfamily. *Science* **240**, 889.
Fink, G. (1979). Feedback actions of target hormones on hypothalamus and pituitary with special reference to gonadal steroids. *Annu. Rev. Physiol.* **41**, 571–585.
Flug, F., Copp, R. P., Casanova, J., Horowitz, Z. D., Janocko, L., Plotnick, M. and Samuels, H. H. (1987). *Cis*-acting elements of the rat growth hormone gene which mediate basal and regulated expression by thyroid hormone. *J. Biol. Chem.* **262**, 6373–6382.
Frisch, S. M. and Ruley, H. E. (1987). Transcription from the stromelysin promoter is induced by interleukin-1 and repressed by dexamethasone. *J. Biol. Chem.* **262**, 16 300–16 304.
Gharib, S. D., Bowers, S. M., Need, L. R. and Chin, W. W. (1986). Regulation of rat luteinizing hormone subunit messenger ribonucleic acids by gonadal steroid hormones. *J. Clin. Invest.* **77**, 582–589.
Giguere, V., Hollenberg, S. M., Rosenfeld, M. G. and Evans, R. M. (1986). Functional domains of the human glucocorticoid receptor. *Cell* **46**, 645–652.
Gill, G. and Ptashne, M. (1988). Negative effect of the transcriptional activator GAL4. *Nature* **334**, 721–724.
Glass, C. K., Franco, R., Weinberger, C., Albert, V. R., Evans, R. M. and Rosenfeld, M. G. (1987). A c-*erb*-A binding site in rat growth hormone gene mediates trans-activation by thyroid hormone. *Nature* **329**, 738–741.
Glass, C. K., Holloway, J. M., Devary, O. V. and Rosenfeld, M. G. (1988). The thyroid hormone receptor binds with opposite transcriptional effects to a common sequence motif in thyroid hormone and estrogen response elements. *Cell* **54**, 313–323.
Glass, C. K., Lipkin, S. M., Devary, O. V. and Rosenfeld, M. G. (1989). Positive and negative regulation of gene transcription by a retinoic acid-thyroid hormone receptor heterodimer. *Cell* **59**, 697–708.

Gonzalez, G. A., Yamamoto, K. K., Fischer, W. H., Darr, D., Menzel, P., Briggs, W., Vale, W. and Montminy, M. (1989). A cluster of phosphorylation sites on the cyclic AMP-regulated nuclear factor CREB predicted by its sequence. *Nature* **337**, 749–752.

Graupner, G., Wills, K. N., Tzukerman, M., Zhang, X.-K. and Pfahl, M. (1989). Dual regulatory role for thyroid hormone receptors allows control of retinoic-acid receptor activity. *Nature* **340**, 653–656.

Guertin, M., LaRue, H., Bernier, D., Wrange, O., Chevrette, M., Gingras, M.-C. and Belanger, L. (1988). Enhancer and promoter elements directing activation and glucocorticoid repression of the α1-fetoprotein gene in hepatocytes. *Mol. Cell. Biol.* **8**, 1398–1407.

Hai, T., Horikoshi, M., Roeder, R. G. and Green, M. R. (1988). Analysis of the role of the transcription factor ATF in the assembly of a functional preinitiation complex. *Cell* **54**, 1043–1051.

Ham, J., Thomson, A., Needham, M., Webb, P. and Parker, M. (1988). Characterization of response elements for androgens, glucocorticoids and progestins in mouse mammary tumour virus. *Nucl. Acids Res.* **16**, 5263–5275.

Han, K., Levine, M. S. and Manley, J. L. (1989). Synergistic acitvation and repression of transcription by *Drosophila* homeobox proteins. *Cell* **56**, 573–583.

Hay, N., Bishop, J. M. and Levens, D. (1987). Regulatory elements that modulate expression of human c-*myc*. *Gene Dev.* **1**, 659–671.

Hodin, R. A., Lazar, M. A., Wintman, B. I., Darling, D. S., Koenig, R. J., Larsen, P. R., Moore, D. D. and Chin, W. W. (1989). Identification of a thyroid hormone receptor that is pituitary specific. *Science* **244**, 76–79.

Hoeffler, J. P., Meyer, T. E., Yun, Y., Jameson, J. L. and Habener, J. F. (1988). Cyclic AMP-responsive DNA-binding protein: structure based on a cloned placental cDNA. *Science* **242**, 1430–1433.

Hollenberg, S. M., Giguere, V., Segui, P. and Evans, R. M. (1987). Colocalization of DNA-binding and transcriptional activation functions in the human glucocorticoid receptor. *Cell* **49**, 39–46.

Horikoshi, M., Hai, T., Lin, Y.-S., Green, M. R. and Roeder, R. G. (1988). Transcription factor ATF interacts with the TATA factor to facilitate establishment of a preinitiation complex. *Cell* **54**, 1033–1042.

Hynes, N. E., Van Ooyen, A., Kennedy, N., Herrlich, P., Ponta, H. and Groner, B. (1983). LTR subfragments cause hormone-responsive expression of mouse mammary tumor virus and of an adjacent gene. *Proc. Nat. Acad. Sci. USA* **80**, 3637–3641.

Ingham, P. W. (1988). The molecular genetics of embryonic pattern formation in *Drosophila*. *Nature* **335**, 25–34.

Israel, A. and Cohen, S. N. (1985). Hormonally mediated negative regulation of human Pro-Opiomelanocortin gene expression after transfection into mouse L cells. *Mol. Cell. Biol.* **5**, 2443–2453.

Izumo, S. and Mahdavi, V. (1988). Thyroid hormone receptor α isoforms generated by alternative splicing differentially activate myosin HC gene transcription. *Nature* **334**, 539–542.

Jameson, J. L., Deutsch, P. J., Gallagher, G. D., Jaffe, R. C. and Habener, J. F. (1987). *Trans*-acting factors interact with a cyclic AMP response element to modulate expression of the human gonadotropin α gene. *Mol. Cell. Biol.* **7**, 3032–3040.

Jameson, J. L., Jaffe, R. C., Deutsch, P. J., Albanese, C. and Habener, J. F. (1988). The gonadotropin α-gene contains multiple protein binding domains that interact to modulate basal and cAMP-responsive transcription. *J. Biol. Chem.* **263**, 9879–9886.

Karin, M., Haslinger, A., Holtgreve, H., Cathala, G., Slater, E. and Baxter, J. D. (1984). Activation of a heterologous promotor in response to dexamethasone and cadmium by metallothionein gene 5'-flanking DNA. *Cell* **36**, 371–379.

Keleher, C. A., Goutte, C. and Johnson, A. D. (1988). The yeast cell-type-specific repressor α2 acts cooperatively with a non-cell-type-specific protein. *Cell* **53**, 927–936.

Koenig, R. J., Lazar, M. A., Hodin, R. A., Brent, G. A., Larsen, P. R., Chin, W. W. and Moore, D. D. (1989). Inhibition of thyroid hormone action by a non-hormone binding c-*erb*A protein generated by alternative mRNA splicing. *Nature* **337**, 659–661.

Larsen, P. R., Harney, J. W. and Moore, D. D. (1986). Sequences required for cell-type specific thyroid hormone regulation of the rat growth hormone promoter activity. *J. Biol. Chem.* **261**, 14 373–14 376.

Lazar, M. A., Hodin, R. A., Darling, D. S. and Chin, W. W. (1988). Identification of a rat c-*erb*-A α-related protein which binds deoxyribonucleic acid but does not bind thyroid hormone. *Mol. Endocrinol.* **2**, 893–901.

Lee, W., Mitchell, P. and Tjian, R. (1987). Purified transcription factor AP-1 interacts with TPA-inducible enhancer elements. *Cell* **49**, 741–752.

Levine, M. and Manley, J. L. (1989). Transcriptional repression of eukaryotic promoters. *Cell* **59**, 405–408.

Maekawa, T., Sakura, H., Kanei-Ishii, C., Sudo, T., Yoshimura, T., Fujisawa, J., Yoshida, M. and Ishii, S. (1989). Leucine zipper structure of the protein CRE-BP1 binding to the cyclic AMP response element in brain. *EMBO J.* **8**, 2023–2028.

Mann, D. R., Evans, D., Edoimioya, F., Kamel, F. and Butterstein, G. M. (1985). A detailed examination of the *in vivo* and *in vitro* effects of ACTH on gonadotropin secretion in the adult rat. *Neuroendocrinology* **40**, 297–302.

Mellon, P. L. and Akerblom, I. E. (1989). Repression of gene expression by glucocorticoid receptor through interference with cAMP responsive enhancers. In "The Steroid/Thyroid Hormone Receptor Family and Gene Regulation", pp. 207–220. Birkhäuser-Verlag, Basel.

Meyer, M.-E., Gronemeyer, H., Turcotte, B., Bocquel, M.-T., Tasset, D. and Chambon, P. (1989). Steroid hormone receptors compete for factors that mediate their enhancer function. *Cell* **37**, 433–442.

Miesfeld, R., Rusconi, S., Godowski, P. J., Maler, B. A., Okret, S., Wikström, A.-C., Gustaffson, J.-A. and Yamamoto, K. R. (1986). Genetic complementation of a glucocorticoid receptor deficiency by expression of cloned receptor cDNA. *Cell* **46**, 389–399.

Miller, J. H. and Reznikoff, W. S. (eds) (1978). "The Operon". Cold Spring Harbor Laboratory, New York.

Mordacq, J. C. and Linzer, D. I. H. (1989). Co-localization of elements required for phorbol ester stimulation and glucocorticoid repression of proliferin gene expression. *Genes Dev.* **3**, 760–769.

Morley, J. E. (1981). Neuroendocrine control of thyrotropin secretion. *Endocr. Rev.* **2**, 396–436.

Nasmyth, K. and Shore, D. (1987). Transcriptional regulation in the yeast life cycle. *Science* **237**, 1162–1170.

8. REPRESSION OF GENE EXPRESSION

Nelson, C., Albert, V. R., Elsholtz, H. P., Lu, L. I.-W. and Rosenfeld, M. G. (1988). Activation of cell-specific expression of rat growth hormone and prolactin genes by a common transcription factor. *Science* **239**, 1400–1405.

Oro, A. E., Hollenberg, S. M. and Evans, R. M. (1988). Transcriptional inhibition by a glucocorticoid receptor-β-galactosidase fusion protein. *Cell* **55**, 1109–1114.

Pierce, J. C. and Parsons, T. F. (1981). Glycoprotein hormones: structure and function. *Annu. Rev. Biochem.* **50**, 465–495.

Rauscher, F. J. I., Cohen, D. R., Curran, T., Bos, T. J., Vogt, P. K., Bohmann, D., Tjian, R. and Franza, B. R. J. (1988). Fos-associated protein p39 is the product of the *jun* proto-oncogene. *Science* **240**, 1010–1016.

Renkawitz, R., Schütz, G., von der Ahe, D. and Beato, M. (1984). Sequences in the promoter region of the chicken lysozyme gene required for steroid regulation and receptor binding. *Cell* **37**, 503–510.

Riggs, A. D. and Bourgeois, S. (1968). On the assay, isolation, and characterization of the *lac* repressor. *J. Mol. Biol.* **34**, 361–364.

Roesler, W. J., Vandenbark, G. R. and Hanson, R. W. (1988). Cyclic AMP and the induction of eukaryotic gene transcription. *J. Biol. Chem.* **263**, 9063–9066.

Sakai, D. D., Helms, S., Carstedt-Duke, J., Gustafsson, J. A., Rottman, F. M. and Yamamoto, K. R. (1988). Hormone-mediated repression of transcription: a negative glucocorticoid response element from the bovine prolactin gene. *Genes Dev.* **2**, 1144–1154.

Sap, J., Munoz, A., Damm, K., Goldberg, Y., Ghysdael, J., Leutz, A., Beug, H. and Vennström, B. (1986). The c-*erb*-A protein is a high-affinity receptor for thyroid hormone. *Nature* **324**, 635–640.

Sassone-Corsi, P., Sisson, J. C. and Verma, I. M. (1988). Transcriptional autoregulation of the proto-oncogene c-*fos*. *Nature* **334**, 314–319.

Scott, M. P. and Carroll, S. B. (1987). The segmentation and homeotic gene network in early *Drosophila* development. *Cell* **51**, 689–698.

Shupnik, M. A., Chin, W. W., Habener, J. F. and Ridgway, E. C. (1985). Transcriptional regulation of the thyrotropin subunit genes by thyroid hormone. *J. Biol. Chem.* **260**, 2900–2903.

Shupnik, M. A., Greenspan, S. L. and Ridgway, E. C. (1986). Transcriptional regulation of thyrotropin subunit genes by thyrotropin-releasing hormone and dopamine in pituitary cell culture. *J. Biol. Chem.* **261**, 12 675–12 679.

Silver, B. J., Bokar, J. A., Virgin, J. B., Vallen, E. A., Milsted, A. and Nilson, J. H. (1987). Cyclic AMP regulation of the human glycoprotein hormone α-subunit gene is mediated by an 18-base pair element. *Proc. Natl. Acad. Sci. USA* **84**, 2198–2202.

Sprague, G. F., Blair, L. C. and Thorner, J. (1983). Cell interactions and regulation of cell type in the yeast *Saccharomyces cerevisiae*. *Annu. Rev. Microbiol.* **37**, 623–660.

Strähle, U., Klock, G. and Schutz, G. (1987). A DNA sequence of 15 base pairs is sufficient to mediate both glucocorticoid and progesterone induction of gene expression. *Proc. Natl. Acad. Sci. USA* **84**, 7871–7875.

Suter, D. E. and Schwartz, N. B. (1985). Effects of glucocorticoids on secretion of luteinizing hormone and follicle-stimulating hormone by female rat pituitary cells *in vitro*. *Endocrinology* **117**, 855–859.

Takimoto, M., Quinn, J. P., Farina, A. R., Staudt, L. M. and Levens, D. (1989). *fos/jun* and Octamer-binding protein interact with a common site in a negative element of the human c-*myc* gene. *J. Biol. Chem.* **265**, 8992–8999.

Tepperman, J. and Tepperman, H. M. (1987) "Metabolic and Endocrine Physiology", 5th edn. Year Book Medical Publishers, Chicago.
Thompson, C. C., Weinberger, C., Lebo, R. and Evans, R. M. (1987). Identification of a novel thyroid hormone receptor expressed in the mammalian central nervous system. *Science* **237**, 1610–1614.
Tremblay, Y., Tretjakoff, I., Peterson, A., Antakly, T., Zhang, C. X. and Drouin, J. (1988). Pituitary specific expression and glucocorticoid regulation of a pro-opiomelanocortin (POMC) fusion gene in transgenic mice. *Proc. Natl. Acad. Sci. USA* **85**, 8890–8894.
Turcotte B., Guertin, M., Chevrette, M., LaRue, H. and Belanger, L. (1985). DNase I hypersensitivity and methylation of the 5′-flanking region of the αl-fetoprotein gene during developmental and glucocorticoid-induced repression of its activity in rat liver. *Nucl. Acids Res.* **14**, 9827–9841.
Umesono, K. and Evans, R. M. (1989). Determinants of target gene specificity for steroid/thyroid hormone receptors. *Cell* **57**, 1139–1146.
Umesono, K., Giguere, V., Glass, C., Rosenfeld, M. G. and Evans, R. M. (1988). Retinoic acid and thyroid hormone induce gene expression through a common responsive element. *Nature* **336**, 262–265.
Vreeburg, J. T. M., de Greef, W. J., Ooms, M. P., van Wouw, P. and Wever, R. F. A. (1984). Effects of adrenocorticotropin and corticosterone on the negative feedback action of testosterone in the adult male rat. *Endocrinology* **115**, 977–983.
Wang, L.-H., Tsai, S. Y., Cook, R. G., Beattie, W. G., Tsai, M.-J. and O'Malley, B. W. (1989). COUP transcription factor is a member of the steroid receptor superfamily. *Nature* **340**, 163–166.
Weinberger, C., Thompson, C. C., Ong, E. S., Lebo, R., Gruol, D. J. and Evans, R. M. (1986). The c-*erb*-A gene encodes a thyroid hormone receptor. *Nature* **324**, 641–646.
Weiner, F. R., Czaja, M. J., Jefferson, D. M., Giambrone, M.-A., Tur-Kaspa, R., Reid, L. M. and Zern, M. A. (1987). The effects of dexamethasone on *in vitro* collagen gene expression. *J. Biol. Chem.* **262**, 6955–6958.
Yamamoto, K. R. (1985). Steroid receptor regulated transcription of specific genes and gene networks. *Annu. Rev. Genet.* **19**, 209–252.

9
Characterization of DNA Receptor Interactions

M. BEATO, D. BARETTINO, U. BRÜGGEMEIER,
G. CHALEPAKIS, R. J. G. HACHÉ, M. KALFF,
B. PIÑA, M. SCHAUER, E. P. SLATER and M. TRUSS

Institut für Molekularbiologie und Tumorforschung, Emil-Mannkopff-Strasse 2, D-3550 Marburg, FRG

9.1 Introduction

Our understanding of how steroid hormones induce the transcription of specific genes has made considerable progress through the discovery of DNA sequences in the vicinity of the regulated promoters that are recognized by the steroid hormone receptors (Beato, 1989). It appears that these nucleotide sequences must be present for the adjacent gene to respond to hormone administration, and therefore they have been named hormone responsive elements (HREs; see Chapters 6.2 and 9.4). Such HREs are found in a variety of hormone-inducible genes including the mouse mammary tumour virus provirus (Chandler *et al.*, 1983; Payvar *et al.*, 1983; Scheidereit *et al.*, 1983), the human metallothionein IIA gene (Karin *et al.*, 1984), the chicken lysozyme gene (Renkawitz *et al.*, 1984; Hecht *et al.*, 1988), the vitellogenin genes from amphibia and birds (Klein-Hitpass *et al.*, 1986, 1989; Seiler-Tuyns *et al.*, 1986; Martinez *et al.*, 1987), the growth hormone genes (Moore *et al.*, 1985; Slater *et al.*, 1985), the rabbit uteroglobin gene (Cato *et al.*, 1984; Bailly *et al.*, 1986; C. Jantzen *et al.*, 1987), the rat tyrosine amino transferase gene (H. M. Jantzen *et al.*, 1987), the rat tryptophane oxygenase gene (Danesh *et al.*, 1987), the rat acidic glycoprotein gene (Klein *et al.*, 1988) and the Moloney murine sarcoma virus (DeFranco and Yamamoto, 1986; Miksicek *et al.*, 1986). A comparison of the nucleotide sequences found in these different regulatory elements yields a consensus sequence that can be used in computer searching programs to predict the existence of regulatory elements for steroid hormones.

In this chapter we will first describe the interaction of purified hormone receptors with the HREs localized on free linear DNA fragments, and describe the influence of individual base pairs within the HREs upon the affinity of the receptors for the corresponding sites. We will then describe the role of DNA topology in the interaction of receptors with the HREs *in vitro*, and in the response to hormone administration in transfection experiments. Then we will address the question of how the packaging of DNA sequences into nucleosomes could influence the accessibility of the HREs to the receptors, as well as the accessibility of binding sites for other transcription factors in the mouse mammary tumour virus (MMTV) promoter. Finally, we will discuss the possible mechanism of transactivation, again using the MMTV sytem as a model. Here, the focus will be on how binding of the hormone receptors facilitates DNA binding of nuclear factor 1 (NF-1), which has been shown to be essetial for transactivation of the MMTV promoter both *in vivo* and *in vitro*. We will conclude by mentioning the lines of research in this field that we expect to be more productive in the next few years.

9.2 Interaction of hormone receptors with hormone responsive elements on linear DNA

The initial identification of DNA sequences specifically recognized by purified hormone receptors was accomplished using linear fragments of the long terminal repeat region of MMTV which had earlier been shown to be important for hormonal induction *in vivo* (Chandler *et al.*, 1983; Payvar *et al.*, 1983; Scheidereit *et al.*, 1983; see Fig. 1). Following identification of the relevant DNA region in nitrocellulose filter binding studies (Geisse *et al.*, 1982), DNase I footprint experiments revealed the existence of several binding sites for the glucocorticoid receptor purified from rat liver (Scheidereit *et al.*, 1983). A comparison of the nucleotide sequences protected by receptor binding against digestion by DNase I revealed a common hexanucleotide motif, 5'-TGTTCT-3', that was soon identified in other HREs (Karin *et al.*, 1984; see Chapters 6.2 and 9.4). Further analysis of methylation protection experiments with dimethyl sulphate served to detect intimate contacts of the receptor with the guanine residues within the hexanucleotide motifs (Scheidereit and Beato, 1984). In addition, contacts with other guanines outside the hexanucleotide motif were detected, in particular contacts with an imperfect palindrome located between −187 and −165 upstream of the initiation site (Scheidereit and Beato, 1984). Similar methylation protection results were obtained in the promoter region of the human metallothionein IIA gene (Karin *et al.*, 1984) and of the chicken lysozyme gene (von der Ahe *et al.*, 1986). The symmetry of the contacts with the guanine residues (Fig. 2) already suggested that the receptor binds to the

HRE imperfect palindrome as a dimer in a head-to-head orientation (Scheidereit *et al.*, 1986). This orientation of the bound receptor homodimers has been confirmed for other members of the steroid receptor family (Kumar and Chambon, 1988; Tsai *et al.*, 1988). It appears that some members of the receptor family can form dimers in solution in the absence of DNA (Wrange *et al.*, 1989; Fawell *et al.*, 1990).

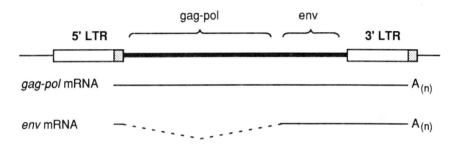

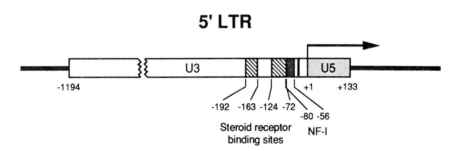

Fig. 1 Schematic representation of mouse mammary tumours provirus and long terminal repeat (LTR). The numbers refer to the distance from the transcription start point.

In the promoter region of the MMTV the array of receptor binding sites is rather complex. Whereas the promoter distal binding site between −187 and −165 corresponds to the canonical imperfect palindrome structure of other HREs, the promoter proximal region between −130 and −75 is composed of three hexanucleotide motifs, two of which lack the complementary half of the palindromic structure (Fig. 3). Whether receptor dimers or monomers bind to each of these three sites remains to be established.

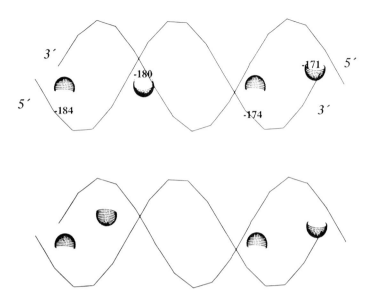

Fig. 2 Contacts between the glucocorticoid receptor and the N-7 positions of guanines within GREs. Computer graphical representation of the promoter distal receptor binding site from the MMTV-HRE (top) and the promoter proximal binding site of the chicken lysozyme gene (bottom). The van der Waals half spheres of the contacted atoms are shown. The numbers refer to distance from the transcription start point of the MMTV-promoter. Data are taken from von der Ahe et al. (1986).

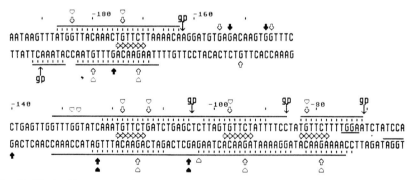

Fig. 3 Nucleotide sequence of MMTV promoter including the HRE. The DNase I footprints generated by the glucocorticoid receptor are indicated by a dotted line (Scheidereit et al., 1983), and the footprint of the progesterone receptor is indicated by a continuous line (Chalepakis et al., 1988a). The exonuclease III stop signals generated by both receptors are indicated (gp). The guanine residues protected by the GR against DMS methylation are shown by the open arrows, and those hypermethylated in the presence of GR are indicated by the black arrows. Similar data for the PR are shown by open and black arrowheads. The positions of the four hexanucleotide motifs TGTTCT are indicated by the horizontal arrows. Numbers refer to distance from the transcription start point.

The same DNA region that is recognized by the rat liver glucocorticoid receptor also binds the rabbit uterus progesterone receptor (von der Ahe et al., 1985). In transfection studies performed with cell lines of mammary epithelial origin, progesterone analogues were more effective than glucocorticoids in inducing promoter activity through the HRE of MMTV (Cato et al., 1986). A detailed comparison of the DNA contacts of glucocorticoid and progesterone receptors in the different sites of the MMTV-HRE show that whereas the promoter distal binding site appears to interact with both receptors in a very similar way, clear differences in interaction are seen in the promoter proximal region (Fig. 3 and Chalepakis et al., 1988a). The main differences occur in the region between -100 and -130, where the progesterone receptor generates a footprint extending 14 bp further upstream than the footprint generated by the glucocorticoid receptor (Chalepakis et al., 1988a).

Oligonucleotide mutation analysis of the HRE region suggests a contribution of the individual TGTTCT motifs to optimal hormone induction. With the exception of the most promoter proximal motif, mutation of any of the other three TGTTCT hexanucleotide motifs has a dramatic effect on hormonal induction, both by progestins and glucocorticoids (Chalepakis et al., 1988a). These results suggest that the receptor molecules bound to the DNA have to interact with each other in order to generate an optimal response. This interaction between receptor molecules is different for the glucocorticoid and the progesterone receptors, as demonstrated by experiments where nucleotide sequences of different lengths were inserted between the promoter distal and the promoter proximal binding regions. Insertion of five base pairs between these two regions improves the response of the corresponding constructions to glucocorticoids, whereas the response to progestins is reduced by 50% (Chalepakis et al., 1988a). Introduction of 10, 20 or 30 bp at the same position does not interfere with the response to progestins, while the response to glucocorticoids is progressively reduced. Only after insertion of 60 bp does the hormone inducibility decrease dramatically, suggesting that the interaction among DNA-bound receptor molecules can occur even when they are separated by five turns of the DNA double helix.

9.3 Role of the hormone ligand

Most experiments with purified steroid hormone receptors are performed with proteins loaded with the corresponding hormone ligand or a synthetic agonist. In order to investigate the role of the ligand in the process of receptor binding to DNA, we have developed a procedure that allows the measurement of specific DNA binding with crude preparations of cytosolic hormone receptor. Using this procedure, we showed that the glucocorticoid receptor present in liver cytosol from adrenalectomized rats is able to bind selectively to the MMTV-HRE in the absence of any ligand (Willmann and Beato, 1986). In fact, binding of the ligand-free glucocorticoid receptor is slightly better than binding observed in the presence of the agonists dexamethasone or triamcinolone acetonide, or of the antagonist RU486 (Willmann and Beato, 1986). These unexpected findings demonstrates that the receptor is able to adopt the conformation required for binding selectively to DNA in the absence of ligand or in the presence of an antagonist. Thus, the question arose as to what the function of the agonist ligand could be.

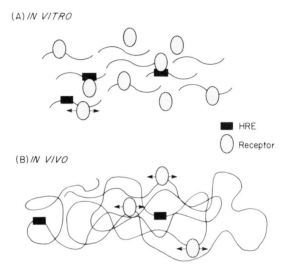

Fig. 4 Binding of the ligand-free hormone receptor to DNA *in vitro* and *in vivo*. (A) *In vitro*, the concentrations of receptor and HRE-containing DNA are sufficiently high so as to allow specific binding even when the on- and off-rates are slow. (B) *In vivo*, the large excess of non-specific genomic DNA may prevent binding of the ligand-free receptor to the few HREs, given the small number of receptor molecules per cell. Addition of hormone accelerates the on- and off-rates and could speed up the scanning process.

In later experiments we investigated the kinetics of the interaction of the receptor with DNA fragments with different nucleotide sequences. We found that binding of a synthetic agonist has a dramatic influence on the rates of association and dissociation of the receptor and DNA (Schauer *et al.*, 1989). Whereas the on-rate is accelerated about five-fold by binding of the ligand, the off-rate is much more dramatically influenced, and is some 20- to 30-fold faster for the liganded receptor than for its ligand-free counterpart (Schauer *et al.*, 1989). This is true not only for the rat liver glucocorticoid receptor in crude cytosol, but also for the rabbit uterine progesterone receptor, suggesting that we are dealing with a general property of steroid hormone receptors. Although the off-rate observed with the non-specific DNA sequences, such as plasmid vector fragments, is much faster than that detected with HRE containing DNA fragments, the effect of the ligand is similar in both cases. Based on these findings, we postulate that the role of the ligand could be to facilitate the scanning of genomic DNA sequences and thus to enable the receptor to find its specific binding sites. The kinetic hypothesis is compatible with the observation that *in vivo* the HRE of the tyrosine amino transferase gene is only occupied by the receptor in the presence of the ligand (Becker *et al.*, 1986). *In vitro*, given the concentrations of DNA and receptor, no kinetic barrier is found by the ligand-free receptor, and specific DNA binding is detected (Fig. 4).

This model does not exclude the possibility that the ligand may have additional functions, such as to facilitate the dissociation of the receptor from the heat-shock protein Hsp90, or even a direct effect on the transactivation function (see Chapters 2.5 and 3.6).

9.4 Similarities and differences between the response elements for various steroid hormones

As mentioned above, the binding sites for the glucocorticoid receptor in the MMTV promoter are also recognized by the progesterone receptor. No clear differences in the sequence recognition mechanism of these two receptors have been conclusively detected. This is probably due to the very high degree of similarity of the DNA binding domains of the two receptors (Beato, 1989). The MMTV-HRE is also recognized by the androgen receptor (Cato *et al.*, 1987) and by the mineralocorticoid receptor (Arizza *et al.*, 1987; Cato and Weinmann, 1988). Using oligonucleotides of various nucleotide sequences it has been found that changes at certain positions within the HRE have a differential influence on androgen and glucocorticoid induction (Ham *et al.*, 1988). However, no DNA-binding data are available for these latter receptors, and therefore no direct comparison of the contacted nucleotides can be performed. Other hormone response elements, such as the ones of the human metallothionein IIA and the rat tyrosine aminotransferase gene, have also been shown to be recognized by both glucocorticoid and progesterone receptors (Danesh *et al.*, 1987; Slater *et al.*, 1988). A comparison of all the sites recognized by these two receptors yields a GRE/PRE consensus sequence: 5'-GGTACANNNTGTYCT-3' (Beato *et al.*, 1989), of which the right half is highly conserved among different sites (see Chapters 6.2 and 9.4).

The consensus sequence for the oestrogen response elements (ERE) of several vitellogenin genes has been found to be 5'-GGTCANNNTGACC-3' (Klein-Hitpass et al., 1989). This sequence appears to be a more perfect palindrome than the above-mentioned GRE/PRE. However, differences in position 3 of the pentanucleotide that forms the half palindrome have been described. For instance, the ERE of the pS2 gene exhibits the sequence TGGCC (Berry et al., 1989), and one half of the ERE of the rabbit uteroglobin promoter is TGCCC (Slater et al., 1990).

A comparison of the consensus sequences for the GRE/PRE and the ERE shows that the half palindromes have several common positions, such as positions 1, 2 and 5, which are T, G and C, respectively, in both cases (Fig. 5). We know that position 6 of the GRE/PRE, although in most cases a conserved T, is not essential for binding *in vitro*, nor for activity *in vivo* (Truss et al., 1990). Therefore, the two elements differ only at positions 3 and 4 of the half palindrome. At position 3 there is always a T in the GRE/PREs, whereas the EREs can accept any base but a T at this position. At position 4, a C can be present in both GRE/PRE and ERE, but EREs will not tolerate a T, the most frequent base in GRE/PREs at this position. Recently, we have been able to develop hybrid HREs that bind oestrogen, progesterone and glucocorticoid receptors, and respond to all three hormones in transfection experiments (M. Truss, G. Chalepakis and M. Beato, unpubl. res.).

Position:	1	2	3	4	5
GRE/PRE:	T	G	T A	T C	C
ERE:	T	G	A C G	C	C

Fig. 5 Comparison of the half palindromes for GRE/PRE and ERE.

The differences in the nucleotide sequences of GRE/PRE and ERE may reflect differences in the DNA-binding domain of the corresponding receptors. In recent experiments it has been shown that three amino acids in the knuckle of the first zinc finger may be responsible for sequence discrimination (Mader et al., 1989; Danielsen et al., 1989; Umesono and Evans, 1989; see Chapters 2.4 and 3.3). In this region, glucocorticoid and progesterone receptors differ from oestrogen receptors (Beato, 1989). Although other differences are found in the second zinc finger, these three positions can be used to classify the nuclear receptors into two subfamilies, the prototype of one family being the glucocorticoid receptor, and the prototype of the other group being the oestrogen receptor. The progesterone receptor, the mineralocorticoid receptor and the androgen receptor belong to the glucocorticoid receptor group, whereas the thyroid hormone receptor, the vitamin D_3 receptor and the retinoic acid receptor belong to the subgroup of the oestrogen receptor. A detailed analysis of different mutants in the DNA-binding domain of the receptors, combined with binding experiments using HREs mutated at individual positions, should help to predict the amino acid side chains that interact with individual bases of the HRE. Ultimately, these predictions will have to be confirmed by elucidating the three-dimensional structure of the complexes between receptors and their DNA-binding sequences.

9.5 Interaction of steroid hormone receptors with circular DNA molecules of different topologies

Most studies on DNA-binding proteins, and in particular with hormone receptors, have been performed with linear DNA fragments. However, since DNA topology could influence the secondary structure of the DNA double helix we wanted to analyse the influence of this parameter on the binding of the hormone receptors to the HRE. To this end, we prepared minicircles containing the HRE of MMTV with DNA ligase and generated different topoisomers by performing the ligation in the presence of different concentrations of ethidium bromide. Analysis of these minicircles on non-denaturing agarose or acrylamide gels reveals a transition in the electrophoretic mobility observed between the topoisomers -2 and -3. Usually, topoisomers migrate faster the higher their degree of supercoiling (Keller, 1975). The MMTV-HRE minicircles show an abnormal behaviour: topoisomer -3 migrates more slowly than topoisomer -2. This change in electrophoretic behaviour correlates with an abrupt increase in the affinity for purified glucocorticoid or progesterone receptors. Whereas relaxed circles, or topoisomers -1 and -2, have relatively low affinity for the hormone receptors, topoisomers -3 and higher have a much higher affinity. We originally detected this behaviour in band retardation assays, but it has been confirmed with DNase I footprinting experiments, and extended to minicircles of different lengths (M. Truss and M. Beato, unpubl. res.). This finding suggests that negative supercoiling has a role in facilitating the interaction among DNA-bound receptor molecules that is required for optimal binding and transactivation.

To determine whether this *in vitro* behaviour has an *in vivo* correlate, we analysed the influence of DNA topology on the expression of transfected plasmids carrying the MMTV-HRE. It has been previously shown that plasmids containing enhancer sequences are expressed much more effectively when transfected in the negative supercoiled form (Weintraub et al., 1986). In agreement with these previous findings, we observed that induction of an MMTV-HRE carrying gene by progestins depends on the initial topology of the transfected plasmid (Fig. 6). Whereas negative supercoiled plasmids are efficiently induced by progestins, their linear counterparts are around seven- to ten-fold less responsive (Piña et al., 1990b). Unexpectedly, induction by glucocorticoids is independent of the initial topology of the transfected DNA; negative supercoiled and linear plasmids both respond efficiently to glucocorticoid administration (Piña et al., 1990b). These findings corroborate the differences found in the mechanisms of transactivation by each hormone, and point to a possible role for negative supercoiling in facilitating optimal activity of the progesterone-dependent MMTV enhancer.

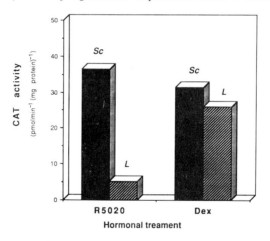

Fig. 6 Influence of DNA topology on hormonal induction of transfected plasmids. Supercoiled (Sc) or linear (L) plasmid MMTV-tk-CAT was transfected in T47D cells, alone or together with the plasmid RSV-GRE (R5020) that expresses the rat glucocorticoid receptor. After transfection the cells were treated with either 10^{-7} M Dexamethasone (Dex) or 10^{-8} M R5020 for 48 h and CAT activity was measured in cell extracts. The data are from Piña et al. (1990a).

To study the mechanism of this topological effect in more detail, we introduced changes in the relative orientation or distance between the different receptor binding sites of the HRE or between the HRE and the binding sites for other transcription factors in the promoter. When the distance between the two sets of receptor binding sites within the HRE is increased, induction by glucocorticoids also becomes dependent on the initial topology of the transfected DNA, as is the induction by progestins in the wild-type constructions (Piña et al., 1990b). Similarly, when the distance between the HRE and the binding sites for the transcription factors in the tk-promoter is increased, a dependence of the glucocorticoid induction on the topology of the transfected plasmid is observed. These data again suggest that negative supercoiling is facilitating not only the interactions among receptor proteins, but also between receptor proteins and other transcription factors.

Since most of the DNA transfected into mammalian cells in culture is linearized immediately after transfection (Alwine, 1985), it was intriguing to find that the initial topology influenced induction by hormones as much as two days after transfection. To understand the mechanism of this topological effect we analysed the DNA remaining in the cell nucleus of the transfected cells at different times after transfection. We found that although the majority of the transfected DNA is rapidly nicked or linearized, a minor fraction of the transfected DNA is organized into chromatin and exhibits a supercoiled structure following purification of the DNA shortly after transfection. Furthermore, this fraction of negative supercoiled DNA is stable for several days and probably represents the fraction of the population of transfected plasmids that is actively engaged in transcription. Thus, very probably the topology of the DNA at the moment of transfection influences its organization into chromatin and thus favours the subsequent response of these plasmids to hormonal induction. Apparently, the wild-type configuration of the MMTV-HRE is such that induction by glucocorticoids is possible even when linear templates are organized into chromatin, but when the relative orientation and the distance between the different elements of the promoter and the enhancer are changed, negative supercoiling is required for optimal induction.

9.6 Interaction of receptors with nucleosomally organized DNA

Studies of binding of regulatory proteins to response elements of DNA are usually performed with free DNA in solution. However, within the cell nucleus DNA is organized into nucleosomes and the influence of DNA topology upon the response of individual plasmids to different hormones suggests that chromatin organization may play a role in the mechanism of hormonal induction. Using bovine papilloma virus minichromosomes, Gordon Hager's group has shown that sequences of the MMTV provirus are organized into nucleosomes in a well-defined manner (Richard-Foy and Hager, 1987; see Chapter 10.3). The long terminal repeat region contains at least six nucleosomes in phase, and a nucleosome-like structure covers the region that is recognized by the steroid hormone receptors. When cells carrying this type of minichromosome are subjected to hormonal induction, the chromatin structure in this region of the minichromosomes changes, and the DNA becomes highly accessible to DNase I (Cordingley et al., 1987). Similar results have been reported for genomically integrated single copies of the MMTV provirus (Zaret and Yamamoto, 1984).

These interesting *in vivo* results prompted us to analyse the behaviour of the relevant sequence of the MMTV in chromatin reconstitution experiments. Short, linear fragments containing the HRE and a few flanking nucleotides, when incubated with rat liver core histones in high salt and subjected to step-wise dialysis, wrap themselves in a precise way on the surface of the histone octamer (Perlman and Wrange, 1988; Piña et al., 1990a). A major population of reconstituted mononucleosomes is obtained with the histone octamer located between position -189 and position -45 of the MMTV-LTR. A detailed analysis of the path of the DNA double helix around the histone octamer suggests that at least two of the receptor binding sites within the HRE, namely the most distal and the most proximal sites, relative to the promoter, are accessible from outside the nucleosome because their major grooves are facing outwards (Fig. 7). In band retardation assays, we find that both glucocorticoid and progesterone receptors bind to the nucleosomally reconstituted MMTV-HRE with an affinity only four- to five-fold lower than that of free DNA (Piña et al., 1990a). A direct interaction of these receptors with the exposed binding sites of the HRE was demonstrated in DNase I footprinting experiments. Therefore, it seems that the DNA sequence of the MMTV-LTR is designed to position itself on the surface of a histone octamer in such a way that two of the four receptor binding sites of the HRE will be exposed.

To exclude the possibility that the nucleosome positioning observed *in vitro* is due to an influence of the free DNA ends, we performed similar experiments with circular DNA molecules. Using microcircles which can accommodate a single nucleosome, or minicircles that could form a dinucleosome, we found that the limits of the histone octamer covering the hormone responsive element are similar to those mentioned above for mononucleosome reconstituted on linear DNA fragments (B. Piña, D. Barettino, M. Truss and M. Beato, unpubl. res.). These findings suggest that intrinsic properties of the DNA sequences, rather than any influence of the free DNA ends, are important for nucleosome positioning. In fact, when these microcircles are digested with DNase I in the absence of histones, the cutting pattern is very similar to that detected in the reconstituted circular mononucleosomes. This strongly suggests that the DNA sequence of the MMTV-HRE has an intrinsic ability to bend and to form a close circular structure in a way that mimics its path on the surface of the histone octamer.

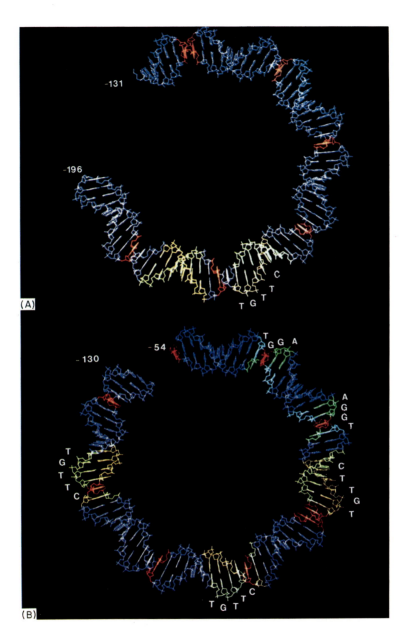

Fig. 7 Computer graphical representation of the superhelix formed by MMTV-HRE DNA around the histone octamere. (A) Sequences between −196 and −131. (B) Sequences between −130 and −54. Those base pairs flanked by the DNase I hypersensitive sites are shown in red. The binding sites for the hormone receptors are shown with the TGTTCT motifs and their palindromic counterparts in yellow. The palindromic sequences recognized by NF-1 are shown in green. The data are from Piña et al. (1990a), and the pictures were prepared by H. Ohlenbush and J. Postma at the EMBL in Heidelberg, FRG, using an E&S EMPS graphics monitor and the SHGRO program (EMBL Computer Graphics Group).

9.7 Mechanism of transcriptional activation

The mechanism by which binding of the steroid hormone receptors to the HRE of the MMTV activates transcription of the promoter seems to imply the binding of nuclear factor 1 (NF-1) to a site located immediately downstream of the most promoter-proximal receptor binding site (Nowock et al., 1985; Miksicek et al., 1987). Mutation of this binding site for NF-1 dramatically reduces transcription from the MMTV promoter after hormonal administration (Miksicek et al., 1987; Buetti et al., 1989). In cells carrying minichromosomes with the MMTV-LTR, binding of NF-1 to the MMTV promoter is only detected after hormonal administration (Cordingley et al., 1987). We have investigated this problem by analysing an oligonucleotide mutant that almost completely eliminates binding of NF-1 to its cognate sequences in vitro. Using this mutant MMTV, a dramatic reduction of the transcription of the MMTV promoter is observed after hormone administration, but a residual hormone-dependent activity can still be detected (Brüggemeier et al., 1990).

One possible mechanism by which binding of the hormone receptors to the HRE could facilitate binding of NF-1 to the MMTV promoter would be a direct protein–protein interaction. We have explored this possibility in DNA-binding experiments with partially purified steroid hormone receptors and NF-1 isolated from pig liver (Meisterernst et al., 1988). Contrary to the prediction of the model, no co-operation or synergism could be detected between the two classes of regulatory proteins for binding to the MMTV promoter (see Chapter 7.5). Instead the proteins compete with each other, apparently because their binding sites on the DNA partially overlap (Brüggemeier et al., 1990).

An alternative explanation for how binding of the receptor facilitates interaction of NF-1 with its cognate sequence on the MMTV promoter involves a change in chromatin structure. In the nucleosome reconstitution experiments mentioned above, we found that the binding site for NF-1 is oriented with the major groove facing inwards and is, therefore, not accessible to proteins from outside the nucleosome (Fig. 7). In band retardation studies and DNase I footprinting experiments, no binding of NF-1 to the nucleosomally organized MMTV could be detected, even at concentrations of NF-1 far higher than those required to obtain a footprint on free DNA (Piña et al., 1990a). It is conceivable that binding of the hormone receptor to the nucleosomally organized MMTV promoter changes the structure of the nucleosome in such a way that the sequence of the NFI-binding site become exposed. Preliminary support for this model has been derived from experiments using exonuclease III digestion of reconstituted nucleosomes. In the absence of receptor, the exonuclease III molecules entering the MMTV-DNA from the 3′-end of the nucleosome yield a main stop signal at position −45. In the presence of either glucocorticoid or progesterone receptor this signal is much weaker or completely absent, indicating that exonuclease III is able to pass this nucleosome border and progress further in its digestion of the DNA inside the nucleosome (Piña et al., 1990a). This suggests that receptor binding to HRE alters the accessibility of the 3′-end of the nucleosomal DNA containing the binding sites for NF-1. However, we have been unable to induce binding of NFI to a nucleosomally organized MMTV-DNA by addition of the hormone receptors. It is probable that other components, related to DNA topology, to histone modifications, or to the mechanism of nucleosome rearrangement, are missing in the reconstitution assay. Alternatively, the binding of the two regulatory proteins to MMTV-DNA in nucleosomes could be sequential rather than simultaneous.

9.8 Cell-free transcription assay

Ultimately, elucidation of the molecular mechanisms underlying transcriptional activation by receptors will require the development of a reconstituted cell-free transcription system in which the individual components can be added in purified form (see Chapter 5.7). As a first step in this direction we have been able to measure the effects of purified progesterone receptor on transcription of the MMTV promoter in a cell-free system derived from HeLa cells (Kalff et al., 1990). Under optimal conditions, in the presence of a progestin agonist and a slight molar excess of receptor, the stimulation of transcription from the MMTV promoter may reach 10-fold. The effect is clearly dependent upon specific binding of the receptor to the MMTV-HRE and can be completed by a synthetic HRE-oligo. In this system, mutation of the binding site for NF-1 reduces the transcriptional efficiency about 20-fold, but the effect of the receptor on transcription initiation is maintained. Thus, although NF-1 clearly is a transcription factor in the MMTV promoter, it is not absolutely required for hormone induction of transcription. When templates containing a mutated NF-1 binding site are used, the effect of the receptor on transcriptional efficiency must be mediated by factors other than NF-1.

There is a quantitative discrepancy between the *in vivo* and *in vitro* effects of hormones on transcription of the MTV promoter. This can be explained if we consider that *in vivo* the MMTV promoter is transcriptionally silent in the absence of hormones, whereas in the cell-free transcription assay the promoter is used very efficiently, even in the absence of receptor. Thus, *in vivo* a repression mechanism seems to be operating that is not efficiently acting in the cell-free system. It is possible that nucleosome organization in the cell nucleus is responsible for the repression observed *in vivo*, and that in the intact cell the main function of the receptor could be to derepress the promoter by allowing binding of NF-1. In addition to this derepression, a positive effect of the receptor on transcription of the MMTV promoter must exist which is detected when the NF-1 binding site is mutated. Fractionation of the cell-free system and reconstitution with the individual components should allow the identification of all factors involved in transcriptional induction by steroid hormone receptors, and may help us to understand their sequential interactions.

Acknowledgements

The experimental work summarized in this paper was supported by grants from the Deutsche Forschungsgemeinschaft and the Fonds der Chemischen Industrie.

References

Akerblom, I. W., Slater, E. P., Beato, M., Baxter, J. D. and Mellon, P. L. (1988). Negative regulation by glucocorticoids through interference with a cAMP responsive enhancer. *Science* **241**, 350–353.

Alwine, J. C. (1985). Transient gene expression control: effects of transfected DNA stability and trans-activation by viral enhancer proteins. *Mol. Cell. Biol.* **5**, 1034–1042.

Arriza, J. L., Weinberger, C., Cerelli, G., Glaser, T. M., Handelin, B. L., Housmann, D. E. and Evans, R. M. (1987). Cloning of human mineralocorticoid receptor cDNA: structural and functional kinship with the glucocorticoid receptor. *Science* **237**, 268–275.

Bailly, A., Le Page, C., Rauch, M. and Milgrom, E. (1986). Sequence-specific DNA binding of the progesterone receptor to the uteroglobin gene: effects of hormone, antihormone and receptor phosphorylation. *EMBO J.* **5**, 3235–3241.

Beato, M. (1989). Gene regulation by steroid hormones. *Cell* **56**, 335–344.

Beato, M., Chalepakis, G., Schauer, M. and Slater, E. P. (1989). DNA regulatory elements for steroid hormones. *J. Steroid. Biochem.* **32**, 737–747.

Becker, P. B., Gloss, B., Schmid, W., Strähle, U. and Schütz, G. (1986). In vivo protein-DNA interactions in a glucocorticoid response element. *Nature* **324**, 686–688.

Berry, M., Nunez, A. M. and Chambon, P. (1989). Estrogen-responsive element of the human pS2 gene is an imperfect palindromic sequence. *Proc. Natl. Acad. Sci. USA* **86**, 1218–1222.

Brüggemeier, U., Rogge, L., Winnacker, E. L. and Beato, M. (1990). Nuclear factor I acts as a transcription factor on the MMTV promoter but competes with steroid hormone receptors for DNA binding. *EMBO J.* **9**, 2233–2239.

Buetti, E., Kühnel, B. and Diggelmann, H. (1989). Dual function of a nuclear factor I binding site in MMTV transcription regulation. *Nucl. Acids Res.* **17**, 3065–3078.

Cato, A. C. B. and Weinmann, J. (1988). Mineralocorticoid regulation of transfected mouse mammary tumour virus DNA in cultured kidney cells. *J. Cell. Biol.* **106**, 2119–2125.

Cato, A. C. B., Geisse, S., Wenz, M., Westphal, H. M. and Beato, M. (1984). The nucleotide sequences recognized by the glucocorticoid receptor in the rabbit uteroglobin gene region are located far upstream from the initiation of transcription. *EMBO J.* **3**, 2731–2736.

Cato, A. C. B., Miksicek, R., Schütz, G., Arnemann, J. and Beato, M. (1986). The hormone regulatory element of mouse mammary tumour virus mediates progesterone induction. *EMBO J.* **5**, 2237–2240.

Cato, A. C. B., Henderson, D. and Ponta, H. (1987). The hormone response element of the mouse mammary tumour virus DNA mediates the progestin and androgen induction of transcription in the proviral long terminal repeat region. *EMBO J.* **6**, 363–368.

Chalepakis, G., Arnemann, J., Slater, E. P., Brüller, H.-J., Gross, B. and Beato, M. (1988a). Differential gene activation by glucocorticoids and progestins through the hormone regulatory element of mouse mammary tumour virus. *Cell* **53**, 371–382.

Chalepakis, G., Postma, J. P. M. and Beato, M. (1988b). A model for hormone receptor binding to the mouse mammary tumour virus regulatory element based on hydroxyl radical footprinting. *Nucl. Acids Res.* **16**, 10 237–10 247.

Chandler, V. L., Maler, B. A. and Yamamoto, K. R. (1983). DNA sequences bound specifically by glucocorticoid receptor *in vitro* render a heterologous promoter hormone responsive *in vivo*. *Cell* **33**, 489–499.

Cordingley, M. G. and Hager, G. L. (1988). Binding of multiple factors to the MMTV promoter in crude and fractionated nuclear extracts. *Nucl. Acids Res.* **16**, 609–630.

Cordingley, M. G., Riegel, A. T. and Hager, G. L. (1987). Steroid-dependent interaction of transcription factors with the inducible promoter of mouse mammary tumor virus *in vivo*. *Cell* **48**, 261–270.

Danesch, U., Gloss, B., Schmid, W., Schütz, G. and Renkawitz, R. (1987). Glucocorticoid induction of the rat tryptophan oxygenase gene is mediated by two widely separated glucocorticoid-responsive elements. *EMBO J.* **6**, 625–630.

Danielsen, M., Hinck, L. and Ringold, G. M. (1989). Two amino acids within the knuckle of the first zinc finger specify DNA response element activation by the glucocorticoid receptor. *Cell* **57**, 1131–1138.

DeFranco, D. and Yamamoto, K. R. (1986). Two different factors act separately or together to specify functionally distinct activities at a single transcriptional enhancer *Mol. Cell. Biol.* **6**, 993–1001.

Fawell, S. E., Lees, J. A., White, R. and Parker, M. G. (1990). Characterization and colocalization of steroid binding and dimerization activities in the mouse estrogen receptor. *Cell* **60**, 953–962.

Freedman, L. P., Yoshinaga, S. K., Vanderbilt, J. N. and Yamamoto, K. R. (1989). *In vitro* transcription enhancement by purified derivatives of the glucocorticoid receptor. *Science* **245**, 298–301.

Geisse, S., Scheidereit, C., Westphal, H. M., Hynes, N. E., Groner, B. and Beato, M. (1982). Glucocorticoid receptors recognize DNA sequences in and around murine mammary tumour virus DNA. *EMBO J.* **1**, 1613–1619.

Ham, J., Thomson, A., Needham, E. and Parker, M. (1988). Characterization of response elements for androgens, glucocorticoids and progestins in mouse mammary tumour virus. *Nucl. Acids Res.* **16**, 5263–5277.

Harland, R. M., Weintraub, H. and McKnight, S. L. (1983). Characterization of response elements for androgens, glucocorticoids and progestins in mouse mammary tumour virus. *Nature* **302**, 38–43.

Hecht, A., Berkenstam, A., Strömstedt, P. E., Gustafsson, J. A. and Sippel, A. E. (1988). A progesterone responsive element maps to the far upstream steroid dependent DNase hypersensitive site of chicken lysozyme chromatin. *EMBO J.* **7**, 2063–2073.

Jantzen, C., Fritton, H. P., Igo-Kemenes, T., Espel, E., Janich, S., Cato, A. C. B., Mugele, K. and Beato, M. (1987). Partial overlapping of binding sequences for steroid hormone receptors and DNaseI hypersensitive sites in the rabbit uteroglobin gene region. *Nucl. Acids Res.* **15**, 4535–4552.

Jantzen, H. M., Strähle, U., Gloss, B., Stewart, F., Schmid, W., Bosshart, M., Miksicek, R. and Schütz, G. (1987). Cooperativity of glucocorticoid response elements located far upstream of the tyrosine aminotransferase gene. *Cell* **49**, 29–38.

Kalff, M., Gross, B. and Beato, M. (1990). Progesterone receptor stimulates transcription of mouse mammary tumour virus in a cell-free system. *Nature* **344**, 360–362.

Karin, M., Haslinger, A., Holtgreve, A., Richards, R. I., Krauter, P., Westphal, H. M. and Beato, M. (1984). Characterization of DNA sequences through which cadmium and glucocorticoid hormones induce human metallothionein IIA. *Nature* **308**, 513–519.

Keller, W. (1975). Characterization of purified DNA-relaxing enzyme from human tissue culture cells. *Proc. Natl. Acad. Sci. USA* **72**, 2550–2554.

Klein, E. S., DiLorenzo, D., Posseckert, G., Beato, M. and Ringold, G. M. (1988). Sequences downstream of the glucocorticoid regulatory element mediate cycloheximide inhibition of steroid induced expression from the rat A1-acid glycoprotein promoter: evidence for a labile transcription fact. *Mol. Endocrinol.* **2**, 1343–1351.

Klein-Hitpass, L., Schorp, M., Wagner, U. and Ryffell, G. U. (1986). An estrogen-responsive element derived from the 5′-flanking region of *Xenopus* vitellogenin A2 gene functions in transfected human cells. *Cell* **46**, 1053–1061.

Klein-Hitpass, L., Tsai, S. Y., Greene, G. L., Clark, J. H., Tsai, M. J. and O'Malley, B. W. (1989). Specific binding of estrogen receptor to the estrogen response element. *Mol. Cell. Biol.* **9**, 43–49.

Kumar, V. and Chambon, P. (1988). The estrogen receptor binds tightly to its responsive element as a ligand-induced homodimer. *Cell* **55**, 145–156.

Kühnel, B., Buetti, E. and Diggelmann, H. (1986). Functional analysis of the glucocorticoid regulatory elements present in the mouse mammary tumor virus long terminal repeat. A synthetic distal binding site can replace the proximal binding domain. *J. Mol. Biol.* **190**, 367–378.

Mader, S., Kumar, V., de Verneuil, H. and Cambon, P. (1989). Three amino acids of the oestrogen receptor are essential to its ability to distinguish an oestrogen from a glucocorticoid-responsive element. *Nature* **338**, 271–274.

Martinez, E., Givel, F. and Wahli, W. (1987). The estrogen-responsive element as an inducible enhancer: DNA sequence requirements and conversion to a glucocorticoid responsive element. *EMBO J.* **6**, 3719–3727.

Meisterernst, M., Rogge, L., Donath, C., Gander, I., Lottspeich, F., Mertz, R., Dobner, T., Föckler, R., Stelzer, G. and Winnacker, E. L. (1988). Isolation and characterization of the porcine nuclear factor I (NFI) gene. *FEBS Lett.* **236**, 27–32.

Miksicek, R., Heber, A., Schmid, W., Danesch, U., Posseckert, G., Beato, M. and Schütz, G. (1986). Glucocorticoid responsiveness of the transcriptional enhancer of Moloney murine sarcoma virus. *Cell* **46**, 283–290.

Miksicek, R., Borgmeyer, U. and Nowock, J. (1987). Interaction of the TGGCA-binding protein with upstream sequences is required for efficient transcription of mouse mammary tumor virus. *EMBO J.* **6**, 1355–1360.

Moore, D. D., Marks, A. R., Buckley, D. I., Kapler, G., Payvar, F. and Goodman, H. M. (1985). The first intron of the human growth hormone gene contains a binding site for glucocorticoid receptor. *Proc. Natl. Acad. Sci. USA* **82**, 699–702.

Nowock, J., Borgmeyer, U., Püschel, A., Rupp, A. W. and Sippel, A. E. (1985). The TGGCA protein binds to the MMTV-LTR, the adenovirus origin of replication, and BK virus. *Nucl. Acids Res.* **13**, 2045–2062.

Payvar, F., DeFranco, D., Firestone, G. L., Edgar, B., Wrange, Ö., Okret, S., Gustafsson, J. A. and Yamamoto, K. R. (1983). Sequence-specific binding of glucocorticoid receptor to MTV DNA at sites within and upstream of the transcribed region. *Cell* **35**, 381–392.

Perlmann, T. and Wrange, Ö. (1988). Specific glucocorticoid receptor binding to DNA reconstituted in a nucleosome. *EMBO J.* **7**, 3073–3079.

Piña, B., Brüggemeier, U. and Beato, M. (1990a). Nucleosome positioning modulates accessibility of regulatory proteins to the mouse mammary tumor virus promoter. *Cell* **60**, 719–731.

Piña, B., Haché, R. J. G., Arnemann, J., Chalepakis, G., Slater, E. P. and Beato, M. (1990b). Hormonal induction of transfected genes depends on DNA topology. *Mol. Cell. Biol.* **10**, 625–633.

Renkawitz, R., Schütz, G., von der Ahe, D. and Beato, M. (1984). Identification of hormone regulatory elements in the promoter region of the chicken lysozyme gene. *Cell* **37**, 503–510.

Richard-Foy, H. and Hager, G. L. (1987). Sequence specific positioning of nucleosomes over the steroid-inducible MMTV promoter. *EMBO J.* **6**, 2321–2328.

Schauer, M., Chalepkais, G., Willmann, T. and Beato, M. (1989). Binding of hormone accelerates the kinetics of glucocorticoid and progesterone receptor binding to DNA. *Proc. Natl. Acad. Sci. USA* **86**, 1123–1127.

Scheidereit, C. and Beato, M. (1984). Contacts between receptor and DNA double helix within a glucocorticoid regulatory element of mouse mammary tumor. *Proc. Natl. Acad. Sci. USA* **81**, 3029–3033.

Scheidereit, C., Geisse, S., Westphal, H. M. and Beato, M. (1983). The glucocorticoid receptor binds to defined nucleotide sequences near the promoter of mouse mammary tumour. *Nature* **304**, 749–752.

Scheidereit, C., Westphal, H. M., Carlson, C., Bosshard, H. and Beato, M. (1986). Molecular model of the interaction between the glucocorticoid receptor and the regulatory elements of inducible genes. *DNA* **5**, 383–391.

Schema, M. and Yamamoto, K. R. (1988). Mammalian glucocorticoid receptor derivatives enhance transcription in yeast. *Science* **241**, 965–968.

Schema, M., Freedman, L. P. and Yamamoto, K. R. (1989). Mutations in the glucocorticoid receptor zinc finger that distinguish interdigitated DNA binding and transcriptional enhancement activities. *Genes & Dev.* **3**, 1590–1601.

Seiler-Tuyns, A., Walker, P., Martinez, E., Merillat, A.-M., Givel, F. and Wahli, W. (1986). Identification of estrogen-responsive DNA sequences by transient expression experiments in a human breast cancer cell line. *Nucl. Acids Res.* **14**, 8755–8770.
Slater, E. P., Rabenau, O., Karin, M., Baxter, J. D. and Beato, M. (1985). Glucocorticoid receptor binding an activation of a heterologous promoter in response to dexamethasone by the first intron of the human growth hormone gene. *Mol. Cell. Biol.* **5**, 2984–2992.
Slater, E. P., Cato, A. C. B., Karin, M., Baxter, J. D. and Beato, M. (1988). Progesterone induction of metallothionein-IIA gene expression. *Mol. Endocrinol.* **2**, 485–491.
Slater, E. P., Redenihl, G., Theis, K., Suske, G. and Beato, M. (1990). The uteroglobin promoter contains a monocanonical estrogen responsive element. *Mol. Endocrinol.* **4**, 604–610.
Strähle, U., Klock, G. and Schütz, G. (1987). A DNA sequence of 15 base pairs is sufficient to mediate both glucocorticoid and progesterone induction of gene expression. *Proc. Natl. Acad. Sci. USA* **84**, 7871–7875.
Strähle, U., Boshart, M., Klock, G., Stewart, F. and Schütz, G. (1989). Glucocorticoid- and progesterone-specific effects are determined by differential expression of the respective hormone receptors. *Nature* **339**, 629–632.
Truss, M., Chalepakis, G. and Beato, M. (1990). Contacts between steroid hormone receptors and thymines in DNA: An interference method. *Proc. Natl. Acad. Sci. USA* **87**, (in press).
Tsai, S. Y., Carlstedt-Duke, J., Weigel, N. L., Dahlman, K., Gustafsson, J. A., Tsai, M. J. and O'Malley, B. W. (1988). Molecular interactions of steroid hormone receptor with its enhancer element: evidence for receptor dimer formation. *Cell* **55**, 361–369.
Tsai, S. Y., Tsai, M. J. and O'Malley, B. W. (1989). Cooperative binding of steroid hormone receptors contributes to transcriptional synergism at target enhancer elements. *Cell* **57**, 443–448.
Umesono, K. and Evans, R. M. (1989). Determinants of target gene specificity for steroid/thyroid hormone receptors. *Cell* **57**, 1139–1146.
von der Ahe, D., Janich, S., Scheidereit, C., Renkawitz, R., Schütz, G. and Beato, M. (1985). Glucocorticoid and progesterone receptors bind to the same sites in two hormonally regulated promoters. *Nature* **313**, 706–709.
von der Ahe, D., Renoir, J. M., Buchou, T., Baulieu, E. E. and Beato, M. (1986). Receptors for glucocorticosteroid and progesterone recognize distinct features of a DNA regulatory element. *Proc. Natl. Acad. Sci. USA* **83**, 2817–2821.
Weintraub, H., Cheng, P. F. and Conrad, K. (1986). Expression of transfected DNA depends on DNA topology. *Cell* **46**, 115–122.
Westphal, H. M., Moldenhauer, G. and Beato, M. (1982). Monoclonal antibodies to the rat liver glucocorticoid receptor. *EMBO J.* **1**, 1467–1471.
Willmann, T. and Beato, M. (1986). Steroid-free glucocorticoid receptor binds specifically to mouse mammary tumour virus DNA. *Nature* **324**, 688–691.
Wrange, Ö., Eriksson, P. and Perlmann, T. (1989). The purified activated glucocorticoid receptor is a homodimer. *J. Biol. Chem.* **264**, 5253–5259.
Zaret, K. S. and Yamamoto, K. R. (1984). Reversible and persistent changes in chromatin structure accompanying activation of a glucocorticoid-dependent enhancer element. *Cell* **38**, 29–38.

10
The Interaction of Steroid Receptors with Chromatin

G. L. HAGER and T. K. ARCHER

Hormone Action and Oncogenesis Section, Laboratory of Experimental Carcinogenesis, National Cancer Institute, N. I. H., Bethesda, MD 20892, USA

10.1 Introduction

A mammalian cell typically contains 2–3 pg of DNA per chromosome set. During the evolution of nucleated cells, a hierarchy of structures has developed to package this extraordinary quantity of polymer into the available volume. This nucleoprotein complex, containing many levels of organization, is collectively referred to as chromatin (see van Holde, 1988 for an excellent overview of chromatin structure). Our current concepts of eukaryotic gene regulation, including control of transcription by steroid receptors, have been developed predominantly in experimental systems where only the interactions between transcription factors, enzymes and the DNA template are considered (see Yamamoto, 1985; Green and Chambon, 1988; Beato, 1989; Evans, 1989; Ham and Parker, 1989 for reviews). Implications of the higher order organization of DNA in chromatin are usually not considered in mechanisms of gene regulation, primarily because of a lack of experimental tools to manipulate these complex structures effectively and test potential functions. The powerful tools of recombinant DNA, applied to the cloning and characterization of soluble transcription enzymes and factors, has led to a rapid expansion in our understanding of these molecules and interactions between them. Technologies for examining and manipulating the nucleoprotein organization of regulatory sequences unfortunately lag far behind.

Despite the inherent difficulties in addressing issues of chromatin function, it is becoming apparent that access of the soluble transcription apparatus to regulatory sequences in DNA can be significantly modulated by nucleoprotein structure. Furthermore, in some model systems our knowledge of chromatin structure is now sufficiently advanced that interactions between soluble factors and regulatory sequences organized in nucleoprotein structure can be profitably pursued. A complete theory of regulation in higher eukaryotes will eventually require an understanding of mechanisms that control the organization of DNA in chromatin, and the extent to which this organization modulates activation and repression of transcription.

10.2 Nucleoprotein organization of eukaryotic DNA

The fundamental structural unit of chromatin is the nucleosome, which contains 146 base pairs of DNA wrapped on a protein macromolecule called the octamer core. This structure contains two copies each of the four histone molecules, H2A, H2B, H3 and H4, and is symmetrical about a dyad axis. The synthesis of these histones is tightly coupled to DNA replication. As DNA is polymerized, parent and daughter strands are organized into linear arrays of nucleosomes, the well-known 'beads on a string'. In addition to the core proteins, the 'mature' nucleosomes in these arrays contain a fifth histone, H1, which is involved both in stabilizing the nucleosome structure, and in facilitating assembly of the array into higher-order structures. The amount of DNA found between succeeding members of the polynucleosome array (the linker region) can vary widely, from as little as 15 nucleotides in close-packed arrays, to greater than 50 base pairs in some regions. These polynucleosome arrays (frequently referred to as the 100 nanometre fibre) are subsequently organized into a more closely packed structure, the 300 nanometre fibre, which is thought to contain six nucleosomes per helical turn. This structure is assembled in turn into higher-order structures, ultimately resulting in the physical structure recognized as a chromosome.

In regions of chromatin where genes are being actively expressed, the higher-order structure described above is reversed, with the nucleoprotein fibres becoming more extended and less defined. This process was first described in the nuclei of insect salivary glands, where active regions of the highly amplified lampbrush chromosomes could be visualized as puffs (see van Holde, 1988 for review). Active transcription was eventually observed in these puff regions, and was associated with the 100 nanometre fibre. More recently, active genes as a class have been found to have a generally increased sensitivity to nucleolytic attack (Weintraub, 1983). This global sensitivity is now thought to reflect a similar loss of high-order structure, that is, a decondensation of chromatin structure much as observed in the puff regions of lampbrush chromosomes. Given the extraordinary packing density of DNA sequences in chromatin, it is not surprising that a breakdown in the level of organization is required for the soluble transcription apparatus to access active genes.

10.3 Receptor interactions with characterized chromatin structures

In the discussion thus far, we have seen that the DNA template in eukaryotic cells is highly complex. However, almost all current experimentation on transactivation by steroid receptors is predicated on models in which complex templates are not considered. In point of fact, little is known about the interaction of receptors and nucleoprotein templates. There is a vast literature in which the binding to DNA of transactivators in general, and receptors in particular, is inferred from regions of hypersensitivity to nucleolytic agents (HSRs) that develop in chromatin (see Elgin, 1981, 1988; Gross and Garrard, 1988 for reviews). Indeed, the development of an HSR has become one of the standard tools for identifying a putative transcription control element at a given regulatory region. However, this is essentially a descriptive literature, and few mechanistic investigations have been attempted. There is a clear need for model systems in which the interaction of receptors and associated soluble components with complex templates can be critically evaluated.

10.3.1 The MMTV Paradigm

In one system, that of the steroid-responsive promoter of mouse mammary tumour virus (MMTV), a body of knowledge has emerged concerning both the soluble factors involved in steroid receptor transactivation, and detailed information on the nucleoprotein structure of the *in vivo* template. Steroid transactivation was first demonstrated in the MMTV system (Huang *et al.*, 1981; Lee *et al.*, 1981), and the MMTV LTR has become a standard steroid response reporter promoter (see Chapter 9.2). Not only has the promoter been extensively used in the mutational analysis of receptors, but it is one of the few cases where the formation of a transcriptional initiation complex has been directly observed *in vivo* (Cordingley *et al.*, 1987).

The MMTV system is unique in that it has been shown to adopt a highly reproducible chromatin structure when integrated as a provirus in cellular DNA (Richard-Foy and Hager, 1987). A series of six nucleosomes are specifically positioned across the LTRs of the virus. The initial findings were obtained with cells in which the viral LTR had been fused to reporter genes [either the *ras* oncogene or the chloramphenicol acetyltransferase (CAT) gene], and the chimaeric fusions mobilized on amplified episomes based on the bovine papilloma virus (BPV) vector. The minichromosome vector system provided two features of critical importance for this investigation; the LTR chimaeras were replicated to sufficient copy number to permit high-resolution analysis of chromatin structure, while maintaining homogeneity of the nucleoprotein template with regard to the transcriptional response (see Cordingley et al., 1987 for discussion of this point). Internucleosomal linker regions were positioned across the promoter and associated regulatory regions by susceptibility to micrococcal nuclease and methidiumpropyl–EDTA–Fe^{2+} (MPE) cleavage. The surprising finding from this investigation is presented schematically in Fig. 1. In all cases examined, nucleosomes were found to be phased at the positions depicted in the figure. The results have subsequently been extended to single integrated copies of the provirus, and single integrated copies of MMTV-ras fusion chimaeras (Richard-Foy and Hager, submitted). Thus the acquisition of highly positioned nucleosomes is an intrinsic feature of MMTV-LTR DNA, not an artifact of extrachromosomal location. These findings are somewhat unexpected for a retrovirus, since these elements integrate at a large number of chromosomal sites, and exhibit a variation in constitutive level of expression (Thompson et al., 1980) (a type of position effect) often suspected to result from variation in chromatin structure at the site of integration.

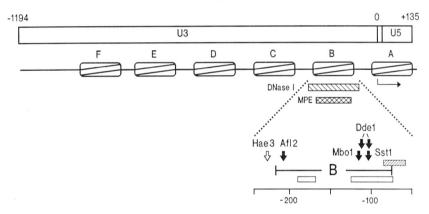

Fig. 1 Chromatin structure of the MMTV-LTR and regulatory elements. Positions are presented schematically for the six-nucleosome phased array. The large hatched box designates the region hypersensitive to DNase I. The cross-hatched box indicates the region hypersensitive to MPE. In the enlarged view of the B region, restriction enzyme sensitivity is indicated, open arrows for enzymes whose access is unaffected by hormone treatment, and closed arrows for enzymes whose access is hormone-dependent. The small hatched box designates the binding site for NF-1, and the grey-filled boxes the binding regions for glucocorticoid receptor.

The feature of greatest interest in the LTR chromatin structure is the coincidence in position between an extended HSR region and the putative core of the second nucleosome (nucleosome B) in the phased array. This region harbours DNA-binding sites not only for the glucocorticoid receptor, but also for the progesterone and androgen receptors. As seen in Fig. 1, hormone-dependent sensitivity to nucleolytic cleavage can be monitored not only with DNase I, but also with MPE and restriction endonucleases. The precise correspondence in position between hypersensitivity to any of the three reagents and the core of nucleosome B indicates that the receptor-dependent HSR results either from displacement, or massive modification in the core structure, of nucleosome B.

The development of HSRs of the approximate size depicted in Fig. 1 has been reported for many genes, and is usually taken as diagnostic of the binding of a regulatory protein and concomitant loss of a nucleosome. Nucleosome B displacement* is unique, however, in that the nucleoprotein modification is an active event that occurs directly in response to receptor binding. In other systems where nucleosome exclusion is thought to result from the presence of a high-affinity DNA-binding protein, it has not been possible to distinguish between actual nucleosome displacement and competition for the site during DNA replication (Schlissel and Brown, 1984). In fact, direct experiments using inhibitors of DNA synthesis have shown that nucleosome B displacement occurs in the absence of replication (Archer et al., 1990a, submitted).

10.3.2 Formation of Initiation Complex

The minichromosome system described above can also be utilized to examine the promoter for binding of factors during transcription activation. Exonuclease III footprinting (Wu, 1985) of promoter chromatin from induced and uninduced cells revealed the appearance of a stable complex on the DNA in response to steroid treatment (Fig. 2A; Cordingley et al., 1987). The large size of the ExoIII footprint (90–100 bp) suggested a multifactorial composition. Fractionation of extracts in fact led to the identification of at least two proteins involved in formation of the initiation complex (Cordingley and Hager, 1988). The upstream boundary of ExoIII resistance resulted from the binding of NF-1 (protecting sequences -80 to -56). The TATA protein (TFIID) was implicated in formation of the 3' boundary (protecting sequences -42 to $+6$). Thus, displacement of nucleosome B from the MMTV phased array is accompanied by the establishment of a stable transcriptional initiation complex composed of at least two factors, NF-1 and TFIID.

---

* The nucleosome B modification will be referred to throughout this discussion as displacement, although formal proof of actual histone removal is not yet available.

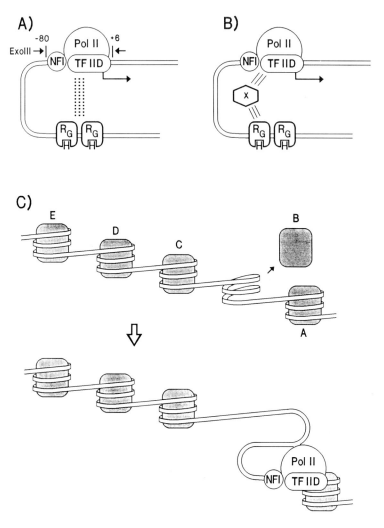

Fig. 2 Models for receptor-mediated transactivation. In panels (A) and (B), putative protein–protein interactions are diagrammed, either direct contact between receptor and the transcription initiation complex (A), or indirect contact mediated by a bridging protein (B). In panel (C), the assembly of the initiation complex is described in the context of the chromatin structure of the locus.

10.3.3 Transactivation by Factor Interactions

Mutational analysis of steroid receptors, coupled with functional analysis in transient transfection assays, has elucidated a multiple domain structure

10. INTERACTION OF STEROID HORMONES WITH CHROMATIN

similar to that derived earlier for transactivating factors of lower eukaryotes (DNA-binding domain, transactivating domain), with the addition of a third domain responsible for hormone binding. Results from a number of studies with several receptors (Kumar *et al.*, 1987; Hollenberg *et al.*, 1987; Miesfield *et al.*, 1987b; see Chapters 2.2 and 3) interpreted in terms of the classic protein–protein contact model (Fig. 2A,B). The essential features are as follows. Receptors bind, as dimers (Kumar and Chambon, 1988; Tsai *et al.*, 1988; Fawell *et al.*, 1990; see Chapter 16.3), to their cognate recognition sites (Fig. 2A,B; see Chapters 6.2 and 9.4). The bound receptor complex interacts with a second protein, which might be a 'bridging' factor, specific to a given promoter (Fig. 2B), or a common member of the promoter initiation complex (Fig. 2A). This interaction might occur as a result of pre-existing contact sites on the receptor, or as a result of allosteric modifications to the receptor upon binding DNA. The net effect is to increase the local concentration of the interacting protein, giving rise to a stable pre-initiation complex (Sawadogo and Roeder, 1985), resulting in an increased rate of transcription initiation. An important feature of this model is the potential for specificity in promoter activation via unique contact sites between receptor and a member of the initiation complex, or between receptor and a specific bridging protein (Tora *et al.*, 1988; see Chapters 2.12 and 7.4).

The critical element of this model for the current discussion is the importance of protein–protein contacts in modulating the rate of initiation complex formation. The concepts underlying this mechanism were developed in prokaryotic systems (Hochschild and Ptashne, 1986). A considerable body of data now exists that is consistent with the general features of this model. Transient transfection studies with several of the known receptors have identified regions of the receptor protein which, when deleted, diminish or eliminate the transactivating properties of the molecule without affecting other critical features, such as hormone binding and DNA binding (Giguere *et al.*, 1986; Hollenberg *et al.*, 1987; Kumar *et al.*, 1987; Miesfeld *et al.*, 1987a; Webster *et al.*, 1988; Lees *et al.*, 1989). Furthermore, modest levels of transactivation have been reported *in vitro* with crude extracts (Corthesy *et al.*, 1988) and partially purified fractions (Freedman *et al.*, 1989; Klein-Hitpass *et al.*, 1990; Kalff *et al.*, 1990) in systems with pure DNA as the template (see Chapter 5.7). One is led to conclude that some type of protein–protein interactions must account for these effects, particularly those observed in cell-free systems.

10.3.4 Factor Exclusion by Chromatin

Two aspects of the results diagrammed in Fig. 2 are of particular interest. First, one of the factors (NF-1) is one of the highest affinity DNA-binding proteins described (Jones et al., 1987). Indeed, binding of this factor to MMTV promoter DNA can be easily detected in simple broken cell extracts (Cordingley et al., 1987). Yet the protein is completely excluded from the MMTV promoter when organized as stable chromatin. Secondly, there is no evidence of the glucocorticoid receptor in the promoter chromatin ExoIII footprints, even though all of the purified receptor DNase I-resistant footprints were scanned in the experiment (Cordingley et al., 1987). In vitro experiments (von der Ahe et al., 1985) have shown that the glucocorticoid receptor is not particularly susceptible to ExoIII digestion. In summary: (1) a protein with very high intrinsic DNA-binding affinity (NF-1) is excluded from chromatin; (2) a protein with relatively low intrinsic DNA-binding affinity (receptor) initiates a process that leads to nucleosome B displacement and NF-1/TFIID binding; and (3) the presence of the initiating protein is not detected in the stable transcription initiation complex. Considering only the results from in vitro transcription experiments (Corthesy et al., 1988; Freedman et al., 1989), one would argue that NF-1/TFIID binding was driven by protein–protein contacts (Fig. 2A,B). Although receptor binds immediately upstream of NF-1, it is unlikely that receptor interacts directly with NF-1. Numerous attempts to detect receptor/NF-1 interactions in vitro have failed; in fact, receptor will compete with NF-1 for binding to MMTV-DNA (Brüggemeier et al., 1990). It is difficult to resolve NF-1 exclusion from promoter chromatin purely on the basis of protein–protein contacts. Not only does the protein have a very high affinity for promoter DNA, we have observed NF-1 dependent, receptor-independent, transcription activation directly with in vitro transcription extracts (M. G. Cordingley and G. L. Hager, unpubl. obs.). It is most likely that NF-1 is excluded from the DNA in vivo by the organization of the template in chromatin (Fig. 2C). This conclusion implies that chromatin is not a neutral, nor transparent, template, but provides selective access to components of the soluble transcription apparatus. This conclusion is further supported by arguments developed below.

10.3.5 Mechanism of Factor Exclusion

We have argued that nucleoprotein organization of the MMTV promoter renders sites for high-affinity DNA-binding proteins selectively accessible. What mechanism could provide for such modulation of site availability? Since we have shown that nucleosomes are specifically positioned over the region of interest, the simplest explanation would be masking of the binding site by octamer core exclusion. Perlmann and Wrange (1988) have reported that octamer cores reconstituted in vitro on an LTR B region DNA fragment

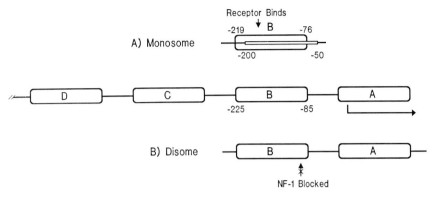

Fig. 3 Reconstitution of nucleosome structures *in vitro*. The *in vivo* nucleosome positions are shown in the centre panel. In panel (A), the monosome described by Perlmann and Wrange (1988) is described. In panel (B), the disome reconstituted by Archer *et al.* (1990a, submitted) is diagrammed.

will adopt a specific position whose boundaries are consistent with the *in vivo* boundaries reported earlier (Fig. 3A; Richard-Foy and Hager, 1987). We have reconstituted a disomic molecule with an A–B region DNA fragment, and find that both A and B nucleosomes are positioned at sites compatible with the *in vivo* results (Fig. 3B; Archer *et al.*, 1990a, submitted). Furthermore, NF-1 is specifically excluded from its binding site in the reconstituted disome. These observations suggest a simple model for *in vivo* exclusion. On a highly-positioned nucleosome, the site for a high-affinity DNA-binding protein can be positioned in such a way that it either remains open to the factor, or becomes inaccessible (Fig. 4A). Perlmann and Wrange (1988) further reported the intriguing observation that glucocorticoid receptor could bind to a B region site even in the presence of the octamer core. Under the model presented in Fig. 4A, the receptor binding site would be facing out from the core.

Although simultaneous occupancy of the B region DNA by the large octamer core and an equally large multimeric receptor complex is difficult to visualize, initial recognition of the site on the nucleosome must take place. Otherwise promoter activation could not occur. We know very little, however, of the precise mechanics of this process. Under all conditions we have examined, activation of the promoter is always accompanied by development of the B region HSR (Sistare *et al.*, 1987; Richard-Foy *et al.*, 1987). That is, occupancy of the B region by receptor *in vivo* appears to be inconsistent with nucleosome placement. We conclude that transactivation in this system is a complex process that includes protein–protein interactions, differential access to sites organized in chromatin, and chromatin remodelling during the assembly of the transcription initiation complex.

A) Factor Access Modulated by Rotational Position

B) Factor Access Modulated by Higher-Order Structure

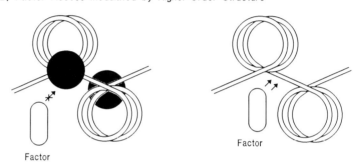

Fig. 4 Models for modulation of factor access by chromatin structure. Panel (A) represents a model wherein factor access is determined primarily by the orientation of the factor DNA-binding site on the octamer core. In panel (B), a model in which factor access is governed by other features of the polynucleosome array, such as the presence of non-octamer core proteins, is shown.

10.3.6 Mechanism of Nucleosome Positioning

There is reason to question the mechanism of nucleosome positioning implied in the *in vitro* phasing results discussed above. It has been argued by several authors that nucleosomes could adopt selective positions due to the intrinsic resistance of DNA to bending through the very small curvature found on the nucleosome surface (see Travers and Klug, 1987 for a review of this concept). In one careful study, Fourier analysis of a large number of sequences isolated from a random preparation of core particles suggested a non-random sequence association with outward-facing minor grooves (Satchwell *et al.*, 1986). These sequences were obtained from preparations of H1-deficient monosomes, however. A more recent analysis of sequences from H1-containing disomes carried out in the same laboratory did not confirm the earlier sequence modalities (Satchwell and Travers, 1989).

There is a notable lack of examples for high-resolution correlation between strong *in vivo* phasing patterns and *in vitro* reconstituted positions. The *Xenopus* 5 S RNA gene will adopt a specific position *in vitro* (Simpson and Stafford, 1983), but an identical 10-bp ladder has not been observed *in vivo*. Only for the heat-shock Hsp26 gene in *Drosophila* (Thomas and Elgin, 1988) has this rigorous correlation been made. Although the MMTV LTR is clearly phased at a resolution of approximately 20 base pairs (Richard-Foy and Hager, 1987), a 10-bp ladder has not been described *in vivo*. The available *in vivo* data is actually consistent with two to three overlapping frames of the position found in reconstituted monosomes or disomes. From this perspective, the intrinsic positioning properties of nucleosome B DNA would dictate a specific rotational frame, but multiple translational positions might be possible (Fig. 5A). Since an *in vivo* 10-bp ladder has not been observed, the possibility of even more complex *in vivo* positioning has not been excluded (multiple translational and rotational positions; Fig. 5B).

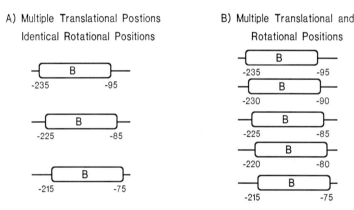

Fig. 5 Models for multiple nucleosome positions on the micro scale. Panel (A) depicts a model in which the rotational frame of the B octamer core is fixed, yet occupancy of multiple frames is possible. In panel (B), the possibility of many translational and rotational positions is presented.

Constraints may be placed on the positioning of nucleosome B during replication *in vivo* that are not present in the *in vitro* reconstitution systems. Histone deposition and octamer core assembly not only occur in concert with DNA replication, but also require the participation of a sophisticated molecular chaperone system. We should be cautious, therefore, in interpreting access experiments with *in vitro* reconstituted nucleosome arrays until direct high-resolution data are available for the *in vivo* structure. It remains possible that selective factor access to sequences within a phased array results from features of the array other than simple site masking by the octamer core (Fig. 4B). In mature nucleosome arrays, histone H1 is positioned on the octamer core where the overlapping strands of DNA enter and leave the nucleosome. The NF-1 site is located at this position on nucleosome B. An alternative model for NF-1 exclusion, therefore, could be based on features of the polynucleosome array other than simple octamer core stereochemistry (see Fig. 4B).

10.3.7 Modulation of Factor Access by Alteration of Nucleosomes

If transcription factor site availability can be significantly altered by nucleoprotein organization, modifications of nucleosome structure or stability might impact transactivation processes in which nucleosome displacement or structural reorganization is involved. To study these issues, we have developed the general approach shown in Fig. 6. The MMTV-LTR promoter, driving an appropriate reporter gene, is established as a multi-copy episome in mouse cells (Ostrowski *et al.*, 1983). The chromatin structure and nucleosome positioning of the LTR region are characterized by standard methodology. The identical promoter and regulatory sequences, driving a different reporter gene, are then introduced transiently in the same cell. Other factors of interest, such as receptors not endogenous to the cell, can also be introduced transiently in appropriate expression vectors. One can then compare the response of the LTR promoter when organized as *stable* chromatin with the response of *transiently* introduced DNA under conditions where both sequences are interacting with the identical soluble transcription machinery. Although transiently transfected DNA becomes associated with histones and acquires some features of chromatin (Cereghini and Yaniv, 1984), it is not known whether this complex reproduces functionally important aspects of nucleoprotein organization. Indeed, we have found in two separate responses that DNA sequences introduced transiently *do not* respond as they do when organized in stable, replicating chromatin.

The first example is illustrated in Fig. 7A. When cells are grown in the presence of sodium butyrate, histones become hyperacetylated due to the inhibition of an enzymatic deacetylase. If the response of the MMTV promoter to glucocorticoid stimulation is examined under conditions where histones are hyperacetylated, we find that the ability of the receptor to activate transcription is strongly inhibited (Bresnick *et al.*, 1990). If the response is monitored in the same nucleoplasm for transiently transfected DNA, however, there is no effect; in fact, activation is slightly more efficient for butyrate-treated cells. Furthermore, nucleosome displacement as monitored by HSR formation and restriction enzyme access is also totally

10. INTERACTION OF STEROID HORMONES WITH CHROMATIN   229

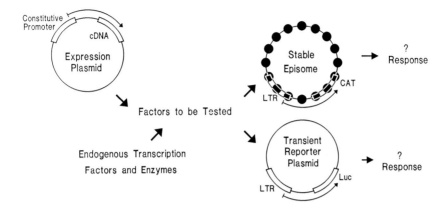

Fig. 6 A system for studying the effect of stable chromatin structure on transactivation. Cell lines contain a responsive regulatory region organized on an episomal element whose chromatin structure is characterized. The identical regulatory sequence is introduced transiently, either alone, or in conjunction with expression vectors to produce other factors whose effects are to be tested.

A) Hyperacetylation blocks displacement

B) Differential receptor access

Fig. 7 Two effects on transactivation mediated by chromatin structure. In panel (A), glucocorticoid receptor activation of the MMTV promoter is shown to require B nucleosome displacement. When histones are hyperacetylated (stippled circles), nucleosome displacement is prevented, and transactivation does not occur. In panel (B), glucocorticoid receptor displacement of nucleosome B readily occurs, while progesterone apparently is inefficient in this displacement process.

inhibited in butyrate-treated cells. It is clear from this series of experiments that hyperacetylation modulates the nucleoprotein structure in some way that prevents receptor-mediated nucleosome displacement and promoter activation, but this inhibitory effect is only observed in appropriately organized chromatin. Hyperacetylation of histones that may or may not be associated with transfected DNA has no effect. We conclude that secondary modification of nucleosomes can have dramatic impact on the ability of high-affinity DNA-binding proteins to induce nucleosome displacement/modification events necessary for transactivation.

10.3.8 Selective Transactivating Factor Access

We have argued above the NF-1 access to MMTV promoter DNA is inhibited by chromatin structure. We do not yet understand the mechanism of NF-1 exclusion, but the NF-1 binding site is uniquely positioned in the phased array, and its sequestration will undoubtedly result from that unique positioning. We have recently discovered a more startling example of selective factor access. Using the same system described in Fig. 6, we characterized the MMTV progesterone response in stable chromatin and transient DNA. The progesterone receptor has been shown to be a more effective activator of MMTV than glucocorticoid receptor, when monitored by classic transfection analysis (Cato *et al.*, 1986). The earlier lack of response of MMTV viral transcription to progesterone was thought to be due simply to a lack of progesterone receptors in most cell types. When we compare the progesterone response in stable chromatin with transient DNA, we find that induction for transfected DNA is much more pronounced with progesterone than glucocorticoid, as reported by Chalepakis *et al.* (1988), whereas MMTV sequences in the stable minichromosomes were essentially refractory to stimulation by progesterone (Archer *et al.*, 1990b, submitted). This result is quite unexpected, particularly, because the DNase I-resistant footprints for the two receptors on pure DNA are quite similar (von der Ahe *et al.*, 1985; see Chapter 9.2). One of two mechanisms could be responsible for this effect. The DNA-binding sites for the two receptors could be differentially positioned in MMTV chromatin in subtle ways that are functionally important. Alternatively, the two receptor proteins could have different properties relative to the functional requirements for nucleosome displacement and transactivation. Whatever the mechanism, it is again clear that nucleoprotein organization of this hormonally responsive promoter can dramatically alter the activation properties of a given transcription factor.

10.4 Summary

Studies of transcriptional regulation in eukaryotic cells have focused almost exclusively on the identification of transcription factors and characterization of domains and mechanisms by which these factors interact with the DNA template and with each other. These interactions can be readily addressed in transient transfection experiments, where both the reporter DNA and a factor of interest can be overexpressed, and mutant versions of a factor can be introduced. This approach has led to the identification of domains in the steroid receptors that are involved in the DNA-binding, transactivation and ligand-binding functions of these molecules.

It now appears that some aspects of the mechanisms involved in hormone transactivation are not faithfully reflected in transient transfection assays. In particular, effects associated with constraints placed on access of soluble factors by the nucleoprotein organization of the DNA template may not be detected in these systems. The two examples discussed above serve as useful demonstrations of this effect. Histone hyperacetylation can prevent a promoter activation process that requires nucleosome displacement, and stable nucleoprotein structure can result in differential access by factors with very similar DNA recognition sites.

The highly reproducible nucleoprotein organization associated with the MMTV-LTR remains a relatively isolated example in terms of promoters that respond to steroid hormone action. Other regulatory elements have been shown to develop hypersensitivity in response to receptor activation, but the extent to which these alterations in nucleoprotein structure reflect chromatin modifications involved mechanistically in the transactivation process remain to be demonstrated. Clearly two avenues of future experimentation are important to evaluate critically the involvement of chromatin in hormone action. First, a detailed understanding of the nucleoprotein organization of a variety of steroid responsive regulatory elements is needed. This should establish the extent to which specific patterns of chromatin organization (such as nucleosome phasing) are frequently or infrequently present in regulated genes. The recent advent of polymerase chain reaction technology, which permits a sensitive analysis of factors and structures associated with diploid genes *in situ* (Pfeifer *et al.*, 1989), may lead to a rapid growth of investigation in this area. Second, systems for the accurate reconstruction of complex chromatin templates *in vitro* are needed to permit experimentation in cell-free transcription extracts, where the concepts discussed here can be critically evaluated.

References

Archer, T. K., Cordingley, M. G., Marsaud, V., Richard-Foy, H. and Hager, G. L. (1989). Steroid transactivation at a promoter organized in a specifically-positioned array of nucleosomes. In "Proceedings: Second International CBT Symposium on the Steroid/Thyroid Receptor Family and Gene Regulation" (eds J. A. Gustafsson, H. Eriksson and J. Carlstedt-Duke), pp. 221–238. Birkhauser-Verlag, Berlin.

Archer, T. K., Cordingley, M. G. and Hager, G. L. (1990a). Active nucleosome displacement during transcription induction by steroid hormone. (Submitted.)

Archer, T. K., Berard, D. S. and Hager, G. L. (1990b). Differential steroid hormone induction of transcription from the mouse mammary tumour virus long terminal repeat. (Submitted.)

Beato, M. (1989). Gene regulation by steroid hormones. *Cell* **56**, 335–344.

Bresnick, E. H., John, S., Berard, D. S., Lefebvre, P. and Hager, G. L. (1990). Glucocorticoid receptor-dependent disruption of a specific nucleosome on the MMTV promoter is prevented by sodium butyrate. *Proc. Natl. Acad. Sci. USA* **87**, 3977–3981.

Brüggemeier, U., Rogge, L., Winnacker, E. L. and Beato, M. (1990). Nuclear factor I acts as a transcription factor on the MMTV promoter but competes with steroid hormone receptors for DNA binding. *EMBO J.* **9**, 2233–2239.

Cato, A. C., Miksicek, R., Schutz, G., Arnemann, J. and Beato, M. (1986). The hormone regulatory element of mouse mammary tumour virus mediates progesterone induction. *EMBO J.* **5**, 2237–2240.

Cereghini, S. and Yaniv, M. (1984). Assembly of transfected DNA into chromatin: structural changes in the origin-promoter-enhancer region upon replication. *EMBO J.* **3**, 1243–1253.

Chalepakis, G., Arnemann, J., Slater, E., Breuller, H. J., Gross, B. and Beato, M. (1988). Differential gene activation by glucocorticoids and progestins through the hormone regulatory element of mouse mammary tumor virus. *Cell* **53**(3), 371–382.

Cordingley, M. G. and Hager, G. L. (1988). Binding of multiple factors to the MMTV promoter in crude and fractionated nuclear extracts. *Nucl. Acids Res.* **16**, 609–628.

Cordingley, M. G., Reigel, A. T. and Hager, G. L. (1987). Steroid-dependent interaction of transcription factors with the inducible promoter of mouse mammary tumor virus *in vivo*. *Cell* **48**, 261–270.

Corthesy, B., Hipskind, R., Theulaz, I. and Wahli, W. (1988). Estrogen-dependent *in vitro* transcription from the vitellogenin promoter in liver nuclear extracts. *Science* **239**, 1137–1139.

Elgin, S. C. R. (1981). DNaseI-hypersensitive sites of chromatin. *Cell* **27**, 413–415.

Elgin, S. C. (1988). The formation and function of DNase I hypersensitive sites in the process of gene activation. *J. Biol. Chem.* **263**, 19 259–19 262.

Evans, R. M. (1989). Molecular characterization of the glucocorticoid receptor. *Recent Progr. Hormone Res.* **45**, 1–22.

Fawell, S. E., Lees, J. A., White, R. and Parker, M. G. (1990). Characterization and colocalization of steroid binding and dimerization activities in the mouse estrogen receptor. *Cell* **60**, 953–962.

Freedman, L. P., Yoshinaga, S. K., Vanderbilt, J. N. and Yamamoto, K. R. (1989). *In vitro* transcription enhancement by purified derivatives of the glucocorticoid receptor. *Science* **245**, 290–301.

Giguere, V., Hollenberg, S. M., Rosenfeld, M. G. and Evans, R. M. (1986). Functional domains of the human glucocorticoid receptor. *Cell* **46**, 645–652.

Green, S. and Chambon, P. (1988). Nuclear receptors enhance our understanding of transcription regulation. *Trends Genet.* **4**, 309–314.

Gross, D. S., and Garrard, W. T. (1988). Nuclease hypersensitive sites in chromatin. *Annu. Rev. Biochem.* **57**, 159–197.

Ham, J. and Parker, M. G. (1989). Regulation of gene expression by nuclear hormone receptors. *Curr. Opinion Cell Biol.* **1**, 503–511.

Hochschild, A. and Ptashne, M. (1986). Cooperative binding of lambda repressors to sites separated by integral turns of the DNA helix. *Cell* **44**, 681–687.

Hollenberg, S. M., Giguere, V., Segui, P. and Evans, R. M. (1987). Colocalization of DNA-binding and transcriptional activation functions in the human glucocorticoid receptor. *Cell* **49**, 39–46.

Huang, A. L., Ostrowski, M. C., Berard, D. and Hager, G. L. (1981). Glucocorticoid regulation of the Ha-MuSV p21 gene conferred by sequences from mouse mammary tumor virus. *Cell* **27**, 245–255.

Jones, K. A., Kadonaga, J. t., Rosenfeld, P. J., Kelly, T. J. and Tjian, R. (1987). A cellular DNA-binding protein that activates eukaryotic transcription and DNA replication. *Cell* **48**, 79–89.

Kalff, M., Gross, B. and Beato, M. (1990). The progesterone receptor stimulates transcription of mouse mammary tumour virus in a cell-free system. *Nature* **344**, 360–362.

Klein-Hitpass, L., Tsai, S., Weigel, N. L., Allen, G. F., Riley, D., Rodriguez, R., Schrader, W. T., Tsai, M-J., and O'Malley, B. W. (1990). The progesterone receptor stimulates cell free transcription by enhancing the formation of a soluble preinitiation complex. *Cell* **60**, 247–257.

Kumar, V. and Chambon, P. (1988). The estrogen receptor binds tightly to its responsive element as a ligand-induced homodimer. *Cell* **55**, 145–156.

Kumar, V., Green, S., Stack, G., Berry, M., Jin, J. R. and Chambon, P. (1987). Functional domains of the human estrogen receptor. *Cell* **51**, 941–951.

Lee, F., Mulligan, R., Berg, P. and Ringold, G. (1981). Glucocorticoids regulate expression of dihydrofolate reductase cDNA in mouse mammary tumour virus chimaeric plasmids. *Nature* **294**, 228–232.

Lees, J. A., Fawell, S. E. and Parker, M. G. (1989). Identification of two transactivation domains in the mouse oestrogen receptor. *Nucl. Acids Res.* **17**, 5477–5488.

Miesfeld R., Godowski, P. J., Maler, B. A., and Yamamoto, K. R. (1987). Glucocorticoid receptor mutants that define a small region sufficient for enhancer activation. *Science* **236**, 423–427.

Ostrowski, M. C., Richard-Foy, H., Wolford, R. G., Berard, D. S. and Hager, G. L. (1983). Glucocorticoid regulation of transcription at an amplified, episomal promoter. *Mol. Cell. Biol.* **3**, 2045–2057.

Perlmann, T. and Wrange, O. (1988). Specific glucocorticoid receptor binding to DNA reconstituted in a nucleosome. *EMBO J.* **7**(10), 3073–3079.

Pfeifer, G. P., Steigerwald, S. D., Mueller, P. R., Wold, B. and Riggs, A. D. (1989). Genomic sequencing and methylation analysis by ligation mediated PCR. *Science* **246**, 810–813.

Richard-Foy, H. and Hager, G. L. (1987). Sequence specific positioning of nucleosomes over the steroid-inducible MMTV promoter. *EMBO J.* **6**, 2321–2328.

Richard-Foy, H. and Hager, G. L. (1990). Nucleosome phasing on the MMTV LTR is preserved on integrated LTR chimeras and endogenous proviruses. (Submitted.)

Richard-Foy, H., Sistare, F. D., Riegel, A. T., Simons, S. S. Jr, and Hager, G. L. (1987). Mechanism of dexamethasone 21-mesylate antiglucocorticoid action: II. Receptor-antiglucocorticoid complexes are unable to interact productively with MMTV LTR chromatin in vivo. *Mol. Endocrinol.* **1**, 659–665.

Satchwell, S. C. and Travers A. A. (1989). Asymmetry and polarity of nucleosomes in chicken erythrocyte chromatin. *EMBO J.* **8**, 229–238.

Satchwell, S. C., Drew, H. R. and Travers A. A. (1986). Sequence periodicities in chicken nucleosome core DNA. *J. Mol. Biol.* **191**, 659–675.

Sawadogo, M. and Roeder, R. G. (1985). Interaction of a gene-specific transcription factor with the adenovirus major late promoter upstream of the TATA box region. *Cell* **43**, 165–175.

Schlissel, M. S. and Brown, D. D. (1984). The transcriptional regulation of *Xenopus* 5S RNA genes in chromatin: the roles of active stable transcription complexes and histone H1. *Cell* **37**, 903–913.

Simpson, R. T. and Stafford, D. W. (1983). Structural features of a phased nucleosome core particle. *Proc. Natl. Acad. Sci. USA* **80**, 51–55.

Sistare, F. D., Hager, G. L. and Simons, S. S. Jr. (1987). Mechanism of dexamethasone 21-mesylate antiglucocorticoid action: I. Receptor-antiglucocorticoid complexes do not competitively inhibit receptor-glucocorticoid complex activation of gene transcription in vivo. *Mol. Endocrinol.* **1**, 648–658.

Thomas, G. H. and Elgin, S. C. (1988). Protein/DNA architecture of the DNase I hypersensitive region of the *Drosophila* hsp26 promoter. *EMBO J.* **7**, 2191–2201.

Thompson, E. B., Venetianer, A., Gelehrter, T. D., Hager, G., Granner, D. K., Norman, M. R., Schmidt, T. J. and Harmon, J. M. (1980). Multiple actions of glucocorticoids studied in cell culture systems. In "Gene Regulation by Steroid Hormones" (eds A. K. Roy and J. H. Clark), pp. 126–152. Springer-Verlag, New York.

Tora, L., Gronemeyer, H., Turcotte, B., Gaub, M. P. and Chambon, P. (1988). The N-terminal region of the chicken progesterone receptor specifies target gene activation. *Nature* **333**, 185–188.

Travers, A. A. and Klug, A. (1987). The bending of DNA in nucleosomes and its wider implications. *Phil. Trans. R. Soc. Lond. (Biol.)* **317**, 537–561.

Tsai, S. Y., Carlstedt-Duke, J., Weigel, N. L., Dahlman, K., Gustafsson, J. A., Tsai, M. J. and O'Malley, B. W. (1988). Molecular interactions of steroid hormone receptor with its enhancer element: evidence for receptor dimer formation. *Cell* **55**, 361–369.

van Holde, K. E. (1988). "Chromatin". Springer-Verlag, Heidelberg.

von der Ahe, D., Janich, S., Scheidereit, C., Renkawitz, R., Schutz, G., and Beato, M. (1985). Glucocorticoid and progesterone receptors bind to the same sites in two hormonally regulated promoters. *Nature* **313**, 706–709.

Webster, N. J. G., Green, S., Jin, J.-R. and Chambon, P. (1988). The hormone binding domains of the estrogen and glucocorticoid receptors contain an inducible transcription activation function. *Cell* **54**, 199–207.

Weintraub, H. (1983). Tissue-specific gene expression and chromatin structure. *Harvey Lect.* **79**, 217–244.

Wu, C. (1985). An exonuclease protection assay reveals heat-shock element and TATA box DNA-binding proteins in crude nuclear extracts. *Nature* **317**, 84–87.

Yamamoto, K. R. (1985). Steroid receptor regulated transcription of specific genes and gene networks. *Annu. Rev. Genet.* **19**, 209–252.

11
Ontogeny of Sex Steroid Receptors in Mammals

G. R. CUNHA[1], P. S. COOKE[2], R. BIGSBY[3], and J. R. BRODY[1]

[1] *Department of Anatomy, University of California, San Francisco, CA 94143, USA*

[2] *Department of Veterinary Biosciences, University of Illinois, 2001 South Lincoln Avenue, Urbana, IL 61801, USA*

[3] *Department of Obstetrics and Gynecology, Indiana University Medical School, Indianapolis, IN 46202-5196, USA*

11.1 Introduction

During specific periods of gestation and early neonatal life the genital tract, mammary glands, and brain are sensitive to the morphogenetic effects of sex steroids. Certain hormones are required for normal sex differentiation growth and function of urogenital organs. Sex differentiation begins during the ambisexual or indifferent stage during which the following structures can be recognized: undifferentiated gonads, mesonephros (mesonephric tubules), Wolffian ducts, Mullerian ducts, and urogenital sinus. These structures constitute the primordia for development of either the male or female genital systems (Fig. 1). Sex differentiation is initiated as the gonads differentiate into either ovaries or testes. In mammals the bias for differentiation of the genital tract is to develop the female phenotype which will spontaneously occur if testes are absent. By contrast, masculine development is imposed via hormones produced by the fetal testes. Leydig cells differentiate within the fetal testes and produce testosterone. In males androgens elicit survival of the WD and stimulate their differentiation into the epididymis, ductus deferens, and seminal vesicle (Fig. 1). Both testosterone (T) and dihydrotestosterone (DHT) appear to be involved in Wolffian duct development. Testosterone is initially involved in WD development since T is effective in preventing programmed cell death of the WD, and during the ambisexual stage of sex differentiation conversion of T to DHT by 5α-reductase is very low in the WD (Wilson and Lasnitzki, 1971; Siiteri and Wilson, 1974; Wilson *et al.*, 1981). However, subsequent morphogenesis of the lower

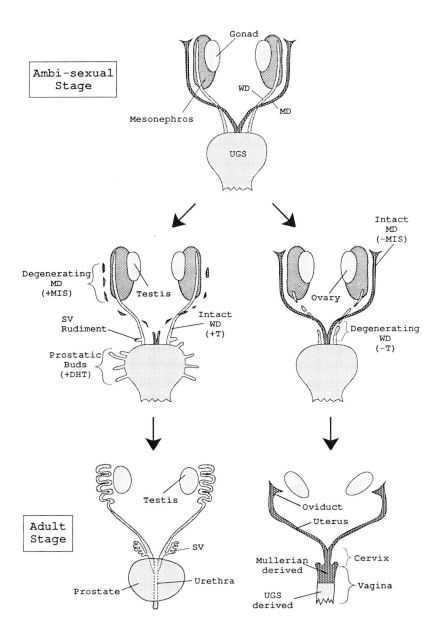

Fig. 1 Process of sex differentiation. WD, Wolffian duct; MD, Mullerian duct; SV, seminal vesicle; + or − T, presence or absence of testosterone; + or − MIS, presence or absence of Mullerian inhibiting substance; + DHT, presence of dihydrotestosterone.

portion of the WD into the seminal vesicle (SV) appears to be dependent upon DHT, since T in the presence of a 5α-reductase inhibitor is completely ineffective in eliciting SV morphogenesis (Shima *et al.*, 1988, 1990). Masculine development of the caudal portions of the genital tract, e.g. urogenital sinus, genital tubercle and genital swellings, is dependent upon DHT produced by 5α-reductase (Wilson and Lasnitzki, 1971; Sulcová *et al.*, 1973; Imperato-McGinley *et al.*, 1974; Wilson *et al.*, 1981; Cooke *et al.*, 1988). The urogenital sinus, under the influence of DHT, develops into the prostate, bulbourethral glands, and most of the male urethra, while the genital tubercle and genital swellings form the penis and scrotum, respectively. A second hormonal substance, Mullerian inhibiting substance (MIS), is produced by Sertoli cells of the fetal testes. This glycoprotein triggers destruction of the Mullerian duct in males (Josso *et al.*, 1977; Donahoe *et al.*, 1982). In some species mammary anlage present initially in both male and female fetuses are eliminated in males through a process of programmed cell death elicited by androgen (Kratochwil, 1971, 1975, 1977; Goldman *et al.*, 1976). Thus, the normal development of the male phenotype is crucially dependent upon androgens and Mullerian inhibiting substance.

In females, due to the absence of steroidal androgens the mesonephric tubules and Wolffian ducts regress while the Mullerian ducts persist because of an absence of MIS. The Mullerian ducts in females give rise to the oviducts, uterus, cervix, and contribute to the upper vagina. The lower vagina, urethra, and bladder are derived from the urogenital sinus in females (Forsberg, 1973; Cunha, 1975) (Fig. 1). Normal development of the female genital tract is thought to be independent of sex steroids. The Mullerian ducts, unlike the Wolffian ducts, are programmed from their inception to survive and differentiate into female genital tract structures. Since the fetal MDs develop normally whether or not ovaries are present (Jost, 1965), and since uterine growth and development are normal during the neonatal period in mice ovariectomized and adrenalectomized at birth (Ogasawara *et al.*, 1983), it appears that oestrogens are not required, at least initially, for normal development of the female genital tract. Nonetheless, prenatal or neonatal exposure to exogenous oestrogen elicits a spectrum of deleterious changes in the developing oviduct, uterus, cervix, vagina, and brain (Greene, 1940; Takasugi, 1976; McLachlan *et al.*, 1980; Forsberg and Kalland, 1981; Branham *et al.*, 1985a; Newbold and McLachlan, 1985; Arai *et al.*, 1988; Brody and Cunha, 1989b). Indeed, a vast literature documents a wide spectrum of teratogenic and carcinogenic changes elicited by exogenous oestrogens in the developing genital tracts of both males and females from several species including man (Forsberg, 1975; McLachlan *et al.*, 1975; Takasugi, 1976; Bern and Talamantes, 1981; Bern *et al.*, 1984; Arai *et al.*, 1988; Mori and Iguchi, 1988; Newbold and McLachlan, 1988). By the same token, when androgens are present at inappropriate times, or in abnormal amounts, they may elicit teratogenic or even carcinogenic change in developing target tissues. Thus, while in male fetuses androgens elicit development of the masculine phenotype, in female fetuses androgens can be considered teratogenic because of their masculinizing effects (Jost, 1953, 1965, 1970; Burns, 1961; Raynaud, 1977). These examples serve to emphasize the sensitivity and/or dependence of fetal tissues to steroidal sex hormones which may be required for normal development or may be teratogenic or carcinogenic.

Needless to say, gestation periods and the degree of maturity at birth vary from species to species. In some species the process of sex differentiation is telescoped into a short period (hamster), while in other species (man) the process occurs over a long time frame (Fig. 2). It should be emphasized that the state of differentiation of the genital tract at birth is radically different in various species.

The biological effects of oestrogens and androgens in developing tissues appear to be elicited via high affinity saturable receptor proteins as in the case for adult tissues. Characterization of hormone receptors in embryonic, fetal, and neonatal tissues presents special problems due to the limited quantities of tissue available and the presence of serum proteins such as α-fetoprotein which, in rodents, binds oestrogens with high affinity. Nonetheless, a variety of techniques have been used successfully to investigate receptors for sex steroids and their role in normal and abormal development.

11.2 Methodological considerations

Three procedures have been used to study steroid receptors in fetal and neonatal animals: biochemical binding assays performed on nuclear or cytosolic extracts, steroid autoradiography, and, more recently, immunohistochemistry. Biochemical studies have been hampered because of limited quantities of tissue. This problem has been circumvented in the past through use of the guinea pig. This animal has a long gestation, attains considerable size and maturity during pre-natal periods, and is sufficiently precocious at birth that it can actually begin eating solid food. In contrast the mouse and rat are very immature at birth. For this reason uteri of guinea pigs are vastly more advanced in their development at birth than uteri of newborn mice or rats. For example, uteri of rats and mice at birth consist of an undifferentiated low columnar epithelial tube surrounded by an undifferentiated mesenchyme (Brody and Cunha, 1989a). Conversely, in the guinea pig the uterine mesenchyme at birth has already differentiated into endometrial stroma and the circular and longitudinal myometrial layers (Gulino *et al.*,

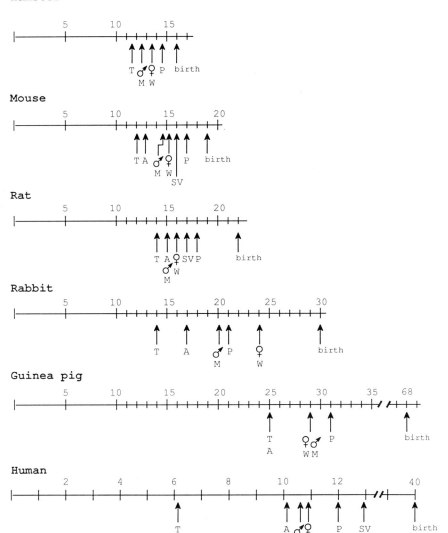

Fig. 2 Comparative sequence of sex differentiation given in days (laboratory animals) and weeks (human). T, testicular differentiation; A, androgen secretion; ♂M, initiation of male Mullerian duct regression; ♀W, initiation of female Wolffian duct regression; SV, appearance of seminal vesicles; P, appearance of prostatic buds. (Adapted from Price and Ortiz, 1965.)

1984). These layers do not become well defined in rats and mice until 10-15 days after birth (Brody and Cunha, 1989a). Thus, the uteri of these two groups are not comparable biological entities at birth and data on hormone receptors in the *fetal* guinea pig are not necessarily applicable to fetuses of other species. At birth the human uterus, though small, has already attained its full morphological complexity.

Steroid autoradiography and immunocytochemistry have also been used to examine steroid receptors in fetal and neonatal tissues. Here problems of limited tissue quantity are eliminated and additional information on the cell and tissue distribution of receptors is achieved. Steroid autoradiography can give results following 3 to 4 weeks of exposure of the autoradiogram (Shannon *et al.*, 1982). With long exposure times (6 months to 2 years) steroid autoradiography is the most sensitive technique for studying steroid receptors. It should be stressed that steroid autoradiography (like biochemical binding assays) is capable of satisfying all diagnostic criteria established by Clark and Peck (Clark and Peck, 1979) for receptor proteins: saturability, high affinity, steroid specificity, tissue specificity, and correlation with biological response (see Cunha *et al.*, 1983b). Monoclonal and polyclonal antibodies to oestrogen, progesterone, and androgen receptors provide for rapid detection of hormone receptors on tissue sections, but do not provide information concerning the ability of the receptor proteins to bind steroid. However, the ease of immunocytochemical methods compared to the more laborious steroid autoradiographic methodology ensures that this technique will become increasingly important in the future. In the few studies in which both autoradiographic and immunocytochemical techniques have been applied to the same tissue specimens, complete congruence has been reported between the two methods.

11.3 Ontogeny of androgen receptors

Biochemical analysis of the urogenital ridge (gonads, Mullerian and Wolffian ducts and mesonephros), urogenital sinus, and genital tubercle of 17- to 21-day rabbit fetuses demonstrated specific uptake of [^{3}H] testosterone (Wilson, 1973; George and Noble, 1984). For this species the presence of androgen receptor correlates precisely with the early periods of sex differentiation during which androgens stabilize the Wolffian ducts and elicit prostatic development from the unrogenital sinus. Similarly, Gupta and Bloch (1976) have reported that high speed supernatants of genital tracts (mesonephros, Wolffian and Mullerian ducts, urogenital sinus, and genital tubercle) of 14.5- to 15-day-old fetal rats (ambisexual stage) contain high affinity ($K_d = 2$ nM) binding proteins. External genitalia of human fetuses also exhibit specific uptake of [³H]testosterone (Sulcová et al., 1973). Biochemical binding assays performed on embryonic mouse mammary anlagen have demonstrated androgen receptors (K_d = 0.4–0.9 nM) that are initially detectable at low levels on the 12th day of gestation. Androgen receptor levels rise sharply in the developing mammary gland and plateau by day 16 of gestation. Responsiveness of the embryonic mammary gland to androgens (which triggers regression of the mammary epithelium) can be demonstrated between days 13 to 15 of gestation (Wasner et al., 1983). After 15 days of gestation androgen fails to elicit regression of the mammary rudiments even though androgen receptor levels remain high (Wasner et al., 1983; Kratochwil, 1987). Thus, biochemical analysis of developing androgen target organs from several species has demonstrated the presence of androgen-binding proteins during stages when androgens masculinize the genital tract and mammary glands.

Biochemical assays of androgen receptors fail to provide information on receptor distribution at the cellular and tissue level. So far only steroid autoradiographic methods have been utilized to gain additional insights on the tissue localization of cells possessing androgen receptors in developing organ rudiments. A theme seen in a broad spectrum of fetal and neonatal organ rudiments is that nuclear androgen-binding sites are initially expressed solely in the mesenchymal cells with the expression of androgen receptors in the epithelium occurring at later stages.

In the rodent urogenital sinus (Shannon and Cunha, 1983; Takeda et al., 1985; Cooke, 1988; Cooke and Cunha, 1990) and Wolffian duct (Cooke, 1988; Cooke and Cunha, 1990) androgen receptors are detected at day 13 and 16 of gestation, respectively, in mesenchymal cells while epithelial cells of either organ rudiment appear to lack androgen receptors at these stages since labelling of the epithelium is indistinguishable from background. The efferent ductules, which are considered to develop from the mesonephric tubules, contain both epithelial and mesenchymal androgen receptors at birth (day 0) in the mouse and at all subsequent time points. Among organs derived from the Wolffian duct (e.g. the seminal vesicle, epididymis and ductus deferens) mesenchymal/stromal androgen receptors are detectable from early fetal through adult stage (Fig. 3). The epididymis and ductus deferens initially express epithelial androgen receptors by day 19 of gestation and are always positive thereafter. The seminal vesicle (which can be recognized as a distinct rudiment on day 15 of gestation in the mouse) expresses mesenchymal androgen receptors at the time of its appearance and at all times thereafter, but epithelial androgen receptors are not expressed until day 2 after birth (Cooke, 1988; Cooke and Cunha, 1990).

The organs derived from the urogenital sinus, the bulbourethral gland and prostate also show a similar pattern of androgen receptor expression. Both

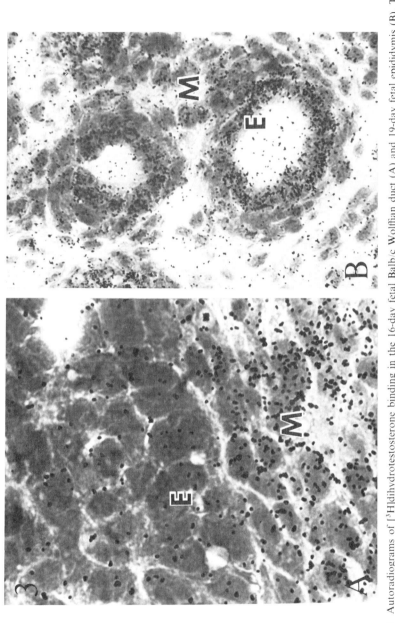

Fig. 3 Autoradiograms of [³H]dihydrotestosterone binding in the 16-day fetal Balb/c Wolffian duct (A) and 19-day fetal epididymis (B). The mesenchymal cells (M) of the Wolffian duct show heavy labelling, while the epithelium (E) shows only background labelling, indicating that only mesenchymal cells have androgen receptors at this stage of development. Conversely, the 19-day epididymis shows labelling in both mesenchymal and epithelial cells, indicating that both of these tissues have androgen receptors. Magnifications: (A) 1420×; (B) 500×.

of these organs contain only mesenchymal androgen receptors at birth and over the first few days of neonatal life. It is not until days 4–6 after birth in the prostate and days 6–8 in the bulbourethral gland that epithelial androgen receptors are detected (Shannon and Cunha, 1983; Cooke, 1988; Cooke and Cunha, 1990). Integumental (skin) derivatives developmentally responsive to androgens such as the mammary glands, preputial glands, and external genitalia of fetal mice also exhibit this common pattern of expression of androgen receptors solely in the mesenchymal elements of these structures during fetal periods (Wasner et al., 1983; Murakami, 1987; G. R. Cunha, unpubl. res.).

In summary, the timing of expression of the epithelial androgen receptor varies from organ to organ, probably is variable from species to species, and may also show slight variation in different strains of mice. In the Balb/c mouse androgen receptor expression in epithelial cells of the male reproductive tract appears to follow a cranial-caudal gradient. Epithelial cells of the epididymis and ductus deferens derived from the upper portion of the Wolffian duct become positive for androgen receptors on day 18 of gestation, while the epithelium of the seminal vesicle (derived from the caudal portion of the Wolffian duct) initially expresses androgen receptors on day 2 after birth. Prostatic epithelium expresses androgen receptors on days 4 to 6 after birth and the bulbourethral gland on day 6 after birth in the Balb/c mouse (Shannon and Cunha, 1983; Cooke, 1988; Cooke and Cunha, 1990).

Androgen receptors are present in homologous structures of the female genital tract and provide the basis for teratogenic effects of androgens in female fetuses. Detailed ontogenic studies of androgen receptor expression are not available for the female.

11.4 Ontogeny of oestrogen receptors in the female genital tract

Biochemical studies have demonstrated the presence of oestrogen receptors during the perinatal period in uteri of rats, mice and guinea pigs. The most extensively studied species is the guinea pig in which oestrogen receptors appear in the urogenital ridge at 34–35 days of gestation, shortly after the Wolffian ducts have regressed in female embryos (Pasqualini et al., 1976; Sumida and Pasqualini, 1979a). At this stage the urogenital ridges contain morphologically differentiated ovaries and Mullerian ducts that are beginning to undergo uterine development (Price and Ortiz, 1965). Oestrogen receptor levels reach a maximum at 50 days of gestation and then decline to lower levels in neonatal guinea pigs (Pasqualini et al., 1976; Sumida and Pasqualini, 1979b). In developing rats oestrogen receptors have been detected biochemically in the developing uterus during late fetal, neonatal, and weanling (21 days) periods (Clark and Gorski, 1970; Somjen et al., 1973; Peleg et al., 1979; Kimmel and Harmon, 1980; Medlock et al., 1981; Sheehan et al., 1981). Biochemical binding assays have also demonstrated specific uptake of [^{3}H]oestrogens in 1- to 5-day-old mice (Eide, 1975a; Bigsby and Cunha, 1986; Korach et al., 1988). In all of these studies the physicochemical properties, K_d, and relative binding affinities for different steroids were found to be similar to those of oestrogen receptors of immature and adult uteri (Clark and Peck, 1979; Pasqualini and Sumida, 1986).

Oestrogen receptors detected during pre- and post-natal periods appear to be biologically active based upon the following observations:

(1) Apparent translocation of receptor from the cytosolic to the nuclear compartment occurs upon exposure to oestrogen (Somjen et al., 1973; Peleg et al., 1979; Sumida and Pasqualini, 1979a; Sumida and Pasqualini, 1980; Medlock et al., 1981; Sheehan et al., 1981).
(2) Oestrogens, whether steroidal or non-steroidal, stimulate cell proliferation and growth in the developing female genital tract of rats, mice and guinea pigs (Eide, 1975b; Forsberg, 1975; Bigsby and Cunha, 1986; Pasqualini and Sumida, 1986).
(3) Oestrogens administered to fetal or neonatal rodents increase levels of ornithine decarboxylase in the developing uterus (Kaye et al., 1973; Kimmel et al., 1981; Sheehan et al., 1981; Branham et al., 1985a).
(4) Induced protein, the BB form of creatine kinase, is stimulated by oestrogen in uteri of neonatal rats (Somjen et al., 1973; Katzenellenbogen and Gregor, 1974).
(5) Progesterone receptor levels in uteri of guinea pig fetuses and neonatal rats are elevated following administration of oestradiol (Raynaud et al., 1980).
(6) RNA and DNA syntheses are stimulated by oestrogen in uteri of fetal and neonatal rodents (Eide, 1975b; Ogasawara et al., 1983; Bigsby and Cunha, 1986; Pasqualini and Sumida, 1986; Taguchi et al., 1988).
(7) Finally, oestrogens administered to pre- and neonatal rats, mice, hamsters, and guinea pigs elicit a variety of cytological, histological and teratogenic changes in the developing female genital tract (Takasugi, 1976; McLachlan et al., 1980; Bern and Talamántes, 1981; Forsberg, 1988; Newbold and McLachlan et al., 1988; Brody and Cunha, 1989b).

Autoradiographic and, quite recently, immunohistochemical studies corroborate and extend the biochemical studies. As is the case for androgen receptors, the initial expression of ER in the fetal mouse occurs in mesenchymal cells of the developing female internal genitalia (Stumpf et al., 1980;

Cunha et al., 1982; Holderegger and Keefer, 1986; Taguchi et al., 1988), mammary glands (Narbaitz et al., 1980a; Haslam, 1989); and larynx (Narbaitz et al., 1980b). As an aside, glucocorticoid receptors are initially detected in the mesenchymal cells of the developing mouse lung on day 11 of gestation and subsequentially are detected in the epithelium on day 14 to 16 of gestation (Beer et al., 1984). Thus, the generalized ontogenetic programme of steroid hormone receptor expression involves an initial expression in mesenchyme followed later by expression of steroid receptors in epithelium. The timing of the initial expression of ER in oestrogen target organs varies from species to species, and from strain to strain, reflecting differing rates of development. In the rat, uterine epithelial oestrogen receptors have been demonstrated autoradiographically as early as 2 days after birth; earlier stages have not been examined (Stumpf and Sar, 1976). Preliminary information on oestrogen receptors in the human fetal genital tract indicates that at the earliest stages examined (10 weeks of gestation) nuclear oestrogen binding sites are present solely in mesenchymal cells of the vagina and urogenital sinus, but as the epithelium differentiates and matures, both the epithelium and mesenchyme express oestrogen receptors (Taguchi et al., 1986).

Strain differences in the ontogeny of uterine epithelial oestrogen receptors has led to some confusion in the literature. Korach and co-workers (Korach et al., 1988; Yamashita and Korach, 1989; Yamashita et al., 1989) suggested that the autoradiographic evidence indicating a lack of oestrogen receptors in the epithelium of the 4-day-old mouse uterus (Bigsby and Cunha, 1986; Taguchi et al., 1988) was erroneous due to a lack of sensitivity of the methods used and that by using immunocytochemical methods they could detect epithelial receptor in 4-day-old animals. However, when this subject was examined systematically using both immunocytochemistry and steroid autoradiography (Bigsby and Cunha, 1987; Bigsby et al., 1990), it was apparent that uterine epithelial oestrogen receptors in the outbred CD-1 mice used by Korach and associates develop at a rate that is 2–3 days ahead of either of the inbred strains of mice (Balb/c or C57B1/J) used in our studies. Indeed, as reported by Yamashita et al. (1989), a small portion of cells within uterine epithelium of CD-1 mice already express ER on day 2 after birth (3 days old). As shown in Fig. 4, uterine epithelial cells of 6-day-old Balb/c mice are devoid of ER immunostaining, while in 5-day-old CD-1 mice the majority of uterine epithelial cells have expressed ER. In NMRI mice oestrogen receptors in uterine epithelial cells are still undetectable at day 8 after birth (Andersson and Forsberg, 1988). In all three strains (Balb/c, C57/10 and CD-1) that we have examined, epithelial cells of the oviduct exhibit oestrogen receptors as early as the day of birth and cervical epithelium is positive in newborn CD-1 mice and 2-day-old Balb/c or C57Bl/

6J mice (Taguchi et al., 1988; Bigsby et al., 1990). The earliest expression of ER in cells within vaginal epithelium was observed in newborn CD-1 mice, 2-day-old Balb/c mice, and 5-day-old C57B1/6J mice (Bigsby, unpubl. res.; Taguchi et al., 1988). Thus, the reason for the discrepancy in reports of ER in mouse uterine epithelium merely resides in differences of developmental rates among different strains of mice. In the female reproductive tract the general pattern of expression of ER in mesenchyme first followed by epithelial expression holds for all developmental studies done to date, in species from mouse to man. This is also true for androgen and glucocorticoid receptors as described above.

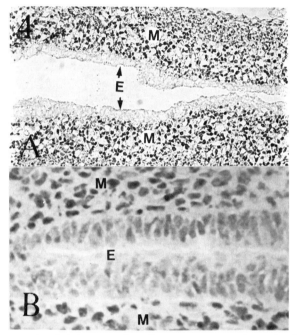

Fig. 4 Immunocytochemical demonstration of oestrogen receptor in uterine tissues of two strains of mice. The reproductive tracts were removed from Balb/c and CD-1 mice of different ages, fixed in neutral buffered formalin for 90 min and frozen as described by Korach et al. (1988). Frozen, longitudinal sections were thaw-mounted and stained with the monoclonal antibody, H222 (supplied by Dr Kris Nolan of Abbott Laboratories) using an avidin–biotin–peroxidase detection method. (A) Uterus of a 6-day-old Balb/c mouse. Note that the uterine epithelium (E) is devoid of staining although nuclei of the mesenchymal cells (M) stain very strongly. Of seven mice examined at this age, six lacked epithelial oestrogen receptor. The epithelium was completely negative in all animals examined as late as 5 days of age in this strain. (B) Uterus of 5-day-old CD-1. A large proportion of the epithelial cells of the uteri of all CD-1 mice examined at this age exhibited staining for oestrogen receptor; oestrogen receptor could be detected in uterine epithelial cells as early as 3 days old in CD-1 mice. Magnifications: (A) 250 × ; (B) 750 × .

11.5 Ontogeny of oestrogen receptors in male mouse reproductive organs

Oestrogens have pronounced deleterious effects on the growth, differentiation, histology and function of the male reproductive tract when administered to neonates, juveniles or adults. Oestrogen receptors (ER) are present in the male reproductive tract, and extensive work has been devoted to characterization and analysis of ER in the juvenile and adult prostate (West et al., 1988). The presence and distribution of ER in other male organs such as the epididymis, testis, etc. have been reported in the adult mouse (Schleicher et al., 1984). However, little information is available on ER distribution in the developing male reproductive tract, although extensive evidence has indicated that the male reproductive tract is maximally sensitive to the deleterious effects of oestrogens during this period. Pre-natal administration of natural or synthetic oestrogens to a variety of animals such as monkeys, guinea pigs, hamsters, mice, rats and humans results in long-term structural and functional abnormalities (for a review, see McLachlan, 1981a). The most extensive information in this area has come from McLachlan and co-workers who have described the deleterious consequences of pre-natal exposure to diethylstilboestrol (DES), a potent synthetic oestrogen, on the male reproductive tract of the mouse. Adult mice exposed to DES pre-natally have a litany of reproductive problems including decreased or abolished fertility, testicular tumours, retention of Mullerian duct remnants, and histological abnormalities such as cysts and squamous metaplasia in the testis, epididymis, seminal vesicle, prostate and coagulating gland (McLachlan, 1981b; Newbold et al., 1987).

ER are present in the urogenital sinus and Wolffian duct of mice as early as 16 and 13 days of gestation, respectively (Stumpf et al., 1980; Holderegger and Keefer, 1986), but ER expression in male organs during the crucial late fetal and neonatal periods has not been addressed. Recent work in our laboratories has sequentially examined ER expression in male mouse reproductive organs from the ambisexual stage of sex differentiation (14 days of gestation) through the neonatal period (day 10 after birth). These observations confirmed that male urogenital sinus and Wolffian duct contain ER in the mesenchyme, but not epithelium at 16 days of gestation. Observations at subsequent time points indicated that organs derived from the urogenital sinus, e.g. the prostate and bulbourethral gland, continue to express mesenchymal ER at the time of their initial morphogenesis (17 days post-coitum) and at all subsequent time points up to day 10 after birth. Interestingly, the frequency of ER-positive mesenchymal cells decreased with advancing age; almost all mesenchymal cells were ER-positive at 16 days of gestation, but the proportion of ER-negative cells increased during development. The epithelium of these organs remained negative for ER throughout the same time period. Likewise, organs derived from the Wolffian duct (the seminal vesicle, epididymis and ductus deferens) exhibited mesenchymal ER from the time of their initial morphological differentiation up to day 10 after birth; in adulthood ER have been detected in stromal cells of these organs (Stumpf and Sar, 1976). The epithelium of the seminal vesicle and ductus deferens remained ER negative during the period of observation (fetal to day 10 after birth). In adulthood epithelial cells of the seminal vesicle and ductus deferens lack ER (Stumpf and Sar, 1976). In contrast, epididymal epithelium became strongly ER positive at day 18 of gestation and remained so at all subsequent time points (Fig. 5). Like the UGS-derived organs such as the prostate and bulbourethral gland, the proportion of ER-positive cells in the mesenchyme declined with increasing age.

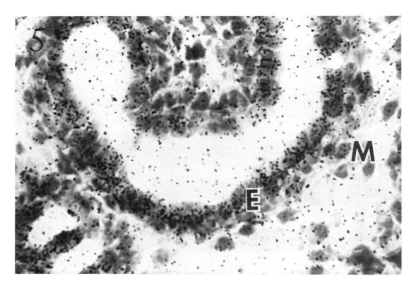

Fig. 5 Autoradiogram of [^{3}H] oestradiol binding in the epididymis of an 18-day Balb/c fetus. The Wolffian duct, from which the epididymis differentiates on day 17 of gestation, contains only mesenchymal oestrogen receptors at 16 days of fetal life. However, by day 18 of gestation, both the mesenchyme (M) and epithelium (E) of the epididymis contain oestrogen receptors. Magnification: 500 ×.

The efferent ductules, whose origin is controversial and may originate from either mesonephric tubules of the primitive kidney or the Wolffian duct itself, also show a pattern of early epithelial ER expression similar to the epididymis. This organ contains both mesenchymal and epithelial ER at the time of birth (the first time point examined) and at all subsequent ages up to day 10 after birth. A possible role of oestrogen in the normal development of the male genital tract has yet to be demonstrated.

In summary, mesenchymal/stromal ER are present in all male reproductive organs examined from day 16 of gestation until day 10 after birth, although the proportion of ER-positive cells declines in these organs with advancing age. The epididymis and efferent ducts begin to express epithelial ER very early, during the late fetal/early neonatal period, but the epithelium of other reproductive organs remains ER-negative up to at least day 10 after birth. It should be emphasized that a wide spectrum of genital tract lesions in both male and female genital tracts are induced by exogenous oestrogen in pre- and neonatal mice during periods before epithelial receptors are detectable, suggesting that putative mediators of oestrogen action could be involved. This is particularly true for the treatment protocol used by the McLachlan group (McLachlan, 1981b; Newbold et al., 1987) in which pregnant mice are treated with DES from day 9 to day 16 of gestation, which is well before the appearance of ER in the epithelia of the genital tract. Evidence from studies of neonatal mice suggests that prolonged administration of exogenous oestrogen accelerates the onset of epithelial receptors by several days (Andersson and Forsberg, 1988; Taguchi et al., 1988). Moreover, administration of oestrogen in an oil vehicle results in oestrogen exposure for several days after the last injection. Thus, for experimental protocols in which DES is injected from day 9 to day 16 of gestation, initial oestrogen exposure occurs during periods when oestrogen receptors are present solely in the mesenchyme. However, at later periods oestrogen receptors may be present in both tissues although this has not been examined specifically.

11.6 Steroid receptors in developing brain and pituitary

Androgens and oestrogens, in addition to their roles in reproductive tract organs, are also important in the development of the brain and pituitary. Previous work has shown that the late fetal and early neonatal periods are critical in rodents for the organizational effects of sex steroids on the brain. These effects are assumed to be mediated by specific hormone receptors, and have led to extensive evaluation of the distribution and ontogeny of oestrogen and androgen receptors during early development (Plapinger and McEwen, 1973; Plapinger et al., 1977; Gorski, 1979; McEwen et al., 1980; Pfaff, 1980).

In mice, oestrogen receptors first appear on day 14 of gestation (Keefer and Holdregger, 1985) in the brain. Labelled cells have been reported in the basal hypothalamus, the pre-optic area, amygdala and midbrain. During the remainder of gestation, the number and labelling intensity of ER positive cells in all of these brain regions increased. Oestrogen receptor positive cells first appear in the anterior pituitary on day 17 of gestation, but they are initially not very numerous and not intensely labelled. Subsequent observations shortly after birth indicated that the number of ER positive cells had increased during the late fetal period, and were more numerous in the early post-natal pituitary (Keefer and Holdregger, 1985).

Oestrogen receptors have also been detected in the fetal guinea pig brain starting at 29–35 days of gestation (Pasqualini *et al.*, 1978). Similarly, oestrogen receptors have been reported in the hypothalamus, pre-optic area, cerebral cortex and anterior pituitary of fetal (135–162 days of gestation) rhesus monkeys (Hochner-Celnikier *et al.*, 1986). Despite the difficulties already mentioned in comparing results obtained from different species, these results suggest that oestrogen receptors may be present in the brains of all species during fetal development, and may play important roles in the development of the pituitary and brain.

Androgen receptors have been reported in the brain and anterior pituitary of fetal (132–162 days of gestation) rhesus monkeys (Pomerantz *et al.*, 1985). In rodents, androgen receptors have not been reported in fetuses. However, these receptors have been detected in the pre-optic area, amygdala and hypothalamus of early neonatal rats and mice (reviewed in Pasqualini and Sumida, 1986). Given the important organizational effects of androgens on the brain during early neonatal development, and the similar times of appearance of androgen and oestrogen receptors in other organ systems, it is possible that androgen receptors are present at earlier developmental periods in rodents even though they have not yet been identified.

In 20-day-old fetal mice, nuclear concentration of a radiolabelled progestin, indicating the presence of receptors, was observed in several areas of the brain, including the pre-optic area, hypothalamus, and central grey area (Shughrue *et al.*, 1989). Additionally, progesterone receptors (PR) were observed in the anterior, intermediate and posterior lobe of the pituitary (Shughrue *et al.*, 1988). These results contrast with earlier results in which attempts to detect PR in fetal rodent brain and pituitary were unsuccessful (Kato and Onouchi, 1983). Progesterone receptors were present in the brain of both males and females, and no sexual dimorphism was noted. The function of these PR is unknown, but Shughrue *et al.* (1989) speculate that PR may be important for brain and pituitary development and expression of female sexual behaviour.

11.7 Progesterone receptors in fetal tissues

Progesterone plays several essential roles in the establishment and maintenance of pregnancy. Progesterone receptors (PR) are found in high levels in the brain (Sar and Stumpf, 1973; Moguilewsky and Raynaud, 1979) and reproductive tract of adult females (Press and Greene, 1988), but recent results using a variety of experimental animals have unexpectedly indicated that PR are expressed in a variety of reproductive and non-reproductive tissues during late fetal life (Pasqualini and Nguyen, 1980; Giannopoulos et al., 1982; Hochner-Celnikier et al., 1986; Nguyen et al., 1986; Shughrue et al., 1988).

Other studies have examined the presence of PR in a variety of body tissues. Progesterone receptors were identified in the oral mucosa, teeth, esophagus, laryngeal muscles, mucosa, dermis, muscle and adipose tissue of skin, abdominal muscles, mesentery and interstitial cells of the kidney medulla of both male and female 20-day-old fetal mice (Shughrue et al., 1988). In females, PR were found in the mesenchyme of the mammary glands, the mesenchyme and epithelium of the oviduct, and the mesenchyme of the uterus. Recent autoradiographic studies detected PR in both the uterine mesenchyme and epithelium of 1-day-old Balb/c mice (R. M. Bigsby, unpubl. res.). In males, the interstitial cells of the testis and the mesenchyme of the ductus deferens and epididymis were found to be PR-positive (Shughrue et al., 1988).

These results with late fetal mice are in agreement with earlier reports that PR are present in the fetal guinea pig uterus, ovary and vagina (Pasqualini and Nguyen, 1980; Nguyen et al., 1986) and a variety of other body tissues such as lung, kidney, brain and heart (Pasqualini and Nguyen, 1980). Similarly, the lung, kidney, small intestine, brain, muscle, heart, skin and liver of the fetal rabbit also express PR (Giannopoulos et al., 1982). The role of the PR in somatic tissues of developing animals is not known, but its ubiquity indicates that it could be involved in growth and development of several tissues.

11.8 Oestrogen induction of progesterone receptors

When examined as whole organ homogenates, progesterone receptor (PR) of reproductive tract organs is increased following oestrogen treatment (Milgrom et al., 1973; Leavitt et al., 1974; Muechler et al., 1976). The involvement of oestrogen in the expression of PR in the fetus appears to be similar, but has not been definitely established. Oestrogen stimulates PR expression in the vagina, ovary and uterus in late gestational guinea pig fetuses, but does not produce increases in PR of non-reproductive tissues (Pasqualini and Nguyen, 1980; Nguyen et al., 1986). However, recent observations in the rat uterus (Ennis and Stumpf, 1988), rabbit oviduct

(Hyde *et al.*, 1989), and monkey uterus (Brenner *et al.*, 1988) indicate that oestrogen regulation of PR expression is cell-type specific. For example, in the immature rat, oestrogen induces increased progestin binding in the stromal and myometrial cells, while it decreases binding in the epithelial cells (Ennis and Stumpf, 1988). Similarly, in the rabbit oviduct, PR immunoreactivity was lost in the epithelium following oestrogen treatment of ovariectomized animals (Hyde *et al.*, 1989). Thus, generalizations made from biochemical analyses of steroid-binding activity in homogenates of whole organs may be misleading. As a case in point, in the previously discussed experiments in which progesterone binding was analysed autoradiographically in tissues of 20-day-old mouse fetuses, oestrogen priming of the mothers was used, presumably to insure enhanced levels of PR expression in reproductive tissues (Shughrue *et al.*, 1988, 1989); the observation that the uterine epithelium of these fetuses was devoid of PR implies that it had not matured to the point of expressing PR. However, we have found that the cells within the uterine epithelium of newborn mice (equivalent in age to the 20-day-old fetuses used by Shughrue *et al.*, 1988) exhibit strong progestin binding (R. M. Bigsby and G. R. Cunha, unpubl. res.); it may be that priming with oestrogen down-regulates epithelial PR in the uterus of fetal/ newborn mice. However, in light of the cellular and tissue differences in the hormonal regulation of PR in adult organs mentioned above, one must be cautious in extending concepts to developing reproductive tract organs until more thorough investigations have been performed at the cellular level.

11.9 Role of cell–cell interactions in the expression of steroid hormone receptors

The expression of androgen receptors in either epithelial or mesenchymal cells is a crucial stage in cellular differentiation that in turn confers new regulatory properties to the cells in question. A fundamental and largely unanswered question is why do androgen receptors (or other types of steroid receptors) appear in the first place? Is androgen receptor expression autonomous or dependent upon interactions with other cells? While the entire picture is still incomplete, it is clear that cell–cell interactions play a crucial role. In the embryonic mouse mammary gland, where androgen receptors are normally localized in a thin layer of mesenchymal cells immediately surrounding the epithelial bud, it has been shown that the mammary epithelium induces androgen receptors in the mesenchyme (Heuberger *et al.*, 1982). Only mammary epithelium has been shown to possess this inductive activity (epidermis and pancreatic epithelium is completely ineffective), but other epithelia that are normally associated with androgen-receptor-positive mesenchyme (preputial gland, genital tubercle, Wolffian duct, or urogenital sinus) were not tested for their inductive ability (Heuberger et al., 1982).

In other systems such as the developing prostate, the initial expression of androgen receptors in the mesenchyme has not been examined. However, the expression of androgen receptors in the epithelium appears to be dependent upon inductive influences from mesenchyme which in a general sense induces the entire process of epithelial differentiation. This conclusion has been derived from studies in which epithelia of the embryonic or adult urinary bladder (BLE) have been grown in association with urogenital sinus mesenchyme (UGM). When grown in intact male hosts, UGM + BLE tissue recombinants undergo prostatic development and the BLE, initially devoid of androgen receptors, forms prostatic ducts whose epithelial cells express androgen receptors (Cunha et al., 1980b; Neubauer et al., 1983; Sugimura et al., 1986). These findings have been recently corroborated in studies in which epithelium of the ureter was induced by seminal vesicle mesenchyme to undergo seminal vesicle differentiation (Cunha et al., 1988). In this experiment, the androgen-receptor-deficient ureteric epithelium is induced to express androgen receptors during its differentiation into seminal vesicle tissue (Cunha et al., 1990). Thus, mesenchyme induces and directs epithelial differentiation, one aspect of which is the expression of androgen receptors, a differentiation marker of the epithelium.

In a fashion similar to that in the developing male genital tract, the initial expression of epithelial oestrogen receptors appears to be dependent upon mesenchymal–epithelial interactions in the female genital tract. This is based upon experiments in which uterine epithelium from 1- to 2-day-old mice was grown in association with mesenchyme from the uterus or vagina (both of which possess oestrogen receptors) or from the urinary bladder or gallbladder (both of which lack oestrogen receptors). Uterine epithelium grown in association with its own mesenchyme (UM + UE) undergoes uterine development, while in association with vaginal mesenchyme the uterine epithelium (VM + UE) undergoes vaginal differentiation (Cunha, 1976). In both cases (UM + UE or VM + UE) the programme of development results in the differentiation of an oestrogen-responsive epithelium (Cooke et al., 1986, 1987), and the expression of epithelial oestrogen receptors. In contrast, growth of UE in association with mesenchyme from the urinary bladder or gallbladder prevents the expression of oestrogen receptors in the uterine epithelium (Cunha and Young, unpubl. res.). Thus, oestrogen receptors in epithelial cells represent a stage in the programme of differentiation whose expression is dependent upon relatively specific mesenchymal–

epithelial interactions. Once epithelial oestrogen receptors are expressed, uterine epithelial cells may come under direct oestrogenic regulation.

11.10 Role of mesenchymal–epithelial interactions in hormonal response

The pattern of androgen receptor expression in the developing male genital tract, being initially confined to mesenchyme, raises the possibility that certain hormonal effects on the epithelium may be elicited indirectly via paracrine influences from mesenchyme. Indeed, in males a wide spectrum of androgen-dependent developmental processes in the male genital tract normally occurs in epithelial cells in which androgen receptors are undetectable. These processes include: (1) Wolffian duct stabilization during the ambisexual stage; (2) appearance of seminal vesicle anlagen; (3) appearance and the initial branching morphogenesis of prostatic ducts; (4) appearance and early branching morphogenesis of the bulbourethral gland; (5) pre-natal development of the external genitalia; and (6) androgen-induced regression of the epithelial mammary anlagen. To explore the role of mesenchymal–epithelial interactions in androgen-dependent development, tissue recombinants have been prepared with epithelium and mesenchyme derived from normal and androgen-insensitive Tfm (testicular feminization) tissues (Fig. 6). Such studies have amply shown that androgen induced development of male accessory sex organs occurs via mesenchymal factors; furthermore, these experiments have shown that androgen-induced epithelial proliferation in the mature prostate is also mediated by the stromal androgen receptors (Cunha *et al.*, 1983a, 1987; Sugimura *et al.*, 1986). The developmental pattern of androgen receptor expression in the mesenchyme prior to epithelial expression during the period of androgen-stimulated growth and differentiation supports the notion of mesenchymally mediated hormonal effects.

Similar to the observations in the developing male reproductive tract, mesenchymal cells of the developing female tract express ER prior to the appearance of ER in the corresponding epithelia. By analogy to the male system, it has been suggested that some oestrogenic responses of the epithelia of the uterus and vagina are mediated by the ER of the underlying mesenchyme or stroma. Using tissue recombinations it has been shown that vaginal or uterine epithelial morphogenesis and hormonal responsiveness is determined by the stroma with which the epithelium is associated (Cunha *et al.*, 1980a, 1983a; Cooke *et al.*, 1986). Although these studies did not definitely point to a specific mesenchymal factor as the mediator of oestrogenic response within the epithelium, a number of paracrine mediators of

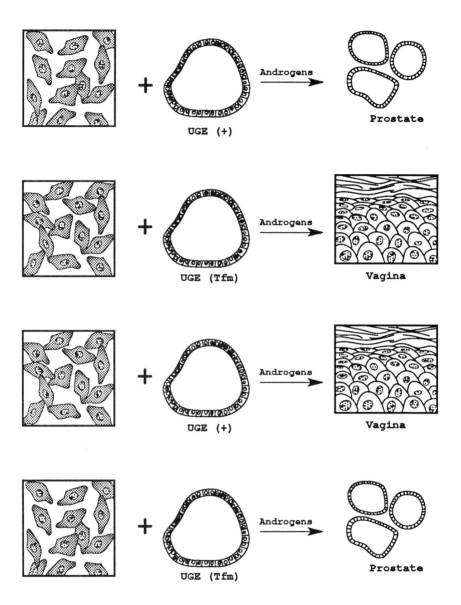

Fig. 6 A summary of recombination experiments between urogenital sinus components from Tfm/Y and wild-type embryos. A positive androgenic response (prostatic morphogenesis) occurs when wild-type mesenchyme is grown in association with either wild-type or Tfm/Y epithelium in male hosts. Conversely, vagina-like differentiation occurs when either wild-type or Tfm/Y epithelium is grown in association with Tfm/Y mesenchyme in male hosts.

mesenchymal effects on epithelium must be considered such as EGF, IGF-1, TGF-B, TGF-α, and KGF (Mukku and Stancel, 1985; Dickson and Lippman, 1987; DiAugustine et al., 1988; Murphy et al., 1988; Finch et al., 1989). In this regard, the uterine epithelium of the early neonatal mouse, while devoid of ER, nonetheless exhibits an increase in the rate of cellular proliferation following oestrogen administration, implicating an indirect mechanism of oestrogen action since the mesenchymal cells do contain ER at this stage (Bigsby and Cunha, 1986; Taguchi et al., 1988; Bigsby et al., 1990).

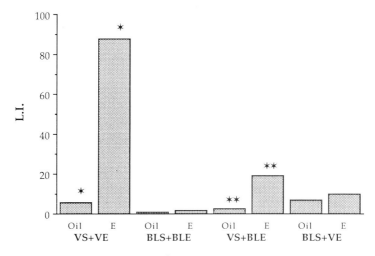

Fig. 7 Epithelial labelling index with [^{3}H]-thymidine of tissue recombinants constructed with epithelium and mesenchyme from urinary bladder and vagina. Epithelium and mesenchyme of neonatal mouse vagina and urinary bladder were separated and recombined into homotypic vaginal (VS + VE) and urinary bladder (BLS + BLE) tissue recombinants or recombined heterotypically (VS + BLE, BLS + VE). Tissue recombinants were grafted into intact female hosts and grown for 3 weeks at which time the hosts were ovariectomized. One week later the hosts were injected with sesame oil or oestradiol (E2) and sacrificed 24 h later. Two hours before sacrifice the hosts were injected with [^{3}H]-thymidine. Following fixation and sectioning epithelial labelling index was determined. Note the oestrogenic responsiveness of VS + VE and VS + BLE recombinants and the lack of responsiveness in BLS + BLE and BLS + VE recombinants. * and **, statistically significant ($P = 0.05$).

The notion that oestrogen-induced proliferation is mediated by stromal factors is further supported by a number of observations. In primary cultures of mammary, uterine, or vaginal epithelial cells a direct mitogenic effect of oestrogen has never been demonstrated (Flaxman et al., 1973; Liszezak et al., 1977; Kirk et al., 1978; Casimiri et al., 1980; Imagawa et al., 1982; Nandi et al., 1982, 1984; Iguchi et al., 1983, 1985; Tomooka et al.,

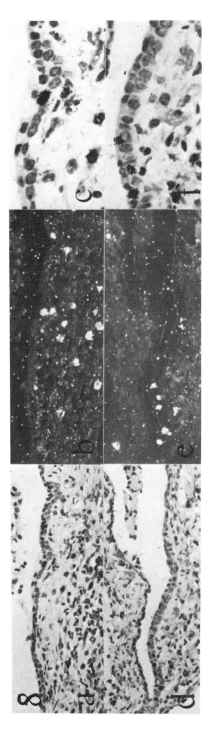

Fig. 8 Histological sections of BLS+VE tissue recombinants. BLS+VE tissue recombinants were prepared as described in Fig. 4 and grown for 3 weeks in intact female hosts which were then ovariectomized. One week later the hosts were injected with sesame oil (a–c) or oestradiol (d–f) and sacrificed 24 hours later. Two hours before sacrifice the hosts received [^{3}H]-thymidine. Note that the epithelium is thin and atrophic and lightly labelled in the presence (b) or absence (e) of oestradiol. Dark-field images (b and e) are identical fields to bright-field images (a and d). Magnifications: (a), (b), (d), (e) = 200×; (c) and (f) = 500×.

1986; Uchima et al., 1987), while in cultures of mixed stromal and epithelial cells oestrogen may stimulate proliferation (McGrath, 1983; Haslam, 1986; Inaba et al., 1988). Furthermore, uterine and vaginal epithelial cells that were unresponsive to oestrogen *in vitro* (Iguchi et al., 1983, 1985; Uchima et al., 1987) exhibited oestrogen-induced proliferation when recombined with their respective stromas and re-introduced into an *in vivo* environment (renal capsule graft) (Cooke et al., 1986). Evidence of indirect oestrogenic action has also been reported for the primate uterus and oviduct. For example, endometrial epithelial cells of the regenerating monkey uterus lack oestrogen receptors at a time when oestrogen induces DNA synthesis (McClellan et al., 1986).

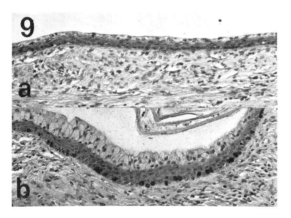

Fig. 9 Histological sections of tissue recombinants composed of vaginal epithelium which has been recovered from BLS + VE and reassociated with fresh vaginal stroma. BLS + VE recombinants were grown for a total of 4 weeks according to the protocol described in Fig. 8 at which time the VE was recovered from the BLS + VE recombinants and reassociated with fresh vaginal stroma (VS). These VS + recovered VE recombinants were grown for 3 weeks in intact female hosts which were then ovarectomized and 1 week later injected with oil (a) or oestradiol (b). As in Fig. 8 the hosts received [^{3}H]-thymidine 2 h before sacrifice. Note that the proliferative response (epithelial labelling) to oestrogen has been recovered in these VS + recovered VE recombinants. Magnification: 500 × .

Conclusions from these negative correlative observations are further supported by experimental tissue recombination studies. Epithelial cells of the urinary bladder, normally unresponsive to oestrogen, exhibit a modest oestrogen-induced increase in epithelial labelling with [^{3}H]thymidine when grown in association with vaginal mesenchyme (Fig. 7) (Cunha and Young, 1991). Conversely, the ability of vaginal epithelial cells to respond proliferatively to oestrogen is abolished when vaginal epithelium is grown in association with mesenchyme of the urinary bladder (BLM + VE recombinants) (Figs 7 and 8). This loss in oestrogenic responsiveness of vaginal epithelium can be reversed by recovering the vaginal epithelium from the BLM + VE tissue recombinants and reassociating it with fresh vaginal stroma (Fig. 9) (Cunha and Young, 1991).

11.11 Conclusions

Target organs of sex steroids within male and female genital tracts are composed of epithelium and mesenchyme and during development are influenced by steroids in profound ways. The effects of androgens, oestrogens and progesterone on developing organs appear to be elicited via high affinity saturable receptor proteins with physico-chemical properties identical to their adult counterparts. The timing of expression of steroid hormone receptors correlates well with hormonal effects elicited during development, even though the expression of hormone receptors occurs initially in the mesenchyme and later in the epithelium. This finding, in conjunction with tissue recombinant analysis of the developmental role of mesenchymal–epithelial interactions, has led to the concept that a variety of hormonal effects expressed in developing epithelial cells are elicited indirectly via the mesenchyme. Indeed, several models have shown that the expression of epithelial androgen and oestrogen receptors occurs as a result of mesenchymally induced differentiation. Finally, there is considerable evidence supporting the idea that certain hormonal effects expressed in receptor-positive adult epithelial cells may be elicited via regulatory influences from the underlying connective tissue.

Acknowledgements

This work was supported by NIH grants CA05388, DK32157, HD17491, HD21919 and CA49996.

References

Andersson, C. and Forsberg, J. G. (1988). Induction of estrogen receptor, peroxidase activity, and epithelial abnormalities in the mouse uterovaginal epithelium after neonatal treatment with diethylstilbestrol. *Teratogenesis Carcinog. Mutagen.* **8**, 347–361.

Arai, Y., Matsumoto, A., Yamanouchi, K. and Nichizuka, M. (1980). Perinatal sex steroid exposure, brain morphology, and neuroendocrine and behavioral functions. *In*: "Toxicity of Hormones in Perinatal Life" (eds T. Mori and H. Nagasawa), pp. 9–20. CRC Press, Inc., Boca Raton, FL.

Beer, D. G., Butley, M. S., Cunha, G. R. and Malkinson, A. M. (1984). Autoradiographic localization of specific [³H]dexamethasone binding in fetal lung. *Devel. Biol.* **105**, 351–364.

Bern, H. A. and Talamantes, F. J. (1981). Neonatal mouse models and their relation to disease in the human female. *In*: "Developmental Effects of Diethylstilbestrol (DES) in pregnancy" (eds A. Herbst and H. A. Bern), pp. 129–147. Thieme Stratton Inc., New York.

Bern, H. A., Mills, K. T., Ostrander, P. I., Schoenrock, B., Graveline, B. and Plapinger, L. (1984). Cervicovaginal abnormalities in BALB/c mice treated neonatally with sex hormones. *Teratology* **30**, 267–274.

Bigsby, R. M. and Cunha, G. R. (1986). Estrogen stimulation of deoxyribonucleic acid synthesis in uterine epithelial cells which lack estrogen receptors. *Endocrinology* **119**, 390–396.

Bigsby, R. M. and Cunha, G. R. (1987). Ontogeny of estrogen receptors in the uterus and vagina of the neonatal mouse: Comparison of autoradiographic and immunohistochemical methods of detection. *Endocrinology* (Suppl.) **120**, 64.

Bigsby, R. M., Aixin, L., Luo, K. and Cunha, G. R. (1990). Strain differences in the ontogeny of estrogen receptors in murine uterine epithelium. *Endocrinology* (in press).

Branham, W. S., Sheehan, D. M., Zehr, D. R., Medlock, K. L., Nelson, C. J. and Ridlon, E. (1985a). Inhibition of rat uterine gland genesis by tamoxifen. *Endocrinology* **117**, 2238–2248.

Branham, W. S., Sheehan, D. M., Zehr, D. R., Ridlon, E. and Nelson, C. J. (1985b). The postnatal ontogeny of rat uterine glands and age-related effects of 17 beta-estradiol. *Endocrinology* **117**, 2229–2237.

Brenner, R. M., McClellan, M. C. and West, N. B. (1988). Immunocytochemistry of estrogen and progestin receptors in the primate reproductive tract. *In*: "Steroid Receptors in Health and Disease" (ed. V. K. Moudgil), pp. 47–70. Plenum, New York.

Brody, J. R. and Cunha, G. R. (1989a). Histologic, morphometric, and immunocytochemical analysis of myometrial development in rats and mice: I. Normal development. *Am. J. Anat.* **186**, 1–20.

Brody, J. R. and Cunha, G. R. (1989b). Histologic, morphometric, and immunocytochemical analysis of myometrial development in rats and mice: II. Effects of DES on development. *Am. J. Anat.* **186**, 21–42.

Burns, R. K. (1961). Role of hormones in the differentiation of sex. *In*: "Sex and Internal Secretions" (ed. W. C. Young), pp. 76–158. Williams and Wilkins, Baltimore.

Casimiri, V., Rath, N. C., Parvez, H. and Psychoyos, A. (1980). Effect of sex steroids on rat endometrial epithelium and stroma cultured separately. *J. Steroid Biochem.* **12**, 293–298.

Clark, J. and Gorski, J. (1970). Ontogeny of the estrogen receptor during early uterine development. *Science* **169**, 76–78.

Clark, J. H. and Peck, E. J. (1979). "Female Sex Steroids: Receptors and Function". Springer-Verlag, New York.

Cooke, P. S. (1988). Ontogeny of androgen receptors in male mouse reproductive organs. *Endocrinology* (Suppl.) **122**, 92.

Cooke, P. S. and Cunha, G. R. (1990). In preparation.

Cooke, P. S., Uchima, F.-D. A., Fujii, D. K., Bern, H. A. and Cunha, G. R. (1986).

Restoration of normal morphology and estrogen responsiveness in cultured vaginal and uterine epithelia transplanted with stroma. *Proc. Natl. Acad. Sci. USA* **83**, 2109–2113.

Cooke, P. S., Fujii, D. K. and Cunha, G. R. (1987). Vaginal and uterine stroma maintain their inductive properties following primary culture. *In Vitro Cell. Dev. Biol.* **23**, 159–166.

Cunha, G. R. (1975). Age-dependent loss of sensitivity of female urogenital sinus to androgenic conditions as a function of the epithelial-stromal interaction. *Endocrinology* **95**, 665–673.

Cunha, G. R. (1976). Stromal induction and specification of morphogenesis and cytodifferentiation of the epithelia of the Mullerian ducts and urogenital sinus during development of the uterus and vagina in mice. *J. Exp. Zool.* **196**, 361–370.

Cunha, G. R. and Young, P. (1991). Role of stroma in estrogen-induced epithelial proliferation. *Epithelial Cell Biol.* (in press).

Cunha, G. R., Chung, L. W. K., Shannon, J. M. and Reese, B. A. (1980a). Stromal-epithelial interactions in sex differentiation. *Biol. Reprod.* **22**, 19–43.

Cunha, G. R., Reese, B. A. and Sekkingstad, M. (1980b). Induction of nuclear androgen-binding sites in epithelium of the embryonic urinary bladder by mesenchyme of the urogenital sinus of embryonic mice. *Endocrinology* **107**, 1767–1770.

Cunha, G. R., Shannon, J. M., Vanderslice, K. D., Sekkingstad, M. and Robboy, S. J. (1982). Autoradiographic analysis of nuclear estrogen binding sites during postnatal development of the genital tract of female mice. *J. Steroid Biochem.* **17**, 281–286.

Cunha, G. R., Chung, L. W. K., Shannon, J. M., Taguchi, O. and Fujii, H. (1983a). Hormone-induced morphogenesis and growth: Role of mesenchymal-epithelial interactions. *Recent Prog. Horm. Res.* **39**, 559–598.

Cunha, G. R., Shannon, J. M., Vanderslice, K. D., McCormick, K. and Bigsby, R. M. (1983b). Autoradiographic demonstration of high affinity nuclear binding and finite binding capacity of 3H-estradiol in mouse vaginal cells. *Endocrinol.* **113**, 1427–1430.

Cunha, G. R., Donjacour, A. A., Cooke, P. S., Mee, S., Bigsby, R. M., Higgins, S. J. and Sugimura, Y. (1987). The endocrinology and developmental biology of the prostate. *Endocrine Rev.* **8**, 338–363.

Cunha, G. R., Higgins, S. J. and Young, P. F. (1988). Seminal vesicle mesenchyme induces the expression of the seminal vesicle functional cytodifferentiation in adult epithelia. *J. Cell Biol.* **107**, 178a.

Cunha, G. R., Young, P., Higgins, S. J. and Cooke, P. S. (1991). Neonatal seminal vesicle mesenchyme induces a new morphological and functional phenotype in the epithelia of adult ureter and ductus deferens. *Development* (in press).

DiAugustine, R. P., Petrusz, P., Bell, G. I., Brown, C. F., Korach, K. S., McLachlan, J. A. and Teng, C. T. (1988). Influence of estrogens on mouse uterine epidermal growth factor precursor protein and messenger ribonucleic acid. *Endocrinology* **122**, 2355–2363.

Dickson, R. B. and Lippman, M. E. (1987). Estrogenic regulation of growth and polypeptide growth factor secretion in human breast carcinoma. *Endocr. Rev.* **8**, 29–43.

Donahoe, P. K., Budzik, G., Trelstad, R., Mudgett-Hunter, M., Fuller, A. F. J., Hutson, J., Ikawa, H., Hayashi, A. and MacLaughlin, D. (1982). Mullerian Inhibiting Substance: An update. *Rec. Progr. Hormone Res.* **38**, 279–330.

Eide, A. (1975a). The effect of oestradiol on the cell kinetics in the uterine and cervical epithelium of neonatal mice. *Cell Tiss. Kinet.* **8**, 249–257.
Eide, A. (1975b). The effect of oestradiol on the DNA synthesis in neonatal mouse uterus and cervix. *Cell Tiss. Res.* **156**, 551–555.
Ennis, B. W. and Stumpf, W. E. (1988). Differential induction of progestin-binding sites in uterine cell types by estrogen and antiestrogen. *Endocrinology* **123**, 1747–1753.
Finch, P. W., Rubin, J. S., Miki, T., Ron, D. and Aaronson, S. A. (1989). Human KGF is FGF-related with properties of a paracrine effector of epithelial cell growth. *Science* **245**, 752–755.
Flaxman, B. A., Chopra, D. P. and Newman, D. (1973). Growth of mouse vaginal epithelial cells *in vitro*. *In Vitro* **9**, 194–201.
Forsberg, J.-G. (1973). Cervicovaginal epithelium: Its origin and development. *Am. J. Obstet. Gynecol.* **115**, 1025–1043.
Forsberg, J.-G. (1975). Late effects in the vaginal and cervical epithelial after injections of diethylstilbestrol into neonatal mice. *Am. J. Obstet. Gynecol.* **121**, 101–104.
Forsberg, J.-G. (1988). Histogenesis of irreversible changes in the female genital tract after perinatal exposure to hormones and related substances. *In* "Toxicity of Hormones in Perinatal Life" (eds T. Mori and H. Nagasawa), pp. 39–62. CRC Press, Boca Raton, FL.
Forsberg, J.-G. and Kalland, T. (1981). Neonatal estrogen treatment and epithelial abnormalities in the cervicovaginal epithelium of adult mice. *Cancer Res.* **41**, 721–734.
George, F. W. and Noble, J. F. (1984). Androgen receptors are similar in fetal and adult rabbits. *Endocrinology* **115**, 1451–1458.
Giannopoulos, G., Phelps, D. S. and Munowitz, P. (1982). Heterogeneity and ontogenesis of progestin receptors in rabbit lung. *J. Steroid Biochem.* **17**, 503–510.
Goldman, A. S., Shapiro, B. H. and Neumann, F. (1976). Role of testosterone and its metabolites in the differentiation of the mammary gland in rats. *Endocrinology* **99**, 1490–1495.
Gorski, R. (1979). The neuroendocrinology of reproduction: an overview. *Biol. Reprod.* **20**, 111–127.
Greene, R. R. (1940). Hormonal factors in sex inversion: the effects of sex hormones on embryonic sexual structures of the rat. *Biol. Symp.* **9**, 105–123.
Gulino, A., Screpanti, I. and Pasqualini, J. R. (1984). Differential estrogen and antiestrogen responsiveness of the uterus during development in the fetal, neonatal and immature guinea pig. *Biol. Reprod.* **31**, 371–381.
Gupta, C. and Bloch, E. (1976). Testosterone-binding protein in reproductive tracts of fetal rats. *Endocrinology* **99**, 389–399.
Haslam, S. Z. (1986). Mammary fibroblast influence on normal mouse mammary epithelial cell responses to estrogen *in vitro*. *Cancer Res.* **45**, 310–316.
Haslam, S. Z. (1989). The ontogeny of mouse mammary gland responsiveness to ovarian steroid hormones. *Endocrinology* **125**, 2766–2772.
Heuberger, B., Fitzka, I., Wasner, G. and Kratochwil, K. (1982). Induction of androgen receptor formation by epithelium-mesenchyme interaction in embryonic mouse mammary gland. *Proc. Natl. Acad. Sci. USA* **79**, 2957–2961.
Hochner-Celnikier, D., Marandici, A., Iohan, F. and Monder, C. (1986). Estrogen and progesterone receptors in the organs of prenatal cynomolgus monkey and laboratory mouse. *Biol. Reprod.* **35**, 633–640.

Holderegger, C. and Keefer, D. (1986). The ontogeny of the mouse estrogen receptor: the pelvic region. *Am. J. Anat.* **177**, 285–297.

Hyde, B. A., Blaustein, J. D. and Black, D. L. (1989). Differential regulation of progestin receptor immunoreactivity in the rabbit oviduct. *Endocrinology* **125**, 1479–1483.

Iguchi, T., Uchima, F. D. A., Ostrander, P. L. and Bern, H. A. (1983). Growth of normal mouse vaginal epithelial cells in and on collagen gels. *Proc. Natl. Acad. Sci. USA* **80**, 3743–3747.

Iguchi, T., Uchima, F.-D. A., Ostrander, P. L., Hamamoto, S. T. and Bern, H. A. (1985). Proliferation of normal mouse uterine luminal epithelial cells in serum-free collagen gel culture. *Proc. Jpn. Acad.* **61**, 292–295.

Imagawa, W., Tomooka, Y. and Nandi, S. (1982). Serum-free growth of normal and tumor mouse mammary epithelial cells in primary culture. *Proc. Natl. Acad. Sci. USA* **79**, 4074–4077.

Imperato-McGinley, J., Guerrero, L., Gautier, T. and Peterson, R. E. (1974). Steroid 5a-reductase deficiency in man: An inherited form of psuedohermaphroditism. *Science* **186**, 1213–1215.

Inaba, T., Weist, W. G., Strickler, R. C. and Mori, J. (1988). Augmentation of the response of mouse uterine epithelial cells to estradiol by uterine stroma. *Endocrinology* **123**, 1253–1258.

Josso, N., Picard, J. Y. and Tran, D. (1977). The anti-Mullerian hormone. *Rec. Progr. Hormone Res.* **33**, 117–160.

Jost, A. (1953). Problems of fetal endocrinology: The gonadal and hypophyseal hormones. *Rec. Progr. Hormone Res.* **8**, 379–418.

Jost, A. (1965). Gonadal hormones in the sex differentiation of the mammalian fetus. *In*: "Organogenesis" (eds R. L. Urpsrung and H. DeHaan), pp. 611–628. Holt, Rinehart and Winston, New York.

Jost, A. (1970). Hormonal factors in the sex differentiation of the mammalian foetus. *Phil. Trans. R. Soc. Lond.* **259**, 119–130.

Kato, J. and Onouchi, T. (1983). Progestin receptors in female rat brain and hypophysis in the development from fetal to postnatal stages. *Endocrinology* **113**, 29–36.

Katzenellenbogen, B. S. and Gregor, N. G. (1974). Ontogeny of uterine responsiveness to estrogen during early development in the rat. *Mol. Cell Endocrinol.* **2**, 31–42.

Kaye, A. M., Icekson, I., Lamprecht, S. A., Gruss, R., Tsafriri, A. and Lindner, H. R. (1973). Steroid receptors in the reproductive system. *Biochemistry* **12**, 3072–3076.

Keefer, D. A. and Holdregger, C. (1985). The ontogeny of estrogen receptors: the brain and pituitary. *Dev. Brain Res.* **19**, 1386–1395.

Kimmel, G. L. and Harmon, J. R. (1980). Characteristics of estrogen binding in uterine cytosol during the perinatal period in the rat. *J. Steroid Biochem.* **12**, 73–75.

Kimmel, G. L., Harmon, J. R., Olson, M. E. and Sheehan, D. M. (1981). Steroid receptors in the reproductive system. *Teratology* **23**, 46A.

Kirk, D., King, R. J. B., Heyes, J., Peachy, L., Hirsch, P. J. and Taylor, R. W. T. (1978). Normal human endometrium in cell culture. *In Vitro* **14**, 651–662.

Korach, K. S., Horigome, Y., Tomooka, Y., Yamashita, S., Newbold, R. R. and McLachlan, J. A. (1988). Immunodetection of estrogen receptor in epithelial and mesenchymal tissues of the neonatal mouse uterus. *Proc. Natl. Acad. Sci. USA* **85**, 3334–3337.

Kratochwil, K. (1971). In vitro analysis of the hormonal basis for the sexual dimorphism in the embryonic development of the mouse mammary gland. *J. Embryol. Exp. Morphol.* **25**, 141–153.

Kratochwil, K. (1975). Experimental analysis of the prenatal development of the mammary gland. *Mod. Probl. Pediat.* **15**, 1–15.

Kratochwil, K. (1977). Development and loss of androgen responsiveness in the embryonic rudiment of the mouse mammary gland. *Devel. Biol.* **61**, 358–365.

Kratochwil, K. (1987). Tissue combination and organ culture studies in the development of the embryonic mammary gland. In "Developmental Biology: A Comprehensive Synthesis" (ed. R. B. L. Gwatkin), pp. 315–334. Plenum Press, New York.

Leavitt, W. W., Toft, D. O., Strott, C. A. and O'Malley, B. W. (1974). *Endocrinology* **94**, 1041–1053.

Liszezak, T. M., Richardson, G. S., MacLaughlin, D. T. and Kornblith, P. L. (1977). Ultrastructure of human endometrial epithelium in monolayer culture with and without steroid hormones. *In Vitro* **3**, 344–356.

McClellan, M., West, N. B. and Brenner, R. M. (1986). Immunocytochemical localization of estrogen receptors in the macaque endometrium during the luteal-follicular transition. *Endocrinology* **119**, 2467–2475.

McEwen, B., Pfaff, D. and Zigmond, R. (1980). Factors influencing sex hormone uptake by rat brain regions. II. Effects of neonatal treatment and hypophysectomy on testosterone uptake. *Brain Res.* **21**, 17–28.

McGrath, C. M. (1983). Augmentation of response of normal mammary epithelial cells to estradiol by mammary stroma. *Cancer Res.* **43**, 1355–1360.

McLachlan, J. A. (1981a). "Estrogens in the Environment". Elsevier/North Holland, New York.

McLachlan, J. A. (1981b). Rodent models for perinatal exposure to diethylstilbestrol and their relation to human disease in the male. In "Developmental Effects of Diethylstilbestrol (DES) in Pregnancy" (eds H. A. Bern and H. Bern), pp. 147–158. Thieme Stratton, New York.

McLachlan, J. A., Newbold, R. R. and Bullock, B. C. (1980). Long-term effects on the female mouse genital tract associated with prenatal exposure to diethylstilbestrol. *Cancer Res.* **40**, 3988–3999.

McLachlan, J. A., Newbold, R. R. and Bullock, B. C. (1975). Reproductive tract lesions in male mice exposed prenatally to diethylstilbestrol. *Science* **190**, 991–992.

Medlock, K. L., Sheehan, D. M. and Branham, W. S. (1981). The postnatal ontogeny of the rat uterine estrogen receptor. *J. Steroid Biochem.* **15**, 283–288.

Milgrom, E., Thi, L., Atger, M. and Baulieu, E. (1973). Mechanisms regulating the concentration and the conformation of progesterone receptor(s) in the uterus. *J. Biol. Chem.* **248**, 6366–6374.

Moguilewsky, M. and Raynaud, J. P. (1979). Estrogen-sensitive progestin binding sites in the female rat brain and pituitary. *Brain Res.* **164**, 165–175.

Mori, T. and Iguchi, T. (1988). Long-term effects of perinatal treatment with sex steroids and related substances on reproductive organs of female mice. In: "Toxicity of Hormones in Perinatal Life" (eds T. Mori and H. Nagasawa), pp. 63–80. CRC Press, Inc., Boca Raton, FL.

Muechler, E., Flickinger, G., Mastroianni, L. and Mikhail, G. (1976). Progesterone binding in rabbit oviduct and uterus. *Proc. Soc. Exp. Biol. Med.* **150**, 275–279.

Mukku, V. R. and Stancel, G. M. (1985). Regulation of epidermal growth factor receptor by estrogen. *J. Biol. Chem.* **260**, 9820–9824.

Murakami, R. (1987). Autoradiographic studies of the localisation of androgen-binding cells in the genital tubercles of fetal rats. *J. Anat.* **151**, 209–219.

Murphy, P. R., Sato, R., Sato, Y. and Friesen, H. G. (1988). Fibroblast growth factor messenger ribonucleic acid expression in a human astrocytoma cell line: regulation by serum and cell density. *Mol. Endocrinol.* **2**, 591–598.

Nandi, S., Imagawa, W., Tomooka, Y., Shiurba, R. and Yang, J. (1982). Mammogenic hormones: possible roles *in vivo*. *In*: "Growth of Cells in Hormonally Defined Media" (eds G. Sato, A. Pardee and D. A. Sirbasku), pp. 779–788. Cold Spring Harbor Laboratory, Cold Spring Harbor.

Nandi, S., Imagawa, W., Tomooka, Y., McGrath, M. F. and Edery, M. (1984). Collagen gel culture system and analysis of estrogen effects on mammary carcinogenesis. *Arch. Toxicol.* **55**, 91–96.

Narbaitz, R., Stumpf, W. E. and Sar, M. (1980a). Estrogen receptors in mammary gland primordia of fetal mouse. *Anat. Embryol. (Berlin)* **158**, 161–166.

Narbaitz, R., Stumpf, W. E. and Sar, M. (1980b). Estrogen target cells in the larynx: autoradiographic studies with ^{3}H-diethylstilbestrol in fetal mice. *Hormone Res.* **12**, 113–117.

Neubauer, B. L., Chung, L. W. K., McCormick, K. A., Taguchi, O., Thompson, T. C. and Cunha, G. R. (1983). Epithelial–mesenchymal interactions in prostatic development. II. Biochemical observations of prostatic induction by urogenital sinus mesenchyme in epithelium of the adult rodent urinary bladder. *J. Cell Biol.* **96**, 1671–1676.

Newbold, R. R. and McLachlan, J. A. (1985). Diethylstilbestrol-associated defects in murine genital tract development. *In* "Estrogens in the Environment II" (ed. J. A. McLachlan), pp. 288–318. Elsevier, New York.

Newbold, R. R. and McLachlan, J. A. (1988). Neoplastic and non-neoplastic lesions in male reproductive organs following perinatal exposure to hormones and related substances. *In* "Toxicity of Hormones in Perinatal Life" (eds T. Mori and H. Nagasawa), pp. 89–109. CRC Press, Inc., Boca Raton, FL.

Newbold, R. R., Bullock, B. C. and McLachlan, J. A. (1987). Mullerian remnants of male mice exposed prenatally to diethylstilbestrol. *Teratogen. Carcinogen. Mutagen.* **7**, 377–389.

Nguyen, B. L., Giambiagi, N., Mayrand, C., Lecerf, F. and Pasqualini, J. R. (1986). Estrogen and progesterone receptors in the fetal and newborn responses to tamoxifen and estradiol. *Endocrinology* **119**, 978–988.

Ogasawara, Y., Okamoto, S., Kitamura, Y. and Matsumoto, K. (1983). Proliferative pattern of uterine cells from birth to adulthood. *Endocrinology* **113**, 582–587.

Pasqualini, J. R. and Nguyen, B. L. (1980). Progesterone receptors in the fetal uterus and ovary of the guinea pig: evolution during fetal development and induction and stimulation in estradiol-primed animals. *Endocrinology* **106**, 1160–1165.

Pasqualini, J. R. and Sumida, C. (1986). Ontogeny of steroid receptors in the reproductive system. *Int. Rev. Cytol.* **101**, 275–315.

Pasqualini, J. R., Sumida, C., Gelly, C. and Nguyen, B.-L. (1976). Specific [^{3}H]estradiol binding in the fetal uterus and testis of guinea pig. Quantitative evolution of [^{3}H]estradiol receptors in the different fetal tissues (kidney, lung, uterus and testis) during fetal development. *J. Steroid Biochem.* **7**, 1031–1038.

Pasqualini, J. R., Sumida, C., Nguyen, B.-L. and Gelly, C. (1978). Quantitative evaluation of cytosol and nuclear [^{3}H]estradiol specific binding in the fetal brain of guinea pig during fetal ontogenesis. *J. Steroid Biochem.* **9**, 443–447.

Peleg, S., DeBoever, J. and Kaye, A. M. (1979). Replenishment and nuclear retention of estradiol-17b receptors in rat uteri during postnatal development. *Biochim. Biophys. Acta* **587**, 67–74.

Pfaff, D. (1980). "Estrogens and Brain Function: Neural Analysis of a Hormone Controlled Mammalian Reproductive Behavior". Springer-Verlag, New York.

Plapinger, L. and McEwen, B. (1973). Ontogeny of estradiol-binding sites in rat brain. I. Appearance of presumptive adult receptors in cytosol and nuclei. *Endocrinology* **93**, 1119–1128.

Plapinger, L., Landau, I., McEwen, B. and Feder, H. (1977). Characteristics of estradiol-binding macromolecules in fetal and adult guinea pig brain cytosols. *Biol. Reprod.* **16**, 586–599.

Pomerantz, S. M., Fox, T. O., Sholl, S. A., Vito, C. C. and Goy, R. W. (1985). Androgen and estrogen receptors in fetal rhesus monkey brain and anterior pituitary. *Endocrinology* **116**, 83–89.

Press, M. F. and Greene, G. L. (1988). Localization of progesterone receptor with monoclonal antibodies to the human progestin receptor. *Endocrinology* **122**, 1165–1175.

Price, D. and Ortiz, E. (1965). The role of fetal androgens in sex differentiation in mammals. *In*: "Organogenesis" (eds R. L. Ursprung and H. DeHaan), pp. 629–652. Holt, Rinehart and Winston, New York.

Raynaud, A. (1977). The development of estrogen-sensitive tissues of the genital tract and mammary gland. *In* "The Ovary" (eds L. Zuckerman and B. J. Weir), pp. 63–100. Academic Press, New York.

Raynaud, J.-P., Moguilewsky, M. and Vannier, B. (1980). *In* "The Development of Responsiveness to Steroid Hormones: Advances in the Biosciences" (eds A. M. Kaye and M. Kaye), pp. 59–75. Pergamon Press, Oxford, UK.

Sar, M. and Stumpf, W. E. (1973). Neurons of the hypothalamus concentrate [^{3}H]-progesterone or its metabolites. *Science* **182**, 1266–1268.

Schleicher, G., Drews, U., Stumpf, W. E. and Sar, M. (1984). Differential distribution of dihydrotestosterone and estradiol binding sites in the epididymis of the mouse. *Histochemistry* **81**, 139–147.

Shannon, J. M., Cunha, G. R., Vanderslice, K. D. and Sekkingstad, M. (1982). Autoradiographic localization of steroid binding in human tissue labeled *in vitro*. *J. Histochem. Cytochem.* **30**, 1059–1065.

Shannon, J. M. and Cunha, G. R. (1983). Autoradiographic localization of androgen binding in the developing mouse prostate. *Prostate* **4**, 367–373.

Sheehan, D. M., Branham, W. S., Medlock, K. L., Olson, M. E. and Zehr, D. R. (1981). Uterine responses to estradiol in the neonatal rat. *Endocrinology* **109**, 76–82.

Shima, H., Young, P. F. and Cunha, G. R. (1988). Postnatal morphogenesis of mouse seminal vesicle is dependent on 5a-dihydrotestosterone. *J. Cell Biol.* **107**, 692a.

Shima, H., Tsuji, M., Young, P. F. and Cunha, G. R. (1990). Postnatal growth of mouse is dependent on 5a-dihydrotestosterone. *Endocrinology* (in press).

Shughrue, P. J., Stumpf, W. E. and Sar, M. (1988). The distribution of progesterone receptor in the 20-day-old fetal mouse: An autoradiographic study with [125I]progestin. *Endocrinology* **123**, 2382–2389.

Shughrue, P. J., Stumpf, W. E., Sar, M., Elger, W. and Schulze, P.-E. (1989). Progestin receptors in brain and pituitary of 20-day-old fetal mice: An autoradiographic study using [125I]progestin. *Endocrinology* **124**, 333–338.

Siiteri, P. K. and Wilson, J. D. (1974). Testosterone formation and metabolism during male sexual differentiation in the human embryo. *J. Clin. Endocrinol. Metab.* **38**, 113–125.

Somjen, D., Somjen, G., King, R. J. B., Kaye, A. M. and Lindner, H. R. (1973). Nuclear binding of oestradiol-17b and induction of protein synthesis in the rat uterus during postnatal development. *Biochem. J.* **136**, 25–33.
Stumpf, W. and Sar, M. (1976). Autoradiographic localization of estrogen, androgen, progestin, and glucocorticosteroid in 'target tissues' and 'non-target tissues'. *In* "Receptors and Mechanism of Action of Steroid Hormones" (ed. J. Pasqualini), pp. 41–84. Marcel Dekker Inc., New York.
Stumpf, W. E., Narbaitz, R. and Sar, M. (1980). Estrogen receptors in the fetal mouse. *J. Steroid Biochem.* **12**, 55–64.
Sugimura, Y., Cunha, G. R. and Bigsby, R. M. (1986). Androgenic induction of deoxyribonucleic acid synthesis in prostatic glands induced in the urothelium of testicular feminized (Tfm/y) mice. *Prostate* **9**, 217–225.
Sulcová, J., Jirásek, J. E. and Stárka, L. (1973). Transformation of testosterone into dihydrotestosterone by the primordia of human genitalia and by the fetal suprascapular skin. *Steroids Lipids Res.* **4**, 129–134.
Sumida, C. and Pasqualini, J. R. (1979a). Determination of cytosol and nuclear estradiol-binding sites in fetal guinea pig uterus by [^{3}H]estradiol exchange. *Endocrinology* **105**, 406–413.
Sumida, C. and Pasqualini, J. R. (1979b). Relationship between cytosol and nuclear oestrogen receptors and oestrogen concentrations in the fetal compartment of guinea-pig. *J. Steroid Biochem.* **11**, 267–272.
Sumida, C. and Pasqualini, J. R. (1980). Dynamic studies on estrogen response in fetal guinea pig uterus: Effect of estradiol administration on estradiol receptor, progesterone receptor and uterine growth. *J. Recep. Res.* **1**, 439–457.
Taguchi, O., Cunha, G. R. and Robboy, S. J. (1986). Expression of nuclear estrogen-binding sites within developing human fetal vagina and urogenital sinus. *Am. J. Anat.* **177**, 473–480.
Taguchi, O., Bigsby, R. M. and Cunha, G. R. (1988). Estrogen responsiveness and the estrogen receptor during development of the murine female reproductive tract. *Growth, Diff. Devel.* **30**, 301–313.
Takasugi, N. (1976). Cytological basis for permanent vaginal changes in mice treated neonatally with steroid hormones. *Int. Rev. Cytol.* **44**, 193–224.
Takeda, H., Mizuno, T. and Lasnitzki, I. (1985). Autoradiographic studies of androgen-binding sites in the rat urogenital sinus and postnatal prostate. *J. Endocrinol.* **104**, 87–92.
Tomooka, Y., DiAugustine, R. P. and McLachlan, J. A. (1986). Proliferation of mouse uterine epithelial cells *in vitro*. *Endocrinology* **118**, 1011–1018.
Uchima, F.-D. A., Edery, M., Iguchi, T., Larson, L. and Bern, H. A. (1987). Growth of mouse vaginal epithelial cells in culture: Functional integrity of the estrogen receptor system and failure of estrogen to induce proliferation. *Cancer Lett.* **35**, 227–235.
Wasner, G., Hennermann, I. and Kratochwil, K. (1983). Ontogeny of mesenchymal androgen receptors in the embryonic mouse mammary gland. *Endocrinology* **113**, 1771–1780.
West, N. B., Roselli, C. E., Resko, J. A., Greene, G. L. and Brenner, R. M. (1988). Estrogen and progestin receptors and aromatase activity in rhesus monkey prostate. *Endocrinology* **123**, 2312–2322.
Wilson, J. D. (1973). Testosterone uptake by the urogenital tract of the rabbit embryo. *Endocrinology* **92**, 1192–1199.

Wilson, J. D., Griffin, J. E., Leshin, M. and George, F. W. (1981). Role of gonadal hormones in development of the sexual phenotypes. *Hum. Genet.* **58**, 78–84.

Wilson, J. D. and Lasnitzki, I. (1971). Dihydrotestosterone formation in fetal tissues of the rabbit and rat. *Endocrinology* **89**, 659–668.

Yamashita, S. and Korach, K. S. (1989). A modified immunohistochemical procedure for the detection of estrogen receptor in mouse tissues. *Histochemistry* **90**, 325–330.

Yamashita, S., Newbold, R. R., McLachlan, J. A. and Korach, K. S. (1989). Developmental pattern of estrogen receptor expression in female mouse genital tracts. *Endocrinology* **125**, 2888–2896.

12
Retinoic Acid Receptors and Vertebrate Limb Morphogenesis

C. W. RAGSDALE, Jr and J. P. BROCKES

The Ludwig Institute for Cancer Research, 91 Riding House St., London W1P 8BT, UK

12.1 Introduction

Retinoids are small lipophilic molecules that share with steroids and thyroxine the ability to affect cell growth and differentiation. They possess, in addition, a much rarer ability—they can produce dose-dependent changes in pattern formation that appear to be alterations in positional specification. These morphogenetic effects of retinoids have provoked particular interest because so little is understood about how position is established in vertebrate pattern formation. The identification of novel members of the hormone nuclear receptor family that respond to retinoic acid (RA) deepens the analogy between steroid and thyroid hormones and retinoids. In addition, it offers a molecular basis for investigating the genetic mechanisms involved in pattern formation in vertebrates. We review here the roles of retinoic acid in differentiation and morphogenesis, and what has been learned to date about the structure and distributions of the retinoic acid receptors (RARs). We focus on morphogenesis in developing and regenerating limbs because it is these systems that currently offer the greatest promise for understanding how RA influences pattern formation in vertebrates.

12.2 Biology of retinoid action

Naturally occurring retinoids are hydrophobic molecules composed of a β-ionone ring conjugated to an unsaturated hydrocarbon side chain with a polar terminal group. Illustrated in Fig. 1 are three retinoids of known biological importance—retinol, retinal and retinoic acid. Retinol is vitamin A. It cannot be synthesised *de novo* by animals and is derived ultimately from plant carotenoids in the diet. It is stored in and released from the liver, and delivered in the circulation as a protein complex (Goodman, 1987). In many tissues retinol can be converted by oxidative metabolism to retinal and, irreversibly, to RA (Napoli, and Race, 1987). The requirements for vitamin A in vision and in reproduction are also satisfied by retinal, but not by RA (Dowling and Wald, 1960; Thompson *et al.*, 1964).

RA, by contrast, has long been recognized as the active retinoid in the differentiation and maintenance of epithelial tissue (Pitt, 1971). More specifically, the action of RA is to promote differentiation of simple (or secretory) epithelia, even when this is at the expense of normal differentiation in squamous epithelia. This effect is referred to as mucous metaplasia. Inappropriate squamous metaplasia is observed in secretory epithelia such as the vagina when RA is absent or in deficiency.

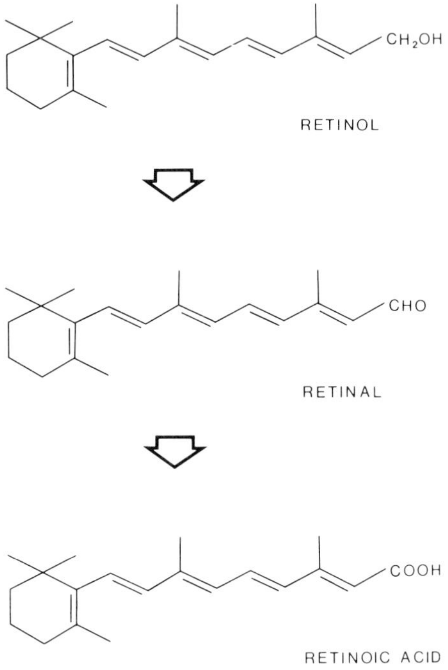

Fig. 1 Structure and biosynthetic relationships of the retinoids vitamin A (retinol), retinal and retinoic acid.

These differentiation effects of RA can be studied in culture with keratinocytes and mesothelial cells (Kim *et al.*, 1987). Moreover, the ability of RA to induce differentiation is not limited to epithelial cells, but can be demonstrated with a variety of cultured cell types. For example, RA is able to promote the differentiation of human HL60 promyelocytic leukaemia cells (Breitman *et al.*, 1980) as well as the conversion of murine F9 teratocarcinoma cells into parietal or visceral endoderm (Strickland and Mahdavi, 1978). In the case of P19 teratocarcinoma cells, the differentiated cell types formed from cell aggregates exposed to RA are dependent on its concentration. Thus, cardiac muscle forms at 10^{-9}M RA, skeletal muscle at 10^{-8}M, while neurons and astroglia predominate at 10^{-7}–10^{-5}M (Edwards and McBurney, 1983).

These culture systems have attracted a great deal of interest, not only because they provide a model for RA-induced differentiation, but also because retinoid differentiation entails anti-neoplastic effects. Promotion of differentiation and suppression of cell division are not, however, the only biological consequences of RA administration. First, RA does not promote differentiation in all cell types. For example, chondrocytes *in vivo* respond to RA treatment with the suppression of cartilage formation (Kochhar, 1977). Such negative effects can also be observed in cell and organ culture of limb bud mesenchyme (Lewis *et al.*, 1978). Second, although the more familiar effect of RA in culture is suppression of cell division, positive effects on proliferation have also been reported, and positive effects on both proliferation and chondrogenesis of chick limb bud cultures have been observed at low concentrations of RA (Ide and Aono, 1988; Paulsen *et al.*, 1988). It seems, then, that RA is able to provoke a variety of different effects that are dependent on both cell-type and concentration. Finally, some of the most striking effects of retinoid treatment, that is, those on pattern formation, do not involve altering cells by their degree or even path of differentiation (Lewis and Wolpert, 1976).

12.3 Effects on pattern formation in limb morphogenesis

The morphogenetic effects of retinoids have been established principally by experiments on the chick limb bud and the regenerating amphibian limb. The chick limb bud develops as an outgrowth of flank mesenchyme jacketed by epithelium. An important influence on the bud comes from an area of mesenchymal cells at its posterior margin termed the polarizing region or the zone of polarizing activity (ZPA). If the ZPA is transplanted to the anterior margin of a host bud, the limb forms with a duplication in its anteroposterior (AP) axis (Saunders, and Gasseling, 1968; Tickle *et al.*, 1975). This results in a digit pattern of 432234 (anterior to posterior; see Fig. 2B) in place of the normal 234 (anterior to posterior; see Fig. 2A) pattern. This and other grafting experiments suggest that the polarizing region either releases a morphogen whose concentration gradient across the AP axis specifies the tissue pattern directly (Tickle *et al.*, 1975), or else influences the function or controls the distribution of primary morphogens, which in turn determine the pattern (Newman and Frisch, 1979; Wolpert and Hornbruch, 1987).

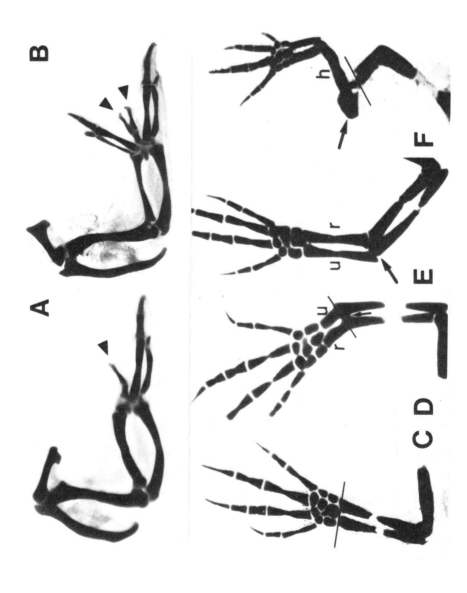

Work by Tickle and her colleagues and by Summerbell has established that the properties of a polarizing region graft can be mimicked by an inert carrier soaked in RA (Tickle *et al.*, 1982; Summerbell, 1983). Moreover, the extent of the duplication, for example 2234 or 432234, produced by a local source of RA implanted at the anterior margin of the limb bud, is dependent on the concentration of RA. This result is consistent with expectations about the relationship between morphogen concentration and the resulting pattern (Tickle *et al.*, 1985). Studies with radioactively labelled analogues of RA released from an ion-exchange bead do in fact indicate that a graded distribution is more effective at inducing duplications than an even one (Eichele *et al.*, 1985), and have led to quite detailed equivalences between the shape of the gradient and the induction of particular digits (Eichele and Thaller, 1987). Thus the concentration of the RA analogue TTNPB required to induce digit 2 can be estimated at 85 nM. These findings have led to the hypothesis that RA is in fact the morphogen released by the ZPA, a possibility to which we return below.

A second important setting for studying morphogenesis is the regeneration of limbs in adult urodele (tailed) amphibians such as the newt and axolotl. If the urodele limb is amputated at any level on the proximodistal (PD) axis (shoulder to fingertip) the wound surface is rapidly covered by migrating epithelial cells, and progenitor cells arise locally at the amputation plane to form a mound of cells known as the blastema (Wallace, 1981). The cells of the blastema give rise to the cartilage, muscle and connective tissue of the regenerate—that is, the mesenchymal tissues which are the substrate for pattern formation. It is an invariant of limb regeneration that the blastema does not give rise to structures proximal to its point of origin on the PD axis. For example, a shoulder blastema forms an arm, but a wrist blastema will only regenerate a hand (Fig. 2C). Retinoids, however, can 'proximalize' the blastema. If a wrist blastema is exposed to RA or its precursor retinol at an early stage in regeneration, cell division is initially inhibited, but the

---

Fig. 2 (Opposite) *Top*: Pattern formation in the developing chick limb is controlled by the ZPA. (A) Skeletal pattern observed in a normal chick wing. (B) Mirror-image pattern induced by a ZPA graft placed at the anterior margin of the chick limb bud early in development. Marked by arrowheads is digit 2, which is anterior in the normal limb. Beads soaked in RA mimic the effects of ZPA grafts on limb pattern formation. Photographs kindly supplied by C. Tickle. *Bottom*: Memory for the position of amputated structures is reset by RA treatment during amphibian limb regeneration. (C) Normal skeletal pattern seen in regenerated axolotl limb following amputation at the wrist and treatment with DMSO. Amputation plane indicated by line segment. (D)–(F) Increasing doses of RA in a DMSO solution progressively proximalize the regenerated limb. In (F), a complete arm including shoulder girdle (arrow) is produced. Arrows in (D) and (E) indicate fusion of the distal ends of the ulna and radius bones in the stump tissue that is formed at the plane of amputation. Abbreviations: h, humerus; r, radius; u, ulna. Photographs kindly supplied by D. Stocum.

blastema eventually gives rise to an entire arm, producing a serial duplication at the point of origin of the blastema (Niazi and Saxena, 1978; Maden, 1982; see Fig. 2F). As in the case of the chick limb, the effects are graded with concentration so that increasing doses progressively respecify the blastema to more proximal locations on the PD axis (Fig. 2D–F).

The effects of RA treatment on limb regeneration are not limited to the PD axis. In regenerating anuran tadpole limbs, mirror-image duplications in the AP axis as well as serial duplications in the PD axis can be elicited by RA treatment (Scadding and Maden, 1986a). In urodele regeneration, AP effects are not normally observed, but they can be demonstrated when surgically constructed double anterior regenerates are exposed to RA (Kim and Stocum, 1986). In summary, while position in the chick system can be respecified by RA only in the AP axis, delivery of RA can regulate the outcome of amphibian limb regeneration in both the PD and the transverse axes, a consideration that becomes important in considering receptor heterogeneity.

An important aspect of the actions of RA on limb development and regeneration are the inhibitory or so-called teratogenic effects (Kochhar, 1977; Alles and Sulik, 1989; Satre and Kochhar, 1989). These range from the deletion of a single skeletal element to severe dysmorphogenesis and the total inhibition of regeneration. A detailed discussion of the teratogenic effects of RA is outside the scope of this review, but ultimately a complete account of RA should 'bridge the void between teratology and molecular biology' (Kochhar, 1989). We wish to make three points about the relationship between the morphogenetic and teratogenic activities of RA as established by studies on amphibian limb development and regeneration (Scadding and Maden, 1986a,b). First, it is possible to evoke deletions in developing limbs with doses of RA that provoke morphogenetic duplications in regenerating limbs. Second, although the most severe inhibitory effects of RA are seen at high doses, there are some weaker effects in the low dose range. Third, deletions of skeletal elements are never seen when morphogenetic duplications are provoked by RA, suggesting that these are mutually exclusive events. A complete account of how RA affects limb morphogenesis must include both these aspects of its actions, not least because it is possible that some of the teratogenic effects could proceed through resetting blastemal position to uninterpretable values—for example, to a body position proximal to the limb field.

Finally, there is great interest in the possibility that RA is deployed in development and regeneration as a morphogen. RA can be extracted from the chick limb bud at an average concentration of 25 nM, whereas its synthetic precursor retinol is present at about 25-fold higher levels (Thaller and Eichele, 1987). When limb buds were split unequally into anterior and posterior fragments, the level of RA was 2.6-fold higher in the posterior tissue after normalization to DNA levels (Thaller and Eichele, 1987). Retinol was uniformly distributed when analysed in this way. These data are consistent with a proposed role for RA as a morphogen released by the polarizing region at the posterior margin, but many uncertainties remain. It is not clear if all or part of the extractable RA is in a form or location that allows it to interact with cells on the developing AP axis. There are several unidentified retinoids in the limb bud (including the so-called A_1 peak which represents a major metabolite of retinol), and their potential roles as morphogens need to be investigated (Thaller and Eichele, 1988). It will also be important to have information about the synthetic and degradative enzymes for the pathway of Fig. 1, and antibody probes to localize and quantitate them.

If there remain significant uncertainties about the endogenous role of RA in the chick limb bud, the ZPA experiments at least suggest a clear hypothesis about the source and role of a morphogen. Such an hypothesis does not exist in the case of limb regeneration, and at present there is no evidence available to support an endogenous role for RA. In fact, it is most unlikely that a gradient exists from the shoulder to fingertip because of the problems of scale (Wolpert, 1969; Crick, 1970), but it is possible that a proximal blastema could secondarily express higher levels of RA than a distal one as a result of events after amputation. Alternatively the levels of retinoid could be constant, and the responsiveness of cells could vary according to position. The conjectural nature of this discussion underlines how much remains to be established. Nonetheless, the ability of RA to respecify positional value may enhance our understanding of the molecular basis of such specification, regardless of whether it is used as an endogenous morphogen. The prospect of understanding the mechanism behind the morphogenetic effects of RA has been greatly improved as a result of the identification of nuclear receptors for RA.

12.4 Retinoic acid receptors

12.4.1 Identification

In 1987 two laboratories reported the identification of human cDNA clones for a new member of the hormone nuclear receptor family that responded to retinoic acid as a ligand (Giguere *et al.*, 1987; Petkovich *et al.*, 1987). The clones were initially isolated by virtue of their reactivity with probes to a region of the DNA-binding domain that is conserved among nuclear receptors (see Chapters 2.4 and 3.3). After the clones were transfected into HeLa or COS-1 cells, cytosolic extracts showed high affinity binding for RA in the nanomolar range, whereas retinol and retinal were at least ten-fold lower in affinity. A more definitive test was to demonstrate that both molecules could function as transactivators of transcription in the presence of RA. At the time, no promoter had been identified that was directly responsive to RA, and the functional test was made by replacing the DNA-binding region of the cDNA clones with the corresponding region of either the oestrogen (Petkovich et al., 1987) or glucocorticoid (Giguere et al., 1987) receptors. When the chimaeric receptors were transfected with reporter genes under the control of oestrogen or glucocorticoid-responsive elements, the activity of the reporters was stimulated by RA in the nanomolar or subnanomolar range. Similar stimulation was elicited by retinol only at 10^2–10^3-fold higher concentrations. It is not clear if this retinol effect is a consequence of conversion to RA in the cultured cells, or represents a direct interaction with the receptor. Ligands such as thyroid hormones, steroids or vitamin D_3 did not result in activity. This original member of the family is now named RAR-α.

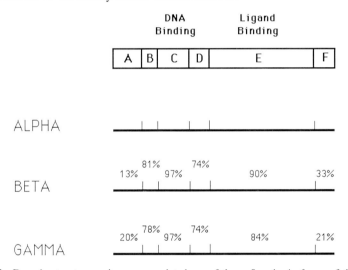

Fig. 3 Domain structure and sequence relatedness of the α, β and γ isoforms of the human RAR. Percentage scores indicate the amino acid sequence identity between each of the regions A–F of RAR-β and RAR-γ with those of RAR-α.

A second member of the family was cloned from the sequences flanking a site of hepatitis B virus integration in a human hepatocellular carcinoma (de The et al., 1987), which appeared to identify a novel member of the hormone nuclear receptor family. Initially, the ligand for this receptor was unknown, but the high sequence identity with RAR-α suggested that it was a second RAR. This was verified by making a chimaeric receptor and demonstrating RA-dependent activation of an oestrogen reporter gene (Brand et al., 1988; Benbrook et al., 1988). This member is referred to as RAR-β. Recently, a third member of the RAR family, RAR-γ, was isolated from a mouse embryo cDNA library and shown to be an RA-dependent transactivator (Zelent et al.,

1989). The human homologue of RAR-γ has also been isolated (Krust et al., 1989), and its variants are considered below. In view of the importance of the urodele blastema for evaluating the morphogenetic effects of RA, it was of great interest to identify the receptors expressed there. To date, homologues for RAR-α (Ragsdale et al., 1989) and RAR-β (Giguere et al., 1989), and also a new member of the family referred to as RAR-δ (Ragsdale et al., 1989), have been identified in the newt.

The relationships among the different members of the human family can be appreciated by reference to a standard representation of the protein structure as comprising six regions A–F (Krust et al., 1986 and Chapter 1.1; see Fig. 3). The most conserved regions between different members of the family are the C domain, mediating DNA binding, and the E domain, mediating RA binding. The homology scores show that the most variable regions of the molecules are the A and F regions (Fig. 3). When the human receptors are compared with RARs α, β and γ that have been isolated in the mouse, it is clear that homologues for a given RAR isoform, for example, the mouse and human α, are significantly closer to each other than they are to the other RAR subtypes identified in the same species (Krust et al., 1989; Zelent et al., 1989). This is further exemplified by a comparison of human and newt RAR-α, which shows impressive conservation in the A and F regions (Ragsdale et al., 1989). The newt RAR-δ is closer to γ than to α or β in regions B–E but it differs markedly in A and F. In view of the splicing of alternate A regions in γ and δ (discussed below), the strongest evidence that δ is not an amphibian homologue of γ is the marked difference in the F region (Ragsdale et al., 1989). This point may only be clarified on isolation of further receptors, for example, an unequivocal newt γ.

We now give a more detailed account of the regions, and the functions mediated by them.

12.4.2 Region A

Studies of human RAR-γ cDNA clones have revealed at least five forms that differ in this region (Krust et al., 1989). Some of these diverge at a point that corresponds to the boundary between regions A and B. Region B is known to be encoded on a separate exon in the human β and newt δ RARs (de The et al., 1987; C. Ragsdale, P. Gates and J. Brockes, unpubl. obs.), and RNase protection analysis of the newt δ receptor provides clear evidence for variable splicing at the A–B junction (see Fig. 4). Interestingly, the major RAR-δ variant appears to predominate in limbs and tails and in their blastemas but not in a variety of other tissues. The minor variant (or variants) appears to be ubiquitously expressed, but it will be necessary to clone and sequence the

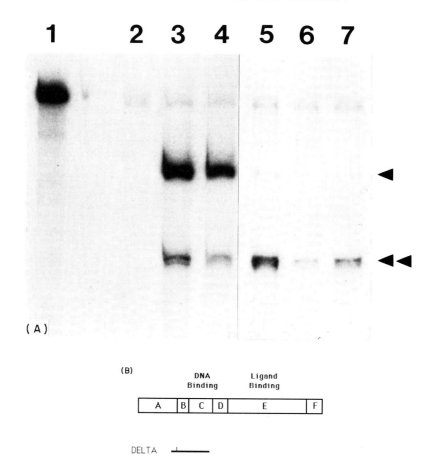

Fig. 4 RNase protection experiments establish that the A region of the newt δ receptor is variably spliced in different tissues. (A) Arrowhead indicates band corresponding to full-length protection of probe illustrated in (B). Double arrowheads note band representing partial probe protection. This band identifies transcripts alternatively spliced at the A–B junction. Lane 1, δ RAR riboprobe; Lane 2, tRNA control; Lane 3, limb blastema; Lane 4, normal limb; Lane 5, heart; Lane 6 liver; Lane 7, kidney. (B) Description of RAR sequence recognized by the antisense riboprobe employed in these experiments. Protection of the probe to the right of the notch is represented by the major band observed in the heart, liver and kidney samples in (A).

alternate A regions in order to clarify this point. Splicing at the A–B boundary is likely to be of functional importance, because evidence from studies of the oestrogen and progesterone receptors suggests that the N-terminal region has a role in determining the spectrum of genes that nuclear receptors activate (Tora et al., 1988; Bocquel et al., 1989).

12.4.3 Region C and Retinoic Acid Response Elements

The C region is responsible for binding to DNA, and is believed to form two zinc-stabilized fingers, of which the N-terminal one, or CI region, appears to confer sequence specificity in DNA binding (Green et al., 1988; Chapters 2.4 and 3.3). The amino acid sequence of CI is identical for α and β, but differs from that of γ and δ by 2 out of 31 residues (see Fig. 5). This raises the issue of the identity of response elements for the receptors (RAREs), and the specific possibility that α/β and γ/δ might recognize distinct targets.

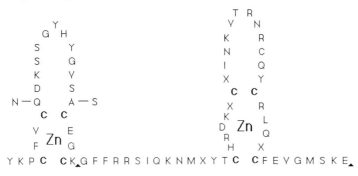

Fig. 5 Proposed zinc finger-like structure of the C region of the RAR. Substitution of two amino acids in the first 'finger' distinguishes the γ and δ from the α and β subtypes. An X is entered at other positions for which the amino acid sequence varies among the RARs cloned to date. Arrowheads note the location of introns identified in the newt RAR-δ by cloning RNA precursors.

The first response element identified was the thyroid hormone response element (TRE), a possibility suggested by the rather high sequence similarity between the C regions of the RARs and thyroid hormone receptors (TRs). When a synthetic palindromic TRE or the natural TREs in the growth hormone gene or the myosin heavy chain gene were juxtaposed to appropriate reporter genes and cotransfected with α and β receptor expression constructs, they mediated RA-dependent activation of the reporter (Umesono et al., 1988; Graupner et al., 1989). The artificially high levels of receptor required for these experiments raise questions about the physiological significance of this interaction, but at least for F9 cells there is sufficient endogenous receptor to activate a transfected TRE containing reporter when differentiation is triggered by RA (Umesono et al., 1988; Graupner et al., 1989). It has recently been shown that RAR-α and TR-β will form heterodimers in solution. Such heterodimers are able to bind to TREs in a concentration range where the RAR alone will not bind (Glass et al., 1989). Furthermore the two receptors will collaborate functionally in mixed transfection experiments. The effect of the heterodimer depends on the structure of the TRE, in that binding to a synthetic palindromic element stimulates transcription, whereas binding to a natural non-palindromic element represses transcription (Glass *et al.*, 1989). The physiological significance of retinoid–thyroid interactions at the receptor and response element levels is unclear, but these experiments illustrate the possibilities that heterodimer formation holds for increasing the range of target genes for a given RAR (see Chapter 16.3).

Specific RAREs have been identified in the promoter of the human RAR-β gene (de The *et al.*, 1990) and that of the laminin B1 chain (Vasios *et al.*, 1989; see Chapter 6.5). Transcription of the β gene is markedly increased by RA treatment (de The *et al.*, 1989; Zelent *et al.*, 1989), and the response element has been characterized as a 27 bp fragment that confers inducibility on a heterologous promoter (de The *et al.*, 1990). This sequence contains a direct repeat of the motif GTTCAC. The fragment binds to RAR-β *in vitro* and binding is not affected by the presence of RA, a property shared with the thyroid hormone receptor which binds to its TRE in the presence and absence of thyroid hormone (Damm *et al.*, 1989; Graupner *et al.*, 1989). There is evidence that both α and β are able to act through this β-promoter RARE, but the possibility of additional RAREs remains, particularly given that the GTTCAC motif was not found in a 46 bp sequence from the laminin B1 chain promoter that confers RA-dependent stimulation (Vasios *et al.*, 1989).

12.4.4 Region D

The D region is positioned as a hinge between the DNA-binding and ligand-binding domains. Its most interesting functional possibility appears to be that of nuclear localization. All of the receptor cDNAs analysed to date contain a lysine-rich sequence (RNDRNKKKK) which is similar to corresponding regions in steroid hormone receptors that are important for this function (Picard and Yamamoto, 1987; Guiochon-Martel *et al.*, 1989). The evidence from antibody staining experiments on cells transfected with cloned receptors is for a predominant, possibly exclusive, nuclear localization for RARs α and β (Gaub *et al.*, 1989). This is supported by both Western blotting (Gaub *et al.*, 1989) and band shift assays (de The *et al.*, 1990) on nuclear and cytoplasmic fractions of such transfectants.

12.4.5 Region E

The E region is responsible for binding ligand, but the relative affinities of the different receptors for RA remain unknown. Brand *et al.* (1988) reported that chimaeric human RARs α and β containing an oestrogen receptor C region could be distinguished by the RA concentration required to produce half-maximal stimulation of reporter activity, with the RAR-β construct responding to 10-fold lower concentrations. A subsequent study by this group (Krust *et al.*, 1989) failed to distinguish among the receptors when they tested intact mouse RARs α, β and γ with a less sensitive synthetic RARE reporter system constructed from a TRE. This point requires further study. Another important uncertainty is whether the RARs can interact with other retinoids, including the A_1 peak of Thaller and Eichele (1988), oxidative derivatives of RA (Strickland, 1979) and the synthetic retinoids that have known morphogenetic activity (Keeble and Maden, 1984).

We expect that, as with other nuclear receptors (see Chapters 2.3 and 3.4) additional E region functions for the RARs will emerge as detailed studies progress. Already, there is evidence that the E region is required to mediate co-operative binding with the thyroid hormone receptor to TREs (Glass *et al.*, 1989).

12.4.6 Region F

The functions of the F region are not understood but the cross-species conservation for a particular RAR subtype is impressive (Krust *et al.*, 1989). Thus for newt and human RAR-α 24 of the last 26 amino acids are identical (Ragsdale *et al.*, 1989).

Our understanding of the functional roles of the six regions of the RAR is largely based on extrapolations from experiments with other members of the hormone nuclear receptor superfamily. It will be necessary to obtain direct evidence by mutagenesis of the RARs, particularly for evaluating the roles of the splice variants of the A region. The recent identification of RAREs in the promoters of responsive genes is an important beginning for the analysis of the transactivation targets of the different RARs. Finally, we must keep in mind that the inventory of RAR subtypes may not be complete, and there could be as yet undiscovered receptors for RA or retinoid metabolites.

12.5 Functions of the retinoic acid receptors

The finding of significant heterogeneity in RAR structure is broadly consistent with the variety of effects obtained by administering RA. As discussed above, these include its action as an inducer of differentiation in epithelia and cultured cell lines, its inhibitory and teratogenic effects on embryos and tissue regeneration, and its morphogenetic effects on developing and regenerating limbs, including the ability to respecify at least two axes in amphibian limbs.

A major challenge for future work is to assign functional roles for the different RARs in mediating these effects. It should be stated that definitive proof that some or all of the effects of RA are mediated by its hormone nuclear receptors and by subsequent changes in gene expression, is lacking. Nonetheless, we are persuaded that the existence of this elaborate machinery provides the most compelling context for future investigation of these issues. While ultimately it will be necessary to manipulate receptor expression *in vivo* and determine the functional consequences, it is likely that the investigation of these effects will be predicated on information about receptor affinity, receptor distribution and regulation, and the identity of genes that are activated by the different receptor subtypes.

12.6 Ligand affinities of the retinoic acid receptors

Scadding and Maden (1986a) in their studies of the effects of retinoid treatment on amphibian limb regeneration suggested that the different effects seen—pattern duplications, skeletal deletions, and suppression of regeneration—showed distinct dose–response profiles. Because these effects were never observed together, it was possible to rank the responses [deletions ≪ duplications ≪ suppression] according to increasing doses of retinoids delivered. Interestingly, these rank orderings may be species-specific. Summerbell (1983) in his study of chick limb development observed [duplications ≪ deletions ≪ suppression] with increasing retinoid treatment. A simple account of these rankings would be that each effect was due to the activation of a different RAR, and that the various RARs have different affinities for RA. With the cloning of multiple RARs in the newt, it should be possible to obtain direct estimates for the relative affinities for RA of the various receptors. Even at this early stage such biochemical measurements might allow for the tentative assignment of function according to the dose–response rankings of Scadding and Maden (1986a) in urodele regeneration.

There are, however, a number of interpretive problems with assigning functional roles on the basis of affinity measurements. Because there are no antagonists for RA, one must depend on agonist affinities. This can be misleading in two ways. First, there could be molecules associated with the receptor *in vivo*, such as heat-shock proteins or bound DNA, that might affect agonist ligand affinity but are absent from purified cloned receptor. Second, agonist affinities are unreliable in predicting dose–response curves. Large effects can result from the occupancy of only a fraction of the receptor population because other factors, such as DNA-binding sites and associated transcription factors, are present in limiting quantities.

Another difficulty is that different receptor subtypes may encounter different RA concentrations according to their spatial and cellular distributions. One molecule, the cytoplasmic retinoic acid binding protein (CRABP), is known to bind RA with high affinity and to be distributed in the target tissues of retinoid action (Goodman, 1987). Maden and his colleagues (1988) have examined by immunohistochemistry the distribution of CRABP in the developing limb bud of the chick (see also Dolle et al., 1989b). They found that the binding protein forms a gradient in the AP axis, being most concentrated anteriorly. Since levels of RA are thought to be highest posteriorly, Maden et al. (1988) speculated that CRABP may act as an RA buffer, increasing the concentration gradient of free RA across the AP axis. Clearly, it will be crucial to know the distributions and levels of molecules that can affect RA concentrations in interpreting the functional implications of ligand affinities for the receptor subtypes. In addition, the possibility that occult retinoid-degrading enzymes may alter the intracellular concentration of RA in a scheme similar to that known for the mineralcorticoid receptor (Funder et al., 1988) cannot be excluded at present.

12.7 Localization of the retinoic acid receptors

The pattern of expression of the three mammalian RARs has been investigated by Northern analysis of RNA samples derived from embryonic and adult tissues. These studies indicate that RAR-α is rather widely distributed and relatively invariant in its level of expression, whereas the distributions of RAR-β and RAR-γ are significantly more variable (Zelent et al., 1989). Particularly impressive are the high levels of RAR-β expression found in brain (see Rees et al., 1989 for similar data in the rat) and RAR-γ in the skin (Krust et al., 1989; Zelent et al., 1989). Neither of these receptors, however, is restricted to these tissues, and in particular, RAR-γ has an extensive but regulated expression in developing neural crest and mesenchyme, suggesting that it may mediate the effects of RA treatment on craniofacial development (Ruberte et al., 1990). Two major transcripts are detected for RAR-α and -β and the basis of this heterogeneity remains unclear.

A more detailed picture of the expression pattern of the receptors has come from recent studies on the mouse embryo with *in situ* hybridization, as illustrated in Fig. 6. The overall impression is, again, that RAR-α is ubiquitously expressed, although at somewhat lower levels in cartilage (Fig. 6). By contrast, RAR-γ and -β show considerable spatial and temporal regulation and tend to be found in non-overlapping distributions (Dolle et al., 1989b; Ruberte et al.,

1990). Thus in the early limb bud (10 days post-coitum) RAR-γ transcripts were homogeneously distributed but by 12.5 days they were restricted in the proximal mesenchyme to pre-cartilaginous condensations (Fig. 6). β transcripts were detected initially only in the most proximal region of the limb bud, but at later ages were concentrated in the interdigital mesenchyme (Fig. 6). While it will be important to extend the analysis of γ transcripts by using probes for distinct A region variants (see above), it already seems likely that this subtype is a strong candidate to participate in retinoid effects on chondrogenesis and on differentiation in squamous and mucous epithelia. There is to date no evidence that RAR transcripts are distributed in gradients along the axes of the developing limb, but it will be important to extend these anlayses to the protein level with antibody probes.

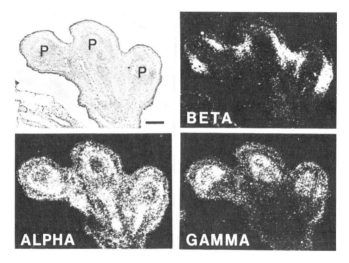

Fig. 6 Transcripts for the α, β and γ RARs are differentially distributed in the developing mouse limb. Illustrated are adjoining sections taken through the distal hindlimb of a 14.5-day-old embryo. See text for description. Message is localized by autoradiography following hybridization of radiolabelled anti-sense riboprobes for the murine α, β and γ RARs, as described in Dolle *et al.* (1989b). Label appears as white grains in dark field photomicrographs. Bright-field photomicrograph in upper left identifies the phalangeal bone anlagen (P). Bar, 160 μm. Photomicrographs kindly supplied by P. Dolle, P. Chambon and their colleagues.

In the newt, both the α and δ receptors are expressed in the adult limb and the limb blastema as determined by RNase protection and Northern blot analysis. The levels of expression of δ in the regenerative tissues of limb and tail, and their blastemas, is an order of magnitude higher than that found in a variety of other tissues. This difference is particularly marked for the major A region splice variant of δ (see Fig. 4), which is expressed in both epithelial and mesenchymal compartments of the limb. It would be particularly interesting if there were a gradient of expression of RAR along the PD axis of the normal limb, as this could be a basis for the level-specific properties inherited by blastemal cells. No such gradient of α or δ RNA was detected in the normal limb after RNase protection analysis, although differences between proximal and distal blastemas may exist. In view of the morphogenetic effects on more than one axis in the urodele, it is important to identify the A region variants of δ, as well as any other subtypes that may be expressed in the blastema, in order to study their distribution on the PD and transverse axes.

12.8 Regulation of gene expression by retinoic acid

The complexity and timing of retinoid effects on cells and tissues suggest not only that a variety of receptor subtypes are involved, but also that there are effects on gene expression not directly mediated by RARs but rather reflecting a chain of causation. It is clear, for example, that RA may regulate RNA levels by post-transcriptional mechanisms (Colberg-Poley et al., 1987); in fact, a post-transcriptional decrease in expression of RNA for the myc gene may play an important role in the ability of RA to induce differentiation of the cell lines F9 and HL60 (Dony et al., 1985; Bentley and Groudine, 1986; Griep and Westphal, 1988). It is not understood at present how the RARs participate in such mechanisms, but they may well act through 'intermediate' steps that involve the expression of other transcriptional regulators. Some effort has already been made to assess this possibility in the case of homeobox genes, and an important recent development is the identification of genes that are directly regulated by the RAR.

In cultures of human hepatoma cells, treatment with RA (10^{-9}M–10^{-6}M) leads to a 10- to 50-fold increase in RAR-β mRNA over 4 h, whereas RAR-α mRNA is not affected. Analysis of nuclear transcript elongation has shown that this effect is at the level of transcription, and has led to the identification of a RARE in the RAR-β promoter (de The et al., 1990; see above). It has recently been observed in the P19 embryonal carcinoma line that RAR-α RNA increased eight-fold after only 2 h of RA treatment (Song and Siu, 1989) and in F9 cultures that cAMP administered in combination with RA will reduce levels of α and γ transcripts and inhibit the β-induction (Hu and Gudas, 1990). Apparently, RAR expression changes following RA treatment are neither simple nor universal, and it will be important to determine if they occur in the context of axis respecification of developing and regenerating limbs. de The and colleagues (1989) have suggested that the β-induction may operate to 'amplify' the rather small gradient of RA found in the chick limb bud. This point, though, needs further study. It is, for example, not clear that β-induction would produce a positive effect on transcription of RAR-β-responsive genes, or that this effect would be non-linear with respect to RA concentration.

In evaluating potential target genes for RARs that might mediate the morphogenetic effects of RA, it is natural to consider genes whose identity and expression patterns make them good candidates to participate in axial specification. At the moment the most attractive possibilities are found in another class of transcriptional activators, the homeobox genes, because of their dramatic temporal and spatial regulation in developing and regenerating limbs and the variety of circumstantial evidence that implicates them in morphogenesis during vertebrate development. For example, in amphibian limb development the homeobox gene XLHBox-1 is present in forelimb mesenchyme, but not that of hindlimb, and in the forelimb bud is expressed in a graded fashion, being concentrated in anterior and proximal regions (Oliver *et al.*, 1988a,b). Similar distributions are also observed during limb regeneration; the newt homologue of this gene, NvHBox-1, is expressed at higher level in a proximal limb blastema than a distal one (Savard *et al.*, 1988; Tabin, 1989).

The most dramatic findings on vertebrate homeobox genes have come from an appreciation of the relationship between the clustered organization on the chromosome of the *Hox* genes, and their pattern of expression in development. Thus, for the murine *Hox*-2 and *Hox*-5 complexes, their order from 5' to 3' on the chromosome parallels their extent of expression along the cranial–caudal axis of the embryo (Graham *et al.*, 1989). This analysis has been adapted to the limb bud in a study of the distribution of transcripts for five contiguous members of the *Hox*-5 complex (Dolle *et al.*, 1989a). The spatial patterns that result appear to reflect the systematic activation of members of the complex at progressively more posterior and distal regions of the bud (see also Oliver *et al.*, 1989). Moreover, this progressive activation observes the order of the genes in the complex. One interpretation of these observations is that the temporal and spatial domains reflect the repression of early activated genes by the late members. Although there is no direct evidence that implicates these genes in the unfolding of positional specification in the limb bud, it seems quite likely that they will be so involved, and the general issue of retinoid regulation of *Hox* genes has therefore attracted attention.

There is now evidence in several of the culture models of RA-induced differentiation for the activation of *Hox* genes, and it has recently been found for the *Hox*-2 complex that the genes at the 3' end of a complex are more responsive to RA treatment than are those at the 5' end (Papalopulu *et al.*, 1990), suggesting the possibility of a sequential scheme of gene activation. Whether any of these changes are mediated by direct action of RARs on a homeobox promoter is unclear. This has been suggested for one of the early RA-response genes (ERA-1) of F9 cells (LaRosa and Gudas, 1988a), a gene that was found to encode the *Hox*-1.6 protein (LaRosa and Gudas, 1988b). Nonetheless, the evidence is not yet compelling and inspection of the promoter sequence as published does not reveal the established β-receptor RARE (de The *et al.*, 1990). Although these cases of induction may reflect indirect mechanisms (that is, mediated by intermediary transcription factors), and could even be due to post-transcriptional events (Papalopulu *et al.*, 1990), they would be no less interesting in terms of understanding how RA could exert its morphogenetic effects.

The findings that, on the one hand, RA can modulate the expression of homeobox genes *in vitro* and, on the other hand, that homeobox genes in limb development are distributed in spatial gradients, raise the question whether RA treatment that produces axis respecification also alters levels of homeobox gene expression. At present there are only results for the amphibian homeobox gene HBox-1 discussed earlier. This gene is expressed at higher levels in a proximal (mid-humerus) versus a distal (mid-radius–ulna) blastema (Savard *et al.*, 1988), and by immunohistochemical staining is known to be concentrated in the proximal and anterior mesenchyme of *Xenopus* and mouse limb buds (Oliver *et al.*, 1988a). When the distal blastemas were proximalized by treatment with RA, no increase in the level of the NvHbox-1 transcript was observed in a midbud stage blastema (Savard *et al.*, 1988; Tabin, 1989). It is possible that the level of expression is unrelated to respecification of the PD axis, or alternatively that RA is acting on a 'downstream' target, for example on the levels of the HBox-1 protein. This finding suggests that it will probably be necessary to study as many of these genes as possible in order to identify those members that are regulated in an expected way, and hence are potential targets of RAR action. The expression pattern of HBox-1 is perhaps not the most appropriate target because it is higher at proximal and anterior positions, and hence would have to be induced by RA on the PD axis and repressed on the AP axis (see discussion above). It is clearly necessary to search for additional potential targets, and in particular to identify genes whose expression pattern presents a gradient in the posteroproximal–anterodistal axis.

12.9 Prospects for functional studies in cells and limbs

The crucial test for the function of RARs will come from the ability to manipulate their levels or activity and then analyse the functional consequences. We therefore consider the possible strategies for genetic manipulation using various receptor constructions, and then the problem of introducing such constructions into cells that are participating in morphogenesis. For a recent study, anti-sense oligonucleotides were synthesized for the start codon and 12

Fig. 7 Expression of β-galactosidase activity establishes that cultured newt limb cells can be transfected with exogenous DNA. Methods: cells cultured from newt limb explants and subsequently passaged were microinjected with a reporter plasmid containing the β-galactosidase gene modified to include a nuclear localization signal and placed under the control of the SV40 early promoter plus enhancer. Enzyme activity was demonstrated by X-gal staining.

bases upstream of the human RAR-α (Cope and Wille, 1989). The addition of these reagents to the culture medium of malignant human keratinocytes resulted in a profound drop in RAR-α mRNA levels, and a block in the induction by retinol of alkaline phosphatase activity, suggesting that further developments of this approach might be fruitful. A second study has investigated the use of dominant negative receptor constructions that derive from our understanding of the modular nature of function in hormone nuclear receptors (Espeseth et al., 1989). F9 cells were stably transfected with a vector expressing the A, B, C and D (partial) regions of the human RAR-α, fused to β-galactosidase as a marker. The fusion protein was expressed in the nucleus, and at particularly high levels in clones that were concomitantly selected for resistance to RA-induced differentiation. The functional consequences of this expression were that the clones no longer exhibited retinoid induction of keratin, laminin B-1 or plasminogen activator mRNA, but did show induction of type IV collagen and *Hox*-1.3 mRNA. The most likely interpretation of this dominant negative effect is that the partial receptor competed with native receptor in dimer formation, or for DNA or transcription activator binding sites. Because the partial receptor lacks an activating ligand-binding region, it served as a repressor that blocked the function of normal receptors in the F9 cells. The basis for the genetic specificity observed with the RAR-α fusion protein is not clear, but one attractive possibility is that the collagen and *Hox*-1.3 induction may be mediated by other receptors, for example RAR-γ, whose function was not compromised by the α construction. In general there appear to be a number of alternative stratagies to explore for constructions that could interfere with function, and we can anticipate substantial progress in this area.

These approaches are immediately applicable to the culture models of retinoid action, but the greater challenge is to extend them to the analysis of developmental events *in vivo*, including the mechanisms of limb morphogenesis. Unfortunately, the development of the limb bud is a relatively late event, and it is likely that many of the genes such as RARs that might be involved in pattern formation are expressed in other locations of the embryo at earlier times. It therefore seems likely that a transgenic approach requires the isolation of specific promoters that can target expression of receptor constructions to the limb. Such an approach should give important insights into the role of retinoic acid in morphogenesis.

The circumstances of limb regeneration have engendered a different approach. It is possible to establish cultures of limb blastemal cells that continue to divide without any obvious crisis or senescence (Ferretti and Brockes, 1988). These cells can be marked by transfection (Fig. 7) and then introduced back into the limb. After amputation the marked cells are recruited into the resulting blastema and contribute to the regenerate. The approach, then, is to analyse the consequence of cotransfecting engineered RAR genes on the location, proliferation, and fate of the marked blastemal cells. Even if cells established from different positions on the PD axis 'forget' their origin after the period in culture, it may still be possible to control this variable with the appropriate nuclear receptor constructions.

12.10 Conclusions

The existence of multiple nuclear receptors for retinoids may reflect the need for distinct genetic mechanisms to implement the diverse effects produced by retinoid treatment. While this complexity may appear to be an encumbrance to research, the fact that it arises at the level of the receptor could prove a great advantage in the long term. We have focused in this review on urodele limb regeneration because it offers special promise for sorting out the functional roles of the various RARs. The full range of effects produced by RA treatment, including pattern duplications, skeletal truncations and deletions, and pronounced histological effects including cartilage breakdown, stimulation of mucous metaplasia in the epidermis and inhibition of cell division in the blastema (Maden, 1983), can be observed in the regenerating limb, and already three functional RARs have been identified in the newt blastema. Both distributional studies and the transgenic limb may help assign candidate functions among the RARs involved in limb regeneration.

Retinoids share with steroid and thyroid hormones classic hormonal effects on development and differentiation as well as multiple receptors within the nuclear hormone superfamily. With the rapid pace of research into the structure, function and targets of the RARs, the greatest uncertainties in retinoid physiology lie in understanding the identity, distribution, synthesis and degradation of the ligands themselves. This research now becomes particularly important given that receptor heterogeneity could be explained in part by multiple retinoid ligands. Finally, the unique properties of RA as a morphogenic agent, including the need for both threshold and concentration gradient responses (Wolpert, 1969), may offer more general insights into novel mechanisms for regulating transcription.

Acknowledgements

We thank Drs P. Chambon, D. Stocum and C. Tickle, who kindly provided us with illustrations of their work. C.W.R. was supported by the National Multiple Sclerosis Society.

References

Alles, A. J. and Sulik, K. K. (1989). Retinoic-acid-induced limb-reduction defects: perturbation of zones of programmed cell death as a pathogenetic mechanism. *Teratology* **40**, 163–171.

Benbrook, D., Lernhardt, E. and Pfahl, M. (1988). A new retinoic acid receptor identified from a hepatocellular carcinoma. *Nature* **333**, 669–672.

Bentley, D. L. and Groudine, M. (1986). A block to elongation is largely responsible for decreased transcription of c-myc in differentiated HL60 cells. *Nature* **321**, 702–706.

Bocquel, M. T., Kumar, V., Stricker, C., Chambon, P. and Gronemeyer, H. (1989). The contribution of the N- and C-terminal regions of steroid receptors to activation of transcription is both receptor and cell-specific. *Nucl. Acids Res.* **17**, 2581–2595.

Brand, N. J., Petkovich, M., Krust, A., Chambon, P., de The, H., Marchio, A., Tiollais, P. and Dejean, A. (1988). Identification of a second human retinoic acid receptor. *Nature* **332**, 850–853.

Breitman, T. R., Selonick, S. E. and Collins, S. J. (1980). Induction of differentiation of the human promyelocytic leukemia cell line (HL-60) by retinoic acid. *Proc. Natl. Acad. Sci. USA* **77**, 2935–2940.

Brockes, J. P. (1989). Retinoids, homeobox genes, and limb morphogenesis, *Neuron* **2**, 1285–1294.

Colberg-Poley, A. M., Puschel, A. W., Dony, C., Voss, S. D. and Gruss, P. (1987). Post-transcriptional regulation of a murine homeobox gene transcript in F9 embryonal carcinoma cells. *Differentiation* **35**, 206–211.

Cope, F. O. and Wille, J. J. (1989). Retinoid receptor antisense DNAs inhibit alkaline phosphatase induction and clonogenicity in malignant keratinocytes. *Proc. Natl. Acad. Sci. USA* **86**, 5590–5594.

Crick, F. H. C. (1970). Diffusion in embryogenesis. *Nature* **225**, 420–422.

Damm, K., Thompson, C. C. and Evans, R. M. (1989). Protein encoded by v-*erb*A functions as a thyroid-hormone receptor antagonist. *Nature* **339**, 593–597.

de The, H., Marchio, A., Tiollais, P. and Dejean, A. (1987). A novel steroid thyroid hormone receptor-related gene inappropriately expressed in human hepatocellular carcinoma. *Nature* **330**, 667–670.

de The, H., Marchio, A., Tiollais, P. and Dejean, A. (1989). Differential expression and ligand regulation of the retinoic acid receptor alpha and beta genes. *EMBO J.* **8**, 429–433.

de The, H., del Mar Vivanco-Ruiz, M., Tiollais, P., Stunnenberg, H. and Dejean, A. (1990). Identification of a retinoic acid responsive element in the retinoic acid receptor B gene. *Nature* **343**, 177–180.

Dolle, P., Izposua-Belmonte, J.-C., Falkenstein, H., Renucci, A. and Duboule, D. (1989a). Coordinate expression of the murine *Hox*-5 complex homeobox-containing genes during limb pattern formation. *Nature* **342**, 767–772.

Dolle, P., Ruberte, E., Kastner, P., Petkovich, M., Stoner, C. M., Gudas, L. J. and Chambon, P. (1989b). Differential expression of genes encoding alpha, beta and gamma retinoic acid receptors and CRABP in the developing limbs of the mouse. *Nature* **342**, 702–705.

Dony, C., Kessel, M. and Gruss, P. (1985). Post-transcriptional control of myc and p53 expression during differentiation of the embryonal carcinoma cell line F9. *Nature* **317**, 636–639.

Dowling, J. E. and Wald, G. (1960). The biological function of vitamin A acid. *Proc. Natl. Acad. Sci. USA* **46**, 587–608.

Edwards, M. K. S. and McBurney, M. W. (1983). The concentration of retinoic acid determines the differentiated cell types formed by a teratocarcinoma cell line. *Dev. Biol.* **98**, 187–191.

Eichele, G. and Thaller, C. (1987). Characterization of concentration gradients of a morphogenetically active retinoid in the chick limb bud. *J. Cell Biol.* **105**, 1917–1923.

Eichele, G., Tickle, C. and Alberts, B. M. (1985). Studies on the mechanism of retinoid-induced duplications in the early chick limb bud: temporal and spatial aspects. *J. Cell Biol.* **101**, 1913–1920.

Espeseth, A. S., Murphy, S. P. and Linney, E. (1989). Retinoic acid receptor expression vector inhibits differentiation of F9 embryonal carcinoma cells. *Genes Dev.* **3**, 1647–1656.

Ferretti, P. and Brockes, J. P. (1988). Culture of newt cells from different tissues and their expression of a regeneration-associated antigen. *J. Exp. Zool.* **247**, 77–91.

Funder, J. W., Pearce, P. T., Smith, R. and Smith, A. I. (1988). Mineralocorticoid action: target tissue specificity is enzyme, not receptor, mediated. *Science* **242**, 583–585.

Gaub, M. P., Lutz, Y., Ruberte, E., Petkovich, M., Brand, N. J. and Chambon, P. (1989). Antibodies specific to the retinoic acid human nuclear receptors alpha and beta. *Proc. Natl. Acad. Sci. USA* **86**, 3089–3093.

Giguere, V., Ong, E. S., Segui, P. and Evans, R. M. (1987). Identification of a receptor for the morphogen retinoic acid. *Nature* **330**, 624–629.

Giguere, V., Ong, E. S., Evans, R. M. and Tabin, C. J. (1989). Spatial and temporal expression of the retinoic acid receptor in the regenerating amphibian limb. *Nature* **337**, 566–569.

Glass, C. K., Lipkin, S. M., Devary, O. V. and Rosenfeld, M. G. (1989). Positive and negative regulation of gene transcription by a retinoic acid-thyroid hormone receptor heterodimer. *Cell* **59**, 697–708.

Goodman, D. S. (1987). Retinoids and retinoid-binding proteins. In "The Harvey Lectures, Series 81", pp. 111–132. Alan R. Liss, New York.

Graham, A., Papalopulu, N. and Krumlauf, R. (1989). The murine and *Drosophila* homeobox gene complexes have common features of organization and expression. *Cell* **57**, 367–378.

Graupner, G., Wills, K., Tzukerman, M., Zhang, X.-K. and Pfahl, M. (1989). Dual regulatory role for thyroid-hormone receptors allows control of retinoic-acid receptor activity. *Nature* **340**, 653–656.

Green, S., Kumar, V., Theulaz, I., Wahli, W. and Chambon, P. (1988). The N-terminal DNA-binding 'zinc finger' of the oestrogen and glucocorticoid receptors determines target gene specificity. *EMBO J.* **7**, 3037–3044.

Griep, A. E. and Westphal, H. (1988). Antisense Myc sequences induce differentiation of F9 cells. *Proc. Natl. Acad. Sci. USA* **85**, 6806–6810.

Guiochon-Martel, A., Loosfelt, H., Lescop, P., Sar, S., Atger, M., Perrot-Applanat, M. and Milgrom, E. (1989). Mechanisms of nuclear localization of the progesterone receptor: evidence for interaction between monomers. *Cell* **57**, 1147–1154.

Hu, L. and Gudas, L. J. (1990). Cyclic AMP analogs and retinoic acid influence the expression of retinoic acid receptor alpha, beta and gamma mRNAs in F9 teratocarcinoma cells. *Mol. Cell. Biol.* **10**, 391–396.

Ide, H. and Aono, H. (1988). Retinoic acid promotes proliferation and chondrogenesis in the distal mesodermal cells of chick limb bud. *Dev. Biol.* **130**, 767–773.

Keeble, S. and Maden, M. (1984). The relationship among retinoid structure, affinity for retinoic acid-binding protein, and ability to respecify pattern in the regenerating axolotl limb. *Dev. Biol.* **132**, 26–34.

Kim, K. H., Stellmack, V., Javors, J. and Fuchs, E. (1987). Regulation of human mesothelial cell differentiation: opposing roles of retinoids and epidermal growth factor in the expression of intermediate filament proteins. *J. Cell Biol.* **105**, 3039–3051.

Kim, W.-S. and Stocum, D. L. (1986). Retinoic acid modifies positional memory in the anteroposterior axis of regenerating axolotl limbs. *Dev. Biol.* **114**, 170–179.

Kochhar, D. M. (1977). Cellular basis of congenital limb deformity induced in mice by vitamin A. In "Morphogenesis and Malformation of the Limb" (eds D. Bergsma and W. Lenz), pp. 111–154. Alan R. Liss, New York.

Kochhar, D. M. (1989). Response to Kalter. *Teratology* **39**, 615–616.

Krust, A., Green, S., Argos, P., Kumar, V., Walter, P., Bornert, J.-M. and Chambon, P. (1986). The chicken oestrogen receptor sequence: homology with v-*erb*A and the human oestrogen and glucocorticoid receptors. *EMBO J.* **5**, 891–897.

Krust, A., Kastner, P., Petkovich, M., Zelent, A. and Chambon, P. (1989). A third human retinoic acid receptor, hRAR-gamma. *Proc. Natl. Acad. Sci. USA* **86**, 5310–5314.

LaRosa, G. J. and Gudas, L. J. (1988a). An early effect of retinoic acid: cloning of an mRNA (ERA-1) exhibiting rapid protein synthesis-independent induction during teratocarcinoma stem cell differentiation. *Proc. Natl. Acad. Sci. USA* **85**, 329–333.

LaRosa, G. J. and Gudas, L. J. (1988b). Early retinoic acid-induced F9 teratocarcinoma stem cell gene ERA-1: alternate splicing creates transcripts for a homeobox-containing protein and one lacking a homeobox. *Mol. Cell. Biol.* **8**, 3906–3917.

Lewis, C. A., Pratt, R. M., Pennypacker, J. P. and Hassel, J. R. (1978). Inhibition of limb chondrogenesis *in vitro* by vitamin A: alterations in cell surface characteristics. *Dev. Biol.* **64**, 31–47.

Lewis, J. H. and Wolpert, L. (1976). The principle of non-equivalence in development. *J. Theoret. Biol.* **62**, 579–590.

Maden, M. (1982). Vitamin A and pattern formation in the regenerating limb. *Nature* **295**, 672–675.

Maden, M. (1983). The effect of vitamin A on the regenerating axolotl limb. *J. Embryol. Exp. Morph.* **77**, 273–295.

Maden, M., Ong, D. E., Summerbell, D. and Chytil, F. (1988). Spatial distribution of cellular protein binding to retinoic acid in the chick limb bud. *Nature* **335**, 733–735.

Napoli, J. L. and Race, K. R. (1987). The biosynthesis of retinoic acid from retinol by rat tissues *in vitro*. *Arch. Biochem. Biophys.* **255**, 95–101.

Newman, S. A. and Frisch, H. L. (1979). Dynamics of skeletal pattern formation in developing chick limb. *Science* **205**, 662–668.

Niazi, I. A. and Saxena, S. (1978). Abnormal hind limb regeneration in tadpoles of the toad, *Bufo andersoni*, exposed to excess vitamin A. *Folia Biol. (Krakow)* **26**, 3–11.

Oliver, G., Wright, C. V. E., Hardwicke, J. and De Robertis, E. M. (1988a). A gradient of homeodomain protein in developing forelimbs of *Xenopus* and mouse embryos. *Cell* **55**, 1017–1024.

Oliver, G., Wright, C. V. E., Hardwicke, J. and De Robertis, E. M. (1988b). Differential antero-posterior expression of two proteins encoded by a homeobox gene in *Xenopus* and mouse embryos. *EMBO J.* **7**, 3199–3209.

Oliver, G., Sidell, N., Fiske, W., Heinzmann, C., Mohandas, T., Sparkes, R. S. and De Robertis, E. M. (1989). Complementary homeo protein gradients in developing limb buds. *Genes Dev.* **3**, 641–650.

Papalopulu, N., Hunt, P., Wilkinson, D., Graham, A. and Krumlauf, R. (1990). Hox-2 homeobox genes and retinoic acid: potential roles in patterning the vertebrate nervous system. In "Advances in Neural Regeneration Research" (ed F. J. Seil). In press. Alan R. Liss, New York.

Paulsen, D. F., Langille, R. M., Dress, V. and Solrush, M. (1988). Selective stimulation of *in vitro* limb-bud chondrogenesis by retinoic acid. *Differentiation* **39**, 123–130.

Petkovich, M., Brand, N. J., Krust, A. and Chambon, P. (1987). A human retinoic acid receptor which belongs to the family of nuclear receptors. *Nature* **330**, 444–450.

Picard, D. and Yamamoto, K. R. (1987). Two signals mediate hormone-dependent nuclear localization of the glucocorticoid receptor. *EMBO J.* **6**, 3333–3340.

Pitt, G. A. J. (1971). Vitamin A. In "Carotenoids" (ed. O. Isler), pp. 717–742. Birkhauser-Verlag, Basel and Stuttgart.

Ragsdale, C. W., Petkovich, M., Gates, P. B., Chambon, P. and Brockes, J. P. (1989). Identification of a novel retinoic acid receptor in regenerative tissues of the newt. *Nature* **341**, 654–657.

Rees, J. L., Daly, A. K. and Redfern, C. P. F. (1989). Differential expression of the alpha and beta retinoic acid receptors in tissues of the rat. *Biochem. J.* **259**, 917–919.

Ruberte, E., Dolle, P., Krust, A., Zelent, A., Morriss-Kay, G. and Chambon, P. (1990). Specific spatial and temporal distribution of retinoic acid receptor gamma transcripts during mouse embryogenesis. *Development* **108**, 213–222.

Satre, M. A. and Kochhar, D. M. (1989). Elevations in the endogenous levels of the putative morphogen retinoic acid in embryonic mouse limb buds associated with limb dysmorphogenesis. *Dev. Biol.* **133**, 529–536.

Saunders, J. W. and Gasseling, M. T. (1968). Ectodermal–mesenchymal interactions in the origin of limb symmetry. *In* "Epithelial–Mesenchymal Interactions" (eds R. Fleischmajer and R. E. Billingham), pp. 78–97. Williams and Wilkins, Baltimore, MD.

Savard, P., Gates, P. B. and Brockes, J. P. (1988). Position dependent expression of a homeobox gene transcript in relation to amphibian limb regeneration. *EMBO J.* **7**, 4275–4282.

Scadding, R. R. and Maden, M. (1986a). Comparison of the effects of vitamin A on limb development and regeneration in the axolotl, *Ambystoma mexicanum*. *J. Embryol. Exp. Morph.* **91**, 19–34.

Scadding, S. R. and Maden, M. (1986b). Comparison of the effects of vitamin A on limb development and regeneration in *Xenopus laevis* tadpoles. *J. Embryol. Exp. Morph.* **91**, 35–53.

Song, S. and Siu, C.-H. (1989). Retinoic acid regulation of the expression of retinoic acid receptors in wild-type and mutant embryonal carcinoma cells. *FEBS Lett.* **256**, 51–54.

Strickland, S. and Mahdavi, V. (1978). The induction of differentiation in teratocarcinoma stem cells by retinoic acid. *Cell* **15**, 393–403.

Strickland, S. (1979). Induction of differentiation by retinoids. *In* "Hormones and Cell Culture" (eds G. H. Sato and R. Ross), pp. 671–676. Cold Spring Harbor Laboratory, Cold Spring Harbor.

Summerbell, D. (1983). The effect of local application of retinoic acid to the anterior margin of the developing chick limb. *J. Embryol. Exp. Morph.* **78**, 269–289.

Tabin, C. J. (1989). Isolation of potential vertebrate limb-identity genes. *Development* **105**, 813–820.

Thaller, C. and Eichele, G. (1987). Identification and spatial distribution of retinoids in the developing chick limb bud. *Nature* **327**, 625–628.

Thaller, C. and Eichele, G. (1988). Characterisation of retinoid metabolism in the developing chick limb bud. *Development* **103**, 473–483.
Thompson, J. N., Howell, J. M. and Pitt, G. A. J. (1964). Vitamin A and reproduction in rats. *Proc. R. Soc. London. B* **159**, 510–535.
Tickle, C., Summerbell, D. and Wolpert, L. (1975). Positional signalling and specification of digits in chick limb morphogenesis. *Nature* **254**, 199–202.
Tickle, C., Alberts, B., Wolpert, L. and Lee, J. (1982). Local application of retinoic acid to the limb bud mimics the action of the polarising region. *Nature* **296**, 564–565.
Tickle, C., Lee, J. and Eichele, G. (1985). A quantitative analysis of the effect of all-trans-retinoic acid on the pattern of chick wing development. *Dev. Biol.* **109**, 82–95.
Tora, L., Gronemeyer, H., Turcotte, B., Gaub, M.-P. and Chambon, P. (1988). The N-terminal region of the chicken progesterone receptor specifies target gene activation. *Nature* **333**, 185–188.
Umesono, K., Giguere, V., Glass, C. K., Rosenfeld, M. G. and Evans, R. M. (1988). Retinoic acid and thyroid hormone induce gene expression through a common responsive element. *Nature* **336**, 262–265.
Vasios, G. W., Gold, J. D., Petkovich, M., Chambon, P. and Gudas, L. J. (1989). A retinoic acid-responsive element is present in the 5' flanking region of the laminin B1 gene. *Proc. Natl. Acad. Sci. USA* **86**, 9099–9103.
Wallace, H. (1981). "Vertebrate Limb Regeneration". John Wiley, Chichester.
Wolpert, L. (1969). Positional information and the spatial pattern of cellular differentiation. *J. Theoret. Biol.* **25**, 1–47.
Wolpert, L. and Hornbruch, A. (1987). Positional signalling and the development of the humerus in the chick limb bud. *Development* **100**, 333–338.
Zelent, A., Krust, A., Petkovich, M., Kastner, P. and Chambon, P. (1989). Cloning of murine retinoic acid receptor alpha and beta cDNAs and of a novel third receptor gamma predominantly expressed in skin. *Nature* **339**, 714–717.

Note added in proof: Mangelsdorf *et al.* (1990, *Nature* **345**, 224–229) have characterized a novel human nuclear receptor, identified as the retinoid X receptor (RXR-α), that is distantly related to the RARs (amino acid sequence identity to the human RAR-α: region C, 61%; region E, 28%). Transfection experiments establish that the RXR-α can mediate RA-dependent transcription activation at high ligand concentrations (half-maximal response at 5×10^{-8} M RA). The synthetic retinoid TTNPB, which is more potent than RA in producing duplications in developing and regenerating limbs, is a very poor transactivator of the human RXR-α, suggesting that this nuclear receptor is not a primary mediator of retinoid effects on position specification. Thaller and Eichele (1990, *Nature* **345**, 815–819) have identified the A_1 peak as 3,4-didehydroretinoic acid. This retinoid is present in the chick limb at six times the concentration of RA and is as effective as RA in inducing AP duplications. Its status as a ligand for the RARs has not been reported. The sequential activation of *Hox*-2 genes by RA has also been described in a human embryonal carcinoma cell line by Simeone *et al.* (1990, *Nature* **346**, 763–766).

13

The Role of Steroid Hormones and Growth Factors in the Control of Normal and Malignant Breast

R. CLARKE, R. B. DICKSON and M. E. LIPPMAN

Lombardi Cancer Research Center, Georgetown University Medical Center, 3800 Reservoir Road NW, Washington, DC 20007, USA

13.1 Introduction

A number of hormones and growth factors have been implicated in mammary development, insulin, the steroids and various pituitary factors being most closely associated with these processes. Steroids can modulate the metabolism of cells by both genomic and non-genomic mechanisms. Non-genomic mechanisms occur in either the absence or presence of a specific receptor protein but within a few minutes of exposure to hormone. These effects are almost exclusively observed at pharmacological concentrations and do not directly involve alteration in gene expression (for an extensive review see Duval *et al.*, 1983). Genomic mediated effects of steroids, including induction of mitogenesis, cannot generally be detected less than 30 min after hormone treatment and follow significant perturbations in the level of expression of a number of genes (Duval *et al.*, 1983). We shall discuss the role of steroid hormones in the genomic growth control of both normal and neoplastic breast tissue, with particular reference to the potential ability of growth factors to function as mediators of steroid hormone action.

13.1.1 Steroid Hormones in Normal and Malignant Breast Tissue

Oestrogen, progesterone and hydrocortisone are amongst the most influential hormones involved in the induction of growth and differentiation in normal mammary tissue. In the human breast, DNA synthesis in epithelial cells appears to reach its maximum during the luteal phase of the menstrual cycle when progesterone levels are highest and oestrogen levels lowest (Ferguson and Anderson, 1981). Oestrogens increase nipple differentiation and induce proliferation of the surrounding mesenchyme in the developing mouse fetus (Raynaud, 1955; see Chapter 11.10). Both oestrogen and progesterone increase alveolar formation and ductal branching (Murr et al., 1974; Bronson et al., 1975). Priming by exposure to both oestrogen and progesterone is required for full lobulo-alveolar development in normal breast tissue (Warner, 1978). The ductal and lobulo-alveolar phases of development are induced by oestradiol in ovariectomized mice (Lyons, 1958; Lieberman et al., 1978).

Oestrogen is the hormone most closely associated with the control of breast tumour growth. Beatson (1896) first reported remissions in premenopausal breast cancer patients following bilateral ovariectomy. Lacassagne (1932) chronically administered an ovarian extract to castrated male mice and subsequently induced mammary tumour formation. Agents which function primarily as competitive inhibitors of oestrogen receptor (ER) function, referred to as anti-oestrogens (see Chapter 2.11 and 16.7), inhibit the growth of breast tumours expressing these receptors (King et al., 1982; Clark and McGuire, 1988). Pharmacological concentrations of anti-oestrogens inhibit, whilst physiological concentrations of oestrogen stimulate, the proliferation in vitro of human breast cancer cells expressing ER (Lippman et al., 1976; Aitken and Lippman, 1982). These cell lines require oestrogen for tumour formation in athymic nude mice (Soule and McGrath, 1980; Shafie and Grantham, 1981; Seibert et al., 1983). In contrast, pharmacological concentrations of oestrogens such as diethylstilboestrol can inhibit some oestrogen-dependent breast tumours, with response rates of up to 38% being achieved in patients not selected for treatment on the basis of the ER content of their tumours (Henderson and Canellos, 1980). The mechanism of action of high concentrations of steroids is unclear but non-specific effects mediated by changes in plasma membrane fluidity appear to be involved (Clarke et al., 1987).

Exposure to oestrogens has been closely associated with endometrial and breast cancers (Jick et al., 1980; Thomas, 1984), vaginal adenocarcinoma (Herbst et al., 1971) and liver tumours (Vana et al., 1979). The precise role of oestrogens in the genesis of breast cancer is unclear. The incidence of rat mammary tumours induced by 7,12-dimethylbenz(a)anthracene is significantly reduced if oestrogens are removed before or shortly after administration of the drug (Dao, 1972). In women, breast cancer occurs almost exclusively after puberty (Akhtar et al., 1983). Primary ovarian failure reduces the incidence of breast cancer to that observed in men. The ability of oestrogens to influence the rate of proliferation, atrophy or differentiation of stem or intermediate cells in normal breast, to stimulate the rate of proliferation of breast cancer cells in vitro, and of ovariectomy and anti-oestrogens to induce remissions in some patients, may reflect their function as tumour promoters rather than mutagens (Thomas, 1984). Progestins may also be involved in mammary carcinogenesis, since the number of mitoses observed in normal breast epithelial cells is greatest during the luteal phase of the menstrual cycle when progesterone levels are highest and oestrogen levels lowest (Ferguson and Anderson, 1981). Whilst progestins and other hormones may contribute to hormonal carcinogenesis, it is generally believed that the mitogenic effects of a cumulative exposure to oestrogens is the primary aetiological factor in breast cancer (Henderson et al., 1988).

13.1.2 Steroid Hormone Receptors in Breast Cancer

The ability of a cell to elicit a response following stimulation with a hormone is largely the result of an interaction between the relevant ligand and a specific, high affinity, low capacity receptor. The presence of receptors may therefore indicate a specific requirement of the cell for a hormone or growth factor. ER and progesterone receptors (PGR) are the most important hormone receptors associated with breast cancer. The majority of breast tumours which respond to anti-oestrogens express ER (Carter, 1981; Clarysse, 1985). However, a combined measurement of both ER and PGR increases the ability to predict initial response to endocrine therapy to greater than 60% (King et al., 1982; Stewart et al., 1982; Clark and McGuire, 1988). Receptors for other steroid hormones have also been demonstrated in breast cancer cells including receptors for glucocorticoids and dihydrotestosterone (Horwitz et al., 1975).

In contrast to its ability to predict response to endocrine manipulation, attempts to correlate ER content with either tumour size (Heuson and Benraad, 1983) or the time of onset or incidence of distant metastases (Campbell et al., 1981; Kamby et al., 1988) have been largely unsuccessful. Low ER levels and a high degree of anaplasia are frequently associated with visceral metastases, whilst tumours expressing high levels of ER often metastasize to bone (Kamby et al., 1988). Histological grade III tumours are more likely to be ER negative than either grade I or grade II tumours (Blanco et al., 1984; Singh et al., 1988; Henry et al., 1988).

A number of investigators have observed a close correlation between ER expression and tumour growth rate. Tumours with low or undetectable levels of ER proliferate more rapidly than tumours expressing high levels of ER (Meyers et al., 1977; Jonat and Maat, 1978; Antoniades and Spector, 1982). The low levels of circulating oestrogens present in post-menopausal women may not be sufficient to provide optimal growth in hormone-dependent tumours. This hypothesis gains support from recent observations obtained with the MCF-7 hormone-dependent breast cancer cell line which requires oestrogen supplementation for growth in ovariectomized athymic nude mice. These mice have levels of circulating oestrogens comparable to post-menopausal women (Seibert et al., 1983). We have established sublines of MCF-7 which are capable of forming proliferating tumours in ovariectomized athymic nude mice (Clarke et al., 1989a,b). Whilst not requiring oestrogen supplementation for growth, proliferation of these tumours is significantly increased by oestrogen supplementation (Clarke et al., 1989b). These cells also retain sensitivity to anti-oestrogens (Clarke et al., 1989c).

13.2 The autocrine hypothesis of steroid hormone function

The post-receptor binding mechanisms through which steroid hormones elicit mitogenic or biological responses in target tissues is unclear, but clearly must involve alterations in gene expression. From the perspective of breast cancer, the induction of PGR synthesis following stimulation with oestradiol is used as an indicator of a functional receptor system. Tumours expressing both ER and PGR elicit a greater response rate to anti-hormonal therapy than tumours expressing neither of these receptors or either receptor alone (King et al., 1982; Clark and McGuire, 1988). Whilst fulfilling an important practical function, induction of PGR synthesis is unlikely to mediate the mitogenic effects of oestradiol. Considerable effort has thus been expended in determining the ability of steroid hormones to regulate the expression of genes associated with cellular proliferation. De Larco and Todaro (1978) have suggested that some tumour cells may produce the factors they require for continued proliferation. These factors could subsequently function in an autostimulatory (autocrine) manner. For hormone-dependent breast cancer, this has included investigation of steroid-induced alterations in the level of expression of mitogenic growth factors, inhibitory growth factors and oncogenes (Lippman et al., 1986). Breast cancer cells possess receptors for a number of mitogenic growth factors and also secrete significant amounts of the appropriate ligands for these growth factor receptors. Hence, an autocrine stimulatory mechanism may control breast tumour proliferation (Fig. 1). This autocrine loop could function externally—the cells could secrete the ligand which could then bind to its receptor on the surface of the cell from which it was secreted (Lippman et al., 1986). An internal autocrine loop may also function with the ligand–receptor interactions occurring intracellularly, perhaps at the endoplasmic reticulum–Golgi complexes or within secretory vesicles (Browder et al., 1989).

For hormone-dependent breast cancer cells, the autocrine hypothesis would predict that the levels of mitogenic growth factors produced would

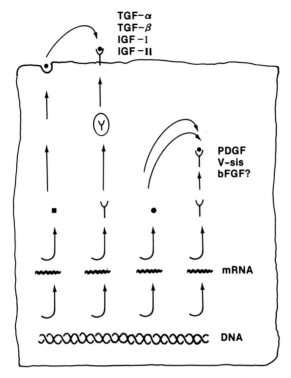

Fig. 1 The autocrine hypothesis of oestrogen stimulation. Both ligand and receptor are produced by the same cell. The receptor–ligand interaction occurs either at the cell surface (e.g. IGF-I) or intracellulary, perhaps at the Golgi or within secretory vesicles (e.g. PDGF).

increase and inhibitory growth factors would decrease on stimulation with oestradiol. Exposure to anti-oestrogens might increase the production of inhibitory factors, whilst decreasing production of mitogens. Hormone-independent cells would constitutively express high levels of mitogenic ligands in the absence of hormone and would not respond to hormonal stimulation. The constitutive expression of the appropriate mitogen or mitogens in a hormone-dependent cell, for example following transfection with plasmid expression vectors, would be expected to induce the hormone-independent phenotype. The ability of oestrogen to regulate the secretion of a number of growth factors has been reported. These include the mitogens epidermal growth factor (EGF), transforming growth factor-α (TGF-α) and other TGF-α-like proteins, the insulin-like growth factors (IGF) and fibroblast growth factors. Direct evidence of the role of these oestrogen-stimulated factors was obtained by Dickson *et al.* (1987), who observed the ability of the conditioned cell culture medium obtained from MCF-7 cells growing *in vitro* to partly support the growth of these oestrogen-dependent cells in castrated athymic nude mice. Regulation of the inhibitory growth factor TGF-β has also been investigated and will be discussed in detail.

13.3 The role of growth factors in normal and malignant breast

13.3.1 Epidermal Growth Factor/Transforming Growth Factor-α

EGF and its naturally occurring homologue TGF-α have both been implicated in the control of normal breast development and in maintaining the malignant phenotype. Both ligands function through their interactions with the same receptor, conventionally designated the EGF receptor (EGF-R). TGF-α and EGF have been isolated from human milk (Zwiebel *et al.*, 1986). Human mammary epithelial cells require EGF for continued growth *in vitro*, especially at low density (Stampfer *et al.*, 1980). Removal of the submaxillary glands, the main source of circulating EGF, reduces milk production in mice. The lactating mice have smaller mammary tissue than sham-operated mice, indicating a possible role for EGF in the development of mammary tissue during pregnancy (Oka *et al.*, 1987). EGF stimulates cell differentiation and division during the proliferative stages of alveolar differentiation (Turkington, 1971). In common with oestrogen and progesterone, both EGF and TGF-α stimulate lobulo-alveolar development *in vitro* but TGF-α appears to be more potent than EGF. TGF-α stimulates breast epithelial growth in C3H/HeN mice without steroid supplementation, although maximal growth is observed in the presence of steroids (Vondehaar, 1988).

Approximately 70% of human breast tumours express TGF-α mRNA (Bates *et al.*, 1988) and a number of breast cancer cell lines have been reported to produce both TGF-α mRNA and protein (Perroteau *et al.*, 1986; Peres *et al.*, 1987; Bates *et al.*, 1988; Clarke *et al.*, 1989d). TGF-α-like material has been isolated from the urine of breast cancer patients (Sherwin *et al.*, 1983; Stromberg *et al.*, 1987). The urines of both normal and tumour-bearing mice also contain TGF-α-like material (Twardzik *et al.*, 1985). The proliferation of neoplastic rodent (Turkington, 1969) and human breast cells is stimulated by EGF/TGF-α (Barnes and Sato, 1978; Osborne *et al.*, 1980; Imai *et al.*, 1982). Sialoadenectomy (removal of the salivary glands), reduces the levels of circulating EGF and can reduce the incidence of mammary tumours in some strains of mice (Oka *et al.*, 1987; Tsutsumi *et al.*, 1987). Infusion of EGF can increase the ability of oestrogen-dependent MCF-7 human breast cancer cells to form small tumours in ovariectomized nude mice (Dickson et al., 1987).

Secretion of TGF-α is regulated by oestrogen in most hormone-dependent human breast cancer cell lines. Induction of TGF-α mRNA expression occurs within hours of treatment with physiological concentrations of oestradiol (Bates et al., 1988). EGF/TGF-α partially reverse the growth inhibitory effects of the anti-oestrogens 4-hydroxytamoxifen, tamoxifen, 4-hydroxyclomiphene and LY117018 (Koga and Sutherland, 1987). Furthermore, EGF-stimulated cell proliferation in MCF-7 cells growing in the absence of oestrogen is inhibited by anti-oestrogens (Vignon et al., 1987). Thus, TGF-α may be involved in mediating the ability of oestrogen to induce a mitogenic response in these cells.

Hormone-independent cells have been reported to constitutively secrete TGF-α (Perroteau et al., 1986; Peres et al., 1987; Bates et al., 1988) and to express high levels of EGF-R (Fitzpatrick et al., 1984; Davidson et al., 1987). There appears to be an inverse relationship between ER and EGF-R expression in malignant breast tissue. Primary breast tumours (Perez et al., 1984; Sainsbury et al., 1985; Cattoretti et al., 1988; Pekonen et al., 1988; Wrba et al., 1988) and human breast cancer cell lines (Fitzpatrick et al., 1984; Davidson et al., 1987) which have either low ER content or have lost the ability to express ER possess high levels of EGF-R (Sainsbury et al., 1985). The up-regulation of EGF-R in tumours which no longer require oestrogen further implicates the EGF-R/TGF-α autocrine loop in the progression to hormone-independence in breast cancer.

13.3.2 Effects of Constitutive Transforming Growth Factor-α Expression in MCF-7 Cells

In an attempt to clarify the role of TGF-alpha in mediating the effects of oestrogens, we transfected hormone-dependent MCF-7 cells with a plasmid expression vector directing the constitutive production of TGF-α (Clarke et al., 1989d). Transfectants secrete levels of biologically active TGF-α equivalent to or greater than both hormone-independent MDA-MB-231 cells and oestrogen-stimulated parental MCF-7 cells. One clone designated H8 secretes sufficient TGF-α to fully down-regulate EGF-R. Detectable levels of EGF-R were observed only following treatment with suramin, a polyanion capable of removing ligand from EGF-R (Coffey et al., 1987). Transfectants retain 'normal' responses to oestrogen as determined by induction of PGR and mitogenesis in vitro. When inoculated subcutaneously into ovariectomized athymic nude mice, TGF-α transfected cells fail to form tumours without oestrogen supplementation (Clarke et al., 1989d). Thus,

TGF-α expression alone is not sufficient to mediate fully the effects of oestrogen in MCF-7 cells growing either *in vitro* or *in vivo*.

Oestrogens and progestins increase EGF-R expression in oestrogen-responsive tissue (Mukku and Stancel, 1985; Leake *et al.*, 1988; Lingham *et al.*, 1988). Oestrogens also increase the levels of secreted ligand, perhaps indicating an additional requirement for elevated numbers of EGF cell surface receptors for mediation of the full effects of TGF-α secretion by an autocrine loop. This hypothesis gains support from the observations that NIH 3T3 cells expressing low levels of EGF-R are not transformed when transfected with the TGF-α cDNA. Cells transfected with the EGF-R cDNA require exogenous EGF for transformation but exhibit the fully transformed phenotype when co-transfected with both EGF receptor and ligand (Di Marco *et al.*, 1989). Thus, for the EGF-R/TGF-α autocrine loop to function in some cell systems, an adequate level of expression of both ligand and receptor is required. The low levels of EGF-R expression in MCF-7 cells, compared with hormone-independent MDA-MB-231 cells, may be responsible for the inability of a constitutive expression of TGF-α to induce the fully hormone-independent phenotype. Alternatively, TGF-α may be necessary but not sufficient when overexpressed alone (or with EGF-R) to induce hormone-independence, requiring a combined action with other oestrogen-regulated mitogenic growth factors, for example, IGF-like and/or FGF-like proteins.

13.3.3 Insulin-like Growth Factor-I and Insulin-like Growth Factor-II

The precise role of IGFs in normal tissue is not yet known, but it appears that the levels of IGF expression are higher in fetal than in adult tissues (Rotwein *et al.*, 1987). IGF-I has been observed in human secretory phase endometrium, where it may function as an autocrine or paracrine growth factor (Rutanen *et al.*, 1988). IGF-II is synthesized in the choroid plexus of rat brain (Hynes *et al.*, 1988) and in rat liver (Froesch *et al.*, 1985), where it appears to be partly under the regulation of glucocorticoids (Levinovitz and Norstedt, 1989). The most important factor controlling the regulation of IGFs in normal tissue is growth hormone (Froesch *et al.*, 1985). Secretion of both IGF-I and IGF-II increases during lactation in bovine mammary tissue (Dehoff *et al.*, 1988). Rosenfeld *et al.* (1983) observed a correlation between changes in breast size in humans and circulating levels of IGF-I. Receptors for IGF-I have been detected in normal human breast tissue (Pekonen *et al.*, 1988) and human benign breast disease (Peyrat *et al.*, 1988a). IGF-I can replace insulin at much lower concentrations for the maintenance of normal rat mammary epithelial cells in culture (Deeks *et al.*, 1988).

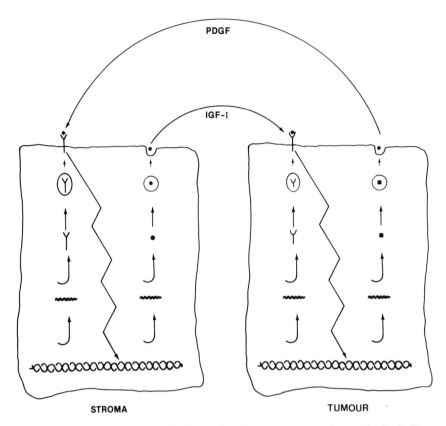

Fig. 2 Oestrogen-induced paracrine interactions between tumour and stromal cells. Following exposure to oestrogen, the tumour cells secrete PDGF which then stimulates the stroma to secrete IGF-I. The IGF-I of stromal origin in turn stimulates the tumour cells to proliferate.

Type-I IGF receptors are widely expressed by malignant breast tissue (Foekens *et al.*, 1986; Peyrat *et al.*, 1988b). The growth of MDA-MB-231 human breast cancer cells both *in vivo* and *in vitro* is inhibited by an antibody which blocks ligand binding to the IGF-I receptor (Rohlik *et al.*, 1987; Arteaga *et al.*, 1989). This antibody also inhibits the growth of a number of human breast cancer cell lines *in vitro* (Arteaga and Osborne, 1989). Whilst the growth of MCF-7 cells is inhibited *in vitro*, growth *in vivo* is not perturbed (Rohlik *et al.*, 1987; Arteaga *et al.*, 1989). Immunocytochemical analysis has revealed the presence of insulin-like material in some human breast tumours (Spring-Mills *et al.*, 1984). Many breast cancer cells produce an IGF-I-like mRNA (Huff *et al.*, 1988). However, this material is not authentic IGF-I, since RNase protection analyses fail to detect IGF-I mRNA in these cells (Yee *et al.*, 1989). IGF-I stimulates the proliferation of some breast cancer cells (Furlanetto and DiCarlo, 1984; Huff et al., 1986) and can induce the formation of hormone-independent MCF-7 tumours in ovariectomized athymic mice (Dickson et al., 1987). IGF-II mRNA is expressed by T47D cells and both IGF-I and IGF-II stimulate the growth of MCF-7 and T47D breast cancer cells in vitro (Cullen et al., 1988). Since the majority of IGF-like mRNA appears to be expressed by the stromal component of many breast tumours (Cullen et al., 1988), IGFs or IGF-like proteins may be both autocrine and paracrine growth factors in breast cancer (Fig. 2).

13.3.4 Fibroblast Growth Factors

COMMA-D normal mouse mammary epithelial cells can be maintained continually in vitro and respond mitogenically to basic fibroblast growth factor (bFGF). In a serum-free cell culture medium which supports the growth of these cells bFGF can be substituted for both EGF and IGF-I (Riss and Sirbasku, 1987). FGFs can stimulate the growth of the androgen-responsive Shionogi mouse mammary carcinoma (Nakmura et al., 1989). The product of the FGF-related oncogene int-2 is mitogenic for C57MG mouse mammary epithelial cells (Dixon et al., 1989). Exogenous bFGF increases the expression of two of the oestrogen-regulated mRNAs, pS2 and cathepsin D (Cavailles et al., 1989). As a result of the presence of an amplicon at chromosome 11q23, coding sequences of both the int-2 and another FGF-related oncogene hst are co-amplified in approximately 17% of human breast cancers. However, overexpression of the respective mRNA species is only apparent for the hst oncogene (Theillet et al., 1989). Whilst the hst gene product can be secreted, interpretation of responses to exogenous FGF can be problematic in the context of autocrine stimulation since many of the FGFs do not appear to be secreted. FGFs may be either retained within the cell to be released on cell lysis or incorporated into the extracellular matrix and subsequently released by tumour-derived heparan sulphate degrading enzymes (Vlodavsky et al., 1987). Alternatively, an internal autocrine loop may function where interaction between ligand and receptor occurs within the Golgi (Browder, 1989), as has been suggested for PDGF (Bejcek et al., 1989; Huang and Huang, 1988). Some human breast cancer cell lines secrete a 60 kD basic fibroblast growth factor-like activity (Swain et al., 1986). Whilst the precise nature of this material remains unclear, its secretion is inhibited by treatment with anti-oestrogens and stimulated by oestrogen (A. Wellstein, pers. commun.). Thus, FGFs may contribute, in conjunction with other mitogenic factors, to the control of proliferation by oestrogens.

13.3.5 Other Mitogenic Growth Factors and Hormones

Stimulation of MCF-7 cells with oestrogen induces the secretion of the mitogenic proteins PDGF (Rozengurt et al., 1985; Bronzert et al., 1987) and a 52 kD cathepsin D (Westley and Rochefort, 1980). Although mitogenic for MCF-7 cells, the cathepsin D appears to be targeted to lysosomes where it requires activation at a very low pH (De Duve, 1984). Its role in mediating the mitogenic effects of oestrogens is unknown. PDGF is a potent mitogen for stromal fibroblasts (Clemmons and Shaw, 1983). However, MCF-7 cells do not express detectable levels of functional receptors, suggesting that PDGF may function as a paracrine stimulator of breast cancer cell proliferation (Bronzert et al., 1987). For example, PDGF can stimulate fibroblasts to secrete IGF-I (Clemmons and Shaw, 1983). The stromal derived IGF may then stimulate the tumour cell which originally produced the PDGF (Fig. 2).

In addition to the growth factor receptors described above, breast cancer cells express a variety of receptors to other growth factors and hormones. For example, functional receptors for glucagon (Cremin et al., 1987) have been reported on a number of breast cancer cell lines, some human breast carcinomas express gonadotropin-releasing hormone receptors (Eidne et al., 1985). T47D cells express receptors for vasoactive intestinal polypeptide (Gespach et al., 1988).

13.4 Inhibitory growth factors and hormones

Transforming growth factor β (TGF-β) is structurally unrelated to TGF-α but acts in conjunction with TGF-α to induce characteristics of the transformed phenotype in normal rat kidney fibroblasts (Carpenter et al., 1983). It stimulates the proliferation of fibroblasts but inhibits epithelial cells (Massague, 1987). In the developing mouse mammary gland, TGF-β induces a reversible inhibition of mammary growth and morphogenesis (Silberstein and Daniel, 1987). Proliferation of normal human mammary epithelial cells in vitro is inhibited by TGF-β (Valverius et al., 1989); however, expression of the milk fat globule antigen is stimulated (Walker-Jones et al., 1989). TGF-β is secreted by a number of breast cancer cell lines (Knabbe et al., 1987; Arteaga et al., 1988) and inhibits the proliferation of these cells (Zugmaier et al., 1988). This secretion is reduced by oestrogen but increased by antioestrogens (Knabbe et al., 1987). Thus, an additional level of control of steroid-induced proliferation may result from altered secretion of potential inhibitory factors. TGF-β reduces TGF-α and EGF binding to EGF-R in normal rat kidney cells (Nugent et al., 1989; Massague, 1985). Thus, TGF-β may function as a direct or indirect inhibitory factor in some cells.

LH-RH analogues have been reported to directly inhibit MCF-7 cells in vitro (Foekens et al., 1986). When administered in combination, PRO 9-LH-RH ethylamide and D SER(t BU)6 Aza GLY 10-LH-RH inhibit the oestrogen-induced stimulation of proliferation in CG-5 cells, a subline of the MCF-7 human breast cancer cell line (Scambia et al., 1988). However, the major mechanism of action of this class of compounds in vivo is more likely to be related to their ability to inhibit gonadotrophin release and the consequent reduction in secreted gonadal steroids (Furr and Nicholson, 1982).

Somatostatin analogues and the parent hormone inhibit the growth of some mammary tumours in vivo (Vuk-Pavlovik et al., 1982; Defeudis and Moreau, 1986) and in vitro (Nelson et al., 1989; Setyono-Han et al., 1987; Cremin et al., 1990). The precise mechanism of action is unknown, but these analogues reduce plasma concentrations of IGF-I (Lamberts et al., 1985) and EGF (Ghirlanda et al., 1983). Thus, inhibition of tumour growth may be the result of a combination of indirect endocrine and direct anti-tumour effects.

13.5. Growth factors as mediators of steroid hormone function

There are a number of restriction sites within the cell cycle which influence the proliferation of non-transformed fibroblasts. For example, EGF is required for early G_1 transit but transition through the remainder of G_1 and subsequent commitment to DNA synthesis require IGF-I (Campisi and Pardee, 1984; Leof et al., 1989). Whilst it is not clear whether analogous restriction sites apply to the proliferation of epithelial cells, insulin and EGF/TGF-α are at least additive in stimulating the growth of MCF-7 cells (Barnes and Sato, 1978; Huff et al., 1988), COMMA-D mouse mammary epithelial cells (Riss and Sirbasku, 1987) and normal rat mammary epithelial cells (Ethier et al., 1987). Some primary rat mammary tumours require EGF and/or insulin for growth in vitro, independence from these growth factors being associated with an increased neoplastic potential in vivo (Ethier an Cundiff, 1987). Both exogenous TGF-α and EGF can increase production of an IGF-I-related material in MCF-7 cells (Huff et al., 1988).

Table 1 shows the major effects of oestrogens and anti-oestrogens on the levels of growth factors. These observations clearly implicate both EGF/TGF-α and either insulin or the IGFs in the stimulation of proliferation in neoplastic breast tissue. Since the secretion of both EGF-like and IGF-like proteins are stimulated by exposure to oestradiol, both of these proteins may be required to modulate the oestrogen-induced stimulation of proliferation in MCF-7 cells. This hypothesis gains further support from the observation that somatostatin and its functional analogues inhibit the growth of some mammary tumours (Vuk-Pavlovic *et al.*, 1982) and reduce plasma concentrations of both IGF-I (Lamberts *et al.*, 1985) and EGF (Ghirlanda *et al.*, 1983). Anti-oestrogens also reduce plasma levels of IGF-I (Colletti *et al.*, 1989). The ability of oestrogen to stimulate the expression of EGF-R and the secretion of other mitogens including bFGF-like proteins, and reduce secretion of inhibitors, *e.g.* TGF-β, implies that the ability of growth factors to mediate the effects of oestrogen may involve complex alterations in the level of expression of various mitogenic and inhibitory ligands and receptors.

Table 1 Effect of hormones on the level of mRNA expression or secretion of growth factor proteins

Growth factor/ receptor	Expression/secretion	Reference
TGF-α	Increased by oestrogen and decreased by anti-oestrogen in breast cancer cells *in vitro*	Bates *et al.* (1988) Dickson *et al.* (1986)
	High levels constitutively expressed by ER negative cells	Bates *et al.* (1988) Perroteau *et al.* (1986)
EGF-R	Increased by oestrogen in rat uterus	Lingham *et al.* (1988) Mukku and Stancel (1985)
TGF-β	Decreased by oestrogen in breast cancer cells *in vitro*	Knabbe *et al.* (1987)
IGF-I	Circulating levels decreased by anti-oestrogens	Colletti *et al.* (1989)
IGF-I-like	Increased by oestrogen in breast cancer cells *in vitro*	Huff *et al.* (1988)
IGF-II	Increased by oestrogen in breast cancer cells *in vitro*	Yee *et al.* (1988)
PDGF	Increased by oestrogen in breast cancer cells *in vitro*	Bronzert *et al.* (1987) Rozengurt *et al.* (1985)
FGF-like	Increased by oestrogen and decreased by anti-oestrogen in breast cancer cells *in vitro*	A. Wellstein *et al.* (pers. commun.)

Oestrogen may influence the growth and differentiation of normal breast tissue by an indirect interaction mediated through stromal tissues. For example, oestrogen only induces epithelial proliferation *in vitro* in the presence of mammary stromal cells (McGrath, 1983; Haslam and Levely, 1985; see Chapter 11.10). In MCF-7 cells, oestrogen induces a small increase in platelet-derived growth factor (PDGF) mRNA production and a larger increase in secreted PDGF biological activity (Bronzert *et al.*, 1987). Whilst these cells do not express PDGF receptors (Bronzert *et al.*, 1987), PDGF can stimulate IGF-I production in human fibroblasts (Clemmons and Shaw, 1983). Thus, an additional paracrine stimulation of breast cancer cells could occur through the effects of tumour derived, oestrogen-stimulated growth factors inducing stromal cells to secrete IGFs or other factors (Fig. 2).

Since treatment with steroids modulates the expression of a number of genes closely associated with the control of proliferation, it seems unlikely that one gene alone is responsible for mediating these effects. The concerted action of many growth factors may be required to induce the full range of biological responses. It is also possible, particularly in neoplastic tissue, that some clones within the same tumour will utilize different combinations of growth factors to achieve a similar hormone-dependent or -independent phenotype.

Whilst there is considerable evidence supporting the role of both mitogenic and inhibitory growth factors in mediating the mitogenic effects of steroids, this evidence remains largely circumstantial. However, it is only a matter of time until the precise role of growth factors in steroid-induced alterations in biological function is elucidated.

References

Aitken, S. C. and Lippman, M. E. (1982). Hormonal regulation of net DNA synthesis in MCF-7 human breast cancer cells in tissue culture. *Cancer Res.* **42**, 1727–1735.

Akhtar, M., Robinson, C., Ali, M. A. and Godwin, J. T. (1983). Secretory carcinoma of the breast. Light and electron microscopic study of three cases and a review of the literature. *Cancer* **51**, 2245–2254.

Antoniades, K. and Spector, H. (1982). Quantitative estrogen receptor values and growth of carcinoma of the breast before surgical intervention. *Cancer* **50**, 793–796.

Arteaga, C. L. and Osborne, C. K. (1989). Growth inhibition of human breast cancer cells *in vitro* with an antibody against the type I somatomedin receptor. *Cancer Res.* **49**, 6237–6241.

Arteaga, C. L., Tandon, A. K., Von Hoff, D. D. and Osborne, C. K. (1988). Transforming growth factor beta: potential autocrine growth inhibitor of estrogen receptor negative human breast cancer cells. *Cancer Res.* **48**, 3898.

Arteaga, C. L., Kitten, L., Coronado, E., Jacobs, S., Kull, F. and Osborne, C. K. (1989). Blockade of the type I somatomedin receptor inhibits growth of human breast cancer cells in athymic mice. *J. Clin. Invest.* **84**, 1418–1423.

Barnes, D. and Sato, G. (1978). Growth of a human mammary tumor cell line in a serum free medium. *Nature* **281**, 388–389.

Bates, S. E., Davidson, N. E., Valverius, E. M., Dickson, R. B., Freter, C. E., Tam, J. P., Kudlow, J. E., Lippman, M. E. and Salomon, S. (1988). Expression of transforming growth factor alpha and its mRNA in human breast cancer: its regulation by estrogen and its possible functional significance. *Mol. Endocrinol.* **2**, 543–545.

Beatson, G. T. (1896). On the treatment of inoperable cases of carcinoma of the mamma: suggestions from a new method of treatment, with illustrative cases. *Lancet* **ii**, 104–107; 162–165.

Bejcek, B. E., Li, D. Y. and Deuel, T. F. (1989). Transformation by *v-sis* occurs by an internal autoactivation mechanism. *Science* **245**, 1496–1499.

Blanco, G., Alavaikko, M., Ojala, A., Collan, Y., Heikkinen, M., Hietanen, T., Aine, R. and Taskinen, P. J. (1984). Estrogen and progesterone receptors in breast cancer: relationships to tumor histopathology and survival of patients. *Anticancer Res.* **4**, 383–390.

Bronson, F. H., Dagg, C. P. and Snell, G. D. (1975). Biology of the laboratory mouse. *Reproduction*, 187–204.

Bronzert, D. A., Pantazis, P., Antoniades, H. N., Kasid, A., Davidson, N., Dickson, R. B. and Lippman, M. E. (1987). Synthesis and secretion of platelet-derived growth factor by human breast cancer cell lines. *Proc. Natl. Acad. Sci. USA* **84**, 5763–5767.

Browder, T. M., Dunbar, C. E. and Nienhuis, A. W. (1989). Private and public autocrine loops in neoplastic cells. *Cancer Cells* **1**, 9–17.

Campbell, F. C., Blamey, R. W., Elston, C. W., Nicholson, R. I., Griffiths, F. C. and Haybittle, J. L. (1981). Oestrogen-receptor status and sites of metastasis in breast cancer. *Br. J. Cancer* **44**, 456–459.

Campisi, J. and Pardee, A. B. (1984) Post-transcriptional control of the onset of DNA synthesis by an insulin-like growth factor. *Mol. Cell Biol.* **4**, 1807–1814.

Carpenter, G., Stoscheck, C. M., Preston, Y. A. and De Larco, L. E. (1983). Antibodies to the epidermal growth factor receptor block the biological activities of sarcoma growth factor. *Proc. Natl. Acad. Sci. USA* **80**, 4684–4688.

Carter, S. K. (1981). The interpretation of trials: combined hormonal therapy and chemotherapy in disseminated breast cancer. *Breast Cancer Res. Treat.* **1**, 43–51.

Cattoretti, G., Andreola, S., Clemente, C., D'Amato, L. and Rilke, F. (1988). Vimentin and P53 expression in epidermal growth factor receptor-positive oestrogen receptor-negative breast carcinomas. *Br. J. Cancer* **57**, 353–357.

Cavailles, V., Garcia, M., and Rochefort, H. (1989). Regulation of cathepsin-D and pS2 gene expression by growth factors in MCF-7 human breast cancer cells. *Mol. Endocrinol.* **3**, 552–558.

Clark, G. M. and McGuire, W. L. (1988). Steroid receptors and other prognostic factors in primary breast cancer. *Semin. Oncol.* **15**, 20–25.

Clarke, R., van den Berg, H. W., Nelson, J. and Murphy, R. F. (1987). Pharmacological and suprapharmacological concentrations of both 17 beta oestradiol (E2) and tamoxifen (TAM) reduce the membrane fluidity of MCF-7 and MDA-MB-436 human breast cancer cells. *Biochem. Soc. Trans.* **15**, 243–244.

Clarke, R., Lippman, M., Dickson, R. B. and Brünner, N. (1989a). *In vivo/in vitro* selection of hormone independent cells from the hormone dependent MCF-7 human breast cancer cell line. In "Immune-Deficient Animals in Experimental Medicine" (eds B. Wu and J. Zheng), pp. 190–195. Karger, Basel.
Clarke, R., Brünner, N., Katzenellenbogen, B. S., Thompson, E. W., Norman, M. J., Koppi, C., Paik, S., Lippman, M. E. and Dickson, R. B. (1989b). Progression from hormone dependent to hormone independent growth in MCF-7 human breast cancer cells. *Proc. Natl. Acad. Sci. USA* **86**, 3649–3653.
Clarke, R., Brünner, N., Thompson, E. W., Glanz, P., Katz, D., Dickson, R. B. and Lippman, M. E. (1989c). The inter-relationships between ovarian-independent growth, antiestrogen resistance and invasiveness in the malignant progression of human breast cancer. *J. Endocrinol.* **122**, 331–340.
Clarke, R., Brünner, N., Katz, D., Glanz, P., Dickson, R. B., Lippman, M. E. and Kern, F. G. (1989d). The effects of a constitutive production of TGF-alpha on the growth of MCF-7 human breast cancer cells *in vitro* and *in vivo*. *Mol. Endocrinol.* **3**, 372–380.
Clarysse, A. (1985). Hormone-induced tumour flare. *Eur. J. Cancer Clin. Oncol.* **21**, 545–547.
Clemmons, D. R. and Shaw, D. S. (1983). Variables controlling somatomedin production by cultured human fibroblasts. *J. Cell Physiol.* **115**, 137–142.
Coffey, R. J., Derynck, R., Wilcox, J. N., Bringman, T. S., Goustin, A. S., Moses, H. L., and Pittelkow, M. R. (1987). Production and autoinduction of transforming growth factor-alpha in human keratinocytes. *Nature* **328**, 317–820.
Cohen, S. (1983). The epidermal growth factor. *Cancer* **51**, 1787–1791.
Colletti, R. B., Roberts, J. D., Devlin, J. T. and Copeland, K. C. (1989). Effect of tamoxifen on plasma insulin-like growth factor I in patients with breast cancer. *Cancer Res.* **49**, 1882–1884.
Cremin, M., Clarke, R., Nelson, J. and Murphy, R. F. (1987). The response of human breast cancer cells to glucagon. *Biochem. Soc. Trans.* **15**, 241–242.
Cremin, M., Clarke, R., Nelson, J. and Murphy, R. F. (1990). Effect of somatostatin and a synthetic analogue SMS-201-995 (Sandostatin) on human breast cancer cells in culture. *Ir. J. Med. Sci* (in press).
Cullen, K. J., Yee, D., Paik, S., Hampton, B., Perdue, J. F., Lippman, M. E. and Rosen, N. (1988). Insulin-like growth factor II expression and activity in human breast cancer. Abstract 947, *Proc. Annu. Meet. Am. Assoc. Cancer Res.* **29**, 238.
Dao, T. L. (1972). Ablation therapy for hormone dependent tumors. *Annu. Rev. Med.* **23**, 1–18.
Davidson, N. E. and Lippman, M. E. (1985). Combined therapy in advanced breast cancer. *Eur. J. Cancer Clin. Oncol.* **21**, 1123–1126.
Davidson, N. E., Gelmann, E. P., Lippman, M. E. and Dickson, R. B. (1987). Epidermal growth factor receptor gene expression in estrogen receptor-positive and negative human breast cancer cell lines. *Mol. Endocrinol.* **1**, 216–223.
De Duve, C. (1984). In "A Guided Tour of the Living Cell", Vol. 1. Scientific American Books Inc.
Deeks, S., Richards, J. and Nandi, S. (1988). Maintenance of normal rat mammary epithelial cells by insulin and insulin-like growth factor 1. *Exp. Cell Res.* **174**, 448–460.
Defeudis, F. V. and Moreau, J. P. (1986). Studies on somatostatin analogues might lead to new therapies for certain types of cancer. *Trends Pharmacol. Sci.* **7**, 384–386.

Dehoff, M. H., Elgin, R. G., Collier, R. J. and Clemmons, D. R. (1988). Both type I and type II insulin-like growth factor receptor binding increase during lactogenesis in bovine mammary tissue. *Endocrinology* **122**, 2412–2417.

De Larco, J. E. and Todaro, G. J. (1978). Growth factors from murine sarcoma virus-transformed cells. *Proc. Natl. Acad. Sci. USA* **75**, 4001–4005.

Dickson, R. B., Huff, K. K., Spencer, E. M. and Lippman, M. E. (1986). Induction of epidermal growth factor-related polypeptides by 17 beta estradiol in MCF-7 human breast cancer cells. *Endocrinology* **118**, 138–142.

Dickson, R. B., McManaway, M. E. and Lippman, M. E. (1987). Estrogen-induced factors of breast cancer cells partially replace estrogen to promote tumor growth. *Science* **232**, 1540–1543.

Di Marco, E., Pierce, J. A., Fleming, T. P., Kraus, M. H., Molloy, C. J., Aaronson, S. A. and Di Fiore, P. P. (1989). Autocrine interaction between TGF alpha and the EGF-receptor: quantitative requirements for induction of the malignant phenotype. *Oncogene* **4**, 831–838.

Dixon, M., Deed, R., Acland, P., Moore, R., Whyte, A., Peters, G. and Dickson, C. (1989). Detection and characterization of the fibroblast growth factor-related oncoprotein INT-2. *Mol. Cell. Biol.* **9**, 4896–4902.

Duval, D., Durant, S. and Homo-Delarche, F. (1983). Non-genomic effects of steroids. Interactions of steroid molecules with membrane structures and functions. *Biochim. Biophys. Acta* **737**, 409–442.

Eidne, K. A., Flanagan, C. A. and Millar, R. P. (1985). Gonadotropin-releasing hormone binding sites in human breast carcinoma. *Science* **229**, 989–991.

Ethier, S. P. and Cundiff, K. C. (1987). Importance of extended growth potential and growth factor independence on *in vivo* neoplastic potential of primary rat mammary carcinoma cells. *Cancer Res.* **47**, 5316–5322.

Ethier, S. P., Kudla, A. and Cundiff, K. C. (1987). Influence of hormone and growth factor interactions on the proliferative potential of normal rat mammary epithelial cells *in vitro*. *J. Cell Physiol.* **132**, 161–167.

Ferguson, D. J. P. and Anderson, T. J. (1981). Morphological evaluation of cell turnover in relation to the menstrual cycle in the nesting human breast. *Br. J. Cancer* **44**, 177–181.

Fitzpatrick, S. L., Brightwell, J., Wittliff, J. L., Barrows, G. H. and Schultz, G. S. (1984). Epidermal growth factor binding by breast tumor biopsies and relationship to estrogen receptor and progestin receptor levels. *Cancer Res.* **44**, 3448–3453.

Froesch, E. R., Schmid, C., Schwander, J. and Zapf, J. (1985). Actions of insulin-like growth factors. *Annu. Rev. Physiol.* **47**, 443–467.

Foekens, J. A., Henkelman, M. S., Fukkink, J. F., Blankenstein, M. A. and Klijn, J. G. M. (1986). Combined effects of buserelin, estradiol and tamoxifen on the growth of MCF-7 human breast cancer cells *in vitro*. *Biochem. Biophys. Res. Commun.* **140**, 550–556.

Furlanetto, R. W. and DiCarlo, J. N. (1984). Somatomedin C receptors and growth effects in human breast cancer cells maintained in long term culture. *Cancer Res.* **44**, 2122–2128.

Furr, B. J. A. and Nicholson, R. I. (1982). Use of analogs of LH-RH for treatment of cancer. *J. Reprod. Fertil.* **64**, 529–539.

Gespach, C., Bawab, W., de Cremoux, P. and Calvo, F. (1988). Pharmacology, molecular identification and functional characteristics of vasoactive intestinal polypeptide receptors in human breast cancer cells. *Cancer Res.* **48**, 5079–5083.

Ghirlanda, G., Uccioli, L., Perri, F., Altomonte, L., Bertoli, A., Manna, R., Frati, L. and Greco, A. V. (1983). Epidermal growth factor, somatostatin, and psoriasis. *Lancet* **i**, 65.

Haslam, S. Z. and Levely, M. L. (1985). Estrogen responsiveness of normal mouse mammary cells in primary cell culture: association of mammary fibroblasts with estrogenic regulation of progesterone receptors. *Endocrinology* **116**, 1835–1844.

Henderson, C. and Canellos, G. P. (1980). Cancer of the breast: the past decade. *New Engl. J. Med.* **302**, 1–17.

Henderson, B. E., Ross, R. and Bernstein, L. (1988). Estrogens as a cause of human cancer. The Richard and Hinda Rosenthal Foundation Award Lecture. *Cancer Res.* **48**, 246–253.

Henry, J. A., Nicholson, S., Farndon, J. R., Westley, B. R. and May, F.E.B. (1988). Measurement of oestrogen mRNA levels in human breast tumours. *Br. J. Cancer* **58**, 600–605.

Herbst, A. L., Ulfelder, H. and Poskanzer, D. C. (1971). Adenocarcinoma of the vagina: association of maternal stilbestrol therapy with tumor appearance in young women. *New Engl. J. Med.* **284**: 878–881.

Heuson, J.-C. and Benraad, T. J. (1983). Biology of breast cancer: receptors; workshop report. *Eur. J. Cancer Clin. Oncol.* **19**, 1687–1692.

Horwitz, K. B., Costlow, M. E. and McGuire, W. L. (1975). MCF-7: A human breast cancer cell line with estrogen, androgen, progesterone and glucocorticoid receptors. *Steroids* **26**, 785–795.

Huang, S. S. and Huang, J. S. (1988). Rapid turnover of the platelet derived growth factor receptor in *sis* transformed cells and reversal by suramin. *J. Biol. Chem.* **263**, 12 608–12 618.

Huff, K. K., Kaufman, D., Gabbay, K. H., Spencer, E. M., Lippman, M. E. and Dickson, R. B. (1986). Human breast cancer cells secrete an insulin-like growth factor-I-related polypeptide. *Cancer Res.* **46**, 4613–4619.

Huff, K. K., Knabbe, C., Lindsey, R., Kaufman, D., Bronzert, D., Lippman, M. E. and Dickson, R. B. (1988). Multihormonal regulation of insulin-like growth factor-I-related protein in MCF-7 human breast cancer cells. *Mol. Endocrinol.* **2**, 200–208.

Hynes, M. A., Brooks, P. J., Van Wyk, J. J. and Lund, P. K. (1988). Insulin-like growth factor II messenger ribonucleic acids are synthesized in the choroid plexus of the rat brain. *Mol. Endocrinol.* **2**, 47–54.

Imai, Y., Leung, C. K. H., Freisen, H. G. and Shiu, R. P. (1982). Epidermal growth factor receptors and effect of epidermal growth factor on growth of human breast cancer cells in long term tissue culture. *Cancer Res.* **42**, 4394–4398.

Jick, H., Walker, A. M. and Rothman, K. J. (1980). The epidemic of endometrial cancer: a commentary. *Am. J. Pub. Hlth* **70**, 264–267.

Jonat, W. and Maat, H. (1978). Some comments on the necessity of receptor determination in human breast cancer. *Cancer Res.* **38**, 4305–4306.

Kamby, C., Andersen, J., Ejlertsen, B., Birkler, N. E., Rytter, L., Zedeler, K., Thorpe, S. M., Norgaard, T. and Rose, C. (1988). Histological grade and steroid receptor content of primary breast cancer-impact on prognosis and possible modes of action. *Br. J. Cancer* **58**, 480–486.

King, R. J. B., Stewart, J. F., Millis, R. R., Rubens, R. D. and Hayward, J. L. (1982). Quantitative comparison of estradiol and progesterone receptor contents of primary and metastic human breast tumors in relation to response to endocrine treatment. *Breast Cancer Res. Treat.* **2**, 339–346.

Kinne, D. W., Ashikari, R., Butler, A., Menendez-Botet, C., Rosen, P. R. and Schwartz, M. (1981). Estrogen receptor protein in breast cancer as a predictor of recurrence. *Cancer* **47**, 2364–2367.
Knabbe, C. K., Lippman, M. E., Wakefield, L. M., Flanders, K. C., Kasid, A., Derynck, R. and Dickson, R. B. (1987). Evidence that transforming growth factor-beta is a hormonally regulated negative growth factor in human breast cancer cells. *Cell* **48**, 417–428.
Koga, M. and Sutherland, R. L. (1987). Epidermal growth factor partially reverses the inhibitory effects of antiestrogens on T47D human breast cancer cell growth. *Biochem. Biophys. Res. Commun.* **146**, 738–745.
Lacassagne, A. (1932). Apparition de cancers de la mamelle chez la souris mâle, soumise à des injections de folliculine. *C. R. Acad. Sci. (Paris)* **195**, 629–632.
Lamberts, S. W. J., Uitterlinden, P., Verschoor, L., van Dongen, K. J. and Del Pozo, E. (1985). Long-term treatment of acromegaly with the somatostatin analogue SMS 201–995. *New Engl. J. Med.* **313**, 1576–1579.
Leake, R. E., George, W. D., Godfrey, D. and Rinaldi, F. (1988). Regulation of epidermal growth factor receptor synthesis in breast cancer cells. *Br. J. Cancer* **58**, 521.
Leof, E. B., Lyons, R. M., Cunningham, M. R. and O'Sullivan, D. (1989). Mid-G1 arrest and epidermal growth factor independence of ras-transfected mouse cells. *Cancer Res.* **49**, 2356–2361, 1989.
Levinovitz, A. and Norstedt, G. (1989). Developmental and steroid hormonal regulation of insulin-like growth factor II expression. *Mol. Endocrinol.* **3**, 797–804.
Lieberman, M. E., Maurer, R. A. and Gorski, J. (1978). Estrogen control of prolactin synthesis *in vitro*. *Proc. Natl. Acad. Sci. USA* **75**, 5946–5949.
Lingham, R. B., Stancel, G. M. and Loose-Mitchel, D. S. (1988). Estrogen regulation of epidermal growth factor messenger ribonucleic acid. *Mol. Endocrinol.* **2**, 230–235.
Lippman, M. E., Bolan, G. and Huff, K. (1976). The effects of estrogens and antiestrogens on hormone responsive human breast cancer cells in long term tissue culture. *Cancer Res.* **36**, 4595–4601.
Lippman, M. E., Dickson, R., Kasid, A., Gelmann, E., Davidson, N., McManaway, M., Huff, K., Bronzert, D., Bates, S., Swain, S. and Knabbe, C. (1986). Autocrine and paracrine growth regulation of human breast cancer. *J. Steroid Biochem.* **24**, 147–154.
Lyons, W. R. (1958). Hormonal synergism in mammary growth. *Proc. R. Soc. (Biol). Lond.* **149**, 303–325.
Massague, J. (1985). Transforming growth factor beta modulates the high-affinity receptors for epidermal growth factor and transforming growth factor alpha. *J. Cell Biol.* **100**, 1508–1514.
Massague, J. (1987). The TGF beta family of growth and differentiation factors. *Cell* **49**, 437–438.
McGrath, C. M. (1983). Augmentation of the response of normal mammary epithelial cells to estradiol by mammary stroma. *Cancer Res.* **43**, 1355–1360.
Meyers, J. S., Rao, B. R., Stevens, S. C. and White, W. L. (1977). Low incidence of estrogen receptor in breast carcinomas with rapid rates of cellular proliferation. *Cancer* **40**, 2290–2298.
Mukku, V. R. and Stancel, G. M. (1985). Regulation of epidermal growth factor receptor by estrogen. *J. Biol. Chem.* **260**, 9820–9824.

Murr, S. M., Bradford, G. E. and Geschwind, I. I. (1974). Plasma leuteinizing hormone, follicle-stimulating hormone and prolactin during pregnancy in the mouse. *Endocrinology* **94**, 112–116.
Nakamura, N., Yamanishi, H., Lu, J., Uchida, N., Nonomura, N., Matsumoto, K. and Sato, B. (1989). Growth-stimulatory effects of androgen, high concentration of glucocorticoid or fibroblast growth factors on a cloned cell line from Shionogi carcinoma 115 cells in a serum-free medium. *J. Steroid Biochem.* **33**, 13–18.
Nelson, J., Cremin, M. and Murphy, R. F. (1989). Synthesis of somatostatin by breast cancer cells and their inhibition by exogenous somatostatin and sandostatin. *Br. J. Cancer* **59**, 739–742.
Nugent, M. A., Lane, E. A., Keski-Oka, J., Moses, H. L. and Newman, M. J. (1989). Growth stimulation, altered regulation of epidermal growth factor receptors, and autocrine transformation of spontaneously transformed normal rat kidney cells by transforming growth factor beta. *Cancer Res.* **49**, 3884–3890.
Oka, T., Kurachi, H., Yoshimura, Y., Tsutsumi, O., Cossu, M. F. and Tag, M. (1987). Study of the growth factors for the mammary gland: Epidermal growth factor and mesenchyme-derived growth factor. *Nucl. Med. Biol.* **14**, 353–360.
Osborne, C. K., Hamilton, B., Titus, G. and Livington, R. B. (1980). Epidermal growth factor stimulation of human breast cancer cells in culture. *Cancer Res.* **40**, 3261–3266.
Pekonen, F., Partanen, S., Makinen, T. and Rutanen, E.-M. (1988). Receptors for epidermal growth factor and insulin-like growth factor and their relation to steroid receptors in human breast cancer. *Cancer Res.* **48**, 1343–1347.
Peres, R., Betsholtz, C., Westermark, B. and Heldin, C.-H. (1987). Frequent expressions of growth factors for mesenchymal cells in human mammary carcinoma cell lines. *Cancer Res.* **47**, 3425–3429.
Perez, R., Pascual, M., Macias, A. and Lage, A. (1984). Epidermal growth factor receptors in human breast cancer. *Breast Cancer Res. Treat.* **4**, 189–193.
Perroteau, I., Salomon, D., DeBortoli, M., Kidwell, W., Hazarika, P., Pardue, R., Dedman, J. and Tam, J. (1986). Immunological detection and quantitation of alpha transforming growth factors in human breast carcinoma cells. *Breast Cancer Res. Treat.* **7**, 201–210.
Peyrat, J. P., Bonneterre, J., Laurent, J. C., Louchez, M. M., Amrani, S., Leroy-Martin, B., Vilain, M. O., Delobells, A. and Demaille, A. (1988a). Presence and characterization of insulin-like growth factor I receptors in human benign breast disease. *Eur. J. Cancer Clin. Oncol.* **24**, 1425–1431.
Peyrat, J. P., Bonneterre, J., Beuscart, R., Djiane, J. and Demaille, A. (1988b). Insulin-like growth factor I receptors in human breast cancer and their relation to estradiol and progesterone receptors. *Cancer Res.* **48**, 6429–6433.
Raynaud, A. (1955). Observations sur les modifications provoquees par les hormones oestrogenes, du mode de developpment des mamelons des foetus de souris. *C. R. Acad. Sci. (Paris)* **240**, 674–676.
Riss, T. L. and Sirbasku, D. A. (1987). Growth and continuous passage of COMMA-D mouse mammary epithelial cells in hormonally defined serum-free medium. *Cancer Res.* **47**, 3776–3782.
Rohlik, Q. T., Adams, D., Kull, F. C. and Jacobs, S. (1987). An antibody to the receptor for insulin-like growth factor I inhibits the growth of MCF-7 cells in tissue culture. *Biochem. Biophys. Res. Commun.* **149**, 276–281.
Rosenfeld, R. I., Furlanetto, R. and Bock, D. (1983). Relationship of somatomedin-C concentration to pubertal changes. *J. Pediatr.* **103**, 723–728.

Rotwein, P., Pollock, K. M., Watson, M. and Milbrandt, J. D. (1987). Insulin-like growth factor gene expression during rat embryonic development. *Endocrinology* **121**, 2141–2144.

Rozengurt, E., Sinnett-Smith, J. and Taylor-Papadimitriou, J. (1985). Production of PDGF-like growth factor by breast cancer cell lines. *Int. J. Cancer* **36**, 247–252.

Rutanen, E.-M., Pekonen, F. and Makinen, T. (1988). Soluble 34K binding protein inhibits the binding of insulin-like-growth factor I to its cell receptors in human secretory phase endometrium: evidence for autocrine/paracrine regulation of growth factor action. *J. Clin. Endocrinol. Metab.* **66**, 173–180.

Sainsbury, J. R. C., Malcolm, A., Appleton, D., Farndon, J. R. and Harris A. L. (1985). Presence of epidermal growth factor receptors as an indicator of poor prognosis in patients with breast cancer. *J. Clin. Pathol.* **38**, 1225–1228.

Scambia, G., Panici, P. B., Baiocchi, G., Perrone, L., Gaggini, C., Iacobelli, S. and Mancuso, S. (1988). Growth inhibitory effect of LH-RH analogs on human breast cancer cells. *Anticancer Res.* **8**, 187–190.

Seibert, K., Shafie, S. M., Triche, T. J., Whang-Peng, J. J., O'Brien, S. J., Toney, J. H., Huff, K. K. and Lippman, M. E. (1983). Clonal variation of MCF-7 breast cancer cells *in vitro* and in athymic nude mice. *Cancer Res.* **43**, 2223–2239.

Setyono-Han, B., Henkelman, M. S., Foekens, J. A. and Klijn, J. G. M. (1987). Direct inhibitory effects of somatostatin (analogs) on the growth of human breast cancer cells. *Cancer Res.* **47**, 1566–1570.

Shafie, S. M. and Grantham, F. H. (1981). Role of hormones in the growth and regression of human breast cancer cells (MCF-7) transplanted into athymic nude mice. *J. Natl. Cancer. Inst.* **67**, 51–56.

Sherwin, S. A., Twardzik, D. R., Bohn, W. H., Cockley, K. D. and Todaro, G. J. (1983). High-molecular-weight transforming growth factor activity in the urine of patients with disseminated cancer. *Cancer Res.* **43**, 403–407.

Silberstein, G. B. and Daniel, C. W. (1987). Reversible inhibition of mammary gland growth by transforming growth factor beta. *Science* **237**, 291–293.

Singh, L., Wilson, A. J., Baum, J., Whimster, W. F., Birch, I. H., Jackson I. M., Lowrey, C. and Palmer, M. K. (1988). The relationship between histological grade, oestrogen receptor status, events and survival at 8 years in the NATO (Nolvadex) trial. *Br. J. Cancer* **57**, 612–614.

Soule, H. D. and McGrath, C. M. (1980). Estrogen responsive proliferation of clonal human breast carcinoma cells in athymic mice. *Cancer Lett.* **10**, 177–189.

Spring-Mills, E. J., Stearns, S. B., Numann, P. J. and Smith, P. H. (1984). Immunocytochemical localization of insulin- and somatostatin-like material in human breast tumors. *Life Sci.* **35**, 185–190.

Stampfer, M., Hallowes, R. C. and Hackett, A. J. (1980). Growth of normal human mammary cells in culture. *In Vitro* **16**, 415–425.

Stewart, J., King, R., Hayward, J. L. and Rubens, R. D. (1982). Estrogen and progesterone receptor: correlation of response rates, site and timing of receptor analysis. *Breast Cancer Res. Treat.* **2**, 243–250.

Stromberg, K., Hudgins, W. R., Dorman, L. S., Henderson, L. E., Sowder, S. C., Sherrell, B. J., Mount, C. D. and Orth, D. N. (1987). Human brain tumor-associated urinary high molecular weight transforming growth factor: a high molecular weight form of epidermal growth factor. *Cancer Res.* **47**, 1190–1196.

Swain, S., Dickson, R. B. and Lippman, M. E. (1986). Anchorage independent epithelial colony stimulating activity in human breast cancer cell lines. *Proc. 77th Annu. Meet. AACR*, Abstract 844: 213.

Theillet, C., Le Roy, X., De Lapeyriere, O., Grosgeorges, J., Adnane, J., Raynaud, S. D., Simony-Lafontaine, J., Goldfarb, M., Escot, C., Birnbaum, D. and Gaudray, P. (1989). Amplification of FGF-related genes in human tumors: possible involvement of HST in breast carcinomas. *Oncogene* **4**, 915–922.

Thomas, D. B. (1984). Do hormones cause breast cancer? *Cancer* **53**, 595–604.

Tsutsumi, O., Tsutsumi, A. and Oka, T. (1987). Importance of epidermal growth factor in implantation and growth of mouse mammary tumors in female nude mice. *Cancer Res.* **47**, 4651–4653.

Turkington, R. W. (1969). Stimulation of mammary cell proliferation by epidermal growth factor *in vitro*. *Cancer Res.* **29**, 1457–1458.

Turkington, R. W. (1971). *In* "Developmental Aspects of the Cell Cycle" (eds I. L. Cameron, G. M. Padilla and A. M. Zimmerman), pp. 315–355. Academic Press, London.

Twardzik, D. R., Kimball, E. S., Sherwin, S. A., Ranchalis, J. E. and Todaro, G. J. (1985). Comparison of growth factors functionally related to epidermal growth factor in the urine of normal and tumor-bearing athymic mice. *Cancer Res.* **45**, 1934–1939.

Valverius, E. M., Walker-Jones, D., Bates, S. E., Stampfer, M., Clark, R., McCormick, F., Dickson, R. B. and Lippman, M. E. (1989). Production of and responsiveness to transforming growth factor beta in normal and oncogene-transformed human mammary epithelial cells. *Cancer Res.* **49**, 6269–6274.

Vana, J., Murphy, G. P., Arnoff, B. L. and Baker, H. W. (1979). Survey of primary liver tumors and oral contraceptive use. *J. Toxicol. Environ. Hlth.* **5**, 255–273.

Vlodavsky, I., Folkman, J., Sullivan, R., Fridman, R., Ishai-Michaeli, R. and Klagsbrun, M. (1987). Endothelial cell-derived basic fibroblast growth factor: synthesis and deposition into subendothelial extracellular matrix. *Proc. Natl. Acad. Sci. USA* **84**, 2292–2296.

Vignon, F., Bouton, M. M. and Rochefort, H. (1987). Antiestrogens inhibit the mitogenic effect of growth factors on breast cancer cells in the total absence of estrogens. *Biochem. Biophys. Res. Commun.* **146**, 1502–1508.

Vonderhaar, B. K. (1988). Regulation of development of the normal mammary gland by hormones and growth factors. *In* "Breast Cancer: Cellular and Molecular Biology" (eds M. E. Lippmann and R. B. Dickson), pp. 251–266. Kluwer Academic Publishers, Boston.

Vuk-Pavlovic, S., Bozikof, V. and Pavelic, K. (1982). Somatostatin reduced proliferation of murine aplastic carcinoma conditioned to diabetes. *Int. J. Cancer* **29**, 683–686.

Walker-Jones, D., Valverius, E. M., Stampfer, M. R., Lippman, M. E. and Dickson, R. B. (1989). Stimulation of epithelial membrane antigen expression by transforming growth factor beta in normal and oncogene-transformed human mammary epithelial cells. *Cancer Res.* **49**, 6407–6411.

Warner, M. R. (1978). Effect of perinatal oestrogen on the pretreatment required for mouse lobular formation *in vitro*. *J. Endocrinol.* **77**, 1–10.

Westley, B. and Rochefort, H. (1980). A secreted glycoprotein induced by estrogen in human breast cancer cell lines. *Cell* **20**, 352–362.

Wrba, F., Reiner, A., Ritzinger, E. and Heinrich, H. (1988). Expression of epidermal growth factor receptors (EGFR) on breast carcinomas in relation to growth fractions, estrogen receptor status and morphological criteria. *Path. Res. Pract.* **183**, 25–29.

Yee, D., Cullen, K. J., Paik, S., Perdue, J. F., Hampton, B., Schwartz, A., Lippman, M. E. and Rosen, N. (1988). Insulin-like growth factor II mRNA expression in human breast cancer. *Cancer Res.* **48**, 6691–6696.

Yee, D., Paik, S., Lebovic, G. S., Marcus, R., Favoni, R. E., Cullen, K. J., Lippman, M. E. and Rosen, N. (1989). Analysis of insulin-like growth factor I gene expression in malignancy: evidence for a paracrine role in human breast cancer. *Mol. Endocrinol.* **3**, 509–517.

Zugmaier, G. C., Knabbe, B., Deschauer, B., Lippman, M. E. and Dickson, R. B. (1988). Response of estrogen receptor negative and estrogen receptor positive human breast cancer cell lines to TGF beta-1 and TGF beta-2. *Proc. Am. Soc. Cancer Res.* **29**, 238.

Zwiebel, J. A., Bano, M., Nexo, E., Salomon, D. S. and Kidwell, W. R. (1986). Partial purification of transforming growth factors from human milk. *Cancer Res.* **46**, 933–939.

14
Genetic Defects of Receptors Involved in Disease

M. R. HUGHES and B. W. O'MALLEY

Institute for Molecular Genetics and the Department of Cell Biology, Baylor College of Medicine, One Baylor Plaza, Houston, TX 77030, USA

14.1 Introduction

Human diseases which we now understand to be associated with defects in steroid-like hormone function have been described since the earliest days of recorded medical history. Defective action of the glucocorticoids, mineralocorticoids, androgens and vitamins have been reported extensively in the literature. The concepts are similar to tissue resistance syndromes that have been recognized in the peptide hormones. Receptor deficiency states have been described for PTH-resistant pseudohypoparathyroidism, growth hormone-resistant dwarfism, insulin-resistant leprechaunism, acanthosis nigricans, and vasopressin-resistant diabetes insipidus, to name just a few. Our focus on naturally occurring mutations of the nuclear receptors is particularly intriguing since these molecules are direct *trans*-acting factors which regulate gene expression.

As described in previous chapters of this text, our knowledge of the mechanism of action of steroid hormone receptors has increased greatly over the past several years. The techniques and technologies of molecular biology have already opened up many of the challenging questions concerning the structure and function of the steroid receptors. Intentional mutagenesis of receptor genes has focused our attention on functional domains of the protein which direct receptor activity in the nucleus. It is apparent from this basic research that naturally occurring mutations in these modulators of gene transcription would have significant phenotypic effects. Genetic mutations in the nuclear receptors are now being described at an increasing pace, and it is fascinating to realize that a single nucleotide alteration in a receptor gene, e.g. the androgen receptor, can result in the classically profound phenotype of testicular feminization. As this information unfolds it is becoming apparent that hormone resistance syndromes are a paradigm of Murphy's Law—what can go wrong will mutate. To review these recent observations involving tissue resistance to the nuclear receptors, we will first discuss the genetic anomalies uncovered in the vitamin D system and then briefly outline similar receptor mutations that have been found in the androgen and glucocorticoid receptor systems.

14.2 The vitamin D receptor

14.2.1 Background

Elucidation of the structure-function relationships of the $1,25\text{-}(OH)_2D_3$ (Vitamin D) receptor has paralleled the studies described throughout this text. The receptor was isolated and purified to homogeneity (Pike et al., 1983) and used as an antigen to raise monoclonal antibodies (Pike, 1984). Monoclonal antibodies were used to recover vitamin D receptor (VDR) cDNA from a chicken intestine expression library (McDonnell et al., 1987) and, subsequently, these clones were utilized to isolate the human full-length cDNA (Baker et al., 1988). The nucleotide sequence of the 4605 bp human cDNA was shown to include 1281 bp of open reading frame, 115 bp of non-coding leader sequence, and 3209 bp of 3'-untranslated sequence. The VDR coding sequence was transfected into COS-1 cells and a single VDR species was produced that was indistinguishable from the native receptor (Baker et al., 1988). Functional analyses of the VDR (McDonnell et al., 1989b) showed that the protein belongs to the gene family of nuclear hormone receptors (Evans, 1988). Kerner and co-workers (Kerner et al., 1989) went on to identify the osteocalcin gene promoter and map the location of the upstream $1,25\text{-}(OH)_2D_3$ response element to approximately 500 bp of the mRNA start site. This sequence was shown to be very similar to the other steroid hormone response class of cis-acting elements and was able to confer $1,25\text{-}(OH)_2D_3$ response to heterologous promoters (see Chapters 6.2, 6.5 and 9.4). Consequently, by all available criteria, it was shown that the VDR functions as a true member of the family of nuclear receptors.

14.2.2 Hereditary Generalized Resistance to 1,25-Dihydroxyvitamin D_3

A genetic lesion in the pathway of vitamin D action has been suspected in the human disorder of hypocalcaemic vitamin D-resistant rickets

(HVDRR). This autosomal recessive disease (McKusick MOM ♯27744) typically has its onset in early infancy with the appearance of severe rickets, hypocalcaemia, secondary hyperparathyroidism, and elevated circulating levels of 1,25-dihydroxyvitamin D_3 (1,25-$(OH)_2D_3$) (McKusick, 1988). Another frequently associated finding in the most severe cases of HVDRR is total absence of hair on head and body (Hochberg et al., 1985; Marx et al., 1986), an unexplained finding perhaps relating to the observation that 1,25-$(OH)_2D_3$ receptors (VDR) have been found in the hair follicle (Berger et al., 1988). Since 1978 when this disorder was first recognized (Brooks et al., 1978), 29 kindreds containing 47 affected individuals with true resistance to 1,25-$(OH)_2D_3$ have been described and characterized by a variety of techniques (for review, see Marx, 1989). Initially progress in unraveling the molecular pathogenesis was hampered due to difficulty in obtaining receptor-containing target tissue (classically intestine and bone) from which analysis could be carried out in this patient population. Once it was shown that skin possessed VDR (Colston et al., 1980; Simpson and DeLuca, 1980) and that cultured human dermal fibroblasts derived from skin biopsies could be used as a model system for the study of the VDR from patients (Feldman et al., 1980), it was demonstrated that defects in this receptor were the likely cause of the HVDRR syndrome. An additional useful finding was that 1,25-$(OH)_2D_3$ hormone could induce the enzyme 25-(OH)vitamin D-24-hydroxylase(24 hydroxylase) in multiple target tissues by a receptor-mediated process (Colston and Feldman, 1982). Using this bioresponse marker in cultured fibroblasts, cells from a variety of patients with HVDRR have been shown, *in vitro*, to be resistant to the action of 1,25-$(OH)_2D_3$ (Feldman et al., 1982; Griffin and Zerwekh, 1983; Chen et al., 1984; Gamblin et al., 1985). Analysis has included assay of high affinity receptors by radiolabelled hormone binding, cellular uptake measurements of 1,25-$(OH)_2D_3$, induction of the 24-hydroxylase, specific monoclonal antibody association, and receptor–DNA interaction. A spectrum of molecular defects has emerged in the VDR of cells from these patients. Absent 1,25-$(OH)_2D_3$ binding is present in certain kindred fibroblasts, whereas others display a quantitative deficiency of receptor or depressed DNA interaction. All of the cell lines show an inability of 1,25-$(OH)_2D_3$ to activate the 24-hydroxylase enzyme. Consistent with the other steroid-like hormone systems, an intact multistep cellular process must be present for successful hormone-receptor regulation of appropriate gene activity, and this effector system is clearly compromised in HVDRR. Thus, this syndrome has provided an excellent model system not only for the analysis of critical gene loci and functional domains of the VDR but, presumably, will provide mechanistic insight into other hormone receptor systems in the steroid–thyroid family.

14.2.3 Families with Defective Receptor–DNA Interaction

(a) HVDRR Patients with Decreased Receptor–DNA Interaction

Two families with tissue resistance to 1,25-$(OH)_2D_3$ have recently been studied at both the protein and DNA level and shown to have a VDR with decreased affinity for DNA and an inability to activate target genes (Ritchie et al., 1989). The first family, identified as the D-kindred, is the result of a consanguineous marriage in a family that includes nine members, five of whom have been evaluated (Hochberg et al., 1984). The parents (denoted D4 and D5) are phenotypically normal, as is one unaffected daughter (D3). Two affected daughters (D1 and D2) display severe rickets and the classic childhood phenotype of HVDRR. The second family, identified as the G-kindred, are Arabs living in the Middle East (Malloy et al., 1989). The parents (G3 and G4) are first cousins and are phenotypically normal without calcium or bone abnormalities. Of their six children, two males display the HVDRR phenotype.

Fibroblasts were obtained from these family members and grown in culture for subsequent analysis of 24-hydroxylase activity, 1,25-$(OH)_2D_3$-binding, DNA-cellulose affinity, Western blot analysis using anti-VDR monoclonal antibody, and karyotyping. When cells were tested for 1,25-$(OH)_2D_3$-responsive 24-hydroxylase induction (Fig. 1), cultures from the G-family parents showed a hormone-dependent increase in enzyme activity. In contrast, the HVDRR cells from the affected children failed to exhibit enzyme induction even when treated with 100 nM 1,25-$(OH)_2D_3$, confirming that these cells were resistant to vitamin D and have a defect in the hormone action pathway (Malloy et al., 1989).

To investigate this further, the functional domains of the defective VDR were each tested for activity (Malloy et al., 1989). Initial studies assessed ligand binding and, as shown in Fig. 2, cells from each individual exhibited 1,25-$(OH)_2[^3H]D_3$ binding that is saturable and of high affinity. Scatchard analysis showed a K_d in the range of 0.02–0.04 nM and an N_{max} of 20–40 fmol (mg protein)$^{-1}$, both parameters being within the normal range.

Since the cells from these patients were unresponsive to hormone treatment but capable of binding 1,25-$(OH)_2D_3$ normally, the VDR was examined for altered physical properties by sucrose gradients and Western blot analysis (Malloy et al., 1989). Sucrose gradients of G1 and G3 VDR revealed a normal sedimentation value of 3.3 S. Western analysis using monoclonal antibody raised against the receptor displayed a protein of 48 000 D which was in excellent agreement with the value of 48 265 D determined from the human VDR cDNA clone (Baker et al., 1988). Because hormone binding was normal and the receptor size was appropriate, it was suspected that a minimal mutation in the receptor gene could represent the underlying genetic defect.

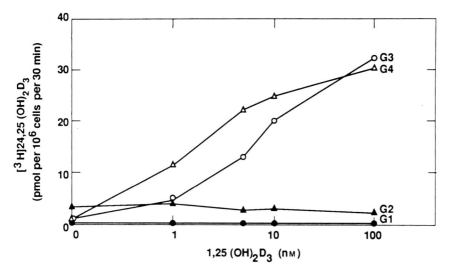

Fig. 1 The induction of 24-hydroxylase activity by vitamin D hormone in fibroblasts from a family with tissue resistance to 1,25-dihydroxyvitamin D. Cultured cells were treated with various concentrations of hormone for 20 h. The level of enzyme activity was measured by the conversion of tritiated 25-hydroxyvitamin D to tritiated 24,25-dihydroxyvitamin D. (From Malloy et al., 1989.)

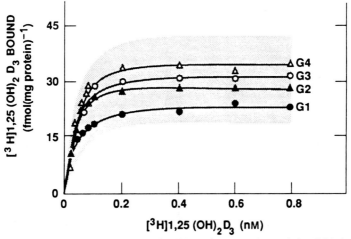

Fig. 2 Radioactive binding to VDR in fibroblasts from members of the G-kindred. The shaded area indicates the mean of five normal unrelated individuals. (From Malloy et al., 1989.)

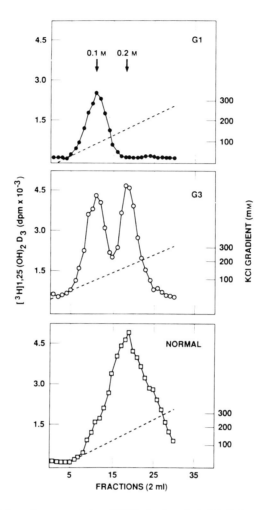

Fig. 3 DNA cellulose chromatography of VDR. Radioactively labelled receptor was applied to DNA cellulose column and eluted with increasing concentrations of KCl. The normal receptor (bottom panel) elutes at 0.2 M KCl, whereas an affected individual (G1, top panel) eluted at one-half this salt concentration of 0.1 M salt. The middle panel shows the chromatography of the father (G3), an obligate heterozygote. (From Malloy et al., 1989.)

The DNA-binding domain of the defective receptors was studied next by analysing the ability of VDR to associate with immobilized calf thymus DNA by DNA-cellulose chromatography, the salt concentration required to elute the receptor being a measure of affinity of protein for DNA. It is well established that the concentration of KCl required to elute wild-type VDR from DNA-cellulose is approximately 0.2 M (Chandler et al., 1984: Hirst et al., 1985; Liberman et al., 1986) and this result was confirmed by the elution pattern shown in Fig. 3, bottom panel. Several kindreds with HVDRR have been described in the literature (Hirst et al., 1985; Liberman et al., 1986) with receptors displaying decreased DNA affinity, the VDR eluting from the column at about 0.1 M KCl. When VDR from G1 was analysed by DNA chromatography (Fig. 3, top panel), a single peak of $1,25\text{-}(OH)_2[^3H]D_3$ eluted at this lower salt concentration (0.1 M KCl), suggesting that the protein possessed markedly depressed DNA affinity (Malloy et al., 1989). Interestingly, when extracts from the parent cells G3 were analysed, a mixed pattern was observed in which half of the receptor eluted from DNA at the normal 0.2 M KCl and half at 0.1 M KCl (Fig. 3, middle panel). Western blots on these column fractions showed that both peaks of VDR contain immunoreactive protein of normal size. Taken together these data suggested that the phenotypic abnormality in these kindreds was likely to result from a defective VDR–DNA interaction in the affected children, with both the abnormal and normal VDR alleles being expressed in the obligate heterozygote parents. Given the biochemical phenotype of defective DNA binding and completely appropriate hormone binding in a protein of normal size, a mutation in the amino-terminal (DNA-binding; see Chapters 2 and 3) portion of VDR was suspected.

(b) Amplification of Receptor DNA

A methodology was needed for rapidly scanning RNA or genomic DNA from a variety of mutational classes, not only for the patients and their families described above, but for additional kindreds with suspected VDR anomalies. Due to the low abundance of receptor mRNA in fibroblasts and transformed lymphoblasts used as the source of the DNA in many of these family studies, the strategy was to utilize the DNA amplification technique of polymerase chain reaction (PCR). The previous molecular cloning of the human vitamin D receptor cDNA (Baker et al., 1988) and the characterization and sequencing of the human chromosomal gene (J. W. Pike, unpubl. res.) allowed the amplification of specific regions of the VDR gene, isolation and sequencing of the amplified products, and interpretation of these data with regard to possible mutations. This approach avoided the tedious methodology of generating a genomic or cDNA library from each family member, followed by screening, mapping, and sequencing the isolated clones.

Figure 4 illustrates the organization of the human VDR chromosomal gene and the two pairs of oligonucleotide primers used to amplify exons 2 and 3 of the nine exons comprising the full length coding region of the gene

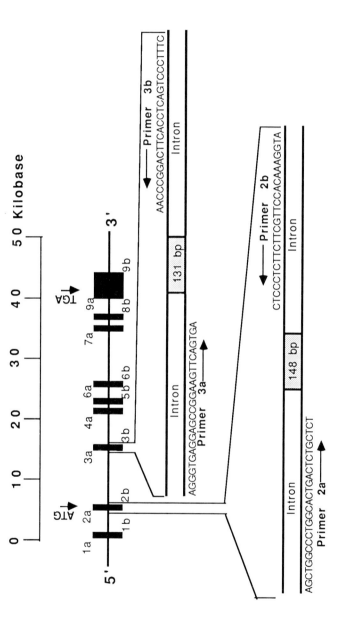

Fig. 4 Organization of the VDR chromosomal gene and polymerase chain reaction for strategy for amplification of exon regions. Oligonucleotide primers complementary to intron flanking region of each vitamin D receptor exon were generated to allow specific amplification of each of the functional domains of the VDR. (From Hughes et al., 1988.)

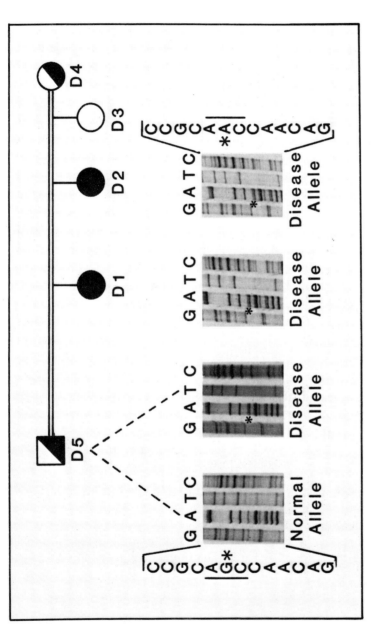

Fig. 5 Partial pedigree and corresponding VDR nucleotide sequence analysis in the D-kindred. Squares represent males, circles indicate females, filled symbols signify obvious clinical disease phenotype, and partially filled symbols indicate heterozygote carrier state for the disease allele. The double bar between the parents indicates consanguineous mating. The normal sequence of exon 3 (left) and the mutant sequence (right) are depicted for this family. A single base substitution (asterisk) of G to A was identified. (From Hughes *et al.*, 1988.)

(J. W. Pike, unpubl. res.). In order to identify potential splice site mutations, primers were designed to anneal to the intron flanking region of each exon such that the *TaqI* DNA polymerase would replicate 10–20 bp of genomic intron sequence prior to crossing the intron–exon boundary and generating the exon of interest. Since exons 4 and 5 as well as exons 7 and 8 were each separated by less than 250 nucleotides of intron, they were amplified together with single pairs of oligonucleotides. In this manner, all exons and limited flanking intervening sequences from the VDR chromosomal gene were amplified and sequenced to identify a potential genetic mutation (Hughes *et al.*, 1988).

Suspecting that the receptor defect may be due to an anomaly of DNA binding, special attention was focused on exons 2 and 3 (Fig. 4) of the VDR gene encoding this functional domain. Together these exons encode the 70 amino acid, cysteine rich, zinc finger region of VDR, one finger generated from each exon. Genomic DNA from cells of each family member was subjected to amplification and the products were sequenced directly from PCR or after subcloning. A reproducible single base mutation (C*G*A- → C*A*A) was identified in the triplet codon for arginine in exon 3 from both affected children of the D-kindred leading to the aberrant substitution of glutamine (Fig. 5). This arginine at residue 70 in the C-terminal DNA-binding finger is positionally conserved throughout all of the steroid receptors sequenced to date (see Chapters 2.4 and 3.3; and Evans, 1988). While only the mutant codon was found in children D1 and D2, parental genomic DNA was heterozygotic and generated both the mutant and the normal gene sequences. Random selection of PCR clones revealed 58% normal and 42% abnormal codons, consistent with the expected equal presence of both alleles in the heterozygote parents. The unaffected sibling (D3) possessed only wild-type sequence and therefore was not a carrier of the disease.

Similar experiments were conducted on members of the G-family (see Fig. 6). While VDR exon 3 was entirely normal in this kindred, a point mutation was identified in exon 2 from the affected children; the triplet codon change of GG*C* → G*A*C resulted in the conversion of glycine to aspartic acid at residue 30 in the N-terminal DNA binding finger. Glycine is a positionally conserved amino acid also in all the steroid (see Chapters 2.4 and 3.3) and thyroid (Chapter 4) receptors, suggesting that it plays a crucial role in protein–DNA interaction. As before, the G-family parents displayed both the normal and disease alleles.

(c) Creation of VDR Mutations with Subsequent Functional Analysis

The point mutations of exon 2 (G-family) and exon 3 (D-family) were the

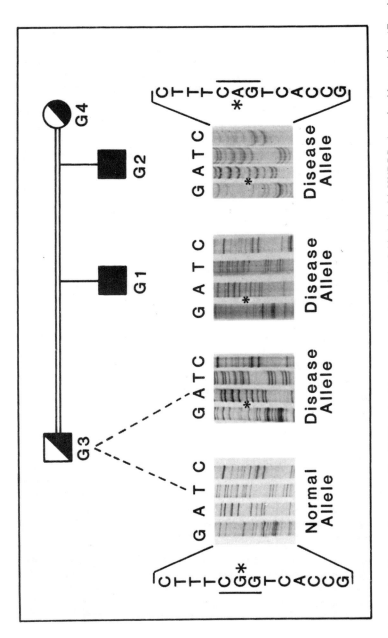

Fig. 6 Partial pedigree and corresponding VDR nucleotide sequences analysis of the G-kindred with HVDRR. A nucleotide transition (G to A) is seen in exon 2 from two affected members of this kindred. The obligate heterozygote parents display both the normal and disease alleles. (From Hughes et al., 1988.)

only VDR sequence variations uncovered in these kindreds. We suspected that these single amino acid substitutions represented the molecular defects responsible for the HVDRR disease phenotype for the following reasons: (1) consistent with an autosomal recessive pattern of inheritance, the mutation predictably segregated in the family, and the parents displayed both the wild-type and mutant alleles; (2) these were the only nucleotide changes found in the coding region for the receptor protein; and (3) the observed mutations were located within the DNA-binding domain of the receptor, consistent with the demonstrated biochemical defect. It was important, however, to confirm that these base changes were actually responsible for the disease phenotype and not simply DNA polymorphisms present in these families. Confirmation was obtained by recreating these nucleotide base changes in wild-type VDR cDNA using oligonucleotide site-directed point mutagenesis. The *in vitro* generated mutant cDNA containing the exact base alterations seen in the HVDRR families was placed into a eukaryotic expression plasmid and transfected into COS-1 (virtually receptor negative) monkey kidney cells. The expressed mutant protein sequence from the D-family contained a glutamine substituted for arginine at residue 70. It was expressed at normal steady-state levels and displayed $1,25\text{-(OH)}_2[^3\text{H}]D_3$ binding of high affinity that was indistinguishable from wild-type (Malloy *et al.*, 1989). As predicted, DNA-binding analysis suggested that the affinity of the mutant receptor was markedly decreased for DNA cellulose (Fig. 7C), consistent with that observed in both the affected patient as well as the heterozygote parent (compare Figs 7A and 7B). The mutant receptor eluted from DNA cellulose at approximately 0.1 M KCl, in contrast to the normal receptor which eluted at 0.2 M KCl.

(d) Transcriptional Activation Capacity of the Mutant Receptors

A question still remained concerning the ability of these mutant receptors to interact functionally with a specific hormone response element in a gene normally regulated by an intact $1,25\text{-(OH)}_2D_3$-receptor complex. Osteocalcin (bone matrix Gla-protein) is encoded by a gene transcriptionally regulated by $1,25\text{-(OH)}_2D_3$ (Zerwekh *et al.*, 1979) and the hormone response element (VDRE) has been localized and sequenced to a site approximately 500 bp upstream of the mRNA start site (Kerner *et al.*, 1989). This sequence confers $1,25\text{-(OH)}_2D_3$ response to a heterologous promoter and strongly resembles the DNA *cis*-acting elements of genes regulated by other members of the steroid-thyroid family of receptors (see Chapter 6.2).

Pike and co-workers (Kerner *et al.*, 1989) have inserted this osteocalcin promoter/enhancer gene fragment containing the VDRE into the chloramphenicol acetyltransferase expression vector, creating a human osteocalcin-CAT reporter plasmid. In addition, normal or mutant VDR cDNA has been cloned into the adenovirus major late promoter-directed expression vector. Cotransfection of both the VDR expression vector and VDRE-CAT reporter plasmid into CV-1 receptorless cells provided a method for assessing the activity of the receptor and various VDR mutants on promoter–enhancer function. Active VDR was measured by the ability of 1,25-$(OH)_2D_3$ hormone to induce CAT activity over the basal uninduced state (Kerner et al., 1989).

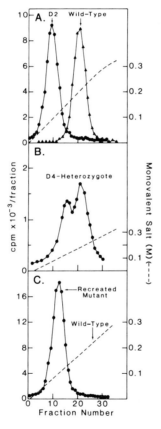

Fig. 7 DNA cellulose elusion profiles of vitamin D receptor from normal individuals, members of the D-kindred, and transiently transvectant mutant receptor. (A) Elution profiles of wild-type VDR (triangles) and receptor from a typcial patient (D2). (B) Profile of VDR isolated from a parent (D4) showing two receptor forms, one eluting at 0.2 M salt coincident with wild-type receptor, and the other at 0.1 M salt coincident with the elution position of VDR from the affected offspring. (C) Profile of deliberately mutated VDR after recreation of the specific mutation found in this family.

This *cis–trans* cotransfection assay was used to monitor transcriptional activity in the receptor mutants from these patients (Stone et al., 1989). As shown in Fig. 8, the wild-type receptor induced strong CAT activity via its transcriptional activation of the osteocalcin promoter in the presence of hormone. This effect was dependent on the amount of expression vector transfected, indicating that the cellular concentration of receptor is capable of driving reporter activity up to 20-fold above background in this experiment. In contrast, when mutant receptors containing the base alterations identified in the HVDRR kindreds were examined with this system, neither the mutant protein from the D-family nor that from the G-family was capable of any significant transcriptional stimulating activity (Fig. 8). Only background activity was observed even when the mutant proteins were significantly overexpressed relative to wild-type receptor. Clearly, these point mutations in the DNA-binding zinc fingers were catastrophic events in terms of receptor transcriptional activity.

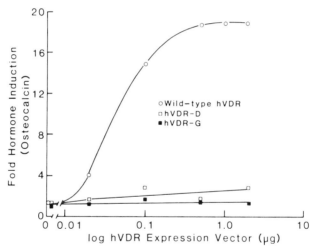

Fig. 8 Cotransfection assay measuring transcriptional activation capacity of normal and mutant vitamin D receptors. Active receptor is determined by the capacity of 1,25-dihydroxyvitamin D (10^{-8} M) to induce the reporter gene (CAT). (From Stone et al., 1989.)

(e) Effect of Hormone on Transcriptional Activation

The effect of 1,25-$(OH)_2D_3$ hormone concentrations on VDR activation and promoter function was tested to determine whether high levels of hormone might be capable of increasing transcriptional activity by the mutant receptors (Stone et al., 1989). This question arose because it is common in clinical practice to treat HVDRR patients with pharmacological doses of 1,25-$(OH)_2D_3$ or other vitamin D metabolites (Balsan et al., 1983; Marx, 1989). The efficacy of this therapeutic approach is far from clear with more than half of these patients, especially those with alopecia, who fail to respond clinically to any treatment short of intravenous calcium infusions (Marx, 1989). We might speculate that this treatment regimen will be shown to be most useful in that subset of patients with the hormone binding class (K_d) of defective vitamin D receptors (Castells et al., 1986). The cotransfection was used to test whether increased hormone occupancy might induce higher transcriptional activity. Osteocalcin promoter activity was found to be markedly elevated in the presence of ligand bound native receptor, with physiological concentrations of 1,25-$(OH)_2D_3$ producing obvious enhancement of transcription. In direct contrast to this wild-type activity, neither mutant receptor was capable of transcriptional induction even with markedly increased hormone occupancy. This underscored the completely inactive nature of the mutant proteins identified in these two families with HVDRR. While they were expressed at steady-state levels and displayed hormone binding capabilities comparable to that of the wild-type VDR, these mutant receptors bound poorly to heterologous DNA and were transcriptionally inactive even when over-expressed or exposed to high levels of hormone. Taken together, these data provided compelling evidence that HVDRR in these two families evolved as a direct result of the inheritance of homozygous receptor gene alleles encoding point mutations which inactivate the 1,25-$(OH)_2D_3$ receptor. Since these receptors were transcriptionally inactive both *in vivo* and *in vitro*, it demonstrated that the major clinical manifestations of this disease, i.e. hypocalcaemia and defective skeletal mineralization, was due to inactivity of the 1,25-$(OH)_2D_3$ receptor at the level of gene activation.

Figure 9 illustrates the VDR amino acid sequence of the DNA binding domain placed into the zinc finger motif. The D-kindred mutation replaces a very basic arginine residue ($pK \simeq 12$) for an uncharged glutamine at a prominent position in the second finger motif. In the G-kindred, the mutation changes a neutral glycine to a negatively charged aspartic acid in the first zinc finger structure. Both arginine and glycine residues are conserved at these positions in all of the nuclear receptors (Chapters 2 and 3), and this evolutionary conservation implies that these amino acids are important to receptor–DNA interaction. It is unclear whether the decreased function of these mutants is because of charge differences or whether the domain conformation in this region of the protein is altered. Related point mutations have been made in this region of the glucorticoid receptor (Chapter 2.4) and oestrogen receptors (Chapter 3.3), and the results suggest that alteration of conserved amino acids at the tip of the fingers generally do

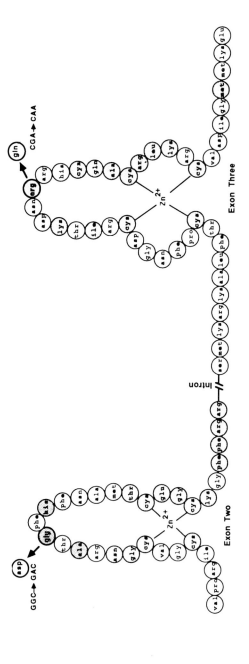

Fig. 9 Amino acid sequences of the VDR DNA-binding domain at the positions of the two mutations that were identified in this region. (From Hughes *et al.*, 1988.)

not compromise DNA binding in those systems. This apparent discrepancy could be due either to some artifact of the *cis–trans* assays used to assess the oestrogen and glucocorticoid receptors, to structure–function differences among the receptors, or to the participation of other critical proteins/factors unique to VDR–DNA interaction. Further clarification will require methods for measuring receptor dimerization and interactions with other proteins (Chapter 7.3). Certainly, crystallization of the receptor protein complexed to the hormone response element will assist in this regard. Until then, the precise mechanism by which these and related mutations affect receptor–DNA interaction will only be conjecture.

(f) Seven Families with a Truncated D-Receptor in the Ligand-binding Domain

A second major group of patients with HVDRR display a different class of molecular defects, namely decreased or absent $1,25\text{-}(OH)_2D_3$ binding. The phenotype may result from either an absence of receptor protein or a dysfunctional hormone-binding domain. This phenotype is the most commonly described in this disorder, and most of the patients display immunoreactive receptor protein but have undetectable high affinity binding of the hormone (Feldman *et al.*, 1982; Liberman *et al.*, 1983; Chen *et al.*, 1984; Gamblin *et al.*, 1985).

An extended kindred with seven related family branches (Fig. 10) having eight children affected with HVDRR has been recently studied at the molecular level. Complex consanguinity was present among the families, there being multiple documented marriages between related individuals, leading to an increased prevalence of this rare disorder. The affected children all clinically presented in a similar manner with early onset rickets, complete hair loss prior to three months of age, dental caries by 2 years, and the aforementioned classic features of HVDRR, including a markedly elevated circulating hormone level. Many of the children have been pharmacologically treated with $1,25\text{-}(OH)_2D_3$ without clinical effect. The obligate heterozygous parents appeared phenotypically normal, displaying no evidence of bone disease, hair loss, or abnormal levels of D-metabolites.

The mutation causing HVDRR in this extended kindred appeared to be the same in all affected children and resulted in undetectable VDR as measured by $1,25\text{-}(OH)_2[^3H]D_3$ binding analysis (Chandler *et al.*, 1980; Chen *et al.*, 1984; Pike *et al.*, 1987). However, Scatchard analysis was normal in cells available from the parents. Western blot analysis using monoclonal antibody with a receptor epitope near the carboxyl side of the VDR zinc finger domain failed to identify the 48-kD $1,25\text{-}(OH)_2D_3$ receptor in extracts of cells from these patients. Careful scrutiny of the immunoblot also failed to

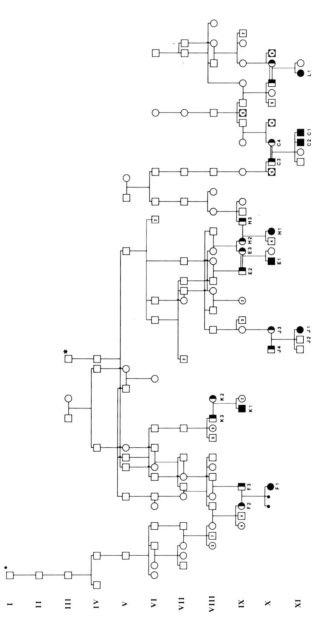

Fig. 10 A pedigree of an extended family with HVDRR. Affected individuals with the classical phenotype are indicated by the solid symbols, whereas the presumed heterozygotic parents are represented by half-solid symbols. Interfamily marriages are shown by double lines.

show any additional band that might represent a truncated receptor. Cells from the heterozygote parents tended to show approximately the same abundance of VDR protein on Western blot as by ligand binding. With functional loss at the C-terminus (hormone-binding domain) of the protein and the apparent absence of antibody binding toward the N-terminus of the receptor, it was hypothesized that either no VDR was synthesized in these cells or an altered VDR protein, caused by a mutation affecting VDR gene transcription or mRNA translation, was synthesized but not detected (Malloy et al., 1990).

Further analysis using Northern blots to measure steady-state levels of VDR mRNA showed no detectable hybridizing bands in many of the patients. However, several cell lines demonstrated very low abundance of a band migrating at the normal 4.4 kb VDR mRNA size, and cells from one patient (F1) clearly displayed an mRNA of normal size but decreased abundance. Confirmation that VDR mRNA was actually present in these patient cells came from PCR amplification of total RNA using oligonucleotide primers complementary to the authentic receptor. The amplified material, when probed with VDR cDNA clearly showed the expected 4.4 kb product (Malloy et al., 1990). These data indicated that perhaps the level of VDR mRNA transcription was depressed or mRNA turnover was increased in cells from these patients. Still, it was not fully apparent that the profound and refractory phenotype of these affected children could be explained solely on decreased VDR mRNA levels.

PCR-based exon amplification of the genomic DNA from selected members of this large kindred was performed as described earlier. Amplification revealed that all of the exons encoding the VDR gene were indeed present and that the sizes of the amplified exons were indistinguishable from those obtained from normal individuals. DNA sequencing of these exons indicated that a point mutation (C→A) was present in exon 7 at position 970 relative to the Cap site of the mRNA. The presence of this mutation changed the normal tyrosine codon (TAC) of amino acid 292 to a termination codon (TAA). The same mutation was found in multiple children (C_1, C_2, E_1, F_1 and H_1) in the family. When translated, this premature termination signal would be predicted to cause the expression of a truncated 291-amino acid receptor with a large portion of the steroid hormone-binding domain deleted.

The genomic DNA of the parents was also amplified and sequenced and showed both the normal (TAC) and mutant (TAA) allelic sequences, confirming their heterozygous state. The remaining seven exons from the patients and parents displayed normal wild-type DNA sequences. Here, the mutant alleles identified were the same in all members of the pedigree. This genetic defect additionally created an altered RSA I restriction enzyme site converting the nucleotide sequence GTAC to GTAA. PCR amplification of exons 7 and 8 followed by RSA I digestion produced an easily identified shifted DNA banding pattern on agarose gel electrophoresis (Malloy et al., 1990). This allowed the rapid screening of other members of the extended family for the defective VDR allele and easily identified the homozygous patients and heterozygous carriers.

The consequences of the premature termination mutation was examined in more detail by reconstructing the mutant receptor from wild-type VDR cDNA using site-directed mutagenesis (Stone et al., 1989). When transfected into COS-1 monkey kidney cells and probed with the monoclonal antibody on Western blots, a prematurely truncated 32-kD protein was expressed (Fig. 11). The receptor fragment was unable to bind $1,25\text{-}(OH)_2[^3H]D_3$ and was expressed in considerably lower amounts than the normal 48-kD receptor. This finding in overexpressing heterologous cultured cells was consistent with the data showing the difficulty in identifying the receptor fragment on Western blots from patient fibroblasts.

The ability of the truncated mutant VDR to initiate transcriptional activity in the presence of $1,25\text{-}(OH)_2D_3$ was tested, as before, using the cotransfection assay. As described earlier in this text (Chapters 2.3, 3.4 and 4.8), numerous laboratories have demonstrated that mutations in the hydrophobic C-terminal region of steroid-like receptors are generally incapable of binding hormone. This premature stop codon (TAA 'ochre' mutation) produced a protein missing more than two-thirds of the ligand-binding domain. While CAT expression was enhanced in the presence of $1,25\text{-}(OH)_2D_3$ when these cells were cotransfected with the wild-type VDR cDNA, it was not surprising that there was no induction of CAT activity in cells transfected with the mutant VDR cDNA lacking the hormone-binding domain. Similar results were demonstrated (McDonnell et al., 1989a) when intentionally truncated VDR peptides of 362 and 281 amino acids were examined for CAT gene expression. In fact, removal of as few as 27 amino acids from the C-terminus of VDR abolished hormone binding. This is clearly different from results obtained for the oestrogen (Kumar et al., 1986), progesterone (Conneely et al., 1988), and glucocorticoid (Chapter 3.4 and Hollenberg et al., 1985) receptors, and underscores some of the heterogeneity of these proteins with respect to their functional ligand-binding domains.

A premature termination codon in an intact VDR mRNA would not, at first glance, be expected to yield the results that were obtained on RNA blots and immunoblots from these HVDRR cells. The almost undetectable 4.4 kb mRNA on Northern blots and the absence of the 32-kD truncated VDR peptide on Western blots suggest that additional mechanisms of control beyond the nonsense mutation may be operative. The mechanisms involved

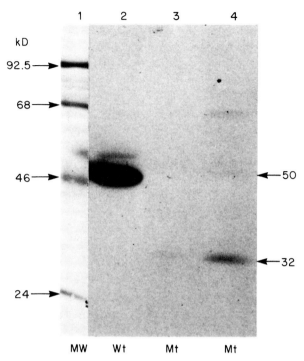

Fig. 11 Western blot analysis of transvected COS-1 cells expressing wild-type and ochre mutant VDR cDNA. Extracts of transfected cells were subjected to SDS/PAGE, transferred to nitrocellulose, and probed with anti-VDR monoclonal antibody 9A7. Lanes: 1, molecular mass standards (kDa); 2, 30 μg of cellular protein containing wild-type VDR; 3 and 4, 30 and 90 μg, respectively, of cellular protein containing ochre VDR mutant. Wild-type VDR are indicated at 50 kD and the mutant VDR at 32 kD.

in mRNA turnover rates are far from understood and it is not clear why the presence of a premature termination signal in the VDR mRNA causes decreased steady-state message. Several other genetic diseases including some forms of β-thalassaemia, Tay Sachs disease, and Duchenne muscular dystrophy have been associated with nonsense mutations and decreased levels of mRNA and the encoded protein gene product (Neufeld, 1989; Sicinski et al., 1989; Orkin, 1990). It is possible that intranuclear mRNA instability or increased turnover of mRNA is responsible for these results. Furthermore, any abnormal protein that is translated might be susceptible to an accelerated turnover rate as well. Whatever the mechanism, the genetic defect in these families was found to be a VDR nonsense mutation which produces lowered steady-state mRNA and a defective protein incapable of binding 1,25-$(OH)_2D_3$, a prerequisite for transcriptional activity. This leads to a 1,25-$(OH)_2D_3$-resistant state and the development of the HVDRR clinical syndrome.

It remains to be determined whether a wide range of VDR mutations are responsible for this disorder in the population, or whether a limited number of defective receptor genotypes, perhaps representing varied ethnic ancestry (e.g. phenylketonuria; Woo, 1989), will be identified. A nonsense mutation producing receptor truncation and mis-sense mutations in the DNA-binding domain are possibly just the beginning of numerous genetic receptor defects to be uncovered in HVDRR. Given the structural and functional similarities between members of the steroid-like receptor gene family, it is not surprising that target-tissue resistance to other hormones such as androgen (Kovacs et al., 1986; Griffin and Wilson, 1989), glucocorticoid (Chrousos et al., 1983a,b), and mineralocorticoid (Armanini et al., 1985) have also been observed.

14.3 Syndrome of androgen insensitivity

Certainly the classic syndrome involving absent or defective steroid response is testicular feminization (Tfm), a condition in which XY genotypic individuals with testis and elevated serum testosterone have tissue unresponsiveness to androgens and thus display a female sexual phenotype. Since it was first described in 1953, this condition has been studied extensively, and now is referred to by the more descriptively correct term androgen insensitivity syndrome (AIS) (for review, see Griffin and Wilson, 1989). The androgen receptor cDNA has been cloned simultaneously by four groups and the gene organization determined (Chang et al., 1988; Lubahn et al., 1988; Trapman et al., 1988; Tilly et al., 1989). The full-length cDNA sequence encodes a protein of approximately 920 amino acids which binds hormone with high affinity and specificity, as well as binding receptor-specific antibodies. The rat cDNA has also been cloned (Chang et al., 1988) and shown to have identical DNA and hormone-binding amino acid sequences with that of the human receptor, with an overall sequence homology of 85%.

The gene, which has been localized to chromosome Xq11–12 (Brown et al., 1989), is over 90 kb in length, but less than 5% of it (divided into 8 exons) is translated into protein. The first exon encodes a large (1589 bp), probably modulatory, amino terminal portion of the receptor. Exons 2 and 3 (152 and 117 bp, respectively) generate the DNA-binding domain, each encoding a zinc finger. The androgen-binding domain is encoded by 5 exons which vary in size from 131 to 288 bp in size. The exon organization of the mammalian gene for androgen receptor is most closely related to the progesterone receptor gene (Huckaby et al., 1987).

Mutations in the androgen receptor have long been suspected to be responsible for complete androgen insensitivity and now this has been supported by direct genetic evidence. Analysis of somatic cell clones for X-inactivation (Meyer *et al.*, 1975; Elawady *et al.*, 1983) and direct linkage analysis (Wieacker *et al.*, 1987) incriminated mutations at Xp11–12 (the map location of the androgen receptor) as the cause of human AIS. Most recently, mutations in the human androgen receptor gene have been identified in families with this syndrome.

Given the heterogeneity in phenotypic expression due to the wide variety of identified androgen receptor protein defects (Griffin and Wilson, 1989), French, Migeon and co-workers (Brown *et al.*, 1988) used cDNA probes to study patients from six unrelated families with complete AIS and undetectable androgen receptor. The rationale was that these would be the most likely patients to carry a receptor gene deletion. The Southern profiles were normal in five of six patients but, in one patient, a partial deletion involving the steroid-binding domain was detected. A second patient has been reported (Pinsky *et al.*, 1989) with complete AIS and, perhaps, two deletions in the androgen gene; one deleted region involves the DNA-binding domain and one-third of the hormone-binding domain, and another eliminates most of the exon 1 modulatory domain. Since the patient also had mental retardation, it was postulated that this may reflect a contiguous gene syndrome. Another patient with AIS and 47,XXY Klinefelter syndrome has been described in which, using DNA markers, it was demonstrated that the supernumerary X chromosome resulted from maternal non-disjunction during meiosis II (Gerli *et al.*, 1979). The error at this stage provided the basis for homozygosity of the mutation at the androgen receptor locus.

Point mutations have most recently been described in the androgen receptor. A particularly interesting base alteration involves a CpG dinucleotide site in exon 6 of the hormone binding domain (Trifiro *et al.*, 1989). Using PCR amplification of total RNA from a patient with complete AIS, a $CGC \rightarrow TCG$ transition was uncovered which alters the sense of non-polymorphic codon 773 from an Arg to Cys, and abolishes a *KpnI* site. Arg_{773} of the androgen receptor steroid-binding domain is evolutionarily conserved at this position in the progesterone, glucocorticoid and mineralocorticoid receptor genes. This same defect was identified in a different patient (relationship of two patients unknown) by Brown and Corden (pers. commun.) who went on to test the mutation in a cotransfection assay using a mouse mammary tumour virus-CAT reporter gene. Induction of CAT activity by this defective receptor was absent.

An additional mis-sense mutation in the androgen receptor has been recently identified by Brown and Corden (pers. commun.) in a subject with complete AIS and absent receptor by hormone and antibody binding. The mutation was found in exon 7 and changed Arg_{831} to a glutamine (C*G*A to C*A*A) destroying a *TaqI* site. This receptor defect in the steroid-binding domain of the androgen receptor was also completely inactive in driving gene transcription in the cotransfection assay.

A nonsense mutation converting the codon TG$G \to$ TGA (an Opal mutation) has been identified by Liao and co-workers (pers. commun.) in exon 4 of the receptor. This premature stop codon was substituted for the normally present Trp_{717} and it truncates approximately 200 amino acids from the carboxy terminus of the protein. This tryptophan is positionally conserved in progesterone, glucocorticoid, mineralocorticoid and oestrogen receptors. The base alteration was present in obligate heterozygote parents and destroyed an *HaeIII* recognition site, making rapid genotyping possible in this particular family.

Finally, Marcelli and co-workers (1990) have identified a single nucleotide substitution in the androgen receptor gene which introduces a premature TGA termination codon. A patient with receptor-negative androgen resistance was examined and found to have a single base change of guanosine to adenosine (TG$G\to$TGA; $Trp_{794}\to$stop) in exon 6 of the androgen receptor cDNA. Analysis of receptor mRNA from the patient fibroblasts using S_1-nuclease protection showed normal quantities of mRNA in fibroblasts from this patient. Truncation of about 125 amino acids from the carboxy terminus (steroid binding domain) of the protein effectively abolished dihydrotestosterone binding both in receptor examined directly from the patient as well as *in vitro* mutated receptor expressed in culture. It is interesting that the premature termination codon described here did not lead to reduced levels of mRNA as was seen in the case of the ochre mutation of the 1,25-$(OH)_2D_3$ receptor (described on p. 337). These early termination signals occur at approximately the same location in the cDNA of VDR and androgen receptors (428 and 375 nucleotides, respectively, from the authentic TGA stop codon), and it is unclear what differences exist in the mechanisms affecting these steady-state message levels.

Androgen resistance is probably the most commonly described hormone insensitivity syndrome. Its frequency may be even larger than realized when one considers a recent report of a family of five men with mild gynecomastia and 'undervirilization', four of whom had fathered children, in a pedigree pattern consistent with X-linked recessive inheritance. Fibroblasts cultured from genital skin displayed normal levels of androgen receptor and receptor affinity for dihydrotestosterone. However, androgen binding in fibroblast monolayers was thermolabile, up-regulation of receptor levels did not occur after prolonged incubation with hormone, and dissociation rates at 37°C were increased with the synthetic androgen mibolerone. In addition, in cytosol preparations the androgen receptor was unstable. It was suggested that this disorder may represent the most subtle functional abnormality of androgen receptor yet characterized. Certainly androgen insensitivity may be causative in many cases of male infertility (Aiman et al., 1979, 1984). It has been estimated that as many as 40% of azoospermic men with otherwise unknown aetiogy for their infertility might have reduced numbers of androgen receptors (Aiman and Griffin, 1982). This alone would make androgen resistance more common than all the other hormone resistance syndromes combined.

There are several possible reasons for this apparently high frequency of androgen resistance syndromes. First, there may be substantial case ascertainment bias. Defects in androgen action often result in abnormal sexual development or in reproductive failure, conditions which are astutely monitored by most individuals. Second, the receptor gene is on the X-chromosome and defects become clinically manifested in the hemizygous (XY) state. Consequently, new mutations in the androgen receptor gene will become clinically obvious immediately. Since described mutations of receptor genes located on autosomes are only manifested clinically in the homozygous state, many new mutations go unnoticed. Finally, expression of androgen activity is necessary for reproduction but not for the life of an individual. As a consequence, those with even the most complete forms of androgen resistance have a normal life-span. Mutations affecting the action of hormones such as $1,25\text{-}(OH)_2D_3$ and aldosterone produce clinical diseases with marked health consequences. Genetic defects have not yet been reported for oestradiol or progesterone, perhaps because those hormones are necessary for blastocyst implantation and early embryogenesis. Profound mutations in thyroid, retinoid or glucocorticoid receptors in many cases might lead also to fetal wastage. Such is not the case, however, for milder defects in glucocorticoid receptor where tissue resistance in cell lines (see Chapter 3.4), patients, and new world primates have now been described.

14.4 Glucocorticoid receptor resistance

(a) Autosomal Recessive Primary Cortisol Resistance

Primary cortisol resistance was first described in 1976 in a family with high circulating levels of plasma cortisol and no evidence of Cushing's syndrome (Vingerhoeds et al., 1976). Since that time, end-organ glucocorticoid resistance has been described in 7 kindreds containing some 20 individuals, and several species of New World primates (Chrousos et al., 1982; Nawata et al., 1984; Iida et al., 1985; Lamberts et al., 1986; Tomita et al. 1986). The best studied kindred has been that of a Dutch family with serum cortisol concentration, cortisol production rate, and ACTH levels greatly increased above normal values (Chrousos et al., 1982; Tomita et al., 1986). The index individual from this family displayed a three-fold reduction in glucocorticoid receptor ligand binding affinity, with variable reduction in affinity in his mildly affected son and nephew. Chrousos and co-workers have studied this family extensively, and recently obtained the 2.3 kb coding region of the glucocorticoid receptor cDNA by reverse transcription of RNA from lymphoblasts of this affected individual (Hurley et al., 1990). PCR amplification of the cDNA from this father showed the amplified product to be of the expected normal length, excluding the possibility that a major deletion was present in the transcript. Subsequent DNA sequencing identified a single base substitution at position 2054. The normal adenosine residue had been replaced by a thymidine ($GAC \to GTC$) changing the codon from Asp_{641} to Val. Both sequences were present in the heterozygous son and nephew and the allele bearing the substitution segregated with the disease phenotype, the severely affected father being homozygous and the mildly affected son and nephew being heterozygous. This mutation was in the hormone-binding domain and altered an amino acid (Asp) that is conserved in glucocorticoid receptors from all species studied to date. It is interesting that an amino acid just three residues away (Cys_{638}) has been shown by dexamethasone 21-mesylate labelling to be within the glucocorticoid receptor ligand-binding cavity and to be intimately involved in hormone binding (Simons et al., 1987). Hence, replacement of the nearby acidic Asp_{641} with a neutral valine provides a plausible mechanism for impaired ligand binding and glucocorticoid resistance in this family.

In the mouse glucocorticoid receptor this adjacent cysteine residue corresponds to Cys_{644}. It is known from receptor-minus mutants of the mouse that two other residues, Glu_{546} and Tyr_{770}, are intimately involved in steroid recognition, since replacement of a glycine at position 546 or an asparagine at 770 markedly increases ligand dissociation rates (Danielsen et al., 1986, 1989; see Chapter 3.4). Both the glutamine and tyrosine are conserved in the human glucocorticoid receptor homologue. Hence, the hormone binding domain must be intricately folded in such a way that murine glucocorticoid receptor residues 546, 644, and 770 are part of the ligand binding pocket. Although significant information can be obtained *in vitro* using oligonucleotide site-directed mutagenesis to intentionally alter amino acid residues, the *in vivo* consequences of these amino acid substitutions are best appreciated in patients and murine models.

(b) Glucocorticoid Resistance in New World Primates

Many of the New World primates exhibit glucocorticoid 'resistance' without apparent pathology. These animals have markedly elevated plasma cortisol and ACTH and their receptors have decreased affinity for glucocorticoids (Brandon et al., 1989). In addition, these receptors are thermolabile and lymphoblasts from the animal are not induced by EB virus transformation to up-regulate receptor concentration, as measured by specific binding and mRNA expression. Western and Nothern blots indicate that both the protein and mRNA are of normal size in these animals (Brandon et al., 1989). Recently, the glucocorticoid receptor has been cloned and sequenced and numerous base substitutions identified, particularly in the hormone-binding domain (G. P. Chrousos, pers. commun.). It remains to be determined which of these amino acid alterations are most responsible for the decreased hormone-binding affinity and generalized receptor lability. It is particularly interesting that these primates also possess generalized oestrogen, progesterone, androgen, aldosterone and $1,25\text{-}(OH)_2D_3$ insensitivity as manifested by high circulating levels of each of these steroid hormones, combined with low concentrations and/or ligand affinity of the respective target tissue receptor. In contrast to the human steroid resistance syndromes, this condition in monkeys appears to be a well-adapted evolutionary change that represents a variant of normal. The selective advantage to the animal is unclear, but it is reasonable to predict that clues into the mechanism of action of these ligand receptors will be gained by comparing and contrasting these biochemically distinct molecules.

14.5 General comments

Elucidation of the pathophysiology of steroid hormone resistance syndromes has been important in providing insight into the normal pathway of nuclear receptor action. Each new type of resistance that is recognized and the mutation responsible identified provides an opportunity for defining the nature of a specific reaction essential for the action of the hormone. How does the affinity of receptor for DNA (Chapter 9) affect transcriptional regulation? How does ligand binding alter receptor conformation and activity at the genome? What additional functional domains and requisite trans-acting factors (Chapters 5.5, and 7.3) are yet to be defined for the receptor molecule? Site-directed point mutagenesis of the various receptors has provided many clues to better identify amino acids which contribute significantly to receptor function. The interpretation of such studies has, in many instances, been difficult owing to limitations in the cis/trans cotransfection assay. In this assay, the receptors are expressed in heterologous cells together with a steroid response element linked to a reporter gene such as CAT or luciferase. Much of the data and most of the structure–function conclusions presented throughout this book have relied on this method of measuring receptor activity. However, a number of biological differences between cotransfected endogenous genes exist. For instance, cotransfected genes exist as episomes and are unlikely to exist in a native chromatin structure. Furthermore, transfection efficiencies are low such that only a fraction of the cells acquire the DNA. Since receptors are often overexpressed in this subpopulation of cells, transfected cells may contain wild-type or mutant receptors in vast excess to that which occurs naturally. Although this type of assay should reflect the gross functional state of the nascent mutagenized protein, more subtle assessments regarding functionality may be difficult to detect.

Further characterization of patients with steroid-resistant disorders should provide important insight into how these proteins function in target cells. In contrast to random *in vitro* mutagenesis, appropriate selection of genetic diseases of hormone resistance in humans and animals should permit the investigation of subtle mutations which result in decreased function under physiological *in vivo* conditions. The study of nature's mutational experiments have been invaluable in understanding many human disease conditions as well as normal physiological processes. It is certain that the mutations producing tissue resistance to nuclear receptors will, when understood, collectively improve out knowledge about the mechanisms by which these proteins regulate cellular transcription.

Acknowledgements

The authors would like to acknowledge our collaboration with the research groups of David Feldman (Stanford) and J. Wesley Pike (Baylor) in generating the vitamin D receptor data described in this chapter. We also thank M. McPhaul, D. Lubahn, F. French, J. Wilson, G. Chrousos and their co-workers for sharing prepublished data.

References

Aiman, J. and Griffin, J. E. (1982). The frequency of androgen receptor deficiency in infertile men. *J. Clin. Endocrinol. Metab.* **54**, 725–732.

Aiman, J., Griffin, J. E., Gazak, J. M., Wilson, J. D. and MacDonald, P. C. (1979). Androgen insensitivity as a cause of infertility in otherwise normal men. *New Engl. J. Med.* **300**, 223–227.

Aiman, J., Griffin, J. E., Gazak, M. M., Wilson, J. D. and MacDonald, P. C. (1984). Androgen insensitivity as a case of infertility in otherwise normal men. *J. Clin. Endocrinol. Metab.* **59**, 672–678.

Armanini, D., Kuhnle, U., Strasser, T., Dorr, H., Butenandt, I., Weber, P. C., Stockigt, J. R., Pearce, P. and Funder, J. W. (1985). Aldosterone-receptor deficiency in pseudohypoaldosteronism. *New Engl. J. Med.* **313**, 1178–1181.

Baker, A. R., McDonnell, D. P., Hughes, M. R., Crisp, T. M., Mangelsdorf, D. J., Haussler, M. R., Pike, J. W., Shine, J. and O'Malley, B. W. (1988). Cloning and expression of full-length cDNA encoding human vitamin D receptor. *Proc. Natl. Acad. Sci. USA* **85**, 3294–3298.

Balsan, S., Garabedian, M., Liberman, U. A., Eil, C., Bourdeau, A., Guillozo, H., Grimberg, R., Le Deunff, M. J., Lieberherr, M., Guimbaud, P., Broyer, M. and Marx, S. J. (1983). Rickets and alopecia with resistance to 1,25-dihydroxyvitamin D: two different clinical courses with two different cellular defects. *J. Clin. Endocrinol. Metab.* **57**, 803–811.

Berger, U., Wilson, P., McClelland, R. A., Colston, K., Haussler, M. R., Pike, J. W. and Coombes, R. C. (1988). Immunocytochemical detection of 1,25-dihydroxyvitamin D receptors in normal human tissues. *J. Clin. Endocrinol. Metab.* **67**, 607–613.

Brandon, D. D., Markwick, A. J., Chrousos, G. P. and Loriaux, D. L. (1989). Glucocorticoid resistance in humans and nonhuman primates. *Cancer Res.* **49**, 2203s–2213s.

Brooks, M. H., Bell, N. H., Love, L., Stern, P. H., Orfei, E., Queener, S. F., Hamstra, A. J. and DeLuca, H. F. (1978). Resistance of target organs to 1,25-dihydroxy-vitamin D. *New Engl. J. Med.* **298**, 996–1000.

Brown, C. J., Goss, S. J., Lubahn, D. B., Joseph, D. R., Wilson, E. M., French, F. S. and Willard, H. F. (1989). Androgen receptor locus on the human X chromosome: regional localization to Xq11-12 and description of a DNa polymorphism. *Am. J. Hum. Genet.* **44**, 264–269.

Brown, T. R., Lubahn, D. B., Wilson, E. M., Joseph, D. R., French, F. S. and Migeon, C. J. (1988). Deletion of the steroid-binding domain of the human androgen receptor gene in one family with complete androgen insensitivity syndrome: evidence for further genetic heterogeneity in this syndrome. *Proc. Natl. Acad. Sci. USA* **85**, 8151–8155.

Castells, S., Greig, F., Fusi, M. A., Finberg, L., Yasumura, S., Liberman, U. A., Eil, C. and Marx, S. J. (1986). Severely deficient binding of 1,25-dihydroxyvitamin D to its receptors in a patient responsive to high doses of this hormone. *J. Clin. Endocrinol. Metab.* **63**, 252–256.

Chandler, J. S., Pike, J. W., Hagan, L. A. and Haussler, M. R. (1980). A sensitive radioreceptor assay for 1 alpha, 25-dihydroxyvitamin D in biological fluids. *Methods Enzymol.* **67**, 522–528.

Chandler, J. S., Chandler, S. K., Pike, J. W. and Haussler, M. R. (1984). 1,25-Dihydroxyvitamin D3 induces 25-hydroxyvitamin D3-24-hydroxylase in a cultured monkey kidney cell line (LLC-MK2) apparently deficient in the high affinity receptor for the hormone. *J. Biol. Chem.* **259**, 2214–2222.

Chang, C., Kikontis, J. and Liao, S. (1988). Molecular cloning of human and rat complementary DNA encoding androgen receptors. *Science* **240**, 324–326.

Chen, T. L., Hirst, M. A., Cone, C. M., Hochberg, Z., Tietze, H. U. and Feldman, D. (1984). 1,25-Dihydroxyvitamin D resistance, rickets, and alopecia: analysis of receptors and bioresponse in cultured fibroblasts from patients and parents. *J. Clin. Endocrinol. Metab.* **59**, 383–388.

Chrousos, G. P., Vingerhoeds, A., Brandon, D., Eil, C., Pugeat, M., DeVroede, M., Loriaux, D. L. and Lipsett, M. B. (1982). Primary cortisol resistance in man: a glucorticoid receptor-mediated disease. *J. Clin. Invest.* **69**, 1261–1269.

Chrousos, G. P., Vingerhoeds, A. C., Loriaux, D. L. and Lipsett, M. B. (1983a). Primary cortisol resistance: a family study. *J. Clin. Endocrinol. Metab.* **56**, 1243–1245.

Chrousos, G. P., Loriaux, D. L., Brandon, D., Tomita, M., Vingerhoeds, A. C., Merriam, G. R., Johnson, E. O. and Lipsett, M. B. (1983b). Primary cortisol resistance: a familial syndrome and an animal model. *J. Steroid. Biochem.* **19**, 567–575.

Colston, K. and Feldman, D. (1982). 1,25-Dihydroxyvitamin D receptors and functions in cultured pig kidney cells (LLCPK1): Regulation of 24,25-dihydroxyvitamin D production. *J. Biol. Chem.* **257**, 2504–2508.

Colston, K., Hirst, M. and Feldman, D. (1980). Organ distribution of the cytoplasmic 1,25-dihydroxycholecalciferol receptor in various mouse tissues. *Endocrinology* **107**, 1916–1922.

Conneely, O. M., Dobson, A. D., Carson, M. A., Maxwell, B. L., Tsai, M. J., Schrader, W. T. and O'Malley, B. W. (1988). Structure–function relationships of the chicken progesterone receptor. *Biochem. Soc. Trans.* **16**, 683–687.

Danielsen, M., Northrop, J. P. and Ringold, G. M. (1986). The mouse glucocorticoid receptor: mapping of functional domains by cloning, sequencing and expression of wild-type and mutant receptor proteins. *EMBO J.* **5**, 2513–2522.

Danielsen, M., Hinck, L. and Ringold, G. M. (1989). Mutational analysis of the mouse glucocorticoid receptor. *Cancer Res.* **49**, 2286s–2291s.

Elawady, M. K., Allman, D. R., Griffin, J. E. and Wilson, J. D. (1983). Expression of a mutant androgen receptor in cloned fibroblasts derived from a heterozygous carrier for the syndrome of testicular feminization. *Am. J. Hum. Genet.* **35**, 376–384.

Evans, R. M. (1988). The steroid and thyroid hormone receptor superfamily. *Science* **240**, 889–895.

Feldman, D., Chen, T., Hirst, M., Colston, K., Karasek, M. and Cone, C. (1980). Demonstration of 1,25-dihydroxyvitamin D3 receptors in human skin biopsies. *J. Clin. Endocrinol. Metab.* **51**, 1463–1465.

Feldman, D., Chen, T., Cone, C., Hirst, M., Shani, S., Benderli, A. and Hochberg, Z. (1982). Vitamin D resistant rickets with alopecia: cultured skin fibroblasts exhibit defective cytoplasmic receptors and unresponsiveness to $1,25(OH)_2D_3$. *J. Clin. Endocrinol. Metab.* **55**, 1020–1022.

Gamblin, G. T., Liberman, U. A., Eil, C., Downs, R. W. Jr., DeGrange, D. A. and Marx, S. J. (1985). Vitamin D-dependent rickets type II. Defective induction of 25-hydroxyvitamin D3-24-hydroxylase by 1,25-dihydroxyvitamin D3 in cultured skin fibroblasts. *J. Clin. Invest.* **75**, 954–960.

Gerli, M., Migliorini, G., Bocchini, V., Venti, G., Ferrarese, R., Donti, E. and Rosi, G. (1979). A case report of complete testicular feminization and 47,XXY karyotype. *J. Med. Genet.* **16**, 480–483.

Griffin, J. E. and Wilson, J. D. (1989). The androgen resistance syndromes. In "The Metabolic Basis of Inherited Disease" (eds C. R. Scriver, A. L. Beaudet, W. S. Sly and D. Valle), pp. 1919–1944. McGraw-Hill, New York.

Griffin, J. E. and Zerwekh, J. E. (1983). Impaired stimulation of 25-hydroxyvitamin D-24-hydroxylase in fibroblasts from a patient with vitamin D-dependent rickets, type II A form of receptor-positive resistance to 1,25-dihydroxyvitamin D3. *J. Clin. Invest.* **72**, 1190–1199.

Hirst, M. A., Hochman, H. I. and Feldman, D. (1985). Vitamin D resistance and alopecia: a kindred with normal 1,25-dihydroxyvitamin D binding, but decreased receptor affinity for deoxyribonucleic acid. *J. Clin. Endocrinol. Metab.* **60**, 490–495.

Hochberg, Z., Benderli, A., Levy, J., Vardi, P., Weisman, Y., Chen, T. and Feldman, D. (1984). 1,25-Dihydroxyvitamin D resistance, rickets, and alopecia. *Am. J. Med.* **77**, 805–811.

Hochberg, Z., Gilhar, A., Haim, S., Friedman-Birnbaum, R., Levy, J. and Benderly, A. (1985). Calcitriol-resistant rickets with alopecia. *Arch. Dermatol.* **121**, 646–647.

Hollenberg, S. M., Weinberger, C., Ong, E. S., Cerelli, G., Oro, A., Lebo, R., Thompson, E. B., Rosenfeld, M. G. and Evans, R. M. (1985). Primary structure and expression of a functional human glucocorticoid receptor cDNA. *Nature* **318**, 635–641.

Huckaby, C. S., Conneely, O. M., Beattie, W. G., Dobson, A. D., Tsai, M. J. and O'Malley, B. W. (1987). Structure of the chromosomal chicken progesterone receptor gene. *Proc. Natl. Acad. Sci. USA* **84**, 8380–8384.

Hughes, M. R., Malloy, P. J., Kieback, D. G., Kesterson, R. A., Pike, J. W., Feldman, D. and O'Malley, B. W. (1988). Point mutations in the human vitamin D receptor gene associated with hypocalcemic rickets. *Science* **242**, 1702–1705.

Hurley, D. M., Accili, D., Vamvakopoulos, M., Statakis, C., Taylor, S. I. and Chrousos, G. P. (1990). Mutation of the human glucorticoid receptor gene in familial glucorticoid resistance—an abstract. *Endocrinology* **71** (Annual Meeting) 284.

Iida, S., Gomi, M., Moriwaki, K., Itoh, Y., Hirobe, K., Matsuzawa, Y., Katagiri, S., Yonezawa, T. and Tami, S. (1985). Primary cortisol resistance accompanied by a reduction in glucorticoid in two members of the same family. *J. Clin. Endocrinol. Metab.* **60**, 967–971.

Kerner, S. A., Scott, R. A. and Pike, J. W. (1989). Sequence elements in the human osteocalcin gene confer basal activation and inducible response to hormonal vitamin D3. *Proc. Natl. Acad. Sci. USA* **86**, 4455–4459.

Kovacs, W. J., Griffin, J. E. and Wilson, J. D. (1986). Androgen resistance in man. *Adv. Exp. Med. Biol.* **196**, 257–267.

Kumar, V., Green, S., Staub, A. and Chambon, P. (1986). Localisation of the oestradiol-binding and putative DNA-binding domains of the human oestrogen receptor. *EMBO J.* **5**, 2231–2236.

Lamberts, S. W. J., Poldermans, D., Zweens, M. and DeJong, F. H. (1986). Familial cortisol resistance: differential diagnostic and therapeutic aspects. *J. Clin. Endocrinol. Metab.* **63**, 1328–1333.

Liberman, U. A., Eil, C. and Marx, S. J. (1983). Resistance to 1,25-dihydroxyvitamin D. *J. Clin. Invest.* **71**, 192–200.

Liberman, U. A., Eil, C. and Marx, S. J. (1986). Receptor-positive hereditary resistance to 1,25-dihydroxyvitamin D: Chromatography of hormone-receptor complexes on deoxyribonucleic acid-cellulose shows two classes of mutation. *J. Clin. Endocrinol. Metab.* **62**, 122–126.

Lubahn, D. B., Joseph, D. R., Sullivan, P. M., Willard, H. F., French, F. S. and Wilson, E. M. (1988). Cloning of human androgen receptor complementary DNA and localization to the X-chromosome. *Science* **240**, 327–330.

Malloy, P. J., Hochberg, Z., Pike, J. W. and Feldman, D. (1989). Abnormal binding of vitamin D receptors to deoxyribonucleic acid in a kindred with vitamin D-dependent rickets, type II. *J. Clin. Endocrinol. Metab.* **68**, 263–269.

Malloy, P. J., Hochberg, Z., Kristjansson, K., Pike J. W., Hughes, M. R. and Feldman, D. (1990). The molecular basis of hereditary 1,25-dihydroxyvitamin D_3 resistant rickets in seven related families. *J. Clin. Invest.* (in press).

Marcelli, M., Tilley, W. D., Wilson, C. M., Wilson, J. D., Griffin, J. E. and McPhaul, M. J. (1990). A single nucleotide substitution introduces a premature termination codon into the androgen receptor gene of a patient with receptor-negative androgen resistance. *J. Clin. Invest.* **85**, 1522–1528.

Marx, S. (1989). Vitamin D and other calciferols. *In* "The Metabolic Basis of Inherited Disease" (eds C. R. Scriver, A. L. Beaudet, W. S. Sly and D. Valle), pp. 2029–2045. McGraw-Hill, New York.

Marx, S. J., Bliziotes, M. M. and Nanes, M. (1986). Analysis of the relation between alopecia and resistance to 1,25-dihydroxyvitamin D. *Clin. Endocrinol. (Oxf.)* **25**, 373–381.

McDonnell, D. P., Mangelsdorf, D. J., Pike, J. W., Haussler, M. R. and O'Malley, B. W. (1987). Molecular cloning of complementary DNA encoding the avian receptor for vitamin D. *Science* **235**, 1214–1217.

McDonnell, D. P., Scott, R. A., Kerner, S. A., O'Malley, B. W. and Pike, J. W. (1989a). Functional domains of the human vitamin D3 receptor regulates osteocalcin gene expression. *Mol. Endocrinol.* **3**, 635–644.

McDonnell, D. P., Pike, J. W., Drutz, D. J., Butt, T. R. and O'Malley, B. W. (1989b). Reconstitution of the vitamin D responsive osteocalcin transcription unit in yeast. *Mol. Endocrinol.* **3**, 635–644.

McKusick, V. A. (1988). "Mendelian Inheritance in Man: Catalogs of Autosomal Dominant, Autosomal Recessive, and X-linked Phenotypes". Eighth Edition. Johns Hopkins University Press, Baltimore, MD.

Meyer, W. J., Migeon, B. R. and Migeon, C. J. (1975). Locus on human X chromosome for dihydrotestosterone receptor and androgen insensitivity. *Proc. Natl. Acad. Sci. USA* **72**, 1469–1472.

Nawata, H., Sekiya, K., Higuchi, K., Yanase, T., Kato, K. and Ibayashi, H. (1984). Decreased deoxyribonucleic acid binding of fibroblasts from a patient with the glucorticoid resistance syndrome. *J. Clin. Endocrinol. Metab.* **65**, 219–226.

Neufeld, E. F. (1989). Natural history and inherited disorders of a lysosomal enzyme, Beta-hexosaminidase. *J. Biol. Chem.* **264**, 10 927–10 930.

Orkin, S. H. (1990). The mutation and polymorphism of the human Beta-globin gene and its surrounding DNA. *Annu. Rev. Genet.* **18**, 131–171.

Pike, J. W., Marion, S. L., Donaldson, C. A. and Haussler, M. R. (1983). Serum and monoclonal antibodies against the chick intestinal receptor for 1,25-dihydroxyvitamin D3. Generation by a preparation enriched in a 64,000-dalton protein. *J. Biol. Chem.* **258**, 1289–1296.

Pike, J. W. (1984). Monoclonal antibodies to chick intestinal receptors for 1,25-dihydroxyvitamin D3. Interaction and effects of binding on receptor function. *J. Biol. Chem.* **259**, 1167–1173.

Pike, J. W., Sleator, N. M. and Haussler, M. R. (1987). Chicken intestinal receptor for 1,25-dihydroxyvitamin D3 immunologic characterization and homogeneous isolation of a 60,000-dalton protein. *J. Biol. Chem.* **262**, 1305–1311.

Pinsky, L., Trifiro, M., Sebbaghian, N., Kaufman, M., Chang, C., Trapman, J., Brinkmann, A. O., Kuiper, G. G. J. M., Ris, C. J., Brown, C. J., Willard, H. F. and Sergovich, F. (1989). A deletional alteration of the androgen receptor gene in a sporadic patient with complete androgen insensitivity who is mentally retarded. *Am. J. Hum. Genet.* **45** (Suppl.), A212.

Ritchie, H. H., Hughes, M. R., Thompson, E. T., Malloy, P. J., Hochberg, Z., Feldman, D., Pike, J. W. and O'Malley, B. W. (1989). An ochre mutation in the vitamin D receptor gene causes hereditary 1,25-dihydroxyvitamin D resistant rickets in three families. *Proc. Natl. Acad. Sci. USA* **86**, 9783–9787.

Sicinski, P., Geng, Y., Ryder-Cook, A. S., Barnard, E. A., Darlinson, M. G. and Barnard, P. J. (1989). The molecular basis of nuclear dystrophy in the mdx mouse: a point mutation. *Science* **244**, 1578–1580.

Simons, S. S. Jr., Pumphrey, J. G., Rudikoff, S. and Eisen, H. J. (1987). Identification of cysteine 656 as the amino acid of hepatoma tissue culture cell glucocorticoid receptors that is covalently labeled by dexamethasone 21-mesylate. *J. Biol. Chem.* **262**, 9676–9680.

Simpson, R. U. and DeLuca, H. F. (1980). Characterization of a receptor-like protein for 1,25-dihydroxyvitamin D3 in rat skin. *Proc. Natl. Acad. Sci. USA* **77**, 5822–5826.

Stone, T., Scott, R. A., Hughes, M. R., Malloy, P. J., Feldman, D., O'Malley, B. W. and Pike, J. W. (1989). Mutant vitamin D receptors which confer hereditary resistance to 1,25-dihydroxyvitamin D3 in humans are transcriptionally inactive in vitro. *J. Biol. Chem.* **264**, 20 230–20 234.

Tilly, W. D., Marcelli, M., Wilson, J. D. and McPhaul, M. J. (1989). Characterization and expression of a cDNA encoding the human androgen receptor. *Proc. Natl. Acad. Sci. USA* **86**, 327–331.

Tomita, M., Brandon, D. D., Chrousos, G. P., Vingerhoeds, A. C., Foster, C. M., Fowler, D., Loriaux, D. L. and Lipsett, M. B. (1986). Glucocorticoid receptors in Epstein–Barr virus-transformed lymphocytes from patients with glucocorticoid resistance and a glucocorticoid-resistant New World primate species. *J. Clin. Endocrinol. Metab.* **62**, 1145–1154.

Trapman, J., Klaassen, P., Kuiper, G. G., van der Korput, J. A., Faber, P. W., van Rooij, H. C., Guerts van, Kessel, A., Voorhorst, M. M., Mulder, E. and Brinkmann, A. O. (1988). Cloning, structure and expression of a cDNA encoding the human androgen receptor. *Biochem. Biophys. Res. Commun.* **153**, 241–248.

Trifiro, M., Prior, L., Pinsky, L, Kaufman, M., Chang, C., Trapman, J., Brinkmann, A. O., Kuiper, G. G. J. M. and Ris, C. (1989). A single transition at an exonic CpG site apparently abolishes androgen receptor binding activity in a family with complete androgen insensitivity. *Am. J. Hum. Genet* **45** (Suppl.), A225.

Vingerhoeds, A. C. M., Thijssen, J. H. H. and Schwartz, F. (1976). Spontaneous hypercortisolism without Cushing's syndrome. *J. Clin. Endocrinol. Metab.* **43**, 1128–1133.

Wieacker, P., Griffin, J. E., Wienker, T., Lopez, J. M., Wilson, J. D. and Breckwoldt, M. (1987). Linkage analysis with RFLPs in families with androgen resistance syndromes: evidence for close linkage between the androgen receptor locus and the DXS1 segment. *Hum. Genet.* **76**, 248–252.

Woo, S. L. (1989). Molecular basis and population genetics of phenylketonuria. *Biochemistry* **28**, 1–7.

Zerwekh, J. E., Glass, K., Jowsey, J. and Pak, C. Y. C. (1979). A unique form of osteomalacia associated with end organ refractoriness to 1,25-dihydroxyvitamin D and apparent defective synthesis of 25-hydroxyvitamin D. *J. Clin. Endocrinol. Metab.* **49**, 171.

15
Avian Erythroleukaemia: Possible Mechanisms Involved in v-erbA Oncogene Function

H. BEUG[1] and B. VENNSTRÖM[2]

[1] *Institute of Molecular Pathology, Dr. Bohrgasse 7, A 1030 Vienna, Austria*

[2] *Karolinska Institute, Department of Molecular Biology CMB, Box 60400, S 10401 Stockholm, Sweden*

15.1 Introduction

The avian erythroblastosis virus (AEV; strains ES4 and R) represents a chicken retrovirus causing lethal erythroleukaemia and sarcomas in chicks (for review see Graf and Beug, 1978, 1983). It contains two oncogenes derived from the chicken genome. The first oncogene (v-*erb*B) represents a virally transduced and truncated form of the chicken EGF receptor, whereas the second (v-*erb*A) is a mutated form of the chicken thyroid hormone receptor α (Downward *et al.*, 1984; Sap *et al.*, 1986; Weinberger *et al.*, 1986).

V-*erb*B is the primary transforming gene of AEV. It is both necessary and sufficient for transformation of fibroblasts and haematopoietic progenitors of the erythroid lineage, the two types of target cells that AEV interacts with *in vitro* and *in vivo* (Frykberg *et al.*, 1983; Fung *et al.*, 1983; Sealy *et al.*, 1983; Yamamoto *et al.*, 1983). The v-*erb*A oncogene, on the other hand, is unable to cause tumours in chickens on its own (Frykberg *et al.*, 1983; Gandrillon *et al.*, 1989). Subsequent work by our laboratories, as well as by many others, has resulted in the following basic understanding of the function and collaboration of the two *erb* oncogenes in transformation of avian cells.

The effects of v-*erb*B in haematopoietic cells and fibroblasts appear to be distinct. In chicken fibroblasts the major effect of v-*erb*B seems to be its ability to abrogate the requirement for mitogens (e.g. EGF, PDGF; Johnsson *et al.*, 1985 and unpublished; Khazaie *et al.*, 1988) and to induce phenotypical changes typical for transformed fibroblasts (Royer-Pokora *et al.*, 1978). In erythroblasts, on the other hand, v-*erb*B (as well as other tyrosine kinase-related oncogenes and v-Ha-*ras*) has two major effects: it induces an abnormal self-renewal in infected erythroid progenitors (perhaps caused by preventing or retarding accumulation of red cell proteins; Knight et al., 1988) and renders them independent of the erythroid growth factor erythropoietin (EPO) for both growth and differentiation (Beug et al., 1982, 1985). While self renewal induction is caused by the unmutated EGF receptor, when overexpressed and activated by ligand, abrogation of EPO dependence is dependent on C-terminal mutations in the c-erbB protein (Khazaie et al., 1988). Since the EGF receptor is not normally expressed in haematopoietic cells (with the possible exception of stem cells) both v-erbB effects may be mediated by interaction with protein substrates not normally affected by the EPO-receptor pathway in erythroblasts.

The v-erbA oncogene, on the other hand, is not overtly transforming in both lineages. In chicken fibroblasts v-erbA elicits only minor effects such as enhancement of agar colony formation and in vitro lifespan. In addition, v-erbA might block or reduce production of extracellular matrix proteins, e.g. fibronectin (Frykberg et al., 1983; Gandrillon et al., 1987; Jansson et al., 1987). v-erbA arrests differentiation of normal erythroid progenitors, but does not appear to alter their lifespan unless it acts in concert with another, primary, oncogene (Frykberg et al., 1983; Schröder et al., 1990). This view is disputed by Gandrillon et al. (1989), who observed the sustained outgrowth of v-erbA-expressing erythroblasts in presence of early hemopoietic growth factors (e.g. anaemic chicken serum containing an Il 3-like activity; Samarut and Bouabdelli, 1980; Gandrillon et al., 1989).

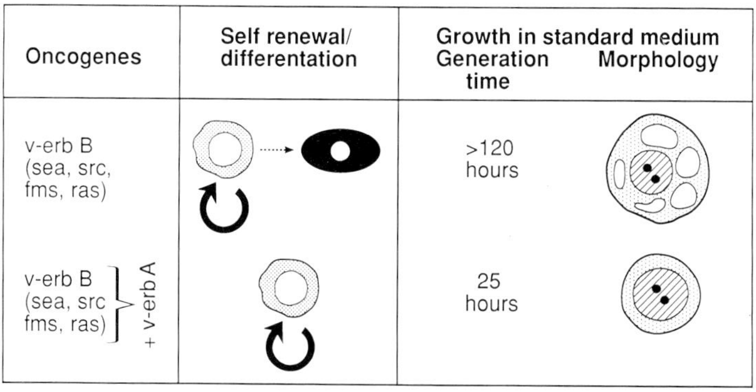

Oncogenes	Self renewal/ differentiation	Growth in standard medium	
		Generation time	Morphology
v-erb B (sea, src, fms, ras)		>120 hours	
v-erb B (sea, src fms, ras) } + v-erbA		25 hours	

Fig. 1 *Contribution of v-erbA to the leukaemic phenotype.* This scheme illustrates how v-erbA co-operates with a variety of tyrosine kinase oncogenes as well as the GTP-binding protein v-Ha-ras in erythroblast transformation. Left panel: v-erbA completely arrests the spontaneous differentiation (right, broken arrow) typical of erythroblasts transformed with one oncogene only, allowing them to self-renew exclusively (broad, circular arrow). Right panel: Erythroblasts transformed by a single oncogene fail to proliferate in standard tissue culture media and disintegrate after extensive vacuolization. In contrast, they proliferate well and do not form vacuoles if they express v-erbA as well.

In transformed erythroblasts, v-*erb*A contributes to the transformed phenotype in two ways. First, v-*erb*A enhances the transforming activity of v-*erb*B by blocking the residual, spontaneous differentiation exhibited by such cells. In addition, v-*erb*A induces certain oncogenes to transform erythroblasts that are unable to do so by themselves (e.g. v-*erb*B mutants and v-*src*; Kahn *et al.*, 1986a; Damm *et al.*, 1987). Second, v-*erb*A enables transformed erythroblasts that require media optimized for pH and HCO_3^-/Na^+ ion concentration to grow in standard tissue culture media (Fig. 1; Kahn *et al.*, 1986b; Damm *et al.*, 1987). These two effects can be quantified, which has aided understanding of the molecular mechanisms by which v-*erb*A exerts its oncogenic activity.

In this review, possible mechanisms of how v-*erb*A arrests differentiation and co-operates with tyrosine kinase-type oncogenes in fibroblast and erythroblast transformation will be discussed in the context of normal thyroid hormone receptor α (TRα) function.

15.2 Avian thyroid hormone receptors

The cellular homologue of v*erb*-A (c-*erb*A) encodes the α-form of thyroid hormone receptors (TRα). TRα belongs to the superfamily of nuclear hormone receptors also comprising the receptors for glucocorticoids, oestrogen, retinoic acid, vitamin D_3, and several receptors with hitherto unidentified ligands (for review, see Evans, 1988).

Four different mammalian and three avain thyroid hormone receptors have been characterized to date (see Chapter 4). Based on their amino acid homologies and chromosomal locations, they have been classified into two groups, α and β (Fig. 2). The mammalian TRα-2 diverges from TRα-1 in domain F due to differential splicing, and neither binds T_3 nor transactivates (Izumo and Mahdavi *et al.*, 1988; Lazar *et al.*, 1988; Nakai *et al.*, 1988; Koenig *et al.*, 1989). The two forms are often co-expressed, and TRα-2 has been shown to repress TRα-1-mediated transactivation. The TRβs differ in the N-terminus, possibly due to differential utilization of adjacent and tissue-specific promoters (Murray *et al.*, 1988; Hodin *et al.*, 1989; Koenig *et al.*, 1989). The situation in the chicken appears to be different: no TRα-2 has been found. However, two N-terminally distinct receptor proteins are made from the same mRNA apparently through differential utilization of AUGs (Sap *et al.*, 1986; Goldberg *et al.*, 1988; Forrest *et al.*, 1990a). The extended N-terminus of the longer protein is phosphorylated at two sites; a casein kinase II and cyclic AMP dependent protein kinase (PKA) consensus site have been found (Goldberg *et al.*, 1988; Glineur *et al.*, 1989).

The v-*erb*A protein, $P75^{gag\text{-}v\text{-}erbA}$ represents a heavily mutated version of

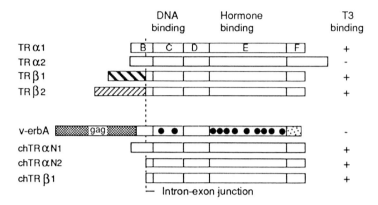

Fig. 2 *Comparison of mammalian and avian thyroid hormone receptors.* The four receptors shown on top represent prototype mammalian receptors, whereas the four below show the structures of the v-*erb*A protein and the three known avian receptors. The stippled or hatched areas indicate regions with no amino acid homology to regions in other receptors. The dots in v-*erb*A show the locations of amino acid replacements.

the TRα. It is fused to the retroviral gag gene with the consequence that the P75$^{gag-v-erbA}$ contains 253 N-terminal amino acids derived from gag, followed by chicken TRα-1 sequences starting at amino acid 13 (Fig. 2). Consequently, P75$^{gag-v-erbA}$ lacks the 12 amino acids that also contain the casein kinase II consensus phosphorylation site. Furthermore, P75$^{gag-v-erbA}$ contains 13 amino acid substitutions, two of which are located between the gag and the DNA-binding domains, two within the DNA-binding domain, and the remaining nine in the ligand-binding region. Finally, P75$^{gag-v-erbA}$ has a 9 amino acid deletion three residues from the C-terminus (Fig. 3A). The first of the amino acid substitutions in the DNA-binding domain (Fig. 3B) is located just after the first zinc finger and represents one of the three amino acids that suffice to change the sequence recognition specificity of nuclear hormone receptors (Mader *et al.*, 1989; see Chapter 2.2). The second amino acid substitution is in-between the two cysteine residues at the base of the second zinc finger (Fig. 3B).

The avian TRα and TRβ bind T_3 with similar affinities; the K_d values have been determined as 0.2–0.4 nM (Sap *et al.*, 1986, 1989; Forrest *et al.*, 1990a; J. Oppenheimer, pers. commun.). In contrast, the v-*erb*A protein fails to bind thyroid hormones (Sap *et al.*, 1986; Muñoz *et al.*, 1988) due to mutations in the region corresponding to the ligand-binding domain of TRα. Consequently, P75$^{gag-v-erbA}$ resembles the mammalian TRα-2, which is also unable to bind thyroid hormones. The fact that several mutations, present in the ultimate C-terminus as well as in internal regions, have accumulated in P75$^{gag-v-erbA}$ and co-operate in abolishing hormone binding, suggests that hormone-independent action by v-*erb*A provided a selective advantage to AEV during the selection for a highly and acutely oncogenic strain of this virus (Rothe-Meyer and Engelbreth-Holm, 1933).

The chicken TRα is expressed in all 14 tissues that have been examined to date, with small variations in levels, and is expressed as early as day 4 of chick embryo development (Forrest *et al.*, 1990a). Expression of TRβ is restricted mainly to brain, eye, lung, kidney and yolk sac and is developmentally controlled; e.g. the level increases 30-fold in brain after hatching. The pattern and developmental control of TRβ expression correlates well with known effects of thyroxine, and it therefore appears as if these classical effects are mediated by TRβ. The functions of TRα are still unclear.

15.3 v-*erb*A, an oncogene acting by aberrant transcriptional regulation?

Since v-*erb*A is a homologue of a thyroid hormone receptor, it seems likely that it might affect transcription of one or several genes important for erythroid cell differentiation. These genes might not be those recognized by the TRα protein, because of the two amino acid changes in the DNA binding region of P75$^{gag\text{-}erbA}$. Thus, when bound to a presumptive regulatory element, P75$^{gag\text{-}erbA}$ could constitutively either up- or down-regulate the transcription of the corresponding gene. If down-regulation is the mechanism by which P75$^{gag\text{-}erbA}$ acts, it seems likely that the expression of the affected erythroid specific gene(s) would be required for proper erythroid differentiation. Conversely, if v-*erb*A up-regulates transcription, the expression of such target genes would inhibit differentiation. In the following, we will first review the evidence (mainly from transient transfection assays in heterologous cells) as to how v-*erb*A and TRα regulate the activity of target genes. In subsequent sections we will describe some recent experiments that illustrate the action of v-*erb*A and TRα on erythrocyte specific genes during erythroid differentiation and leukaemia induction.

15.4 Transcriptional regulation by v-*erb*A and TRα

The question of how v-*erb*A and TRα act on artificial reporter genes carrying various thyroid hormone responsive elements (TREs) was addressed by Damm *et al.* (1989) and Sap *et al.* (1989). They reported that TRα repressed transcription of a CAT reporter gene driven by a T_3-

A

```
                                                      gag
    1                                                                           MEQK   TRα
  185   AGAGEQGQGGDTPRGAEQPRAEPGHAGQAPGPALTDWARIREELASTGPPVVAMPVVIKT                   v-erbA
                                                                                       TRβ


    1   PSTLDPLSEPEDTRWLDGKRKRKSSQCLVKSSMSGYIPSYLDKDEQCVVCGDKATGYHYR                   TRα
  245   EGPAWTPIEPEDTRWLDGKHKRKSSQCLVKSSMSGYIPSQLDKDEQCVVCGDKATGYHYR                   v-erbA
    1                             MSGYIPSYLDKDELCVVCGDKATGYHYR                         TRβ

                      DNA binding region
   65   CITCEGCKGFFRRTIQKNLHPTYSCKYDGCCVIDKITRNQCQLCRFKKCISVGMAMDLVL                   TRα
  305   CITCEGCKSFFRRTIQKNLHPTYSQIYDGCCVIDKITRNQCQLCRFKKCISVGMAMDLVL                   v-erbA
   29   CITCEGCKGFFRRTIQKNLHPTYSCKYEGKCVIDKMTRNQCQECRFKKCIEVGMATDLVL                   TRβ


  125   DDSKRVAKRKLIEENRERRRKEEMIKSLQHRPSPSAEEWELIHVVTEAHRSTNAQGSHWK                   TRα
  365   DDSKRVAKRKLIEENRERRRKEEMIKSLQHRPSPSAEEWELIHVVTEAHRSTNAQGSHWK                   v-erbA
   89   DDSKRIAKRKLIEENREKRRREEELQKTMGHKPEHTDEEWELIKIVTEAHVATNAQGSHWK                  TRβ


  185   QKRKFLPEDIGQSPMASMPDGDKVDLEAFSEFTKIITPAITRVVDFAKKLPMFSELPCED                   TRα
  425   QBRKFLIEDIGQSPMASMLDGDKVDLEAFSEFTKIITPAITRVVDFAKNLPMFSELPCED                   v-erbA
  149   QKRKFLPEDIGQAPIVNAPEGGKVDLEAFSQFTKIITPAITRVVDFAKKLPMFQELPCED                   TRβ


  245   QIILLKGCCMEIMSLRAAVRYDPESETLTLSGEMAVKREQLKNGGLGVVSDAIFDLGKSL                   TRα
  485   QIILLKGCCMEIMSLRAAVRYDPESETLTLSGEMAVKREQLKNGGLGVVSDAIFDLGKSL                   v-erbA
  209   QIILLKGCCMEIMSLRAAVRYDPESETLTINGEMAVIHGQLKNGGLGVVSDAIFDLGMSL                   TRβ


  305   SAFNLDDTEVALLQAVLLMSSDRTGLICVDKIEKCQETYLLAFEHYINYRKHNIPHFWPK                   TRα
  545   SAFNLDDTEVALLQAVLLMSSDRTGLICVDKIEKCQESYLLAFEHYINYRKHNIPHFWSK                   v-erbA
  269   SSFNLDDTEVALLQAVLLMSSDREGIMCVERIEKCQEGFLLAFEHYINYRKHHVAHEWPK                   TRβ


  365   LLMKVTDLRMIGACHASRFLHMKVECPTELFPPLFLEVFEDQEV                                   TRα
  605   LLMKVADLRMIGAYHASRFLHMKVECPTEISP---------QEV                                   v-erbA
  329   LLMKVTDLRMIGACHASRFLHMKVECPTELFPPLFLEVFED                                      TRβ
```

Fig. 3 *Amino acid differences between avian erbA proteins.* (A) The amino acid sequences of the TRα, TRβ, and v-*erb*A are compared in the one-letter code. Differences are indicated by boxes; deletions by —.

responsive thymidine kinase (tk) promoter in absence of thyroid hormone (T_3), but strongly transactivated in the presence of T_3. Interestingly v-*erb*A was found to interfere with this hormone-induced reporter gene expression by constitutively repressing transcription of genes containing a TRE (Damm et al., 1989), or blocking normal TR-mediated hormonal induction (Damm et al., 1989; Sap et al., 1989). It has therefore been proposed that v-*erb*A exemplifies a novel type of oncogene, exhibiting a repressor activity that is unaffected by ligand *and* dominant over the normal thyroid hormone receptor. The fact that thyroid hormone receptors can form heterodimers

B

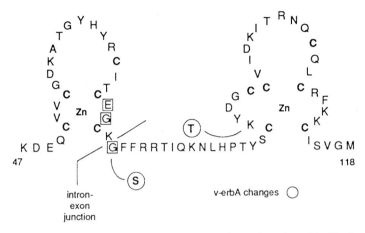

Fig. 3 (B) Close-up of changes in the DNA-binding region of v-*erb*A. The boxes indicate amino acids that determine recognition specificity for DNA sequences as described by Mader *et al.* (1989).

with retinoic acid receptors (Forman *et al.*, 1989; Glass *et al.*, 1989) and that in such a complex unliganded TRs may repress induction by retinoic acid as well (Graupner *et al.*, 1989) raises the question of whether or not the v-*erb*A protein (P75$^{gag\text{-}v\text{-}erbA}$) possesses similar capabilities.

The binding of the p75$^{gag\text{-}v\text{-}erbA}$ to thyroid hormone response elements (TRE) is dependent on the particular TRE tested. In contrast to TRα which binds well, v-*erb*A binds very poorly to the major TRE element in the rat growth hormone gene (Sap *et al.*, 1990). Also, its binding to a TRE in the MoMLV LTR is reduced by ~70% (Sap *et al.*, 1989). In contrast, P75$^{gag\text{-}v\text{-}erbA}$ binds equally well as TRα to a TRE that represents a perfect palindrome with the sequence TCAGGTCATGACCTGA (Damm *et al.*, 1989). Thus caution has to be applied in extending these findings to other T_3-regulated genes, i.e. those actually affected by TRα and v-*erb*A in AEV-transformed erythroleukaemic cells.

15.5 Regulation of erythrocyte differentiation and erythrocyte gene expression by v-*erb*A, TRα and chimaeras thereof

15.5.1 v-*erb*A

Since erythroblasts expressing v-*erb*A alone could not be obtained in sufficient quantities to allow biochemical characterization or a detailed analysis of their phenotype, an alternative approach was taken to study the effects of v-*erb*A on erythroid differentiation (Fig. 4). This approach consists of combining the *erb*A oncogene to be analysed with a temperature-sensitive tyrosine kinase oncogene (e.g. ts v-*sea*, Knight *et al.*, 1988) in retrovirus constructs. On its own, ts v-*sea* transforms erythroblasts at the permissive temperature (37°C) which can then be grown to the desired cell numbers. These transformed erythroblasts can subsequently be induced to differentiate synchronously into mature erythrocytes by a shift to the non-permissive temperature (42°C, Fig. 4). If erythroblasts are transformed instead with a v-*erb*A-ts v-*sea* retrovirus, a shift to 42°C allows us to analyse the effects of v-*erb*A on proliferation, differentiation and erythrocyte gene expression (Fig. 4).

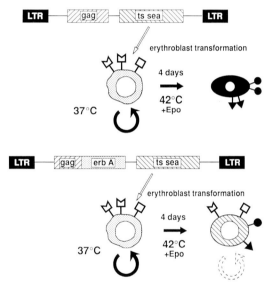

Fig. 4 *Using ts oncogene containing retroviral vectors to study v-*erbA* oncogene function.* Top panel: A simplifed diagram of the ts v-*sea* retrovirus containing the gag, Δ pol (not shown) and env-ts *sea* (ts *sea*) genes and the differentiation phenotypes of ts v-*sea* transformed erythroblasts obtained at the permissive temperature (37°C) and mature red cells obtained after 4 days at 42°C is shown. The bottom panel shows the v-*erb*A-ts v-*sea* retrovirus used to study v-*erb*A function and the aberrant erythroblast phenotype obtained after inactivation of the ts v-*sea* oncogene at 42°C. Broad circular arrow, self-renewal at 37°C; broken circular arrow, limited proliferation of v-*erb*A-ts v-*sea* erythroblasts at 42°C; straight arrow, differentiation at 42°C. Open symbols at cell surface, erythroblast differentiation antigens; closed symbols, erythrocyte differentiation antigens (Zenke *et al.*, 1988, 1990; Schröder *et al.*, 1990).

These studies revealed that v-*erb*A induces an aberrant, largely immature phenotype in erythroblasts, characterized by the coexpression of erythroblast and erythrocyte differentiation antigens (Fig. 4). Similar to erythroblasts infected by v-*erb*A only, erythroblasts expressing v-*erb*A together with a temperature-inactivated ts v-*sea* oncogene had a limited proliferation ability and were dependent on EPO for survival and growth (Fig. 4; Schröder *et al.*, 1990).

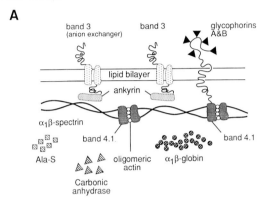

Fig. 5 *Differential effect of v-*erbA* on erythrocyte gene expression.* (A) Diagram of the organization and subcellular localization of the erythrocyte-specific proteins whose possible repression by v-*erb*A was studied. (B) The expression of the genes encoding these erythrocyte proteins in erythroblasts containing or lacking v-*erb*A after shift to 42°C for 3 days, as analysed by protein immunoprecipitation, measurement of total mRNA levels and nuclear run-on analysis (data from Zenke *et al.*, 1988 and 1990).

To try and understand the molecular mechanism responsible for the differentiation arrest caused by v-*erb*A, we studied the effects of this oncogene on the expression of various erythrocyte-specific genes, e.g. globins, red cell skeleton proteins and erythrocyte enzymes (Fig. 5), using the ts oncogene approach described above (Fig. 4). v-*erb*A did *not* act like an erythroid 'master' control gene which would affect expression of a large number of red cell genes. Instead, v-*erb*A selectively suppressed the activity of three erythrocyte genes: the erythrocyte-specific carbonic anhydrase II

(CAII) gene, the erythrocyte anion transporter gene (Band 3) and the erythrocyte version of the δ-aminolevulinic acid synthethase (δALA-S) gene (Fig. 5). This suppression occurred at the transcriptional level, as demonstrated by nuclear run-on analysis (Fig. 5; Zenke et al., 1988). In contrast v-erbA had little or no effect on transcription of other erythrocyte genes, such as encoding globins and other red cell skeleton proteins (Fig. 5).

15.5.2 TRα

To study possible effects of the normal TRα (c-erbA) on erythroid differentiation, we overexpressed it as a gag fusion protein in erythroblasts transformed by a ts v-sea oncogene (Figs 4 and 6). The cells were then induced to differentiate at 42°C in the presence or absence of thyroid hormone (T_3) and analysed for their phenotype and expression of erythrocyte-specific genes. In the absence of T_3, gag-c-erbA arrested differentiation and expressed reduced amounts of CAII, Band 3 and δALA-S m-RNA. In the presence of T_3, however, the erythroblasts underwent abnormal differentiation followed by disintegration of the partially mature red cells formed and expressed elevated levels of the three mRNAs (Fig. 6; Zenke et al., 1990). In contrast, the endogeneous TRα (present in very low amounts in ts v-sea erythroblasts lacking exogeneous erbA) did not severely affect erythroid differentiation and erythrocyte gene expression either in the presence or absence of T_3 (Zenke et al., 1990).

15.5.3 Mutations that Activate TRα as an Oncogene

In accord with the results of the transient transfection experiments described above, the following picture emerges of how c- and v-erbA function in erythroblasts. In the absence of T_3, overexpressed c-erbA may act as a repressor of those erythrocyte genes which contain a suitable TRE-like sequence. This repression is relieved in the presence of hormone, allowing the respective genes to be expressed at maximum efficiency as regulated by other (e.g. erythroid specific) transcription factors. v-erbA, on the other hand, is thought to act as a dominant repressor of these genes having lost the ability to bind and thus to be regulated by T_3 (Damm et al., 1989; Sap et al., 1989; Zenke et al., 1990).

Two types of experiments were performed to substantiate this idea. First, chimaeric v/c-erbA genes have been constructed and inserted into ts v-sea-containing retroviruses to study the role of the various mutations in v-erbA for oncogenic activation. Combination of the DNA-binding domain of v-erbA with the hormone-binding domain of TRα resulted in a protein (v_2-

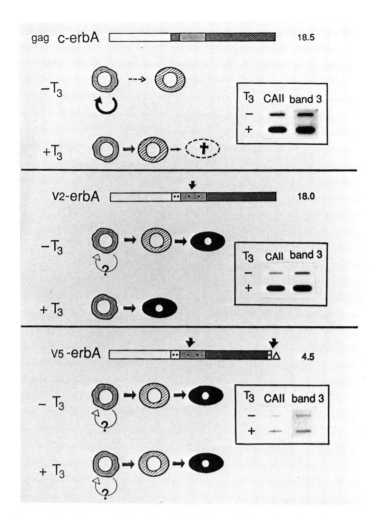

Fig. 6 *Hormone-dependent regulation of erythroid differentiation and erythrocyte gene expression by chimaeric* erbA *proteins*. This diagram shows (in a highly simplified manner) the gag-c-erbA protein and v/c-erbA chimaeric proteins derived from it, and the differentiation phenotypes of erythroblasts expressing these *erbA* proteins in the presence of a temperature-inactivated ts v-*sea* gene. The effects of the various *erbA* proteins on carbonic anhydrase II (CAII) and erythrocyte anion transporter (Band 3) mRNA expression are shown in the insets. Protein diagrams: open regions derived from gag-v-*erbA*; stippled regions derived from c-*erbA*. The DNA-binding region is marked by light stippling in all constructs. Dots, points mutations, see Figs 2 and 3; Δ, C-terminal deletion of v-*erbA*. The relative ability to bind thyroid hormone (fold binding over non-specific background; Muñoz *et al.*, 1988) is indicated by numerals. Diagrams describing the differentiation programmes of erythroblasts expressing the respective *erbA* protein in the absence ($-T_3$) and presence ($+T_3$) of thyroid hormone (T_3) are shown below the protein diagrams. Round cells with stippled cytoplasm, erythroblasts; oval cells with hatched cytoplasm, partially mature cells (early to late reticulocytes; Beug *et al.*, 1982); oval cells with small nucleus and black cytoplasm, erythrocytes. Circular arrows, self-renewal; straight arrows, differentiation. Circular arrows with ?, differentiation retarded, but not blocked in a sustained fashion. The insets show slot blots of total RNA extracted from the respective erythroblasts after 48 h at 42°C probed with CAII and Band 3 (data from Zenke *et al.*, 1990).

erbA) which bound T_3 with affinity similar to that of the gag-c-erbA protein and led to a similar T_3-dependent regulation of the three erythrocyte-specific genes (Fig. 6). However, the phenotypes induced by v_2-erbA in erythroblasts were distinct: v_2-erbA caused only a partial inhibition of differentiation in the absence of hormone, while permitting normal or even accelerated differentiation into red cells in the presence of T_3. Thus the amino acid substitutions in the v-erbA DNA-binding region clearly affect its activity (Zenke et al., 1990), as is also indicated by the fact that the v_2-erbA binds to TREs with the same low capacity as that of $P75^{gag-v-erbA}$, whereas the gag-c-erbA protein exhibited the same binding behaviour as TRα (Sap et al., 1989).

The role of the C-terminal deletion in v-erbA was studied in another chimaeric protein, v_5-erbA (Fig. 6). This protein, carrying the v-erbA C-terminus on a v_2-erbA background, was able to bind thyroid hormone, although with reduced affinity. V_5-erbA inhibited differentiation and erythrocyte gene expression in a fashion similar to v_2-erbA, but had completely lost its ability to induce differentiation and up-regulate erythrocyte gene transcription in presence of hormone (Fig. 6). This was not due to the reduced T_3 binding since other chimaeric proteins containing the v-erbA hormone-binding domain with a TRα C-terminus (v_6; v_7-erbA; Zenke et al., 1990) bound T_3 with a similarly reduced affinity, but still regulated differentiation and gene expression in a hormone-dependent fashion.

Inspection of the sequence deleted in v- and v_5-erbA revealed, that it was highly conserved among avian and mammalian TRs and similar motifs were even present in the retinoic acid receptor (RAR)-family, although not at the very C-terminus (Fig. 7A). In addition, this domain could be predicted to form an amphipathic helix (Fig. 7B) and thus to play a role in trans-activation by TRα and/or its interaction with other transcription factors (Zenke et al., 1990).

15.5.4 Repression of Erythrocyte Genes by v-erbA: Direct or Indirect Mechanism?

In a second approach to confirm that v-erbA acts as a dominant repressor of erythrocyte genes, we tried to elucidate whether v-erbA and TRα directly affected CAII, Band 3 and δALA-S transcription, and whether TRE-like sequences could be identified in the promotors of these genes. While induction of CAII gene transcription by TRα and v_2-erbA in the presence of T_3 was rapid (2–4 h) and not affected by pre-incubation with cycloheximide (CHO), T_3 induction of the Band 3 gene under the same conditions was slower (6–10 h) and partially sensitive to CHO (C. Disela et al., 1990). Subsequently, we were able to identify a TRE-like element in the 5' region of the CAII gene (GGTGAGTGAACGT) which was specifically bound by TRα and (less efficiently) by v-*erb*A in DNase 1 protection assays. Preliminary results with firefly luciferase reporter gene constructs carrying this element in front of a thymidine kinase promoter suggest, that this element might be involved in T_3-dependent regulation of the CAII gene in erythroblasts (Disela *et al.*, 1990). Trials to detect TRE-like elements in the Band 3 and δALA-S genes are in progress.

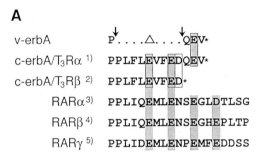

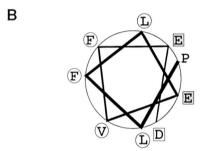

Fig. 7 *Structure of a potential amphipathic α-helix at the* erbA *carboxy-terminus.* (A) Alignment of the carboxy-terminal amino acid residues (actual C-termini marked by asterisks) of the v-*erb*A oncoprotein with those of the avian and mammalian thyroid hormone receptors type α and β and its homology to internal sequences of the retinoic acid receptor protein type α, β and γ.
(B) Helical wheel projection of the C-terminal *erb*A protein domain deleted in v-*erb* A (arrows in A) showing the amphipathic α-helical structure predicted according to Chou and Fasman (1978). The highly negatively charged amino acids present in c-*erb*A are boxed as shown in (A). Hydrophobic residues are circled (data from Zenke *et al.*, 1990).

15.6 Significance of erythrocyte gene repression for the leukaemic erythroblast phenotype

It is unclear how much the v-*erb*A-induced repression of CAII, Band 3 and δALA-S gene transcription contributes to the differentiation phenotype and physiology of AEV-transformed erythroblasts, since v-*erb*A could act on other, so far undetected genes. Experiments using inhibitors of Band 3 and δALA-S function in v-*erb*A-expressing erythroblasts have suggested that repression of these genes may contribute to the v-*erb*A phenotype (Schmidt *et al.*, 1986; Zenke *et al.*, 1988). However, the toxicity of these drugs has prohibited more detailed experiments. To clearly define the role of Band 3 we have therefore introduced a complete avian Band 3 cDNA (Kim *et al.*, 1988) into a suitable retrovirus vector containing a neo-resistance marker. When used to infect fibroblasts, high levels of the Band 3 protein were ectopically expressed at the plasma membrane of these cells. The Band 3-expressing fibroblasts displayed numerous large vacuoles when grown in standard medium; however these vacuoles disappeared when the cells were incubated in media of more alkaline pH (Zenke *et al.*, 1988) or treated with DIDS, a drug that inhibits the HCO_3^-/Cl^- antiport function of the Band 3 protein. This indicates that the retrovirus-expressed Band 3 protein is clearly functional (Fürstenberg *et al.*, 1990). We are currently trying to introduce the Band 3 gene into v-*erb*A-ts v-*sea* erythroblasts (in which the endogenous Band 3 gene is repressed) to study whether re-expression of this gene can abolish at least some of the phenotypic effects caused by v-*erb*A in erythroblasts. This approach might also allow a clarification of the two effects of the v-*erb*A in erythroblasts: it blocks differentiation and enables the cells to grow in standard tissue culture media. A similar approach will be applied to analyse the contribution of CAII and δ-ALA-S gene repression to the leukaemic phenotype.

15.7 Modulation of v-*erb*A activity by phosphorylation

Previous work by J. Ghysdael's group in Lille, France had established, that both v-*erb*A and TRα proteins are phosphorylated on serine (Ser 28, Ser 29) by either cAMP-dependent kinases or protein kinase C, both *in vitro* and *in vivo* (Goldberg *et al.*, 1988). To test the biological significance of v-*erb*A phosphorylation at these sites, both were changed into alanine residues. When introduced into erythroblasts using ts v-*sea*-containing retrovirus vectors, the mutant AA-v-*erb*A protein was non-phosphorylated and essentially devoid of biological activity, causing neither an arrest of erythroid differentiation nor a suppression of erythrocyte-specific gene transcription. In contrast, a v-*erb*A protein bearing threonine residues at positions 28 and 29 (TT) was again phosphorylated in cells and exhibited biological activity, albeit to a lesser extent than the wild-type v-*erb*A protein (Fig. 8; Glineur *et al.*, 1990).

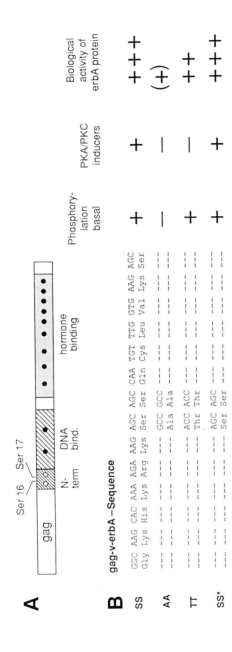

Fig. 8 *Biological activity of v-erbA phosphorylation mutants.* The scheme in (A) shows the localization of the two serine residues (Ser 28,29) in the N-terminal domain of the v-erbA protein that are phosphorylated by PKA and PKC *in vivo* and *in vitro*. (B) The site-directed mutations introduced into v-erbA (SS*, backmutant of AA-v-erbA to wild-type sequence): their phosphorylation status *in vivo* (basal phosphorylation, phosphorylation in absence of drugs; PKA/PKC inducers, enhanced phosphorylation after treatment with Forskolin or phorbol esters) and their ability to arrest differentiation when introduced into erythroblasts together with a ts- kinase oncogene (see Fig. 4) are shown. +++, complete inhibition of differentiation and erythrocyte gene transcription, growth in standard medium (Fig. 1); ++, strong but incomplete inhibition of differentiation and erythrocyte gene transcription, no or slow growth and vacuole formation in standard medium (Fig. 1); (+), no detectable effect on erythroid differentiation, erythrocyte gene transcription and medium requirements (data taken from Glineur *et al.*, 1990).

To prove independently that phosphorylation at Ser 28 and 29 was indeed required for biological activity of v-*erb*A, the protein kinase C inhibitor H7 was employed. At concentrations sufficient to completely prevent *erb*A phosphorylation in erythroblasts, H7 was able to strongly reduce the biological activity of v-*erb*A and TT-*erb*A in erythroblasts, inducing both differentiation and the requirement for complex growth media. In contrast, H7 did not affect proliferation or differentiation of erythroblasts containing either no v-*erb*A or the non-phosphorylated AA-v-*erb*A protein. These results indicate that phosphorylation of v-*erb*A does indeed modulate its function as a transcriptional repressor (Glineur *et al.*, 1990). Work is in progress to determine if repression and/or transactivation by TRα is also modulated by phosphorylation. Furthermore, the possible role of TRα phosphorylation at Ser 12 by casein kinase II (Glineur *et al.*, 1989) is also being investigated, since this site, unlike Ser 28/29, is conserved in all avian and mammalian TRs (Forrest *et al.*, 1990a).

15.8 Conclusions and open questions

The work summarized in this review leads to one major conclusion: the accumulation of the v-*erb*B and v-*erb*A oncogenes into one retrovirus does *not* represent a random combination of two diverse oncogenes that at most would have additive effects on the leukaemic phenotype. Rather, the intense selection of the original AEV strains for a short latency period *in vivo* carried out during isolation of the AEV R and ES4 strains (Rothe-Meyer and Engelbreth-Holm, 1933) seems to have favoured the combination of two oncogenes that act in a highly synergistic fashion. v-*erb*B causes abnormal self-renewal in infected erythroid progenitors and renders them independent of erythroid growth/differentiation hormones. For two reasons, however, v-*erb*B alone is only weakly leukaemogenic. These reasons are: (1) only a proportion of the transformed erythroblasts are arrested in erythroid differentiation, while the remainder spontaneously enter the differentiation pathway and thus escape the transforming action of v-*erb*B (Beug *et al.*, 1982); and (2) v-*erb*B reduces, but not abolishes, the expression of two proteins (CAII, Band 3, Knight *et al.*, 1988) that are essential for erythrocyte function (they allow erythrocytes to respond to external changes in CO_2 partial pressure with an altered internal pH and thus a changed affinity of haemoglobin for oxygen). Since erythroblasts probably need to control their intracellular pH as tightly as fibroblasts, expression of these two erythrocyte proteins in proliferating erythroblasts may interfere with their regulation of intracellular pH and thus cause inhibition of proliferation, unless the cells are grown in special media allowing them to attain the correct intracellular pH in the absence of such regulation (Kahn et al., 1986b).

These two shortcomings in the transforming potential of v-*erb*B, i.e. ineffective arrest of differentiation and thus expression of erythrocyte proteins deleterious for growth in an environment of changing CO_2 concentrations (e.g. the peripheral blood), are effectively corrected by the action of v-*erb*A. First, v-*erb*A completely arrests the spontaneous differentiation of v-*erb*B-transformed erythroblasts, thus preventing loss of leukaemic cells through maturation into erythrocytes. And second, v-*erb*A completely prevents expression of CAII and Band 3 proteins, the production of which in v-*erb*B erythroblasts is thought to render these cells unable to tolerate a wide range of external CO_2 concentrations.

Two major questions remain unresolved at this point. First it remains unclear whether suppression of CAII, Band 3 and δ-ALA-S is sufficient to cause all phenotypic changes induced in erythroblasts by v-*erb*A, or whether this oncogene represses (or even activates) additional genes. The fact that v-*erb*A subtly alters the growth properties of fibroblasts (Gandrillon et al., 1987) and that mutants of v-*erb*A, which have lost their ability to block differentiation, to repress erythrocyte genes and to alter medium requirements, can still increase the growth rate of v-*erb*B transformed erythroblasts (Forrest et al., 1990b) argue for the latter possibility. We are currently testing erythroblast specific genes (that are down-regulated during erythroid differentiation) and erythroid transcription factors (e.g. NF-E1) for their potential regulation by v-*erb*A.

Second, it is unknown at present if the erythrocyte-specific genes suppressed by v-*erb*A and regulated by overexpressed, gag-fused TRα (Zenke et al., 1990) are subject to regulation by endogeneous TRα at some time during erythroid development, or whether they are regulated via related receptors that recognize similar elements on the DNA (e.g. retinoid acid receptor, vitamin D_3 receptor). Although we have failed to detect clear-cut effects of T_3 on the differentiation of erythroblasts into erythrocytes (using mainly ts-oncogene transformed cells and normal bone marrow preparations (Zenke et al., 1990; Schroeder et al., unpubl. res.), we cannot exclude a function of T_3 in embryonic erythrocytes or more mature red cells (Hentzen et al., 1987, Bigler and Eisenman, 1988; H. Beug, unpubl. obs.). Since TRs can obviously form heterodimers with RARs (Glass et al., 1989; Foreman et al., 1989) a synergistic action of T_3 and retinoid acid should also be considered. It is therefore tempting to speculate, that alterations in the specificity of protein–DNA interaction of the normal TRα (presumably caused by overexpression of v-*erb*A and/or mutations in its DNA-binding domain) may provide the basis for the v-*erb*A-induced abnormal pattern of erythrocyte gene regulation in avian erythroleukaemia.

References

Beug, H. and Graf, T. (1989). Cooperation between viral oncogenes in avian erythroid and myeloid leukaemia. *Eur. J. Clin Invest.* **19**, 491–502.
Beug, H., Palmieri, S., Freudenstein, C., Zentgraf, H. and Graf, T. (1982). Hormone-dependent terminal differentiation *in vitro* of chicken erythroleukemia cells transformed by *ts* mutants of avian erythroblastosis virus. *Cell* **28**, 907–919.
Beug, H., Kahn, P., Döderlein, G., Hayman, M. J. and Graf, T. (1985). Characterization of hematopoietic cells transformed *in vitro* by AEV-H, an *erb*-containing avian erythroblastosis virus. In "Modern Trends in Human Leukaemia VI" (eds R. Neth, R. Gallo, M. Greaves and K. Janko), Vol. 29, pp. 290–297.
Bigler, J. and Eisenman, R. N. (1988). c-*erb*A encodes multiple proteins in chicken erythroid cells. *Mol. Cell Biol.* **8**, 4155–4161.
Chou, P. Y. and Fasman, G. D. (1978). Prediction of the secondary structure of proteins from their amino acid sequence. *Adv. Enzymol.* **47**, 45–147.
Damm, K., Beug, H., Graf, T. and Vennström, B. (1987). A single point mutation in *erb*A restores the erythroid transforming potential of a mutant avian erythroblastosis virus (AEV) defective in both *erb*A and *erb*B oncogenes. *EMBO J.* **6**, 375–382.
Damm, K., Thompson, C. C. and Evans, R. M. (1989). Protein encoded by v-*erb*A functions as a thyroid-hormone receptor antagonist. *Nature* **339**, 593–597.
Disela, C., Dodgson, J., Beug, H. and Zenke, M. (1990). The carbonic anhydrase gene promoter is a direct target for transcriptional regulation by *erb*A proteins. *Oncogene* (in preparation).
Downward, J., Yarden, Y., Mayes, E., Scrace, G., Totty, N., Stockwell, P., Ullrich, A., Schlessinger, J. and Waterfield, M. D. (1984). Close similarity of epidermal growth factor receptor and v-*erb*B oncogene protein sequences. *Nature* **307**, 521–527.
Evans, R. M. (1988). The steroid and thyroid hormone receptor superfamily. *Science* **240**, 889–895.
Forman, B. M., Yang, C. R., Au, M., Casanova, J., Ghysdael, J. and Samuels, H. H. (1989). A domain containing leucine-zipper-like motifs mediate novel *in vivo* interactions between the thyroid hormone and retinoic receptors. *Mol. Endocrinol.* **3**, 1610–1626.
Forrest, D., Sjöberg, M. and Vennström, B. (1990a). Contrasting developmental and tissue-specific expression of a and b thyroid hormone receptors. *EMBO J.* **9**, 1519–1528.
Forrest, D., Munnoz, A., Raynoschek, C., Vennström, B. and Beug, H. (1990b). Requirement for the C-terminal domain of the v-*erb* A oncogene protein for biological function and transcriptional repression. *Oncogene* **5**, 309–316.
Frykberg, L., Palmieri, S., Beug, H., Graf, T., Hayman, M. J. and Vennström B. (1983). Transforming capacities of avian erythroblastosis virus mutants deleted in the *erb*A or *erb*B oncogenes. *Cell* **32**, 227–238.
Fung, Y. K. T., Lewis, W. G., Kung, H. J. and Crittenden, L. B. (1983). Activation of the cellular oncogene c-*erb*B by LTR insertion: molecular basis for induction of erythroblastosis by avian leukosis virus. *Cell* **33**, 357–368.
Fürstenberg, S., Beug, H., Introna, M., Khazaie, K., Muñoz, A., Ness, S., Nordström, K., Sap, J., Stanley, I., Zenke, M. and Vennström, B. (1990). Ectopic expression of the erythrocyte band 3 anion exchange protein using a new avian retrovirus vector. *J. Virol.* (in press).

Gandrillon, O., Jurdic, P., Benchaibi, M., Xiao, J.-H., Ghysdael, J. and Samarut, J. (1987). Expression of the v-*erb*A oncogene in chicken embryo fibroblasts stimulates their proliferation *in vitro* and enhance tumor growth *in vivo*. *Cell* **49**, 687–697.

Gandrillon, O., Jurdic, P., Pain, B., Desbois, C., Madjar, J. J., Moscovici, M. G., Moscovici, C. and Samarut, J. (1989). Expression of the v-*erb*A product, an altered nuclear hormone receptor, is sufficient to transform erythroid cells *in vitro*. *Cell* **58**, 115–121.

Glass, C. K., Lipkin, S. M., Devary, S. M. and Rosenfeld, M. G. (1989). Positive and negative regulation of gene transcription by retinoic acid–thyroid hormone receptor heterodimer. *Cell* **59**, 697–708.

Glineur, C., Bailly, M. and Ghysdael, J. (1989). The c-*erb*A alpha-encoded thyroid hormone receptor is phosphorylated in its amino terminal domain by casein kinase II. *Oncogene* **4**, 1247–1254.

Glineur, C., Zenke, M., Beug, H. and Ghysdael, J. (1990). Phosphorylation of the v-*erb*A protein at serines 16 and 17 is required for its function as an oncogene. *Genes Dev.* (in press).

Goldberg, Y., Glineur, C., Gesquiere, J. C., Ricouart, A., Sap, J., Vennström, B. and Ghysdael, J. (1988). Activation of protein kinase C or cAMP-dependent kinase increases phosphorylation of the c-*erb*A-encoded thyroid hormone receptor and of the v-*erb*A encoded protein. *EMBO J.* **7**, 2425–2433.

Graf, T. and Beug, H. (1978). Avian leukemia viruses: interaction with their target cells *in vivo* and *in vitro*. *BBA Rev. Cancer* **516**, 269–299.

Graf, T. and Beug, H. (1983). Role of the v-*erb*A and v-*erb*B oncogenes of avian erythroblastosis virus in erythroid cell transformation. *Cell* **34**, 7–9.

Graupner, G., Wills, K. N., Tzukerman, M., Zhang, X. K. and Pfahl, M. (1989). Dual regulatory role for thyroid-hormone receptors allows control of retinoic acid receptor activity. *Nature* **340**, 653–656.

Hentzen, D., Renucci, A., leGuellec, D., Benchaibi, M., Jurdic, P., Gandrillon, O. and Samarut, J. (1987). The chicken c-*erb*A proto-oncogene is preferentially expressed in erythrocytic cells during late stages of differentiation. *Mol. Cell. Biol.* **7**, 2416–2424.

Hodin, R. A., Lazar, M. A., Wintman, B. I., Darling, D. S., Koenig, R. J., Larsen, P. R., Moore, D. D. and Chin, W. W. (1989). Identification of a thyroid hormone receptor that is pituitary-specific. *Science* **244**, 76–79.

Izumo, S. and Mahdavi, V. (1988). Thyroid hormone receptor alpha isoforms generated by alternative splicing differentially activate myosin HC gene transcription. *Nature* **334**, 539–542.

Jansson, M., Beug, H., Gray, C., Graf, T. and Vennström, B. (1987). Defective v-*erb*B genes can be complemented by v-*erb*A in erythroblast and fibroblast transformation. *Oncogene* **1**, 167–173.

Johnsson, A., Betsholtz, C., Heldin, C. H. and Westermark, B. (1985). Antibodies against platelet-derived growth factor inhibit acute transformation by simian sarcoma virus. *Nature* **317**, 438–440.

Kahn, P., Frykberg, L., Brady, C., Stanley, I. J., Beug, H., Vennström, B. and Graf, T. (1986a). V-*erb*A cooperates with sarcoma oncogenes in leukemic cell transformation. *Cell* **100**, 349–356.

Kahn, P., Frykberg, L., Graf, T., Vennström, B. and Beug, H. (1986b). Cooperativity between v-*erb*A and v-*src*-related oncogenes in erythroid cell transformation. *In*: "XII Symposium for Comparative Research on Leukaemia and Related Diseases" (ed. F. Deinhard), pp. 41–50. Springer-Verlag, Heidelberg.

Khazaie, K., Dull, T. J., Graf, T., Schlessinger, J., Ullrich, A., Beug, H., and Vennström, B. (1988). Truncation of the human EGF receptor leads to differential transforming potentials in primary avian fibroblasts and erythroblasts. *EMBO J.* **7**, 3061–3071.

Kim, H.-R. C., Yew, W., Ansorge, H., Voss, C., Schwager, B., Vennström, B., Zenke, M. and Engel, J. D. (1988). Two different mRNAs are transcribed from a single genetic locus encoding the chicken erythrocyte anion transport protein (Band 3). *Mol. Cell. Biol.* **8**, 4416–4424.

Knight, J., Zenke, M., Disela, Ch., Kowenz, E., Vogt, P., Engel, D., Hayman, M. J. and Beug, H. (1988). Ts v-*sea* transformed erythroblasts: A model system to study gene expression during erythroid differentiation. *Genes Dev.* **2**, 247–258.

Koenig, R. J., Lazar, M. A., Hodin, R. A., Brent, G. A., Larsen, P. R., Chin, W. W. and Moore, D. D. (1989). Inhibition of thyroid hormone action by a non-hormone binding c-*erb*A protein generated by alternative mRNA splicing. *Nature* **337**, 659–661.

Lazar, M. A. and Chin, W. W. (1988). Regulation of two c-*erb*A messenger ribonucleic acids in rat GH3 cells by thyroid hormone. *Mol. Endocrinol.* **2**, 479–484.

Lazar, M. A., Hodin, R. A., Darling, D. S. and Chin, W. W. (1988). Identification of a rat c-*erb*A alpha-related protein which binds deoxyribonucleic acid but does not bind thyroid hormone. *Mol. Endocrinol.* **2**, 893–901.

Mader, S., Kumar, V., de Verneuil, H. and Chambon, P. (1989). Three amino acids of the oestrogen receptor are essential to its ability to distinguish an oestrogen from a glucocorticoid-responsive element. *Nature* **338**, 271–274.

Muñoz A., Zenke, M., Gehring, U., Sap, J., Beug, H. and Vennström B. (1988). Characterization of the hormone-binding domain of the chicken c-*erb*A/thyroid hormone receptor protein. *EMBO J.* **7**, 155–159.

Murray, M. B., Zilz, N. D., McCreary, N. L., MacDonald, M. J. and Towle, H. C. (1988). Isolation and characterization of rat cDNA clones for two distinct thyroid hormone receptors. *J. Biol. Chem.* **263**, 12 770–12 777.

Nakai, A., Sakurai, A., Bell, G. I. and DeGroot, L. J. (1988). Characterization of a third human thyroid hormone receptor coexpressed with other thyroid hormone receptors in several tissues. *Mol. Endocrinol.* **2**, 1087–1092.

Rothe-Meyer, A. and Engelbreth-Holm, J. (1933). Experimentelle Studien über die Beziehungen zwischen Hühnerleukose und Sarkom an der Hand eines Stammes von übertragbaren Leukose-Sarkom-Kombinationen. *Acta Pathol. Microbiol. Scand.* **10**, 380–427.

Royer-Pokora, B., Beug, H., Claviez, M., Winkhardt, H. J., Friis, R. R. and Graf, T. (1978). Transformation parameters in chicken fibroblasts transformed by AEV and MC29 avian leukaemia viruses. *Cell* **13**, 751–760.

Samarut, J. and Bouabdelli, M. (1980). *In vitro* development of CFU-E and BFU-E in cultures of embryonic and post embryonic chicken hematopoietic cells. *J. Cell Phys.* **105**, 553–563.

Sap, J., Muñoz, A., Damm, K., Goldberg, Y., Ghysdael, J., Leutz, A., Beug, H. and Vennström, B. (1986). The c-*erb*A protein is a high-affinity receptor for thyroid hormone. *Nature* **324**, 635–640.

Sap, J., Muñoz, A., Schmitt, J., Stunnenberg, H. and Vennström, B. (1989). Repression of transcription mediated at a thyroid hormone response element by the v-*erb*-A oncogene product. *Nature* **340**, 242–244.

Sap, J., de Magistris, L., Stunnenberg, H. and Vennström, B. (1990). A major thyroid hormone response element in the third intron of the rat growth hormone gene. *EMBO J.* **9**, 887–896.

Schmidt, J. A., Marshall, J., Hayman, M. J., Ponka, P. and Beug, H. (1986). Control of erythroid differentiation: possible role of the transferrin cycle. *Cell* **46**, 41–51.

Schroeder, C., Raynoschek, C., Fuhrmann, U., Damm, K., Vennström, B. and Beug, H. (1990). The v-*erb*A oncogene causes repression of erythrocyte-specific genes and an immature, aberrant differentiation phenotype in normal erythroid progenitors. *Oncogene* (in press).

Sealy, L., Privalsky, M. L., Moscovici, G., Moscovici, C. and Bishop, J. M. (1983). Site-specific mutagenesis of avian erythroblastosis virus: *erb*B is required for oncogenicity. *Virology* **130**, 155–178.

Weinberger, C., Thompson, C., Ong, E. S., Lebo, R., Gruol, D. and Evans, R. (1986). The c-*erb*A gene encodes a thyroid hormone receptor. *Nature* **324**, 641–646.

Yamamoto, T., Hihara, H., Nishida, T., Kawai, S. and Toyoshima, K. (1983). A new avian erythroblastosis virus, AEV-H, carries *erb* B gene responsible for both erythroblastosis and sarcomas. *Cell* **34**, 225–232.

Zenke, M., Kahn, P., Disela, Ch., Vennström, B., Leutz, A., Keegan, K., Hayman, M., Choi, H. R., Yew, N., Engel, J. D. and Beug, H. (1988). V-*erb*A specifically suppresses transcription of the avian erythrocyte anion transporter gene. *Cell* **52**, 107–119.

Zenke, M., Muñoz, A., Sap, J., Vennström, B. and Beug, H. (1990). V-erb A oncogene activation entails loss of hormone dependent regulatory activity of c-*erb*A. *Cell* **61**, 1035–1049.

16
Nuclear Hormone Receptors: Concluding Remarks

M. G. PARKER and O. BAKKER

Molecular Endocrinology Laboratory, Imperial Cancer Research Fund, Lincoln's Inn Fields, London WC2A 3PX, UK

16.1 Introduction

The nuclear receptors comprise a large family of transcription factors whose activity depends on the binding of a hormonal ligand. Prior to the isolation of cDNA clones it was unforeseen that steroid receptors would be related not just to one another, but also to the thyroid hormone and retinoic acid receptors, so the discovery of so many additional related proteins was completely unexpected (see Chapters 1.1 and 5.6). The second major surprise was that the DNA-binding sites for these receptors were rather similar (see Chapters 6.2 and 9.4). However, many other transcription factors have now been shown to share both of these features. For example, the *Fos/Jun* family of transcription factors, which consists of at least six proteins, and the ATF/CREB family with at least seven proteins, each bind to a single DNA recognition site (Jones, 1990). The members of these two families function as dimers and may form heterodimers as well as homodimers in combinations which give rise to markedly different DNA-binding properties. It is now emerging that some nuclear hormone receptors may also function in a similar manner to generate complex patterns of transcriptional control. In this final Chapter, our objective is to review very briefly the major advances in our understanding of the molecular mechanisms of nuclear hormone receptors and to discuss a number of topics that have not been dealt with elsewhere in this book.

16.2 Receptor gene structure and function

The genes for the nuclear hormone receptors can be grouped into one large family, each of which consists of eight coding exons. However, despite this similarity, they vary in size from 15 to 200 kb and are located on different

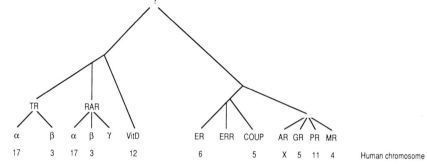

Fig. 1. Possible evolutionary relationships of the nuclear hormone receptors. The evolutionary relationships are derived from a comparison of the amino acid sequences of the different receptors. Shown below the different receptors are their locations on human chromosomes where known (for references see Evans 1988; AR, Lubahn *et al.*, 1988; VitD and COUP, B. O'Malley, pers. commun.).

chromosomes (Fig. 1). On the basis of amino acid sequence information an evolutionary tree can be drawn for the nuclear hormone receptor family (Fig. 1). One possibility is that a progenitor gene gave rise to the ancestors of both the steroid receptor and the thyroid hormone and retinoic acid receptor families by duplication and divergence. The individual receptor genes were then derived by subsequent duplications. This is supported by the finding that genes for different forms of both the thyroid hormone and the retinoic acid receptors are located near to one another on each of the human chromosomes 3 and 17 and, in addition, their exon 2/intron boundary differs from that of the steroid receptor genes (Green and Chambon, 1988).

The receptors can be divided into several structural domains, based on results from protease digestion experiments and from functional studies with chimaeric receptors (for reviews see Yamamoto, 1985; Gehring, 1988; and Chapters 2 and 3) and this is partly reflected in the exon/intron organization of their genes (Fig. 2). The N-terminal domain is largely encoded by one exon, the DNA-binding domain by two exons—one for each zinc finger—and the remaining five coding exons give rise to the hormone-binding domain which is responsible for a number of different receptor functions (Fig. 2).

While the exon arrangement of the coding region of the receptors is highly conserved, the organization of their promoters is significantly different. For example, the genes for the human and mouse oestrogen receptors differ at their 5' ends insofar as translation is initiated in the first exon in human (Green *et al.*, 1986) and in the second exon in mouse (White *et al.*, 1987). The progesterone receptor gene has two promoters in both human and chicken (Kastner *et al.*, 1990a) which generate distinct transcripts that encode

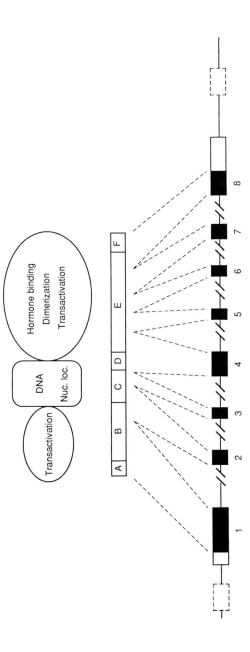

Fig. 2. Domain structure and gene organization of the nuclear hormone receptors. The different domains of the nuclear hormone receptors and their function as identified by mutagenesis are outlined schematically. The nuclear receptors have been divided into three structural domains (Yamamoto, 1985; Gehring, 1988) and their functions have been determined by mutagenesis (Chapters 2 and 3). All receptor genes characterized to date contain 8 coding exons (black boxes) but some contain additional non-coding exons (dashed boxes), for example the mouse oestrogen receptor has one extra 5′ exon (White *et al.*, 1987) and the human thyroid hormone receptor has extra 5′ and 3′ non-coding exons (Zahraoui and Cuny, 1987; V. Laudet, pers. commun.).

receptors with different N-terminal sequences (Conneeley *et al.*, 1989; Jeltsch *et al.*, 1990). The expression of the thyroid hormone and retinoic acid receptors is even more complex, as they are encoded by multiple genes with alternative promoter and splicing patterns that give rise to different forms of receptor (Lazar *et al.*, 1989; Miyajima *et al.*, 1989; Kastner *et al.*, 1990b; see Chapter 4). To date there is very little information about any receptor gene promoters but variations in their activities presumably result in the complex tissue-specific and developmentally controlled receptor expression.

The expression of all receptor genes investigated to date is subject to autoregulation. This appears to depend on a combination of regulatory mechanisms involving gene transcription and the turnover of mRNA and protein. In many types of cells steroid receptors are down-regulated by their respective hormones. The best characterized example is the expression of the glucocorticoid receptor whose autoregulation is achieved partly by a reduction in mRNA levels (Okret *et al.*, 1986) brought about by a reduction in its rate of gene transcription (Rosewicz *et al.*, 1988; Dong *et al.*, 1988) and partly by a decrease in the half-life of the receptor protein itself (Cidlowski and Cidlowski, 1981; Svec and Rudis, 1981; McIntyre and Samuels, 1985). Expression is also down-regulated in the case of the oestrogen receptor in chicken oviduct (Maxwell *et al.*, 1987) and in the human breast cancer cell line MCF-7 (Eckert *et al.*, 1984; Saceda *et al.*, 1988), the progesterone receptor in breast cancer cell lines (Mullick and Katzenellenbogen, 1986; Wei *et al.*, 1988) and the androgen receptor in rat prostate (Quarmby *et al.*, 1990). In contrast, oestrogen stimulates the expression of its receptor in male *Xenopus* liver (Barton and Shapiro, 1988). Whether autoregulation is positive or negative, therefore, may depend on the tissue type and/or its developmental stage.

16.3 Hormone binding and receptor dimerization

Steroid receptors are inactive in the absence of their respective hormone and may be isolated as part of a large protein complex that includes the heat-shock protein Hsp90 (see Chapter 3.6). They are induced to activate transcription upon steroid binding by forming receptor dimers which interact with specific sites on DNA (see Chapters 6.2 and 9.4). Recently a region that is involved in the dimerization of the mouse oestrogen receptor has been identified which overlaps a region essential for oestradiol binding (Fawell *et al.*, 1990a). Mutagenesis experiments suggest that critical amino acids for dimerization include arginine 507, leucine 511 and isoleucine 518 and a peptide encompassing these residues, namely 501–522, was sufficient to restore partial DNA binding to a mutant receptor defective in dimerization (Lees et al., 1990). A sequence alignment of the related proteins of the nuclear receptor superfamily revealed that this region is associated with a conserved heptad repeat of hydrophobic residues that contains additional hydrophobic amino acids in the intermediate positions (Fig. 3). The corresponding region in the thyroid hormones and retinoic acid receptors has been implicated in the formation of heterodimers (Forman et al., 1989; Glass et al., 1989; see Chapter 4.9) although there is no direct evidence as yet that they form homodimers. Therefore, although the sequence of this region in individual receptors is different, it is conveivable that its overall structure may be conserved and that the divergent sequences might contribute to the specificity of interaction between receptor monomers.

We have suggested (Fawell et al., 1990a) that the structure of the region may resemble that involved in the dimerization of uteroglobin. Uteroglobin is a dimeric progesterone-binding protein (Beato and Baier, 1975) whose structure has been determined by X-ray diffraction analysis (Mornon et al., 1980; Morize et al., 1981). The dimer forms a hydrophobic cavity in which the steroid is proposed to bind (Bally and Delettré, 1989). In steroid receptors the location of a hormone-binding pocket at the dimer interface may present a means for steroid-stabilized receptor dimerization based on hydrophobic shielding.

It is beginning to emerge that certain receptors, like a number of other transcription factors, may be able to form heterodimers as well as homodimers. This generates a number of possible regulatory mechanisms, as shown for certain leucine zipper and helix–loop–helix proteins. First, heterodimers may bind to DNA with higher affinity than homodimers of either protein as, for example, *Jun* and *Fos* (for a review see Kouzarides and Ziff, 1989). The difference in DNA binding reflects differences in the stability of *Jun–Jun* homodimers versus *Jun–Fos* heterodimers. The ability of the thyroid hormone receptors, β-1 and to a lesser extent α-1, to synergise with the retinoic acid receptor is proposed to involve the formation of heterodimers and may involve a similar mechanism (Glass et al., 1989). In this context it is of interest that a splicing variant of the thyroid hormone receptor α-1, c-*erb*A α-2, in which the putative dimerization and ligand-binding domain is disrupted by alternative C-terminal sequences (see Chapter 4.4), fails to interact co-operatively with the retinoic acid receptor.

Secondly, the formation of heterodimers may represent a mechanism for repressing transcriptional activity by a process referred to as dominant negative regulation (Herskowitz, 1987). The formation of inactive heterodimers has been shown to be important in two developmental programmes, namely the peripheral nervous system in *Drosophila melanogaster* (Rushlow et al., 1989) and muscle cell differentiation (Benezra et al., 1990). In both cases the activity of certain genes is controlled by transcription factors containing the helix–loop–helix motif. One set of factors, *achaete-scute* complex, *daughterless* and *Myo* D, play a positive role whereas another set, *hairy*, and *extracrochaetae* and Id, play a negative role. In both cases the negative regulators, in which the basic DNA-binding region is either missing or defective, repress the activity of positive regulators by forming heterodimers that are unable to bind DNA.

Certain nuclear receptors may also function as dominant negative regulators by heterodimerization. For example, the two proteins c-*erb*Aα-2 and v-*erb*A, which are both capable of inhibiting thyroid hormone action (Damm *et al.*, 1989; Koenig *et al.*, 1989; Sap *et al.*, 1989) may be acting in this way (see Chapters 4 and 15). The possibility that the α-2 protein may generate heterodimers with other forms of the thyroid hormone receptor, but not with the retinoic acid receptor, is not necessarily inconsistent in view of the behaviour of the transcription factor, *Fos*, which forms dimers with *Jun* but not with itself. Nevertheless, inhibitory effects of receptors could be explained in other ways such as competition for DNA-binding sites and therefore more work is needed to distinguish between these mechanisms. Another candidate for a dominant negative regulator is a receptor-like protein induced by the insect hormone, ecdysone, in the salivary glands of Drosphila melanogaster (Segraves and Hogness, 1990). It has long been known that ecdysone stimulates the appearance of 'puffs', sites of gene activity, in two phases: 'early' transcription puffs, which are autoregulatory, and 'late' puffs, which are dependent upon proteins produced during the primary response (Clever and Karlson, 1960; Clever, 1964; Ashburner, 1972). The products of two 'early' genes E74 and E75, are both DNA-binding proteins, which is consistent with their predicted role in the repressing 'early' and activating 'late' gene expression. Interestingly, both genes give rise to two mRNAs that encode proteins with common C-terminal sequences but different N-terminal sequences. The E75 proteins are members of the nuclear receptor family, although their ligand has yet to be identified. Surprisingly E75A contains two zinc fingers whereas E75B contains only one. The E75A protein binds to the 74 puff site, probably to repress its activity, but the function of E75B is unknown. Could it be, by analogy with the mutant helix–loop–helix proteins described above, a dominant negative regulatory protein repressing the activity of the E75A receptor by heterodimer formation? If so, dominant negative regulation may be a general mechanism by which nuclear receptors regulate various developmental pathways.

A third consequence of dimerization may be to influence DNA binding specificity. For example, *Jun*A binds to the AP-1 site both as a homodimer or as a heterodimer with *Fos*, but to the CRE site as a heterodimer with one

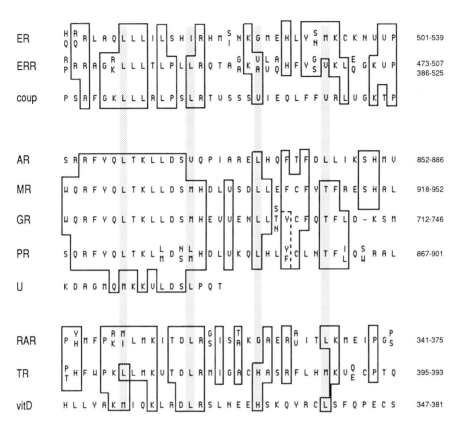

Fig. 3. A potential dimerization region in all members of the nuclear receptor family. The amino acid sequences of members of the nuclear receptor family were aligned by computer (Genalign) or manually and a conserved region within the ligand-binding domain is shown. The position of the conserved region in the human protein is, where appropriate, indicated by the numbers to the right of the sequence. The repeat of hydrophobic residues (see text) is shaded and further homologies are boxed. The sequence of each class of receptor represents the sequence of a number of different species or receptor subclasses used in the alignment and divergent residues are shown. ER, estrogen receptor: human (Green et al., 1986), mouse (White et al., 1987), chicken (Krust et al., 1986), rat (Koike et al., 1987), Xenopus (Weiler et al., 1987). ERR 1 & 2: human (Giguère et al., 1988). COUP: chicken (Wang et al., 1989). AR, androgen receptor: human and rat (Chang et al., 1988). MR, mineralocorticoid receptor: human (Arriza et al., 1987). GR, glucocorticoid receptor: human (Hollenberg et al., 1985), mouse (Danielsen et al., 1986), rat (Miesfield et al., 1986). PR, progesterone receptor: human (Mishrahi et al., 1987), rabbit (Loosfelt et al., 1986), chicken (Gronemeyer et al., 1987). U, uteroglobin: rabbit (Mornon et al., 1980). RAR, retinoic acid receptor: human (Petkovich et al., 1987; Giguère et al., 1987; Brand et al., 1988; Benbrook et al., 1988), mouse α, β, γ forms (Zelent et al., 1989). TR, thyroid hormone receptor: human (Weinberger et al., 1986), chicken (Sap et al., 1986), rat (Thompson et al., 1987). VIT D, vitamin D receptor: human (Baker et al., 1988).

member of the ATF/CREB family of transcription factors (Benbrook and Jones, 1990). It has yet to be established whether binding specificity of nuclear receptors can be altered in a similar way by the formation of heterodimers but retinoic acid and thyroid hormone receptors are potential candidates.

16.4 DNA binding of nuclear hormone receptors

The identification and characterization of the DNA-binding domain and the binding sites for receptors has been comprehensively reviewed by several workers in the field (see Chapters 2.4, 3.3, 6 and 9). Therefore in this section we will only mention a number of details not previously covered. First, it has been inferred from studies of steroid receptors that other nuclear receptors also bind to DNA as dimers. This may not, however, be the case for all receptors, since the thyroid hormone and retinoic acid receptors are able to function by binding to either direct or inverted repeats (see Chapters 4.5, 6.2 and 9.4). These receptors may bind as dimers to inverted repeats but as monomers to direct repeats, stable DNA binding being achieved by protein–protein contacts with other nuclear proteins (Murray and Towle, 1989; Burnside et al., 1990), or between monomers formed after the initial binding. Certainly it would not be too surprising if various receptors exhibit differences in their monomer/dimer equilibrium.

Secondly, the three-dimensional structure of the DNA-binding domain of the glucocorticoid receptor has been determined by nuclear magnetic resonance spectroscopy (NMR) (Härd et al., 1990). The two zinc finger motifs are found to be folded into a single domain (see front cover) whose structure is distinct from that determined for a zinc finger in Xfin (Lee et al., 1989) or GAL4 (Pan and Coleman, 1990). The residues near the first zinc co-ordination site, which have been shown to be responsible for the discrimination between glucocorticoid and oestrogen response elements (see Chapters 2 and 3; Danielsen et al., 1989; Mader et al., 1989; Umesono and Evans, 1989) form part of a helical structure analogous to the recognition α-helix found in helix–turn–helix proteins (Berg, 1989, 1990). Residues near the second zinc co-ordination site, which are involved in the discrimination of different spacings between the half-sites (Umesono and Evans, 1989), form loop structures that may be involved in protein–protein interactions. A model is proposed (Härd et al., 1990) in which the DNA-binding domain interacts with the major groove of the glucocorticoid response element but confirmation of this will require further NMR and/or X-ray diffraction analysis.

Finally, by analogy with the leucine zipper and helix–loop–helix transcription factors, the DNA-binding domain of nuclear receptors may consist not only of the zinc finger region but also of the dimerization region. Although these regions are separated by about 250 amino acids in the primary sequence of the receptors (Fawell *et al.*, 1990a), they could be spatially juxtaposed in the tertiary structure of the protein to form a discrete domain. In view of this, endocrine disorders that arise from defective DNA binding of receptors may be caused by mutations in either the dimerization or zinc finger region.

16.5 Transcriptional activation

A number of models have been described to explain the mechanism by which eukaryotic transcription factors modulate rates of gene transcription. The first, developed initially for prokaryotic promoters (Ptashne 1986, 1988) and mentioned in several chapters, suggests that the formation of a transcription initiation complex is dependent on protein–protein contacts between transcription factors. Some of the factors may not bind directly to DNA in which case they are called 'bridging' proteins. A second model, based primarily on the mouse mammary tumour virus (MMTV) promoter, emphasizes the importance of chromatin structure and, in particular, the availability of binding sites for transcription factors (see Archer and Hager in Chapter 10.3). Finally, the state of DNA supercoiling may also be important (see Beato and co-workers, Chapter 9.5). These proposals are not mutually exclusive but it is fair to say that the vast majority of experiments are capable of testing only one of them.

The different approaches used to investigate the mechanism of transcriptional activation are described in detail in Chapters 2.10, 5.5, 5.7, 7, 9.7 and 10.3. A model involving protein–protein contacts in the formation of a stable transcription initiation complex is favoured by most workers in the field. A popular hypothesis is that receptors interact directly with the TATA box binding protein and possibly with one or more additional transcription factors. Indirect evidence to support this is provided by several observations. First, optimum transcription depends on the stereo-specific alignment of a receptor and transcription factor (see Chapter 7.4). Secondly, in the mouse mammary tumour virus (MMTV) promoter, mutations in the NF-1 binding site reduce the stimulatory effects of hormone (see Chapters 7.3 and 9.7); and third, studies of template commitment suggest that the progesterone receptor interacts directly with TFIID and TFIIE/F (see Chapter 5.7).

Results from at least two types of study of the MMTV promoter, however, are not entirely consistent with the suggestion that receptors form stable interactions with transcription factors. First, Hager's group (see Chapter 10.3) has obtained convincing evidence to suggest that the role of the glucocorticoid receptor is to displace a specific nucleosome, thereby allowing NF-1 and TFIID to bind to sites previously masked. In this example, the receptor could not be detected in the transcription complex once transcription had been initiated by steroid treatment. Secondly, a number of groups have attempted to demonstrate an interaction between receptor and NF-1 *in vitro* but without success to date (see Chapters 7.5 and 9.7). Thus Archer and Hager (Chapter 10.3) conclude that the reorganization of chromatin to allow transcription factor binding is an important step in the hormonal control of MMTV promoter activity. They suggest that protein–protein interactions may also be required without specifying what they might be. It would seem that they may be only transient and could involve nucleosomal proteins as well as specific transcription factors. Further analysis using cell-free transcription systems may help to unravel some of these apparent contradictions.

Finally, one type of protein–protein interaction for which there is good evidence involves a bridging protein. Ironically, before COUP was discovered to be a member of the nuclear receptor superfamily, it was found to require a protein called S300-II to stimulate transcription of the ovalbumin gene. This protein is unable to bind to DNA but may serve as a transcriptional activation domain by binding to COUP (see Chapter 5.5). To date transcriptional activation domains are rather poorly characterized but an amphipathic helix has been implicated in the glucocorticoid receptor (Hollenberg and Evans, 1988). Sequences with the potential to form such a structure are also important for transcriptional activity of the thyroid hormone receptor (Zenke *et al.*, 1990; see Berg and Vennström, Chapter 15.5) and the mouse oestrogen receptor (Lees *et al.*, 1989). Interestingly, this sequence appears to be conserved in other members of the nuclear receptor family with the notable exception of COUP (R. White and P. Danielian, pers. commun.).

16.6 Hormonal effects on mRNA stability

Numerous studies describe indirect evidence to suggest that hormones regulate gene expression, at least in part, by modulating mRNA stability. Progress in elucidating the mechanisms involved has been slow but some discussion is essential because mRNA stability is undoubtedly an important aspect of hormone action.

The oestrogen regulation of egg-yolk protein synthesis, particularly vitellogenin, has provided a model system to study post-transcriptional as well as transcriptional control of gene expression (Jost et al., 1978; Wiskocil et al., 1980; Brock and Shapiro, 1983; Noteborn et al., 1986). In *Xenopus laevis* the half-life of the vitellogenin mRNA is decreased dramatically upon oestrogen withdrawal (Brock and Shapiro, 1983), whereas albumin and γ-fibrinogen mRNAs are de-stabilized in the presence of hormone (Wolffe et al., 1985; Kazmaier et al., 1985; Riegel et al., 1987; Pastori et al., 1990). To explain the effect of oestrogen on the vitellogenin mRNA, Shapiro and co-workers have suggested that there may be a specific protein in *Xenopus* liver cytosol that might bind the mRNA and modulate its stability. Furthermore, a mini-vitellogenin gene containing the 5' and 3' untranslated regions and a small proportion of the adjacent coding regions was shown to contain sufficient sequence information for the stabilization of the RNA (Nielsen and Shapiro, 1990). Evidence supporting such a model can be obtained from the post-transcriptional regulation of transferrin receptor mRNAs by iron. A number of so-called iron response elements which form characteristic stem-loop structures at the 3'-end of the mRNAs are responsible for RNA de-stabilization (Casey et al., 1988). A cytosolic protein has now been identified which binds specifically to these response elements and may therefore be involved in the regulation of mRNA stability (Leibold et al., 1990).

Analysis of a number of mRNAs that are subject to rapid turnover has identified AU-rich sequences near the 3'-end which are implicated as recognition signals for processing (Shaw and Kamen, 1986; Brewer and Ross, 1988). Interestingly, upon oestrogen withdrawal, the apoVLDLII mRNA is cleaved at specific sites in its 3' untranslated region in the vicinity of AU-rich sequences which occur in bulges and loops in the predicted secondary structure (Bakker et al., 1988; Binder et al., 1989; Cochrane and Deeley, 1989). To demonstrate a role for these sequences it will be necessary to analyse the expression of mutant RNAs as was described for vitellogenin mRNA. By combining this approach with studies on RNA-protein interactions it should be possible to elucidate the mechanisms by which hormones regulate mRNA stability in much the same way as promoters are characterized currently.

16.7 Mechanism of action of hormone antagonists

Anti-hormones can be divided into two classes, so-called pure antagonists, and those with partial agonist activity (Jordan, 1984; Raynaud and Ojasoo, 1986). Work on oestrogen receptors has provided insight into the mechanism of action of both types. Tamoxifen, a well-characterized partial antagonist of oestrogen action (Jordan, 1984) promotes DNA binding but fails to induce the transcriptional activity of the hormone-binding domain (Webster et al., 1988). The agonist effects of tamoxifen appear to be derived from the N-terminal transactivation domain, whose activity is dependent on the organism, the target gene and the cell type in which it is expressed (Tora et al., 1989; see Chapter 2.11). In contrast, ICI 164 384, which seems to behave as a pure anti-oestrogen (Wakeling and Bowler, 1988), inhibits DNA binding (Fawell et al., 1990b). This seems to result from disruption of receptor dimerization, since an anti-receptor antibody which is capable of restoring DNA-binding activity to receptors with a defective dimerization domain is similarly able to overcome the inhibitory effect of ICI 164 384. Since the dimerization and steroid-binding regions partially overlap in the oestrogen receptor, Fawell et al. (1990b) proposed that when ICI 164 384 binds to receptor the large 7α-alkylamide side chain might interfere sterically with dimerization of the protein. In the absence of dimerization and DNA binding neither the N- nor C-terminal transactivation domain would be expected to be active. In other studies the anti-glucocorticoid, RU 486, was shown to function by a mechanism similar to that described for tamoxifen (Webster et al., 1988).

It will be important to determine whether other antagonists function by one or other of these mechanisms or by alternative means because a better understanding of their of action is likely to lead to more efficacious treatment of a number of endocrine disorders. For example, tamoxifen is currently used successfully for the treatment of hormone-dependent breast cancer with up to 60% of patients responding to treatment. Nevertheless, in time, tamoxifen-resistant tumours invariably arise in patients. Since tamoxifen is a partial agonist it is conveivable that a pure anti-oestrogen such as ICI 164 384 might be more effective. The initial choice of treatment might still be tamoxifen since few side-effects have been observed with this drug, rather than ICI 164 384, which may adversely affect conditions such as osteoporosis. However, a subset of tamoxifen-resistant patients may well benefit from treatment with a pure anti-oestrogen with a different mechanism of action.

16.8 Final remarks

Clearly during the last few years there has been dramatic progress in our fundamental knowledge of the structure and function of nuclear receptors. These studies should form the basis of future work to elucidate the multihormone control of gene activity in target cells and to unravel their role in endocrine disorders and certain common cancers. As this book goes to press, it is emerging that nuclear hormones may communicate with alternate signalling pathways since the activity of the glucocorticoid receptor can be regulated by *Fos* and *Jun* and vice versa (Diamond *et al.*, 1990; Lucibello *et al.*, 1990; P. Herrlich, M. Karin and R. Evans, personal communications). These interactions may be extremely important in complex physiological processes such as cell growth and differentiation and there is no doubt that they will be the focus of much work in the future. Therefore, while much has been achieved since Elwood Jensen first described the two-step mechanism for steroid hormone action, there is still much to be done.

References

Albrecht, G., Krowczynska, A. and Brawerman, G. (1984). Configuration of β-globin messenger RNA in rabbit reticulocytes. *J. Mol. Biol.* **178**, 881–896.

Arriza, J. L., Weinberger, C., Cerelli, G., Glaser, T. M., Handelin, B. L., Housman, D. E. and Evans, R. M. (1987). Cloning of human mineralocorticoid receptor cDNA: structural and functional kinship with the glucocorticoid receptor. *Science* **237**, 268–275.

Ashburner, M. (1972). Patterns of puffing activity in the salivary gland chromosomes of *Drosophila*. *Chromosoma* **38**, 255–281.

Baker, A. R., McDonnell, D. P., Hughes, M., Crisp, T. M., Mangelsdorf, D. J., Haussler, M. R., Pike, J. W., Shine, J. and O'Malley, B. W. (1988). Cloning and expression of full length cDNA encoding human vitamin D receptor. *Proc. Natl. Acad. Sci. USA* **85**, 3294–3298.

Bakker, O., Arnberg, A. C., Noteborn, M. H. M., Winter, A. J. and AB, G. (1988). Turnover products of the apo very low density lipoprotein II messenger RNA from chicken liver. *Nucl. Acids Res.* **16**, 10 109–10 118.

Bally, R. and Delettré, J. (1989). Structure and refinement of the oxidized P21 form of uteroglobin at 1.64 Å resolution. *J. Mol. Biol.* **206**, 153–170.

Barton and Shapiro, J. D. (1988). Transient administration of estradiol-17β establishes an autoregulatory loop permanently inducing estrogen receptor mRNA. *Proc. Natl. Acad. Sci. USA* **85**, 7119–7123.

Beato, M. and Baier, R. (1975). Binding of progesterone to the proteins of the uterine luminal fluid. Identification of uteroglobin as the binding protein. *Biochim. Biophys. Acta* **392**, 346–356.

Benbrook, D., Lernhardt, E. and Pfahl, M. (1988). A new retinoic acid receptor identified from a hepatocellular carcinoma. *Nature* **333**, 669–672.

Benbrook, D. M. and Jones, N. C. (1990). The heterodimerization of CREB and Jun. *Oncogene* **5**, 295–302.

Benezra, R., Davis, R. L., Lockshon, D., Turner, D. L. and Weintraub, H. (1990). The protein Id: a negative regulator of helix–loop–helix DNA binding proteins. *Cell* **61**, 49–59.

Berg, J. M. (1989). DNA binding specificity of steroid receptors. *Cell* **57**, 1065–1068.

Berg, J. M. (1990). Zinc fingers and other DNA binding domains. *J. Biol. Chem.* **265**, 6513–6516.

Binder, R., Hwang, S.-P. L., Ratnasabapathy, R. and Williams, D. L. (1989). Degradation of apolipoproteinII mRNA occurs via endonucleolytic cleavage at 5'-AAU-3'/5'-UAA-3' elements in single-stranded loop domains of the 3'-noncoding region. *J. Biol. Chem.* **264**, 16 910–16 918.

Brand, N., Petkovich, M., Krust, A., Chambon, P., de The, H., Marchio, A., Tiollais, P. and Dejean, A. (1988). Identification of a second human retinoic acid receptor. *Nature* **332**, 850–853.

Brewer, G. and Ross, J. (1988). Poly(A) shortening and degradation of the 3' A + U rich sequences of the human c-myc mRNA in a cell-free system. *Mol. Cell. Biol.* **8**, 1697–1708.

Brock, M. L. and Shapiro, D. J. (1983). Estrogen stabilizes vitellogenin mRNA against cytoplasmic degradation. *Cell* **34**, 207–214.

Burnside, J., Darling, D. S. and Chin, W. W. (1990). A nuclear factor enhances binding of thyroid hormone receptors to thyroid hormone response elements. *J. Biol. Chem.* **265**, 2500–2504.

Casey, J. L., Hentze, M. W., Koeller, D. M., Caughman, S. W., Rouault, T. A., Klausner, R. D. and Harford, J. B. (1988). Iron-responsive elements: regulatory RNA sequences that control mRNA levels and translation. *Science* **240**, 924–928.

Chang, C., Kokontis, J. and Liao, S. (1988). Structural analysis of complementary DNA and amino acid sequences of human and rat androgen receptors. *Proc. Natl. Acad. Sci. USA* **85**, 7211–7215.

Cidlowski, J. A. and Cidlowski, N. B. (1981). Regulation of glucocorticoid receptors by glucocorticoids in cultured HeLa S3 cells. *Endocrinology* **109**, 1975–1982.

Clever, U. (1964). Actinimycin and puromycin: effects on sequential gene activation by ecdysone. *Science* **146**, 794–795.

Clever, U. and Karlson, P. (1960). Induktion von Puff-veränderungen in den Speicheldrüsenchromosomen von Chironomus tentans durch Ecdyson. *Exp. Cell. Res.* **20**, 623–626.

Cochrane, A. and Deeley, R. G. (1989). Detection and characterization of degradative intermediates of avian apo very low density lipoprotein II mRNA present in estrogen-treated birds and following destabilization by hormone withdrawal. *J. Biol. Chem.* **264**, 6495–6503.

Conneely, O. M., Kettelberger, D. M., Tsai, M. J., Schrader, W. T. and O'Malley, B. W. (1989). The chicken progesterone receptor A and B isoforms are products of an alternate translation initiation event. *J. Biol. Chem.* **264**, 14 062–14 064.

Damm, K., Thompson, C. C. and Evans, R. M. (1989). Protein encoded by v-erb-A functions as a thyroid hormone receptor antagonist. *Nature* **339**, 593–597.

Danielsen, M., Northrop, J. P. and Ringold, G. M. (1986). The mouse glucocorticoid receptor: mapping of functional domains by cloning, sequencing and expression of wild-type and mutant receptor proteins. *EMBO J.* **5**, 2513–2522.

Diamond, M. I., Miner, J. N., Yoshinaga, S. K. and Yamamoto, K. R. (1990). Transcription factor interactions: selectors of positive or negative regulation from a single DNA element. *Science* **249**, 1266–1272.

Dong, Y., Poellinger, L., Gustafsson, J.-Å. and Okret, S. (1988). Regulation of glucocorticoid receptor expression: Evidence for transcriptional and postranslational mechanisms. *Mol. Endocrinol.* **2**, 1256–1264.

Eckert, R. L., Mullick, A., Rorke, E. A. and Katzenellenbogen, B. S. (1984). Estrogen receptor synthesis and turnover in MCF-7 breast cancer cells measured by a density shift technique. *Endocrinology* **114**, 629–637.

Evans, R. M. (1988). The steroid and thyroid hormone receptor superfamily. *Science* **240**, 889–895.

Fawell, S. E., Lees, J. A., White, R. and Parker, M. G. (1990a). Characterization and localization of steroid binding and dimerization activities in the mouse estrogen receptor. *Cell* **60**, 953–962.

Fawell, S. E., White, R., Hoare, S., Sydenham, M., Page, M. and Parker, M. G. (1990b). Inhibition of estrogen receptor DNA binding by the pure antioestrogen ICI 164 384 appears to be mediated by impaired receptor dimerization. *Proc. Natl. Acad. Sci. USA* **87**, 6883–6887.

Forman, B. M., Yang, C., Au, M., Casanova, J., Ghysdael, J. and Samuels, H. H. (1989). A domain containing a leucine-zipper-like motif mediates novel *in vivo* interactions between the thyroid hormone and retinoic acid receptors. *Mol. Endocrinol.* **3**, 1610–1626.

Gehring, U. (1988). Steroid hormone receptors: biochemistry, genetics, and molecular biology. *Trends Biochem. Sci.* **12**, 399–402.

Giguere, V., Ong, E. S., Segui, P. and Evans, R. M. (1987). Identification of a receptor for the morphogen retinoic acid. *Nature* **330**, 624–629.

Giguere, V., Yang, N., Segui, P. and Evans, R. M. (1988). Identification of a new class of steroid hormone receptors. *Nature* **331**, 91–94.

Glass, C. K., Lipkin, S. M., Devary, O. V. and Rosenfeld, M. G. (1989). Positive and negative regulation of gene transcription by a retinoic acid-thyroid hormone receptor heterodimer. *Cell* **59**, 697–708.

Gordon, D. A., Shelness, G. S., Nicosia, M. and Williams, D. L. (1988). Estrogen-induced destabilization of yolk precursor protein mRNAs in avian liver. *J. Biol. Chem.* **263**, 2625–2631.

Green, S. and Chambon, P. (1988). Nuclear receptors enhance our understanding of transcription regulation. *Trends Genet.* **4**, 309–314.

Green, S., Walter, P., Kumar, V., Krust, A., Bornert, J. M., Argos, P. and Chambon, P. (1986). Human oestrogen receptor cDNA: sequence, expression and homology to v-erb-A. *Nature* **320**, 134–139.

Gronemeyer, H., Turcotte, B., Quirin-Stricker, C., Bocquel, M. T., Meyer, M. E., Krozowski, Z., Jeltsch, J. M., Lerouge, T., Garnier, J. M. and Chambon, P. (1987). The chicken progesterone receptor: sequence, expression and functional analysis. *EMBO J.* **6**, 3985–3994.

Härd, T., Kellenbach, E., Boelens, R., Maler, B. A., Dahlman, K., Freedman, L. P., Carlstedt-Duke, J., Yamamoto, K. R., Gustafsson, J.-A. and Kaptein, R. (1990). Solution structure of the glucocorticoid receptor DNA-binding domain. *Science* **249**, 157–160.

Herskowitz, I. (1987). Functional inactivation of genes by dominant negative mutations. *Nature* **329**, 219–222.

Hollenberg, S. M. and Evans, R. M. (1988). Multiple and cooperative transactivation domains of the human glucocorticoid receptor. *Cell* **55**, 899–906.

Hollenberg, S. M., Weinberger, C., Ong, E. S., Carelli, G., Oro, A., Lebo, R., Thompson, E. B. and Evans, R. M. (1985). Primary structure and expression of a functional human glucocorticoid receptor. *Nature* **318**, 635–641.

Huckaby, C. S., Conneely, O. M., Beattie, W. G., Dobson, A. D. W., Tsai, M.-J. and O'Malley, B. W. (1987). Structure of the chromosomal chicken progesterone receptor gene. *Proc. Natl. Acad. Sci. USA* **84**, 8380–8384.

Jeltsch, J.-M., Turcotte, B., Garnier, J.-M., Lerouge, T., Krozowksi, Z., Gronemeyer, H. and Chambon, P. (1990). Characterization of multiple mRNAs originating from the chicken progesterone receptor genes. *J. Biol. Chem.* **265**, 3967–3974.

Jones, N. (1990). Transcriptional regulation by dimerization: two sides to an incestuous relationship. *Cell* **61**, 9–11.

Jordan, V. C. (1984). Biochemical pharmacology of antiestrogen action. *Pharmacol. Rev.* **36**, 245–276.

Jost, J.-P., Ohno, T., Panyim, S. and Schuerch, A. R. (1978). Appearance of vitellogenin mRNA sequences and rate of vitellogenin synthesis in chicken liver following primary and secondary stimulation by 17β-estradiol. *Eur. J. Biochem.* **84**, 355–361.

Kastner, P., Krust, A., Mendelsohn, C., Garnier, J. M., Zelent, A., Leroy, P., Staub, A. and Chambon, P. (1990a). Murine isoforms of retinoic acid receptor gamma with specific patterns of expression. *Proc. Natl. Acad. Sci. USA* **87**, 2700–2704.

Kastner, P., Krust, A., Turcotte, B., Stropp, U., Tora, L., Gronemeyer, H. and Chambon, P. (1990b). Two distinct estrogen-regulated promoters generate transcripts encoding the two functionally different human progesterone receptor forms A and B. *EMBO J.* **9**, 1603–1614.

Kazmaier, M., Bruning, E. and Ryffel, G. U. (1985). Post-transcriptional regulation of albumin gene expression in *Xenopus* liver. *EMBO J.* **4**, 1261–1266.

Koenig, R. J., Lazar, M. A., Hodin, R. A., Brent, G. A., Larsen, P. R., Chin, W. W. and Moore, D. D. (1989). Inhibition of thyroid hormone action by a non-hormone binding c-erb-A protein generated by alternative mRNA splicing. *Nature* **337**, 659–661.

Koike, S., Sakai, M and Muramatsu, M. (1987). Molecular cloning and characterization of rat estrogen receptor cDNA. *Nucl. Acids Res.* **15**, 2499–2513.

Kouzarides, T. and Ziff, E. (1988). The role of the leucine zipper in the fos-jun interaction. *Nature* **336**, 646–651.

Krust, A., Green, S., Argos, P., Kumar, V., Walter, P., Bornert, J.-M. and Chambon, P. (1986). The chicken oestrogen receptor sequence: homology with v-erb-A and the human oestrogen and glucocorticoid receptors. *EMBO J.* **5**, 891–897.

Kuiper, G. G. J. M., Faber, P. W., van Rooij, H. C. J., van der Korput, J. A. M., Ris-Stalpers, C., Klaassen, P., Trapman, J. and Brinkman, A. O. (1989). Structural organization of the human androgen receptor gene. *J. Mol. Endocrinol.* **2**, R1–R4.

Lazar, M. A., Hodin, R. A., Darling, D. S. and Chin, W. W. (1989). A novel member of the thyroid/steroid hormone receptor family is encoded by the opposite strand of the c-erb-A α transcriptional unit. *Mol. Cell. Biol.* **9**, 1128–1136.

Lee, M. S., Gippert, G. P., Soman, K. V., Case, D. A. and Wright, P. E. (1989). Three-dimensional solution structure of a single zinc finger DNA-binding domain. *Science* **245**, 635–637.

Lees, J. A., Fawell, S. E. and Parker, M. G. (1989). Identification of two transactivation domains in the mouse oestrogen receptor. *Nucl. Acids Res.* **17**, 5477–5488.

Lees, J. A., Fawell, S. E., White, R. and Parker, M. G. (1990). A 22 amino acid peptide is sufficient to restore dimerization to a mutant receptor. *Mol. Cell Biol.* **10**, 5529–5531.

Leibold, E. A., Laudano, A. and Yu, Y. (1990). Structural requirements of iron-responsive elements for binding of the protein involved in both transferin receptor and ferritin mRNA post-transcriptional regulation. *Nucl. Acids Res.* **18**, 1819–1824.

Loosfelt, H., Atger, M., Misrahi, M., Guiochon-Mantelo, A., Meriel, C., Logeat, F., Benarous, R. and Milgrom, E. (1986). Cloning and sequence analysis of rabbit progesterone-receptor cDNA. *Proc. Natl. Acad. Sci. USA* **83**, 9045–9049.

Lubahn, D. B., Joseph, D. R., Sullivan, P. M., Willard, H. F., French, F. S. and Wilson, E. M. (1988). Cloning of human androgen receptor complementary DNA and localization to the X-chromosome. *Science* **240**, 327–330.

Lucibello, F. C., Slater, E. P., Jooss, K. U., Beato, M. and Müller, R. (1990). Mutual transrepression of Fos and the glucocorticoid receptor: involvement of a functional domain in Fos which is absent in FosB. *EMBO J.* **9**, 2827–2834.

Maxwell, B. L., McDonnell, D. P., Conneely, O. M., Schulz, T. Z., Greene, G. L. and O'Malley, B. W. (1987). Structural organization and regulation of the chicken estrogen receptor. *Mol. Endocrinol.* **1**, 25–35.

McIntyre, W. R. and Samuels, H. H. (1985). Triamcinolone acetonide regulates glucocorticoid-receptor levels by decreasing the half-life of the activated nuclear receptor form. *J. Biol. Chem.* **260**, 418–427.

Miesfeld, R., Rusconi, S., Godowski, P. J., Maler, B. A., Okret, S., Wikstrom, A.-C., Gustafsson, J.-Å. and Yamamoto, K. R. (1986). Genetic complementation of a glucocorticoid deficiency by expression of cloned receptor cDNA. *Cell* **46**, 389–399.

Misrahi, M., Atger, M., D'Auriol, L., Loosfelt, H., Meriel, C., Fridlansky, F., Guiochon-Mantel, A., Galibert, F. and Milgrom, E. (1987). Complete amino acid sequence of the human progesterone receptor deduced from cloned cDNA. *Biochem. Biophys. Res. Commun.* **143**, 740–748.

Miyajima, N., Horiuchi, R., Shibuya, Y., Fukushige, S., Matsubara, K., Toyoshima, K. and Yamamoto, T. (1989). Two erb A homologs encoding proteins with different T3 binding capacities are transcribed from opposite DNA strands of the same genetic locus, *Cell* **57**, 31–39.

Morize, I., Surcouf, E., Vaney, M. C., Epelboin, Y., Buehner, M., Fridlansky, F., Milgrom, E. and Mornon, J. P. (1981). Refinement of the C2221 crystal form of oxidized uteroglobin at 1.34 Å resolution. *J. Mol. Biol.* **194**, 725–739.

Mornon, J. P., Fridlansky, F., Bally, R. and Milgrom, E. (1980). X-ray crystallographic analysis of a progesterone-binding protein. The C2221 crystal form of oxidized uteroglobin at 2.2 Å resolution. *J. Mol. Biol.* **137**, 415–429.

Mullick, A. and Katzenellenbogen, B. S. (1986). Progesterone receptor synthesis and degradation in MCF-7 human breast cancer cells as studied by dense amino acid incorporation. Evidence for a non-hormone binding receptor precursor. *J. Biol. Chem.* **261**, 13 236–13 246.

Murray, M. B. and Towle, H. C. (1989). Identification of nuclear factors that enhance binding of the thyroid hormone receptor to a thyroid hormone response element. *Mol. Endocrinol.* **3**, 1434–1442.

Nielsen, D. A. and Shapiro, D. J. (1990). Estradiol and estrogen receptor-dependent stabilization of a mini vitellogenin mRNA lacking 5100 nucleotides of coding sequence. *Mol. Cell. Biol.* **10**, 371–376.

Noteborn, M. H. M., Bakker, O., de Jonge, M. A. W., Gruber, M. and AB, G. (1986). Differential estrogen responsiveness of the vitellogenin and apo Very Low Density Lipoprotein II genes from rooster liver. *J. Steroid Biochem.* **24**, 281–284.

Okret, S., Poellinger, L., Dong, Y. and Gustafsson, J. Å. (1986). Down-regulation of glucocorticoid receptor mRNA by glucocorticoid hormones and recognition by the receptor of a specific binding sequence within a receptor cDNA clone. *Proc. Natl. Acad. Sci. USA* **83**, 5899–5903.

Pan, T. and Coleman, J. E. (1990). The DNA binding domain of GAL4 forms a binuclear metal ion complex. *Biochemistry* **29**, 3023–3029.

Pastori, R. L., Moskaitis, J. E., Smith, Jr, L. H. and Schoenberg, D. R. (1990). Estrogen regulation of *Xenopus laevis* γ-fibrinogen gene expression. (In press).

Petkovich, M., Brand, N. J., Krust, A. and Chambon, P. (1987). A human retinoic acid receptor which belongs to the family of nuclear receptors. *Nature* **330**, 444–450.

Ponglikitmongkol, M., Green, S. and Chambon, P. (1988). Genomic organization of the human oestrogen receptor gene. *EMBO J.* **7**, 3385–3388.

Ptashne, M. (1986). Gene regulation by proteins acting nearby and at a distance. *Nature* **322**, 697–701.

Ptashne, M. (1988). How eukaryotic transcriptional activators work. *Nature* **335**, 683–689.

Quarmby, V. E., Yarbrough, W. G., Lubahn, D. B., French, F. S. and Wilson, E. M. (1990). Autologous down-regulation of androgen receptor messenger ribonucleic acid. *Mol. Endocrinol.* **4**, 22–28.

Raynaud, J.-P. and Ojasoo, T. (1986). The design and use of sex-steroid antagonists. *J. Steroid Biochem.* **25**, 811–833.

Riegel, A. T., Aitken, S. C., Martin, M. B. and Schoenberg, D. R. (1987). Posttranscriptional regulation of albumin expression in *Xenopus* liver: evidence for an estrogen receptor-dependent mechanism. *Mol. Endocrinol.* **1**, 160–167.

Rosewicz, S., McDonald, A. R., Maddux, B. A., Goldfine, J. D., Miesfeld, R. L. and Logsdon, C. D. (1988). Mechanism of glucocorticoid receptor down-regulation by glucocorticoids. *J. Biol. Chem.* **263**, 2581–2584.

Ross, J. and Kobs, G. (1986). H4 histone messenger RNA decay in cell-free extracts initiates at or near 3' terminus and proceeds 3' to 5'. *J. Mol. Biol.* **188**, 579–593.

Rushlow, C. A., Hogan, A., Pinchin, S. M., Howe, K. M., Lardelli, M. and Ish-Horowicz, D. (1989). The *Drosophila hairy* protein acts in both segmentation and bristle patterning and shows homology to N-*myc*. *EMBO J.* **8**, 3095–3103.

Saceda, M., Lippman, M. E., Chambon, P., Lindsey, R. L., Ponglikitmongkol, M., Puente, M. and Martin, M. B. (1988). Regulation of estrogen receptor in MCF-7 cells by estradiol. *Mol. Endocrinol.* **2**, 1157–1162.

Sap, J., Muñoz, A., Damm, K., Goldberg, Y., Ghysdael, J., Leutz, A., Beug, H. and Vennström, B. (1986). The c-*erb*-A protein is a high-affinity receptor for thyroid hormone. *Nature* **324**, 635–640.

Sap, J., Munoy, A., Schmitt, J., Stunnenberg, W. and Vennström, B. (1989). Repression of transcription mediated at a thyroid hormone response element by the v-*erb*-A oncogene product. *Nature* **340**, 242–244.

Segraves, W. A. and Hogness, D. S. (1990). The E75 ecdysone-inducible gene responsible for the 75B early puff in *Drosophila* encodes two members of the steroid receptor superfamily. *Genes Dev.* **4**, 204–219.

Shaw, G. and Kamen, R. (1986). A conserved AU sequence from the 3'-untranslated region of GM-CSF mRNA mediates selective mRNA degradation. *Cell* **46**, 659–669.

Svec, F. and Rudis, M. (1981). Glucocorticoids regulate the glucocorticoid receptor in the AtT-20 cells. *J. Biol. Chem.* **256**, 5984–5987.

Thompson, C. C., Weinberger, C., Lebo, R. and Evans, R. M. (1987). Identification of a novel thyroid hormone receptor expressed in the mammalian central nervous system. *Science* **237**, 1610–1614.

Tora, L., White, J., Brou, C., Tasset, D., Webster, N., Scheer, E. and Chambon, P. (1989). The human estrogen receptor has two independent nonacidic transcriptional activation functions. *Cell* **59**, 477–487.

Umesono, K. and Evans, R. M. (1989). Determinants of target gene specificity for steroid/thyroid hormone receptors. *Cell* **57**, 1139–1146.

Wakeling, A. E. and Bowler, J. (1988). Biology and mode of action of pure antioestrogens. *J. Steroid Biochem.* **30**, 141–147.
Wang, L.-H., Tsai, S. Y., Cook, R. G., Beattie, W. G., Tsai, M.-J. and O'Malley, B. W. (1989). COUP transcription factor is a member of the steroid receptor superfamily. *Nature* **340**, 163–166.
Webster, N. J. G., Green, S., Jin, J. R. and Chambon, P. (1988). The hormone binding domains of the estrogen and glucocorticoid receptors contain an inducible transcription activation function. *Cell* **54**, 199–207.
Wei, L. L., Krett, N. L., Francis, M. D., Gordon, D. F., Wood, W. M., O'Malley, B. W. and Horwitz, K. B. (1988). Multiple human progesterone receptor messenger ribonucleic acids and their autoregulation by progestin agonists and antagonists in breast cancer cells. *Endocrinology* **2**, 62–72.
Weiler, I. J., Lew, D. and Shapiro, D. J. (1987). The *Xenopus laevis* estrogen receptor: sequence homology with human and avian receptor and identification of multiple estrogen messenger ribonucleic acids. *Mol. Endocrinol.* **1**, 355–362.
Weinberger, C., Thompson, C. C., Ong, E. S., Lebo, R., Gruol, D. J. and Evans, R. M. (1986). The c-*erb*-A gene encodes a thyroid hormone receptor. *Nature* **324**, 641–646.
White, R., Lees, J. A., Needham, M., Ham, J. and Parker, M. (1987). Structural organization and expression of the mouse estrogen receptor. *Mol. Endocrinol.* **1**, 735–744.
Wiskocil, R., Bensky, P., Dower, W., Goldberger, R. F., Gordon, J. I. and Deeley, R. G. (1990). Coordinate regulation of two estrogen-dependent genes in avian liver. *Proc. Natl. Acad. Sci. USA* **77**, 4474–4478.
Wolffe, A. P., Glover, J. F., Martin, S. C., Tenniswood, M. P. R., Williams, J. L. and Tata, J. R. (1985). Deinduction of transcription of *Xenopus* 74-kDa albumin genes and destabilization of mRNA by estrogen *in vivo* and in hepatocyte cultures. *Eur. J. Biochem.* **146**, 489–496.
Yamamoto, K. R. (1985). Steroid receptor regulated transcription of specific genes and gene networks. *Ann. Rev. Genet.* **19**, 209–252.
Zelent, A., Krust, A., Petkovich, M., Kastner, P. and Chambon, P. (1989). Cloning of murine α and β retinoic acid receptors and a novel receptor γ predominantly expressed in skin. *Nature* **339**, 714–717.
Zahraoui, A. and Cuny, G. (1987). Nucleotide sequence of the chicken proto-oncogene c-*erb*A corresponding to domain 1 of v-*erb*A. *Eur. J. Biochem.* **166**, 63–69.
Zenke, M., Muñoz, A., Sap, J., Vennstrom, B. and Beug, H. (1990). V-erb-A oncogene activation entails the loss of hormone-dependent regulator activity of c-*erb*A. *Cell* **61**, 1035–1049.

Index

Alpha-fetoprotein gene, 184
Alpha-myosin heavy chain, rat, TREs, localization, 129
Androgen insensitivity, testicular feminization, 342–345
Androgen receptors
  expression,
    cell–cell interactions, 252–254
    mesenchymal–epithelial interactions, 254–259
    ontogeny, 240–243
Androgen response elements, genes, experimental data, 127
Anti-hormones, mechanism of action, 29–32, 387–388
Avian erythroleukaemia
  possible mechanisms, 355–357
  summary and open questions, 370–371
  v-*erb*A oncogene, aberrant transcriptional regulation, 359
Avian thyroid hormone receptors, 357–359

Brain, and pituitary, development, and sex steroid receptors, 249–250
Breast cancer
  epidermal growth factors, 302–303
  oestrogen receptor-positive, treatment, 16
  role of steroid hormones, 297–299
  role of tamoxifen, 29–32
  steroid hormone receptors, 299–300
  transforming growth factor-α, 302–303
Breast tissue
  MCF-7 cells, constitutive transforming growth factor-α, 303–304
  normal,
    epidermal growth factors, 302–303
    role of steroid hormones, 297–299

Breast tissue (*cont.*)
  transforming growth factor-α, 302–303
Bridging transcription factors, 33

Carcinogenicity, sex differentiation process, 235–238
Cathepsin D, 307
Chicken apo-VLDL-II, EREs, localization, 128
Chicken lysozyme
  GREs, localization, 126
  PREs, localization, 127
Chicken oestrogen receptor, amino acid sequences, 17
Chicken ovalbumin, EREs, localization, 128
Chicken ovalbumin upstream promoter (COUP) transcription factor
  cloning, 106
  identification and characterization, 104–105
  and orphan receptors, 108–110
  S300-II requirement, 108
  specific binding to 'dissimilar' promoter sequences, 105
  as steroid receptor superfamily member, 106–107
Chicken vitellogenin II
  EREs, localization, 128
  GREs, localization, 126
Chromatin
  interaction with steroid receptors, 217–231
  factor exclusion, 223–226
  formation of initiation complex, 221–222
  MMTV paradigm, 219–221
  nucleosome alteration, 228–230
  nucleosome positioning, 226–228

Chromatin (cont.)
  selective transacting factor access, 230
  summary, 231
  transactivation by factor interactions, 222–223
Co-operative transactivation of steroid receptors
  co-operation between HREs, 156–157
  and other binding sites, 157–161
  functional details, 161–163
  functional synergism and DNA-binding activity, 163–164
  synergistic domains of the receptor, 164–168
  summary, 168–170
Cortisol, autosomal recessive primary resistance, 345–346
COUP-TF see Chicken ovalbumin upstream promoter transcription factor
Cytosol protein, conversion to nuclear receptor, 4–6

1,25-Dihydroxyvitamin D_3, hereditary generalized resistance, 322–323
DNA-binding of receptors, 18–21, 41–51, 384–385
DNA, circular, interaction of hormone response elements with hormone receptors, 205–207
DNA, linear, interaction of hormone response elements with hormone receptors, 198–201
DNA, nucleosomally organized, interaction of hormone receptors, 207–209
DNA-receptor interactions
  cell-free transcription assay, 210–211
  interactions, characterization, 197–211
  linear DNA, hormone response elements, interaction with hormone receptors, 198–201
  mechanism of transcriptional activation, 209–210
  role of hormone ligand, 201–203
Dominant negative effects of receptors, 89, 381–382

Drosophila Hsp22, 23, 27, EcdREs, localization, 129
Drosophila pSC7, GREs, localization, 127

Ecdysteroids
  moulting hormones, role, 131
  response elements, genes, experimental data, 129
  untransformed receptors, 6
Enucleation experiments, receptor localization, 7
Epidermal growth factor
  effects on mRNA/secretion of growth factor proteins, 309
  role in normal breast tissue, 302–303
ER see Oestrogen receptor
ERE see Oestrogen response elements
Erythroleukaemia, avian, 355–357

Fetal tissues, progesterone receptors, 250–251
α-Fetoprotein gene, 184
Fibroblast growth factors, in breast tissue, 306
Fibroblast-like growth factors, effects on mRNA/secretion of growth factor proteins, 309

Glucocorticoid receptor
  autosomal recessive primary cortisol resistance, 345–346
  DNA-binding domain, 41–52
    mutations, 41–45
    nuclear localization, role of basic region in, 51–52
    specificity, 45–51
  heat-shock proteins,
    functional role, 63
    hormone binding, 66–67
    Hsp90, 63–65
    p56/p59, 67
    receptor activation, 65–66
  hormone-binding domain,

Glucocorticoid receptor (*cont.*)
  affinity labelling studies, 57
  characterization, hormone-binding deficient receptors, 58–59
  limits of site, 52–57
  transcriptional activation, 59
  N-terminal domain, 39–41
  phosphorylation, 60–62
    and activation, 61–62
    hormone-induced, *in vivo*, 62
    sites, 60–61
  resistance in New World primates, 346–347
  transcriptional activation 59, 209–211
  'zinc fingers', 18–20
Glucocorticoid response elements
  genes, experimental data, 126
  negative, 185–186
  potential transcription factor binding sites, 159
Glycoprotein hormones, α-subunit gene, 179–183
GREs *see* Glucocorticoid response elements
Growth factors
  epidermal growth factor, 302–303
  fibroblast growth factors, 306
  inhibitory growth factors, 307–308
  insulin-like growth factor-I and II, 304–306
  mediators of steroid hormone function, 308–310
  mitogenic, 307
  transforming growth factor-α, 302–304
  transforming growth factor-β, 307–309
Growth hormone
  human, GREs, localization, 126
  rat, TREs, localization, 128

Heat-shock protein
  and glucocorticoid receptor, 63–67
  interaction with oestrogen receptor, 21–22
  micromolecular factors, association with conversion of cytosol protein to nuclear receptor, 5–6
  p56/p59, 67

Hormone antagonists, mechanism of action, 29–32, 387–388
Hormone resistance diseases, 321–348
Hormone response elements
  characterization, 125–146, 203–205
  consensus sequences, 132
  co-operation, 156–157
  correlation with two subfamilies of nuclear receptors, 142–144
  functional co-operativity with other transcription factor binding sites, 157–161
  as hormone-inducible enhancers, 133–134
  and other transcription factor binding sites, 157–160
  palindromic DNA sequences, 125–133
  problem of specificity, 136–145
  summary and conclusions, 145–146
  synergistic action, 26–27, 113–115, 134–136, 161–162
  see also Hormone-receptor complexes; Sex steroid receptors
Hormone-inducible enhancers, 133–134
Hormone–receptor complexes, 3–4
  hormone binding, 17–18, 52–59, 115–117, 201–203, 380–381
  interaction in target cells, 4–8
  receptor localization, 6–8
  receptor transformation, 4–6
  two-step mechanism, 4, 8–10
  see also Hormone response elements, Steroid hormone receptors
HREs *see* Hormone response elements
Hsp90 *see* Heat-shock protein (90 kD)
Human growth hormone, GREs, localization, 126
Human MCF-7 cells, EREs, localization, 128
Human metallothionine IIA
  GREs, localization, 126
  PREs, localization, 127
Human osteocalcin, VDRE, localization, 129
Human retinoic acid receptor β, RRE, localization, 129
Hypocalcaemic vitamin D-resistant rickets, 322–323
  amplification of receptor DNA, 327–330

HVDRR (*cont.*)
  creation of VDR mutations with subsequent functional analysis, 330–332
  patients with decreased receptor–DNA interaction, 324–327
  seven families with truncated D-receptor, ligand-binding domain, 337–342
  transcriptional activation,
    capacity of mutant receptors, 332–334
    effect of hormone, 334–337
  see also Vitamin D receptor

ICI 164 384, action, 29, 387–388
Inhibitory growth factors
  effects on mRNA/secretion of growth factor proteins, 309
  TGF-β, 307–308
Insect EcdRE
  relationship to oestrogen response elements, 139
  see also Ecdysteroids
Insulin-like growth factor-I and II
  in breast tissue, 304–306
  effects on mRNA/secretion of growth factor proteins, 309
Intracellular receptors
  primary structure, 2
  see also Nuclear receptors

Laminin B1, mouse RREs, localization, 129
Limb morphogenesis
  effects of pattern formation, 271–275
  prospects for functional studies, 287–290
Luteinizing hormone β, rat, EREs, localization, 128

Metallothionine IIA, human, GREs, localization, 126

Mineralocorticoid response elements
  genes, experimental data, 127
  GRE consensus sequence, 130
Mineralocorticoids, untransformed receptors, 6
Mitogenic growth factors, 307
Mitogenic protein PDGF, 307
Moloney murine leukaemia virus, TREs, localization, 128
Moloney murine sarcoma virus, GREs, localization, 126
Morphogenesis, vertebrate limb, retinoic acid receptors, 269–290
Mouse laminin B1, RREs, localization, 129
Mouse mammary tumour virus
  AREs, MREs, PREs, localization, 127
  GREs, 156–157
    localization, 126
  LTR, chromatin structure, 207–208, 220
  nucleoprotein template structure, 219–221
  steroid receptor binding sites, 198–201
  topology of DNA, 205–207
mRNA stability, hormonal effects, 386–387

New World primates, resistance to glucocorticoid receptor, 346–347
Nuclear hormone receptors
  defined, 1
  DNA binding, 18–21, 41–51, 384–385
  gene structure and function, 377–380
  hormone antagonists, mechanism of action, 387–388
  hormone binding, 17–18, 52–59, 115–117, 201–203, 380–381
  hormone–receptor complexes, 3–4
  mRNA stability, hormonal effects, 386–387
  overview, 1–10
  primary structure, 2
  receptor dimerization, 24–26, 380–384
  summary and final remarks, 388

Nuclear hormone receptors (*cont.*)
  transcriptional activation, 385–386
  see also Intracellular receptors
Nucleoproteins, eukaryotes, organization, 218

Oestrogen receptor
  binding as a palindromic dimer, 24–26
  and breast tumours, 16
  deletion mutants, 17
  DNA-binding domain, 18–21
  expression,
    cell–cell interactions, 252–254
    mesenchymal–epithelial interactions, 254–259
  historical perspective, 15–16
  hormone-binding domain, 17–18, 380–384
  how is gene transcription activated?, 32–33
  interaction with 90-kD heat-shock protein, 21–22
  mechanisms of oestrogen antagonism, 29–32, 387–388
  nuclear localization signal, 22–23
  ontogeny,
    in female genital tract, 243–246
    male mouse reproductive organs, 246–249
  primary amino acid sequence, 16–17
  response elements, palindromes, 23–24, 125–133
  synergy, other transcription factors, 26–27
  and thyroid hormone receptors, 189
  transcriptional activation, 27–29, 385–386
Oestrogen response elements
  conversion to GRE, 137
  first identification, 130
  genes, experimental data, 128
  hER binding to palindromic ERE, 25
  location, 15–16
  relationship to insect EcdRE, 139
  sequences, 24
Oestrogens
  autocrine hypothesis, 300–302
  untransformed receptors, 6

Osteocalcin, vitamin D response elements, 129

Phosphoenol pyruvate carboxykinase, rat, GREs, localization, 127
Pituitary, developing, and sex steroid receptors, 249–250
Pro-opiomelanocortin gene, 179
Progesterone receptor
  fetal tissues, 250–251
  oestrogen induction, 251–252
  steroid receptor superfamily, 110–112
  synergism with adjacent transcription factors, 160
  transactivation of target gene expression, 110–112
Progesterone response elements, genes, experimental data, 127
Progestins, untransformed receptors, 6
Prolactin, rat, EREs, localization, 128
Prolactin gene, 184–185
  oestrogen regulation, 187
Proliferin gene, glucocorticoid regulation, 187–188
Proto-oncogene c-*erb*A, encoding thyroid hormone receptors, 81–82

Rabbit uteroglobin
  GREs, localization, 126
  PREs, localization, 127
Rat α-myosin heavy chain, RREs, TREs, localization, 129
Rat growth hormone, RREs, TREs, localization, 128, 129
Rat luteinizing hormone β, EREs, localization, 128
Rat phosphoenol pyruvate carboxykinase, GREs, localization, 127
Rat prolactin, EREs, localization, 128
Rat T-kininogen, GREs, localization, 127
Rat tryptophan oxygenase, GREs, localization, 127
Rat tyrosine aminotransferase
  GREs, localization, 126
  PREs, localization, 127

Retinoic acid receptors
 functions, 281–282
 identification,
  regions A–F, 277–281
  retinoic acid response elements, 279–280
 ligand affinities, 282–283
 localization, 283–285
 prospects for functional studies, cells and limbs, 287–290
 regulation of gene expression, 285–287
 summary, 290
 and thyroid hormone receptors, 188–189
 vertebrate limb morphogenesis, 269–290
Retinoic acid response elements, 130–131, 279–280
 genes, experimental data, 129
 synthetic palindromic TRE oligonucleotide, localization, 129
Retinoids, biology, 269–271
Rickets see Hypocalcaemic vitamin D-resistant rickets
RREs see Retinoic acid response elements

S300-II, requirement, chicken ovalbumin upstream promoter transcription factor, 108
Sex differentiation process, 236
Sex steroid receptors
 expression,
  cell–cell interactions, 252–254
  mesenchymal–epithelial interactions, 254–259
  summary, 259
 ontogeny, 235–259
  androgen receptors, 240–243
  developing brain and pituitary, 249–250
  methodological considerations, 238–240
  oestrogen induction of progesterone receptors, 251–252
  oestrogen receptors in female genital tract, 243–246

Sex steroid receptors (cont.)
  oestrogen receptors in male mouse reproductive organs, 246–249
  progesterone receptors in fetal tissues, 250–251
 see also Steroid hormone receptors
SREs see Hormone response elements
Steroid hormone receptor superfamily
 chicken ovalbumin upstream promoter transcription factor, 104–110
 progesterone receptor, 110–112
 stimulation of target gene expression, 103–104
 summary and conclusions, 117–118
Steroid hormone receptors
 breast cancer, 299–300
 competition between, 188–190
 conversion of cytosol protein to nuclear receptor, 4–6
 function, role of ligand, 115–117
 hormone binding, 17–18, 52–59, 115–117, 201–203, 380–381
 interaction with chromatin, 217–231
 interaction with circular DNA, various topologies, 205–207
 receptor transformation, 5–6
 summary and conclusions, 168–170, 190
 synergistic action, 26–27, 113–115, 155–170
 transcription factor binding sites, 157–161
 transcriptional activation, 27–29, 39–40, 108, 110–112, 209–211, 359–361, 385–386
 two-step mechanism, 4, 8–10
 see also Hormone response elements; Nuclear hormone receptors; Sex steroid receptors; Steroid hormones
Steroid hormone resistance syndromes, 321–348
Steroid hormone response elements see Hormone response elements
Steroid hormones
 autocrine hypothesis, 300–302
 function, mediation by growth factors, 308–310

Steroid hormones (*cont.*)
  interaction of receptors with other transcription factors, no DNA-binding, 186–188
  negative regulatory functions, 175–190
    DNA-binding competition, receptors and transcription factors, 178–186
    molecular mechanisms, 176–178
    repression of gene expression, 175–190
    see also Hormone response elements; Steroid hormone receptors; Sex steroid hormones

Tamoxifen, action, 29–32, 387–388
Teratogenicity, sex differentiation process, 235–238
Testicular feminization, syndrome of androgen insensitivity, 342–345
TFIIIA 'zinc fingers', 18, 19
Thyroid hormone receptors
  biochemistry, 80–81
  clinical syndromes, 94–95
  encoding by proto-oncogene c-*erb*A, 81–82, 357–360
  function, 88–89, 361–367
  functional domains, 86–88
  heterodimer formation, 90
  major types, and related isoforms, 82–85, 357–359
  multiple forms, competition between, 189–190
  negative regulaton, 89–90, 359–361
  and oestrogen receptors, 189
  other aspects, 90–94
  and retinoic acid receptor, 188–189
  summary, 95–96
  see v-*erb*A
Thyroid hormone response elements, 85–86
  first identification, 130
  genes, experimental data, 128–129
  synthetic palindromic TRE oligonucleotide, localization, 128
Thyroid hormones, structures of major hormones, 79

TRα oncogene
  mutations activating, 364–366
  regulation of erythrocyte differentiation and gene expression, 364
  transcriptional regulation, 359–361
  see v-*erb*A
Transcription factor binding sites
  combination of GRE with another factor, 164
  functional co-operativity, 161–162
    with HREs, 157–161
  functional synergism and DNA-binding affinity, two receptor binding sites, 163–164
  natural genes, 157–160
  potential, 159
  stereo-specific alignment, 162
  synthetic genes, 160–161
Transcription factors, bridging, 33
Transcriptional activation, nuclear hormone receptors, 385–386
Transforming growth factor-α
  effects on mRNA/secretion of growth factor proteins, 309
  MCF-7 cells, 303–304
  role in normal breast tissue, 302–303
Transforming growth factor-β
  effects on mRNA/secretion of growth factor proteins, 309
  inhibitory growth factor, 307–308
TREs *see* Thyroid hormone response elements
Tryptophan oxygenase, rat, GREs, 157
  localization, 127
Tyrosine aminotransferase, rat
  GREs, localization, 126
  PREs, localization, 127

Uteroglobin, rabbit
  GREs, localization, 126
  PREs, localization, 127

v-*erb*A oncogene
  avian erythroleukaemia, 359
  modification of activity by phosphorylation, 368–370

v-*erb*A oncogene (*cont.*)
  regulation of erythrocyte differentiation and gene expression, 361–364
  repression of erythrocyte genes, 366–367
  and leukaemic erythroblast phenotype, 367–368
  transcriptional regulation, 359–361
Vertebrate limb morphogenesis,
  prospects for functional studies, 287–290
  retinoic acid receptors, 269–290
Vitamin D receptor, 322–342
  cytosol, 6, 8
  hereditary generalized resistance, 322–323

Vitamin D receptor (*cont.*)
  see also Hypocalcaemic vitamin D-resistant rickets
Vitamin D response elements, human osteocalcin, 129
Vitellogenins, EREs, localization, 128

Xenopus vitellogenin, EREs, localization, 128

'Zinc fingers'
  oestrogen receptor, 18, 19
  representation, 143